Applied Probability and Statistics (Continued)

continued overleaf

Stochastic Analysis

IN loving remembrance of ROLLO DAVIDSON, and in gratitude for his vivid imagination, his great clarity and integrity of mind, the gentle courtesy of his spirit, and the affection this engendered in all who knew him

Preface

This volume, and the companion work *Stochastic Geometry*, have been compiled by the friends of Rollo Davidson and by one or two others who had not met him but who hoped to become his friends. The two books are very closely linked, and those who wish to explore either one in depth will almost certainly need to consult the other. The degree of unity within each single book is, however, much greater than that which could have been attained by bringing them together, and we believe that most people will find the two-volume arrangement a great convenience.

We have taken some pains to try to avoid the 'miscellaneous' character often to be found in such cooperative works; both books contain much that is new, even to the specialist, but we believe that each will be found valuable as an introduction to the field named in its title, especially by those interested in the possibility of undertaking research in either area and daunted by the scattered nature of the periodical literature and the absence of any comparable synthesis of it.

Each book is furnished with an introductory chapter surveying the field, and setting the stage for the specialist articles which follow it.

This volume contains a very large part of the periodical literature on the theory of Delphic semigroups, and is in fact the first book in any language dealing with that subject. It also presents a large number of new results in the theory of Kingman's p-functions which characterize regenerative phenomena, and thus it complements his recent monograph *Regenerative Phenomena* (John Wiley & Sons, London, 1972). Other papers presented here for the first time discuss various special classes of stochastic processes, and a number of surveys are also included, dealing for example with the sample-path properties of additive processes, and with Doob's 'theory of versions'.

We are especially grateful for the trouble which has been taken by Mr D. S. Griffeath, Professor J. F. C. Kingman, and Professor G. E. H. Reuter in preparing for publication one of Davidson's most remarkable unpublished manuscripts from a rough draft found among his papers. We also wish to thank Dr G. K. Eagleson and Dr D. N. Shanbhag for

their assistance with another Davidson manuscript, and Miss Mary
Brooks and Miss Madeleine Wuidart who drew the pictures and made the
index.

Chapters 2.1–2.5 were first published by Springer–Verlag; 2.6 by the
Cambridge Philosophical Society; and 2.7 by the Academy of Sciences,
Paris. To all these bodies we are most grateful for their kindness in making
possible the complete coverage attempted in this book.

D. G. K.

E. F. H.

Contents

1
INTRODUCTION

1.1

An Introduction to Stochastic Analysis

D. G. KENDALL

(1) Stochastic analysis is the field of interest of the members of a loosely knit body called the Stochastic Analysis Group, which was formed in Oxford in December 1961 to promote interest in the analytical aspects of probability theory among mathematicians and statisticians in the United Kingdom. Very much has been achieved since then, and the primary purpose of the Group may be said to have been attained, as the flourishing state of the two magnificent journals (*Journal of Applied Probability*, *Advances in Applied Probability*) published from Sheffield University under the editorship of Professor Gani bear witness. (Most of the published work emanating from the Group has appeared in their pages, or in those of the journal *Zeitschrift für Wahrscheinlichkeitstheorie*, which was founded during the same period, and has been edited with such distinction by Professor Schmetterer.) There are, however, still some who need an occasional reminder that the fashionable trees of statistics and operational research draw some of their nourishment from mathematical roots, as well as from the photosynthetic activities of practical consultation. In this review, intended primarily for mathematicians, we shall therefore take the opportunity not only to say what stochastic analysts do, and why what they do is useful, but also to give some indication of the mathematical foundations of the subject, and of its links with other branches of mathematics.

(2) We shall begin by recalling that a real-valued random variable X is a measurable mapping from a probability-space $(\Omega, \mathscr{F}, \mathrm{pr})$ (Ω any non-vacuous set, $\mathscr{F}$ a σ-algebra of subsets thereof and pr a non-negative measure on $\mathscr{F}$ of total mass 1) into the real line R endowed with Borel sets. When this random variable is considered in isolation from others, what is important is not the mapping X itself but rather the probability

3

measure P_X defined on the Borel subsets B of the line by

(1) $$P_X(B) = \mathrm{pr}\,(X^{-1}B).$$

This measure P_X is called the 'distribution' of X, and it tells us the probability with which the realized 'value' $X(\omega)$ $(\omega \in \Omega)$ will fall in the generic Borel set B. Obviously many random variables, perhaps defined over different probability-spaces, will have the same distribution, and we can regard the probability-space

$$(R, \mathscr{B}(R), P_X)$$

(with the identity-mapping) as supplying a uniquely defined 'canonical' model for all of them. Here Ω has been replaced by the real line R, $\mathscr{B}(R)$ denotes the σ-algebra of Borel sets in R and P_X is the distribution just defined. The identity-mapping of R into R *is* the random variable. For some purposes an equally well-defined and minor adjustment of the model is convenient, in which $\mathscr{B}(R)$ is replaced by the smallest σ-algebra $\mathscr{B}^+(R)$ containing both all the Borel sets and also all subsets of Borel sets having P_X-measure zero. That there exists a unique consistent extension of P_X from $\mathscr{B}(R)$ to $\mathscr{B}^+(R)$ is a familiar fact, and we shall often use this and similar 'completion' procedures in the pages which follow. (The passage from Borel sets to Lebesgue-measurable sets on the line is an instance of the same technical device, and indeed a special case of the procedure described here if we replace R by the unit segment and P_X by Lebesgue measure.) Notice, however, that in the canonical model,

$$\text{identity}: R \to R,$$

while we can use either $\mathscr{B}(R)$ or $\mathscr{B}^+(R)$ in the measurability condition on the left-hand side, we must always use $\mathscr{B}(R)$ on the right-hand side.

(3) When two random variables X and Y are of interest, defined over the same probability-space, their two distributions P_X and P_Y do not suffice for their study save in the very exceptional case when X and Y are 'independent'; the pair (X, Y) is a measurable mapping from the probability-space into the plane R^2, and we define the 'joint distribution' $P_{X,Y}$ by

(2) $$P_{X,Y}(B) = \mathrm{pr}\,((X, Y)^{-1}B)$$

for all planar Borel sets B. Here (and in the higher but finite-dimensional analogues) it is obvious what the analogous canonical models (using the σ-algebra of Borel sets, or its completion) should be. Nor (as we shall see in more detail below) is there any difficulty in extending these definitions to random variables which take their values in an arbitrary second countable compact Hausdorff space Z, and the specific examples just mentioned can be included within such a generalization by an appropriate compactification of $R, R^2, \ldots$, etc. The Alexandrov one-point compactification is convenient save in the case of R, and here the two-point compactification $R \cup \{-\infty, \infty\}$ can be used instead, if desired, and has certain obvious advantages. (There are also some special problems connected with Markov processes where quite sophisticated compactifications may be appropriate.) We indicate such compactifications by a bar; thus $\bar{R}$ will denote the compactified real line.

(4) Now let us generalize these ideas to *an indexed family* X of component random variables X_α, where α ranges through some arbitrary index-set A, called the 'parameter-set'; each one of the random variables takes its values in R, compactified to $\bar{R}$, or more generally in any fixed second countable compact Hausdorff space Z which is called the 'state-space'. Thus, if ω in Ω is fixed, then $X(\omega)$ maps A into Z, while if α in A is fixed, then X_α maps Ω into Z. In such a situation we speak of the whole family X in association with the probability-space $(\Omega, \mathscr{F}, \mathrm{pr})$ as a 'stochastic process'. The terminology harks back to the days when the parameter-set A was invariably the real line, the non-negative half-line, the integers or the non-negative integers, and was thought of as the time-axis; in those more special situations we can think of $X_\alpha(\omega)$ as specifying the realized 'state' of a randomly developing system as 'time' α 'proceeds' through A, the identification of ω in Ω having fixed all the chance contributions to this development. More generally, when ω is free, we can think of X as a generic element of the space Z^A consisting of all Z-valued functions over A; any one such function is called a 'path' or 'trajectory', and for given ω the particular function $X(\omega)$ is called the 'sample path'. In fact there is no distinction now between a stochastic process and a random function; the domain of the function can be arbitrary, and a very wide range of choices is available for its range Z. A detail which will be obvious, but which requires emphasis, is that we do not have a stochastic process X unless the component random variables X_α are all defined over the same probability-space. The first step will be to seek the proper analogue to the formulae (1) and (2), and the instinctive solution, to accept equation (1) as it stands with $X = \{X_\alpha : \alpha \in A\}$ and with B a Borel set, turns out to be *the wrong one* (save for rather special parameter-sets A).

Now the system of Borel sets on the compactified line $\bar{R}$ can usefully be thought of as the smallest σ-algebra containing the open intervals and at least one compactification-point, or alternatively we can describe it as the smallest σ-algebra containing all the half-open intervals $(x', x'']$, where x' and x'' are extended real numbers and $x' \leqslant x''$. It is entirely reasonable to use this σ-algebra on $\bar{R}$ because

(3) $x' < X \leqslant x''$

typifies a practically possible observation on a real-valued random variable X. If we follow up this idea we see that a typical practically possible observation on a real-valued *stochastic process* X will take the form

(4) $x'_{\alpha_j} < X_{\alpha_j} \leqslant x''_{\alpha_j}$ for $j = 1, 2, ..., n,$

where n is any positive integer, the αs are in A, and the xs are extended real numbers. If we ask what is the smallest σ-algebra of subsets of $\bar{R}^A$ which contains all sets of the form (4), where now X_α is thought of as the αth coordinate of a generic point in that function-space, the answer is *the σ-algebra $\mathscr{B}_0$ of Baire sets*. This is *not* the same thing as the σ-algebra $\mathscr{B}$ of Borel sets, and to make this clear it will suffice to say that here (and also more generally) the σ-algebra of Borel sets is that which contains all the compact sets, and is otherwise minimal, while the σ-algebra of Baire sets is that which contains all those compact sets *which happen to be expressible as countable intersections of open sets*, and is otherwise minimal; these definitions work for any compact Hausdorff space, and show that in general $\mathscr{B}$ is bigger than $\mathscr{B}_0$. Of course $\bar{R}^A$ is compact Hausdorff, and so we can use these definitions in our work. (A good general reference for Baire and Borel sets is Halmos [29], but the reader should be prepared to find his account considerably more complicated; this is due to the fact that he does not confine his attention to the compact case.)

(5) It is obviously important to know when the Borel and Baire σ-algebras coincide, and there are just two facts concerning this matter which we shall need to use here:

(*i*) If the space is not merely compact Hausdorff, but is in addition *second countable*, then the two σ-algebras will be identical. We shall assume throughout that our *component* random variables range through state-spaces Z of this sort, and so in the definition of such a random variable, where we require the mapping

(5) $X_\alpha : \Omega \to Z$

to be measurable, we can use the Borel or Baire σ-algebras indifferently on the right-hand side, for they are the same.

(*ii*) If the space is merely compact Hausdorff, and not necessarily second countable, then in general $\mathscr{B}$ can be properly bigger than $\mathscr{B}_0$. In particular this can happen if we are talking about the space Z^A; Z^A is a Cartesian product of copies of the space Z, and so inherits both its compact and Hausdorff properties, but it will inherit its second countable character if and only if the parameter-set A is countable. We can in fact be more precise than this: *in Z^A the systems of Borel and Baire sets coincide if and only if A is countable.* (A useful fact to bear in mind is that the singleton sets in Z^A are always Borel sets, because they are compact, but they are Baire sets if and only if A is countable.)

(6) Let us now apply (*i*) and (*ii*) above to the most general situation we shall want to consider; suppose in fact that we have a probability-space $(\Omega, \mathscr{F}, \mathrm{pr})$ and that we have associated with it a family of component mappings (for $\alpha \in A$)

$$(6) \qquad X_\alpha: \Omega \to Z,$$

which we can also think of as a combined mapping

$$(7) \qquad X: \Omega \to Z^A$$

into the Cartesian-product space. In order to call $X = \{X_\alpha: \alpha \in A\}$ a stochastic process, we want each mapping (6) to be measurable (it being understood that Ω carries the σ-algebra $\mathscr{F}$). We have already remarked at (*i*) in the preceding section that we can use the Baire or Borel σ-algebra indifferently in Z, and so the question of what we mean by the random-variable status of X_α is not in question; we simply mean that $X_\alpha^{-1} B \in \mathscr{F}$ whenever B is a Baire (= Borel) set in Z.

Now let us look at the mapping (7); the only kind of observation we can imagine actually making on such a process would be of the form

$$(8) \qquad X_{\alpha_j} \in B_j \quad \text{for } j = 1, 2, \ldots, n,$$

where n is a positive integer, the αs are in A and the Bs are Baire (= Borel) sets in Z. It is obvious, therefore, that we must require (8) to determine a measurable subset of Z^A for every choice of n, the αs and the Bs, and the smallest σ-algebra which contains all these sets as members is *the Baire σ-algebra $\mathscr{B}_0$ on Z^A.* We can therefore call X, defined over $(\Omega, \mathscr{F}, \mathrm{pr})$, a stochastic process when and only when

$$X^{-1} B \in \mathscr{F} \quad \text{for every Baire set } B \text{ in } Z^A.$$

There is another approach which has some interest and leads to the same conclusion. Let Φ denote any continuous real-valued function over Z^A. Then as a minimal desideratum for the stochastic process (7) we might reasonably demand that the composed mapping

$$\Phi \circ X: \Omega \to R$$

should determine a real-valued random variable, and this immediately suggests that we should use, as the σ-algebra on Z^A, the smallest one which makes the mapping

$$(9) \qquad\qquad \Phi: Z^A \to R$$

measurable for every continuous Φ (R as usual carrying the Borel sets, i.e. the sets in $\mathscr{B}(R)$). But this, once again, turns out to be the Baire σ-algebra on Z^A.

(7) So much being agreed, it will be seen that the correct analogue to equations (1) and (2) will be as follows: the *distribution* (or, as I shall prefer to call it, the *name*) of the stochastic process

$$(10) \qquad\qquad (\Omega, \mathscr{F}, \mathrm{pr}\,; Z, A\,; X)$$

is the probability measure P_X defined by

$$(11) \qquad\qquad P_X(B) = \mathrm{pr}\,(X^{-1} B)$$

on the Baire σ-algebra $\mathscr{B}_0$ for Z^A. Our terminology here is dictated by the fact that, when Z and A have been given, the only property of the structure (10) which is of the slightest practical importance is the measure P_X on $\mathscr{B}_0$. If two variants of the structure (10) have the same Z, the same A and the same P_X, then there is absolutely no reason for using one rather than the other apart from questions of aesthetics or analytical expediency. Thus, Z and A being fixed, P_X characterizes an equivalence class of structures (10) which there can be no *practical* reason to refine further. If we write $\mu = P_X$, then for practical purposes (Z, A, μ) defines the process, and this is why (Z and A being normally fixed in any such discussion) we call μ the 'name' of the process. Any structure (10) having μ as its name will here be called a *version* of the process; the collection of all versions with name μ will be called the *name-class* of μ. The methodology of a good deal of stochastic analysis consists in the shrewd choice of an appropriate version for the particular analytical exercise one has in view. As a glance at the literature will show, versions proliferate at an alarming rate, and we shall make a serious attempt here to 'platonize' the situation by making precise what we mean by those versions, which we shall then call *models*, which are *canonical*.

(8) Before proceeding to this matter, let us note one immediate corollary to the discussion so far. We have observed that we are at liberty to confound the Borel and Baire sets in Z^A if and only if A is countable, and the resulting simplifications in the interaction between the measure theory and the topology are so useful that there is an unavoidable methodological gulf between countable-A problems and uncountable-A problems. For

example, this is why in the more classical parts of the theory 'discrete-time' stochastic processes (with $A = \{0, 1, 2, ...\}$) are so much more easy to deal with than 'continuous-time' stochastic processes (with $A = \{\alpha: 0 \leqslant \alpha < \infty\}$). It will be obvious that one would find it intolerable to have to sacrifice continuous-time stochastic processes altogether, so that the Baire–Borel contrast cannot in general be shirked.

It might perhaps be thought that a drastic simplification in the nature of the state-space Z would have a healing effect on the breach, but this is not so; the difficulties are fully present even in the apparently trivial case when Z contains just two points.

(9) A large number of important questions now pose themselves almost automatically. One, to which we shall return later, is of considerable difficulty. Suppose that we want to talk about random functions having some special *property* Γ; for example we may have in mind the Bachelier–Lévy–Wiener theory of Brownian motion, where $Z = \bar{R}$ and $A = R$, and we may want to be able to speak of this random motion as a random *continuous* motion. Expressed in the notation we are using here, this means that we are dealing with a random path X which (for any fixed ω) is a point in $Z^A = \bar{R}^R$, and we want to be able to say, perhaps with probability one, that X lies in the subset $C(A) = C(R)$ of $Z^A = \bar{R}^R$. Many of the 'nice properties' one would like a sample path to have are of this character; they amount to a requirement of the form $X(\omega) \in \Gamma \subset Z^A$, but often, as here, the portion Γ of path-space to which we should like to restrict ourselves is not a Baire set, and so immediately we are in trouble. We shall see in due course how this difficulty can be turned, but for the moment we record only the following very useful 'rule-of-thumb'. If the specification of the subset Γ involves *essential* reference to uncountably many values of the parameter α, then Γ cannot possibly be a Baire set. If, on the other hand, only countably many αs are involved in the specification of Γ, then Γ *may* be a Baire set. This is perhaps also a good point at which to mention that not all 'nice properties' can be thrown into the form '$X(\omega) \in \Gamma$', and those which cannot be so expressed (e.g. process-measurability) raise further difficulties of a different kind.

We now leave this matter for the moment, and turn to two other problems which have, happily, been fully solved. We have chosen to characterize a name-class of stochastic processes by a rather complicated object: a probability measure μ on the Baire sets of Z^A. Can we simplify this characterization, and can we do so in such a way that the new formulation preserves both the unicity and existence properties? We want to be sure that, when we have learned how to 'spell' the 'name' more simply, it still characterizes the name-class of processes *uniquely*, and also that a name-class always *exists* for it to characterize.

(10) The simplification of the 'name' can be carried out in various ways, and it would be undesirable to go into much detail on this matter here, beyond saying that all amount to specifying in one way or another the system of *finite-dimensional distributions*,

(12) $\operatorname{pr}_{X_{\alpha_1}, X_{\alpha_2}, \ldots, X_{\alpha_n}}$ on $\mathscr{B}(Z^n)$,

where n runs through the positive integers and the αs through A. (Note that $\mathscr{B}(Z^n)$ in the expression (12) is the same as $\mathscr{B}_0(Z^n)$.) Clearly the 'name' determines each of the finite-dimensional distributions (which are in fact just the 'names' of all the finite subprocesses). It is quite easy to prove (by monotone-class or by Dynkin's π/λ arguments—see [21] for the latter) that if two stochastic processes determine the same finite-dimensional distributions (12), then they must belong to the same name-class, and conversely. So we are merely left with the question, which are the systems of finite-dimensional distributions that can be associated with name-classes of stochastic processes? This question of *existence* lies a lot deeper than that of *unicity*, but fortunately (with the topological assumption on Z which we assume throughout) the answer is fully known, and is agreeably simple. It was given by P. J. Daniell in 1918 [14, 15] and later, but more definitively, by A. N. Kolmogorov in 1933 [52], and it can be expressed thus: the finite-dimensional distributions correspond to a name-class of stochastic processes when, and only when, they satisfy two trivially obvious consistency conditions associated with (*a*) dropping one of the αs and (*b*) permuting the αs, respectively. Though the theorem is easy to state, it is not so easy to prove. The modern proof uses functional analysis (Stone–Weierstrass plus Riesz) and is associated with the names of Bourbaki and Nelson. (See Nelson [63] or Meyer [61].)

The Daniell–Kolmogorov theorem proves existence by constructing a special member of the name-class which is deservedly called *the* canonical model,

(13) $(Z^A, \mathscr{B}_0(Z^A), \mu; Z, A; X)$.

Here we have written out $\mathscr{B}_0(Z^A)$ in full, but in future we shall just call this $\mathscr{B}_0$ (and similarly for $\mathscr{B}_0^+, \mathscr{B}$, etc.). In (13) μ denotes the 'name' of the name-class of which the model (13) is to be the canonical representative, and X denotes, here and henceforth, the family of 'coordinate-mappings' from Z^A into Z:

(14) $(.)_\alpha \colon Z^A \to Z \quad (\alpha \in A)$.

Note that the assemblage of these mappings (properly labelled with α) is *both* the stochastic process X *and also* the generic point X in Z^A (comparable with the generic point ω in Ω). The notational difficulties here are awkward but unavoidable.

(11) The model (13) is canonical because of an important property which we shall now elucidate. Let

(15) $$(\Omega, \mathscr{F}, \text{pr}; Z, A; Y)$$

be some stochastic process with parameter-set A and state-space Z which is in the name-class containing and characterized by the model (13). Suppose that the version (15) has 'sample paths' possessing the 'nice properties' $\{\Gamma_\lambda : \lambda \in \Lambda\}$, where the λth item of this collection describes the 'nice property',

$$\text{the sample-path } (\dots, Y_\alpha(\omega), \dots) \in \Gamma_\lambda \subset Z^A,$$

(16)   *for all ω outside an $\mathscr{F}$-set N_λ of* pr*-measure zero.*

There are no restrictions here on the size of the collection labelled by $\lambda \in \Lambda$ (i.e. on Card(Λ)), and in fact we could if we wished suppose that we have added to it, appropriately labelled by a λ, *every* Γ for which the property (16) is true. Notice (and this is very important) that the collection $\{\Gamma_\lambda : \lambda \in \Lambda\}$ then has the following property

$$\text{if } L \text{ is a countable subset of } \Lambda, \text{ then } \lambda(L) \in \Lambda$$

(17)   *exists such that* $\Gamma_{\lambda(L)} = \bigcap_{\lambda \in L} \Gamma_\lambda.$

We should also note that if we are only interested in a specific (maybe very large) collection of Γ_λs (in that only these determine properties of the sample path that we consider quite 'nice'), then we can still assert the property (17) without loss of generality by adjoining to the collection all countable intersections of its original elements.

We now introduce the *canonical mapping*

(18) $$T: (\Omega, \mathscr{F}, \text{pr}) \to (Z^A, \mathscr{B}_0, \mu)$$

defined by

(19) $$T\omega = (\dots, Y_\alpha(\omega), \dots),$$

which will have the properties

(20) $$T^{-1}\mathscr{B}_0 \subset \mathscr{F}, \quad \text{pr } T^{-1} = \mu \quad \text{and} \quad Y = X \circ T.$$

It may happen that Γ_λ is a Baire set, and in that case properties (16) and

(20) tell us that $T^{-1}\Gamma_\lambda$ covers $\Omega\backslash N_\lambda$, and as this last is an $\mathscr{F}$-set of pr-measure 1, we must have $\mu(\Gamma_\lambda) = 1$. If, however, Γ_λ is *not* a Baire set, then we can still say that

$$T^{-1}B_0 \supset T^{-1}\Gamma_\lambda \supset \Omega\backslash N_\lambda$$

whenever B_0 is a Baire set which covers Γ_λ, and then as before we shall have $\mu(B_0) = \text{pr}\,(T^{-1}B_0) = 1$; that is, Γ_λ has outer measure 1 with respect to $\mathscr{B}_0$ and μ, or more succinctly,

(21) Γ_λ is $(\mathscr{B}_0, \mu)$-*thick for each* $\lambda \in \Lambda$.

We now investigate the consequences of property (21) *when taken together with property* (17). We begin by defining $\mathscr{B}_0[\Gamma_\lambda: \lambda \in \Lambda]$, or $B_0[\Lambda]$ for short, to be *the smallest σ-algebra which contains all Baire sets and all the sets* Γ_λ $(\lambda \in \Lambda)$. This of course extends $\mathscr{B}_0$, and we readily see that there is *at most one* extension of μ from $\mathscr{B}_0$ to $\mathscr{B}_0[\Lambda]$ which gives measure 1 to every Γ_λ. For $\mathscr{B}_0[\Lambda]$ is generated by all the sets of the form

(22) B_0, Γ_λ and $B_0 \cap \Gamma_\lambda$,

and it is clear that any two extensions of μ which make all Γ_λs 'full' (i.e. give them measure 1) must agree on each generating set. The system (22), however, is closed under pairwise intersection (a weak consequence of property (17)) and so two such extensions must agree over the whole of $\mathscr{B}_0[\Lambda]$ (by a monotone class argument, or by the π/λ theorem).

It is not obvious, but it is true, that such a (necessarily unique) extension of μ *always exists*, in virtue of properties (17) and (21). The proof will be found in Meyer [61; Ch. II. Theorem 27] (see also Neveu [66]); it depends on the following construction and argument. First complete $\mathscr{B}_0$ to $\mathscr{B}_0^+$ under μ, and then consider the system of all subsets of Z^A which are of the form

(23) $M_\lambda \cup (A \cap \Gamma_\lambda)$,   where $\lambda \in \Lambda$, $M_\lambda \subset \Omega\backslash\Gamma_\lambda$ and $A \in \mathscr{B}_0^+$.

A few simple calculations show that the sets at (23) form a σ-algebra. Next, if we write

(24) $\mu(M_\lambda \cup (A \cap \Gamma_\lambda)) = \mu(A)$,

then this defines $\mu(.)$ uniquely over the sets at (23), because if

(25) $M_{\lambda'}' \cup (A' \cap \Gamma_{\lambda'}) = M_{\lambda''}'' \cup (A'' \cap \Gamma_{\lambda''})$,

and if

$$\Gamma_{\lambda'} \cap \Gamma_{\lambda''} = \Gamma_\lambda,$$

by an appeal to property (17), then we can rewrite equation (25) successively in the forms

$$M'_\lambda \cup (A' \cap \Gamma_\lambda) = M''_\lambda \cup (A'' \cap \Gamma_\lambda) \, (M'_\lambda, M''_\lambda \text{ disjoint with } \Gamma_\lambda),$$
$$M'_\lambda \triangle (A' \cap \Gamma_\lambda) = M''_\lambda \triangle (A'' \cap \Gamma_\lambda)$$

and

$$(A' \triangle A'') \cap \Gamma_\lambda = M'_\lambda \triangle M''_\lambda \subset \Omega \backslash \Gamma_\lambda,$$

so that

$$(A' \triangle A'') \cap \Gamma_\lambda = 0.$$

Thus $A' \triangle A''$ is a completed Baire set disjoint with Γ_λ, and so has μ-measure zero because of property (21); that is, $\mu(A') = \mu(A'')$. (Note that we do not alter the class of $(\mathscr{B}_0, \mu)$-thick sets if we replace $\mathscr{B}_0$ by $\mathscr{B}_0^+$.) Thirdly, μ so extended to the sets at (23), which form what we shall call the σ-algebra $\mathscr{B}_0^+[\Lambda]$, is a probability measure over $\mathscr{B}_0^+[\Lambda]$ (which obviously extends μ from $\mathscr{B}_0$ and makes each Γ_λ full). The only tricky part of the proof is that which establishes the countable additivity of the extended measure, but this, though complicated, is still technically elementary. Finally, we observe that $\mathscr{B}_0[\Lambda]$ is a subalgebra of $\mathscr{B}_0^+[\Lambda]$, and so the existence proof follows on retracting the extended μ to $\mathscr{B}_0[\Lambda]$. To round the story off it is worth noticing that $\mathscr{B}_0^+[\Lambda]$ is, as the notation suggests, complete under μ, and that it is the smallest μ-complete σ-algebra which extends $\mathscr{B}_0[\Lambda]$, so that it is in fact the completion of the latter.

On fitting together the various pieces of the argument it will be found that we have reached the following highly satisfactory conclusion. *Given any member* (15) *in the name-class determined by the canonical model* (13) *and any countable-intersection-closed system* $\{\Gamma_\lambda : \lambda \in \Lambda\}$ *of 'nice' properties which the sample paths of the stochastic process* (15) *possess, there is a unique minimal* (Λ)*-extension,*

$$(26) \qquad (Z^A, \mathscr{B}_0[\Lambda], \mu_{[\Lambda]}; Z, A; X),$$

of the canonical model which makes each Γ_λ *measurable and gives to it measure* 1.

(12) It is important to notice that we *must* attach a suffix $[\Lambda]$ to μ in the expression (26), because two such *'canonical extensions'* may both make measurable a set Γ which is thick under $(\mathscr{B}_0, \mu)$ and yet disagree about its measure. To see that this can happen we have only to observe that there can exist thick sets which are also co-thick (i.e. have thick complements). Such a set could have measure 1 in one canonical extension and measure 0 in another. Here is an example. Take a point z in Z and let ζ denote the point in Z^A all of whose components are equal to z. This is a singleton-set, so a Borel set. Construct the measure on Borel sets which

gives mass 1 to this singleton and mass 0 to its complement, and then retract this measure to the Baire sets, $\mathscr{B}_0$. Obviously $\Gamma = \{\zeta\}$ is $(\mathscr{B}_0, \mu)$-thick, and the complement of Γ is also $(\mathscr{B}_0, \mu)$-thick *if A is uncountable* because then the only Baire subset of the singleton is the empty set.

This example shows quite clearly that there is *no* 'universal' canonical extension which extends all the others. An interesting problem, which we will not examine further here, is whether there can exist 'maximal' canonical extensions. At least it is evident that Zorn's lemma cannot be invoked to produce them, for the Zorn condition cannot accommodate property (17).

One becomes reconciled to this situation when one reflects that there is, presumably, no 'universal' definition of 'nice'.

We now see why it will often be inadvisable to accept *all* sets Γ satisfying property (16) as Γ_λs. Some of these, though describing properties possessed by the Ω-version, may not be features of that version to which we attach any value, and indeed some, far from being 'nice', may be downright 'nasty'. Also the larger we make the collection $\{\Gamma_\lambda : \lambda \in \Lambda\}$, which in the canonical extension is to become a collection of sets that are each measurable and full, the more difficult it will be to adjoin to this collection some other 'nice' property which we know to be thick because it holds for yet another version; the difficulty here alluded to arises, of course, from property (17). In fact property (17) is essentially necessary and sufficient for it to be possible to make each member of a collection of $(\mathscr{B}_0, \mu)$-thick sets Γ_λ simultaneously measurable and full; more exactly, *the necessary and sufficient condition is that all countable intersections of members of the family should again be $(\mathscr{B}_0, \mu)$-thick.*

Thus, to take a very extreme case, if we have a vast family of versions like the version (15), the *versions themselves* now being labelled by λ, and if each such version has just one property Γ_λ which we consider 'nice', then we can build a canonical extension which makes them all measurable and full, and is otherwise minimal, provided that the intersection condition just mentioned is satisfied, and not otherwise. (There will of course be no necessity to check the thickness of the individual Γ_λs; this is assured by the mere fact that there exist versions which possess them.)

(13) We have now justified the adjective 'canonical' which we attached to the Daniell–Kolmogorov model (13), because we have shown that, given any rival 'version' in the same name-class, there is a uniquely defined minimal extension of the canonical model which is 'just as nice'.

Whether we use the canonical model and its canonical extensions, associated with countable-intersection-closed classes of $(\mathscr{B}_0, \mu)$-thick sets, or whether we prefer to work with the completions of these, is a matter of taste and convenience. It is, after all, very easy to pass from one to the other.

It is important to add a warning that the canonical extensions of the canonical model do not necessarily enable us to deal directly with 'nice' properties of the more sophisticated kind which cannot be expressed in the form $\mu(\{X \in \Gamma\}) = 1$. (What is rather confusingly called stochastic-process-measurability appears to be an instance of this, but according to Doob ([19] Ch. II, Theorem 2.9) the difficulty can be circumvented, at least when $A = R$.)

Notice that if there is *any* extension of the canonical model (13) which makes some sets Γ_λ $(\lambda \in \Lambda)$ measurable and full, then these must have formed part of a countable-intersection-closed family of $(\mathscr{B}_0, \mu)$-thick sets to start with, and so no such extension will do more for us than a canonical extension would do. (This follows easily on applying the preceding argument, with T replaced by the identity mapping.)

Our 'canonical extensions' are very close in spirit to the 'standard extensions' used in Doob [18, 19]. It would be interesting to examine these, and also 'standard modifications' and 'Kakutani extensions', and the use of all these devices to obtain (e.g.) 'separable' and 'measurable' versions, from the point of view adopted here. We shall do so in this book [39] and in its sister-volume *Stochastic Geometry* (in the article on 'random sets' [38]), but for the present we must move on to other matters.

(14) Let us now return to distributions of single random variables. P_X at (1) is quite a complicated object; to state it in full requires a list as long as the list of Borel sets, so it is convenient that we can identify it by naming a much simpler object called (in these Schwartzian days somewhat misleadingly) the *distribution function* of X (or, less ambiguously, the *law* of X). This is the real-valued function F_X of a real variable defined by

$$(27) \qquad F_X(x) = P_X((-\infty, x]) = \mathrm{pr}\,(\{\omega: X(\omega) \leqslant x\}).$$

In the early days of probability theory the most adventurous thing probabilists did with random variables was to add together independent ones (often copies of the same one), and the corresponding operation for laws is *convolution*:

$$F_{X+Y}(z) = \int_R F_Y(z-x)\,dF_X(x) \quad (-\infty < z < \infty)$$

if X and Y are independent. But convolution is a messy operation, and so it is convenient that there exists an isomorphism from the set of laws into another set of functions, under which convolution corresponds to simple multiplication. It is obvious what this must be, for by deciding to add together the values of random variables we have invoked the translation group, whose 'characters' now appear. We define, by

$$\Phi_X(s) = \mathbf{E}(e^{isX}) = \int_\Omega e^{isX}\, d\,\mathrm{pr} = \int_R e^{isx} P_X(dx) = \int_R e^{isx}\, dF_X(x)$$

(28) $(-\infty < s < \infty)$,

the characteristic function of X. This is a double misnomer, first because 'characteristic function' is now used in so many ways that it conveys nothing to the mind, and secondly because Φ_X characterizes not X itself, but rather the law of X. I shall speak of it as 'the cf of the law of X'. The basic facts are then:

(*i*) the cf of the law of X determines that law uniquely;

(*ii*) there exists a (Fourier-type) inversion formula;

(*iii*) we know exactly which complex-valued functions of a real variable can arise as cfs of laws;

(*iv*) for laws and their corresponding cfs, $F_n \to F$ on a dense set if and only if $\Phi_n \to \Phi$ everywhere (and then $\Phi_n \to \Phi$ uniformly on compact sets);

(*v*) if X and Y are independent, then

(29) $$\Phi_{X+Y}(s) = \Phi_X(s)\Phi_Y(s).$$

The last property, coupled with the fact that $\Phi_{aX}(s) = \Phi_X(as)$, shows that the cf is ideal for the analytical treatment of problems involving linear combinations of independent random variables, although (as has been stressed recently by Kolmogorov) there do exist such problems for which it is relatively powerless, and where direct methods have the advantage [54].

Property (*ii*) is *almost* useless, and in practice is normally replaced by property (*i*) plus guesswork. Property (*iii*) is Bochner's theorem: the Φs are exactly the continuous positive-definite functions taking the value 1 at $s = 0$.

Property (*iv*) has recently given rise to an interesting question. We know that uniform convergence on compact sets is just convergence with respect to the compact-open (metric!) topology, while 'convergence everywhere' is just convergence with respect to the topology of simple convergence (and this is *not* metric). The convergences are equivalent, so we may ask, are the topologies the same? The answer is, they are *not* (Kendall and Lamperti [40]). This question, curiously long overlooked, was suggested to us by a surprising result of Davidson to which we shall return.

(15) When the random variable X is non-negative in value we may as well allow it to be possibly infinite. The cf is then appropriately replaced by

$$(30) \qquad \Phi_X(s) = \mathbf{E}(e^{-sX}) = \int_{[0,\infty]} e^{-sX} \, dF(x) \quad (0 < s < \infty).$$

We shall call this a Laplace cf. It can be extended to a complex-valued function of a complex variable s, regular in the half-plane $\text{Re}(s) > 0$, and continuous and bounded in the closed half-plane $\text{Re}(s) \geqslant 0$, and a (normally useless) complex inversion formula of Laplace type is available. Moving in this way onto the imaginary axis, we recover equation (28), but the (Fourier) cf thus obtained is rather special among such Fourier cfs, because its Fourier transform is supported by a half-line (such functions have been extensively studied by Paley and Wiener [67]). Normally it is more natural to stay on the positive real axis, and then there is another inversion formula involving the high derivatives of Φ at very large values of s (see e.g. Widder [75] or Feller [24]); this, too, is normally useless, and practical inversion is again replaced by the equivalent of (*i*) plus guesswork (the only counter-example of this which occurs to me is Daniels [16]).

Laplacian analogues of (*i*)–(*v*) exist but will not be set out here, save for the analogue of (*iii*); this says that the Φs are now exactly the completely monotonic real functions of a positive real variable which are bounded above by unity. This, of course, is a famous analytical theorem of Bernstein (Widder [75]). Complete monotony can be defined in the customary way; Φ is required to be infinitely differentiable, and its kth derivative is required to have the sign of $(-1)^k$ for every k. Alternatively we may prefer an equivalent definition in which Φ is merely required to be continuous on $(0, \infty)$ and to have kth differences with the sign of $(-1)^k$. We shall meet a set-theoretic equivalent of this second definition, important also in potential theory, in our work on random sets (for which see *Stochastic Geometry*, 6.2).

The second definition shows that Φ is the cf of a non-negative random variable if and only if it is a continuous mapping from $(0, \infty)$ into R which satisfies an infinite number of *weak* inequalities each one of which restricts the value of Φ at only a finite number of s-values. (The inequalities are weak because 'has the sign of' is to be interpreted as 'has the sign of, if not zero'.) With this definition, the Φs in question form a positive convex cone in the space of continuous functions; the cone has nice topological properties which enable us to recognize this version of Bernstein's theorem as a typical instance of the theorems of Choquet, sharpening that of Krein–Milman, on the representation of the rays of a convex cone as barycentres of the extreme rays of the cone. For a proof of the theorem from this standpoint see Meyer ([61], pp. 237–238).

We can use the Laplace cf to deal with problems involving linear combinations of non-negative independent random variables, provided that the coefficients in the linear form are all non-negative. We are deprived of this option, however, as soon as we interest ourselves in the *difference* of two independent non-negative random variables, and then we must use the Fourier cf instead. Operational research studies concerned with queues and waiting times have raised the following question: given a Fourier cf, when is it associated with a random variable Z which can be written as the difference $X - Y$ of two independent non-negative random variables X and Y? The answer is unknown.

(16) The cf of (any real-valued) random variable is just one of many 'characterizers' of its law; others can be built up using

$$(31) \qquad (f, P) = \mathbf{E}(f(X)) = \int_R f(x) \, dF_X(x)$$

for a suitably wide family of *bounded continuous* real-valued functions f. (We get the classical cf if we take the trigonometric functions as the functions f.) If we are not specially interested in adding independent random variables, but rather in quite general limit theorems (where the limit is to be the limit *of the laws* and not *of the random variables*), we may find it useful to work with the class of all such fs. In fact, if the laws F_n converge to a law F (in the sense that $F_n(x) \to F(x)$ on a dense set of xs) then

$$(32) \qquad (f, P_n) \to (f, P) \quad \text{for all bounded continuous } f,$$

and (trivially, in view of (*iv*)) conversely. We refer to such a situation by saying that *the law of X_n converges weakly to the law of X*.

Now definition (32), if we take it as a definition rather than as part of a theorem, is clearly capable of a very much wider application, and it can in fact be used when the P_ns and P each refer to a stochastic process rather than a random variable, although now of course the f in definition (32) will be a continuous bounded real-valued function defined over the product-topologized function-space $\bar{R}^A$, say. Thus we can speak of the *weak convergence of stochastic processes*, and this has proved to be a fruitful concept in recent years. A very beautiful theory exists, for an account of which see Billingsley [4], and it is now having enormously valuable practical applications, especially in operational research and (potentially) in biometry. In our brief discussion of this theory we shall deliberately side-step some of its technicalities, for which see [4].

The classic example of weak convergence relates random walks of smaller and smaller steps asymptotically to continuous Brownian motion, and such so-called 'invariance principles' (another misnomer) have three uses. First, it turns out that the fine details of the random walks (e.g. the statistical distributions of the step-lengths) are not very important; we get the continuous Brownian motion as limit in any (suitable) case. Now it may be that we are specially interested in (f, P_n), for large n (small steps) when the step-lengths are distributed in some particular and very complicated way, and it may also be the case that (f, P_n') can readily be computed when the $P_n's$ describe random walks with very simple steps, say instances of the classical Pólya walk. We can then use definition (32) in the form,

$$(33) \qquad (f, P_n) \sim (f, P_n') \quad \text{as } n \to \infty,$$

in order to obtain information for the complicated walk in terms of results for the elementary walk.

Next, it may happen that (f, P), where P describes continuous Brownian motion, is easy to compute. We can then make asymptotic statements about (f, P_n) true for a very large class of random walks.

Lastly, it may happen that (f, P) for the Brownian motion is the whole object of our enquiry and that (f, P_n) is easily calculable for some particular elementary random walk. We can then evaluate the sought-for (f, P) as $\lim (f, P_n)$, where the $P_n s$ now relate to that elementary walk.

(17) It turns out that this three-pronged use of weak convergence is very valuable in queueing theory. In all queueing, waiting and other congestion problems a controlling role is played by a parameter ρ, called the *traffic intensity*; this measures the ability of the capacity of the system to respond to the volume of the 'input' (rate of arrival of aircraft, or whatever), and $\rho = 1$ corresponds to a perfect match between the two, while $\rho > 1$ means that the input is too big.

Now $\rho = 1$ is in fact *not* perfect, in a stochastic situation, for stochastic fluctuations will then inevitably lead to some form of breakdown. Thus real systems must be run in such a way that $\rho < 1$, or, if ρ is time-dependent, in such a way that in a time-averaged sense they are subcritical (as at a nuclear power station). However, if ρ is very much less than unity then

economic considerations become important, and it will often be observed
that ρ could be increased with an acceptable extra risk of breakdown and
with a greatly improved performance. The outcome of these two factors
in the real situation is that, in practice, ρ is always (on average) less than
unity, *but 'only just'*. In any study of such a system, therefore, the
asymptotic behaviour of any quantity of interest, in the limit as ρ
approaches unity from below, is well worth having.

The theory of weak convergence might well have been invented to deal
with exactly such questions, and is currently being applied in this area of
operational research with great energy and success by Iglehart and his
school (for a recent review, see Whitt [73]). Just as for the random walk,
we can only solve queueing problems explicitly in a very small number of
really simple (and grossly oversimplified) special cases, but in what is
called the *heavy traffic limit* (i.e. when $\rho \uparrow 1$) it can sometimes be proved
that the solution for the oversimplified case is valid asymptotically for
the realistic case as well. The recognition of the importance of heavy
traffic approximations is due to Kingman [43, 44], while Prohorov and
his collaborators were the first to see that weak convergence provides
the appropriate tool for developing the full potentialities of the situation.

This is not the only limiting situation in queueing theory where weak
convergence concepts are applicable, however. Another structural parameter which is often present is the equivalent of 'the number of servers'
s (in practice it may be 'the number of airstrips'), and it is meaningful
to perform asymptotics as $s \to \infty$, with ρ held fixed; theorems of this
kind seem first to have been obtained by the probabilists in the U.S.S.R.
(see e.g. [27], and the papers cited there). More ambitious asymptotics
of the weak convergence type may involve a *balanced* double limit, with
$s \to \infty$ and $\rho \uparrow 1$, and there will usually be many distinct ways of 'striking
the balance', with corresponding differences in the limit theorems obtained.

(18) It has recently become clear that the programme for queueing
theory outlined above is very widely applicable and may represent an
entirely new and fruitful approach to all situations where stochastic
modelling is appropriate. Fields in biometry in which it has been shown to
be of positive value are (*a*) epidemiology and (*b*) the theory of branching
processes.

In mathematical models of epidemics there is a parameter analogous
to the traffic intensity, called the *threshold*; this, in the simplest case of a
closed epidemic with infection and recovery as the two component
mechanisms, is just the critical number of susceptibles for a trace of
infection to yield a big epidemic. 'Balanced' double limits, as the ratio of

the number of susceptibles to the threshold approaches unity, and the size of the whole population simultaneously tends to infinity, have been studied by several writers (e.g. Nagaev and Starcev [62], Daniels [17], Barbour [1]), for several different types of 'balance'. (Care in studying the literature is needed because the implied 'balance' is not always explicitly stated.)

In the theory of branching processes (which includes topics in population genetics, in molecular biology and in the control of nuclear power stations) every schoolboy is now familiar with the criticality concept, and 'heavy traffic' approximations (i.e. asymptotics for the near critical situation) were first given by A. M. Yaglom [78] and are being pursued now by Fahady, Quine and Vere-Jones [22].

In both of these applications the advantages of a weak convergence approach are evident; complicated and insoluble systems can be approximated by oversimplified but manageable ones. It is worth noticing that these very important practical applications are only available because stochastic analysts have taken the trouble to pursue, for its own sake, what may initially have seemed a somewhat abstract target—'convergence in law' for probability distributions over function-space. It should also be noticed that 'heavy traffic' theory has not been the only practical reward for these activities; in queueing theory (and potentially in other applications also) an important 'continuity principle' has emerged which is very valuable in quantifying the approximation of complicated systems by simpler ones (Kennedy [41], Whitt [74]).

(19) Before leaving 'weak convergence' we should mention a recent and exciting development; the possibility of 'strong' limit theorems of this kind. The distributions P_n which appear in the weak convergence theorems correspond to stochastic processes defined over *different* probability spaces. This is the force of the adjective 'weak' in this context. We speak of a 'strong' limit theorem if all the approximand processes *and* the limit process are defined over the *same* probability space; for example, Knight [51] constructed continuous Brownian paths as limits of random walk trajectories. It seems likely that 'strong' limit theorems will have a big role to play in the future, permitting, as they do, asymptotic statements about the 'history' of the approximand process as a whole.

(20) We now return to equation (29) and recall the beautiful 'arithmetical' theory of cfs developed by Khinchin [42] and Paul Lévy [57]; see also Linnik [59] and Lukacs [60]. This is based on the observation that cfs form a commutative associative semigroup under multiplication, as do the positive integers, and the question is posed, how far can the

2

parallel be carried? The semigroup of cfs certainly has some features which would seem strange in number theory; for example, there exist many cfs Φ which are *infinitely divisible* in the sense that, for each positive integer k, there exists a cf Φ_k such that

$$(34) \qquad\qquad (\Phi_k)^k = \Phi.$$

Examples are

$$\Phi(s) = e^{-s^2/2}, \quad \Phi(s) = e^{-|s|}, \quad \Phi(s) = (1-is)^{-\lambda} \quad (\lambda > 0).$$

A less obvious example is: all mixtures of negative-exponential laws (Goldie [28], Steutel [71]).

One of the main results of this theory, given by Khinchin and Lévy independently, is a unique representation for the infinitely divisible Φs in the form

$$(35) \qquad\qquad \Phi = \exp(\lambda\Psi),$$

where Ψ is uniquely expressed as an integral of a kernel with respect to an arbitrary probability measure Q. (In a variant of equation (35) which we shall mention below, $\Psi(s)$ consists of a quadratic in s, corresponding to a translation and a Gaussian component, added to an integral-mixture of Poisson distributions, where the mixing-measure involved in the mixture has no atom at the scale-origin. The (unique) support of this measure is called the *Poisson spectrum* of the infinitely divisible law.) This representation can again be recognized as a Choquet theorem and has been proved as such by Johansen [31]; a Choquet proof of the simpler result for infinitely divisible laws of non-negative random variables will be found in [34].

Apart from the above, the main results of the arithmetical study of cfs are as follows:

(*A*) *Either* Φ is itself 'prime' or contains a 'prime' factor, *or* it is infinitely divisible and all its factors are infinitely divisible (in which case it is said to belong to *class I_0*).

(*B*) Every cf Φ can be expressed as a convergent product of countably many factors,

$$(36) \qquad\qquad \Phi = \Phi_0 \Phi_1^{m_1} \Phi_2^{m_2}\ldots,$$

where Φ_0 is of class I_0, and $\Phi_1, \Phi_2, \ldots$ are distinct 'prime' cfs.

(*C*) If

$$\Phi_{11}$$
$$\Phi_{21}, \quad \Phi_{22}$$
$$\Phi_{31}, \quad \Phi_{32}, \quad \Phi_{33}$$
$$\ldots, \quad \ldots, \quad \ldots, \quad \ldots$$

is a 'convergent null triangular array', then the limit Φ of the array is infinitely divisible, and each infinitely divisible Φ can be represented as a limit in this way.

Here we say that Φ_2 is a 'factor' of Φ_1 if both are cfs and if there exists a cf Φ_3 such that $\Phi_1 = \Phi_2 \Phi_3$, and we say that Φ is 'prime' if, for any such factorization of Φ, one at least of the components is a 'unit', e.g. $\Phi = \Phi_1 \Phi_2$ and

(37) $$\Phi_1(s) = e^{ics}, \quad \Phi_2(s) = e^{-ics} \Phi(s).$$

Here is a prime:

$$\Phi(s) = (1 - s^2) e^{-s^2/2}.$$

The terminology at (C) calls for some explanation. Given a triangular array, we can form the row-products

$$\Phi_{11}, \ \Phi_{21} \Phi_{22}, \ \Phi_{31} \Phi_{32} \Phi_{33}, \ ...,$$

and the array is said to *converge* (say to Φ) when the row-products do so. The array is said to be *null* if, uniformly within rows, its elements are asymptotically unitary as the rank of the row tends to infinity; more precisely if, given $\varepsilon > 0$ and $\sigma > 0$, then

$$|1 - \Phi_{ij}(s)| < \varepsilon \quad (-\sigma < s < \sigma)$$

when $i \geqslant I(\varepsilon, \sigma)$, for all j such that $1 \leqslant j \leqslant i$. A neater formulation of this can be given by using the Khinchin functionals defined by

$$N_a(\Phi) = -\int_0^a \log|\Phi(s)| \, ds;$$

notice that $N_a(\Phi_1 \Phi_2) = N_a(\Phi_1) + N_a(\Phi_2)$.

The interest of these theorems is twofold. First, (C) gives us a procedure for finding all the infinitely divisible cfs (but it does not help us to identify the class I_0). Secondly, (B) is near (and just about as near as we can get) to an analogue of the unique decomposability of the integers into primes.

We cannot expect to improve much on (B). For example, we can have

$$\Phi_1^2 = \Phi_2 \Phi_3,$$

where Φ_1 and Φ_2 are primes, and Φ_3 is infinitely divisible (and in fact Gaussian).

These results show that the theory runs into a quicksand of pathologies very soon, in almost every direction. It would be very desirable to identify the 'primes' and the members of the class I_0, because (B) shows that these are basic bricks out of which all cfs can be built by countable operations, even though we know that there is in general no unicity in such a representation. Unfortunately neither class has been identified as yet. Here are some members of the class I_0:

(*a*) the cf of a Gaussian law;

(*b*) the cf of a Poisson law;

(*c*) the product of a cf of type (*a*) and one of type (*b*);

(*d*) any infinitely divisible cf Φ without Gaussian component which is such that

$$\text{Poisson spectrum } (\Phi) \subseteq [\alpha, 2\alpha],$$

where $\alpha > 0$.

If Φ_α momentarily denotes the cf of a Poisson law of parameter 1 and scale-factor α, then

$$\Phi_\alpha \Phi_\beta \Phi_\gamma$$

must of course be infinitely divisible; for some choices of (α, β, γ) it *is* in the class I_0, and for others it is *not*! Some recent work (see e.g. [60]) shows that class I_0 is 'big' because *any* infinitely divisible cf can be expressed as a countable product of cfs in class I_0.

(21) If a cf Φ corresponds to a law with df F, then F may or may not be absolutely continuous as a real-valued function of a real variable; if it is, then the corresponding distribution P_X will be absolutely continuous (as a measure) with respect to Lebesgue measure, and this is of course also a sufficient condition. When the condition holds then F has almost everywhere a finite non-negative summable derivative f, and the equivalence class of summable functions differing from f at most on a set of Lebesgue measure zero is a (weakly) positive element of the Lebesgue function space $L = L(-\infty, \infty)$. L is a semigroup under the appropriate version of the convolution operation, and the non-negative part of this is evidently isomorphic to a sub-semigroup of the Lévy semigroup with which we have been dealing. We may therefore ask similar arithmetic questions about this sub-semigroup, which we shall call $L_+(-\infty, \infty)$.

If the restriction to positivity is ignored, and if indeed we admit complex-valued as well as real-valued summable functions, then there is a remarkable theorem of Rudin [69] which tells us that *L has no primes*; every element admits a proper decomposition.

Cohen [12] has studied the real-and-positive case (the only one which is probabilistically relevant) under the restriction that the carrier is a compact group; e.g. L_+(circle). Here he showed that if an f in this space has a continuous strictly positive version, then it always admits a proper decomposition into convolution-factors which are real and admit strictly positive versions. If, however, we merely know that $f \in L_+$(circle) and that it is continuous and non-negative, then he showed by an example that it *could* be prime; i.e. a proper decomposition into real convolution-factors admitting non-negative versions may then be impossible.

Recently Lewis [58] brought these theorems into contact with the applications by proving the long-conjectured result that if a random variable X has the uniform distribution, i.e. if for some $-\infty < a < b < \infty$ its law is absolutely continuous and has the density

$$f(x) = 1/(b-a) \quad (a < x < b),$$
$$f(x) = 0 \quad (x < a, x > b),$$

then X cannot be expressed as the sum of two independent random variables Y and Z each with absolutely continuous laws. This therefore gives a (non-negative, not continuous) prime in $L_+(-\infty, \infty)$. Important extensions to multi-dimensional situations have recently been made by Basterfield [2].

(22) It is clear that the theory of the arithmetic of cfs could be extended in various immediately obvious ways; for example, to take the simplest possibility, we could look at probability distributions on the perimeter of the circle, for which the analogue of the cf is the sequence of 'Fourier coefficients' $c = \{\dots, c_n, \dots\}$, where

(38) $$c_n = \mathbf{E}(e^{inX}) \quad (-\infty < n < \infty),$$

but rather strangely a full-scale attack on this problem had to wait until a much less obvious extension had been investigated. This leads us to introduce the theory of Delphic semigroups, of which three of Davidson's papers (reprinted here) represent the most complete and sophisticated development to date. To appreciate them properly a little background is required, and we shall try to give that now.

(23) Kingman in [46] drew attention to the curious fact (already mentioned in Daley [13]) that the so-called *renewal sequences* form a commutative associative semigroup. For our purposes we can introduce a renewal sequence as follows. Let X be a non-negative infinite-or-integer-valued random variable, and let $X_1, X_2, \dots$ be independent copies of it. Write

(39) $$S_0 = 0, \quad S_1 = X_1, \quad S_2 = X_1 + X_2, \dots.$$

Then write $u = (u_0, u_1, u_2, \dots)$, where

(40) $$u_n = \mathrm{pr}(S_j = n \text{ for some } j) \quad (n = 0, 1, 2, \dots).$$

Evidently $0 \leqslant u_n \leqslant 1$ for all n, and from Feller's work (synthesized in [24]) we know in principle how to construct all such renewal sequences u. The prescription is, take a (perhaps deficient) probability distribution $(f_1, f_2, ...)$ over the non-negative integers and write $F(s) = f_1 s + f_2 s^2 + ... \; (|s| < 1)$. Then

(41)
$$U(s) = \frac{1}{1 - F(s)}$$

defines a complex-valued function of a complex variable s which is regular over the open disk $\{s : |s| < 1\}$, and so can be expanded as a power-series at $s = 0$. The coefficients u_n of this power-series give us the most general possible renewal sequence, and it is uniquely characterized by the sequence of fs. Kingman's binary operation under which the class $\mathcal{R}$ of renewal sequences becomes a commutative associative semigroup is $u = u' \otimes u''$, where

(42)
$$(u' \otimes u'')_n = u'_n u''_n \quad (n = 0, 1, 2, ...).$$

It is very difficult to prove that Kingman's circle-operation has semigroup character using any of the known analytical definitions of a renewal sequence, but very easy to establish it by a probabilistic argument, as follows.

Let us say that 'phenomenon Φ' happens at 'time n' if $S_j = n$ for some j. Thus it surely happens at time 0, and it happens at each of the times $S_1, S_2, ...$, and at no other time. Because of the independence of the Xs, the process 'forgets its past and starts again just as if it were at $t = 0$, at any epoch at which Φ occurs'; we refrain from formulating this statement precisely, because we are only concerned here to communicate the intuitive content of the subject. Such a 'phenomenon' is called *regenerative*, and in discrete time the most general regenerative phenomenon arises from a sequence of cumulative sums of non-negative integer-or-infinite identically distributed random variables as at (39) above, so is fully characterized distributionally by the associated renewal sequence.

Given two renewal sequences, let us set up (on distinct probability spaces) a model for each one of the two associated regenerative phenomena Φ' and Φ''. Take the Cartesian product of these two probability spaces; this enables us to speak of Φ' and Φ'' in the same breath. Now introduce a new phenomenon Φ by agreeing that

Φ *happens at time n if and only if* Φ' *and* Φ'' *both do so.*

Then Φ is regenerative, and $u' \otimes u''$ is its renewal sequence.

Kingman also remarked on the same occasion that $\mathcal{R}$ is topologically as well as algebraically 'nice'; in fact, given a sequence of renewal sequences $\{u^{(k)}: k = 1, 2, ...\}$ which converges to a sequence v in the componentwise sense, then v must also be in $\mathscr{R}$. (*This* property, however, is best proved analytically.) These remarks by Kingman provoked me into an attempt to investigate the 'arithmetic of renewal sequences' [36].

Kingman's observations about $\mathscr{R}$ were preceded by two other very important contributions to the arithmetical study of $\mathscr{R}$ and of the analogous class of renewal *densities*; thus Daley [13] had considered the questions, if each of u_1 and u_2 is a renewal density when is this true of $u_1 u_2$, or of cu_1, where c is a scalar multiplier? Also Lamperti [55] had used a result of Kaluza [32] to show that if $u = (u_0, u_1, u_2, ...)$ is a bounded sequence of non-negative real numbers such that $u_0 = 1$ and

$$(43) \qquad u_{n-1} u_{n+1} \geqslant u_n^2 \quad (n = 1, 2, ...),$$

then $u \in \mathscr{R}$. This is a charming result because it enables us to identify renewal sequences which could not otherwise so readily be seen to be such:

$$(44) \qquad u_n = \frac{1}{n+1} \quad (n = 0, 1, 2, ...)$$

is a notable example.

In [36], presented at a Symposium held at Loutraki in the Gulf of Corinth, I gave an account of the arithmetic of the so-called *aperiodic* renewal sequences (forming a class which we shall call $\mathscr{R}_{\mathrm{ap}}$) defined by the condition

$$(45) \qquad \text{g.c.d.} \{n: u_n > 0\} = 1;$$

it is well known that very little generality is lost by considering $\mathscr{R}_{\mathrm{ap}}$ rather than $\mathscr{R}$ itself. It was first shown that there is a close analogue of Theorem (*C*) which goes as follows.

Suppose that we have a convergent triangular array of renewal sequences (aperiodic or not) which is 'null' in the following sense:

$$(46) \qquad \lim_{i \to \infty} u_1^{(i,j)} = 1, \quad \text{uniformly with respect to } j;$$

here i is the row-index and j is the column-index for the array. Then the limit is infinitely divisible, and is either everywhere positive (so aperiodic) or is the sequence $(1, 0, 0, 0, ...)$. Also, every aperiodic infinitely divisible renewal sequence can be constructed in this way.

This result, coupled with the identification of all the aperiodic infinitely divisible renewal sequences, may be called 'the central limit theorem' for renewal sequences, by analogy with the triangular array limit theorems in classical probability theory.

The aperiodic (= positive) infinitely divisible renewal sequences were identified in [36] in two different ways, intrinsically and constructively. The intrinsic characterization neatly says that *they are just the Kaluza sequences such that $u_1 > 0$* (which implies the positivity of all the u_ns).

The constructive characterization is of Choquet type and asserts essentially that *we get the most general aperiodic infinitely divisible renewal sequence by forming componentwise products of the 'extreme' sequences of the form*

$$u_n = e^{-\lambda n} \quad (n = 0, 1, ..., k),$$

(47) $$u_n = e^{-\lambda k} \quad (n > k).$$

Here the full set of 'extreme' sequences is obtained by taking $k = 1, 2, ..., \infty$.

After this it was natural to look for analogues to Theorems (*A*) and (*B*) for $\mathcal{R}_{\mathrm{ap}}$, and it turned out [36] that *they both hold as they stand for $\mathcal{R}_{\mathrm{ap}}$*.

In a later paper [37], worked out at Delphi and Loutraki, an attempt was made to build an abstract structure which might come near to containing arithmetical theorems concerning both cfs and renewal sequences as special cases, and to this end *Delphic semigroups* were defined as follows.

DSG1:   *G is a commutative associative semigroup with (unique) neutral element e, topologized so that G is Hausdorff and $(u, v) \to u \otimes v$ is continuous. Also G is provided with a continuous homomorphism Δ to the additive semigroup of non-negative real numbers, such that $\Delta(u) = 0$ if and only if $u = e$;*

DSG2:   *for each u in G, the factors of u form a compact set;*

DSG3:   *each convergent triangular array of elements of G, null in the sense that*

$$\lim_{i \to \infty} \Delta(u^{(i,j)}) = 0, \quad \textit{uniformly with respect to } j,$$

converges to an infinitely divisible element of G.

For Delphic semigroups it was proved that (*i*) *every infinitely divisible element can be represented as at* DSG3; (*ii*) *the exact analogue of Theorem (A) holds; and* (*iii*) *the exact analogue of Theorem (B) holds.*

$\mathcal{R}_{\mathrm{ap}}$ does not, as it stands, satisfy these axioms, but if S is an additive semigroup of non-negative integers with g.c.d. unity, and if $\mathcal{R}(S)$ denotes all those renewal sequences which are positive on S, then $\mathcal{R}(S)$ was shown to satisfy the Delphic axioms for every such S. The results about $\mathcal{R}_{\mathrm{ap}}$ can then be deduced from the abstract theorems via the semigroups $\mathcal{R}(S)$.

The problem of squeezing the cf-arithmetic foot into the Delphic shoe was then solved by J. G. Basterfield [2], and later with more detail by

Davidson (see his third 'Delphic' paper, Section 5). This, like the theory for $\mathscr{R}$, presents (worse) complications, and I hoped that a nice clean non-trivial example of the Delphic axioms might be found in the system $\mathscr{P}$ of Kingman's (standard) p-functions; this hope was fulfilled [37]. Kingman p-functions appear in many places in the present volume, and this introductory chapter would be incomplete without an account of them, so to this we now turn.

(24) The definitive account of p-functions is Kingman's book [50]. His definition (given first in [45]) is as follows. In some probability space let E_t be measurable for each $t \geqslant 0$, and suppose that

(48) $\quad \text{pr}(E_{t_1} \cap E_{t_2} \cap ... \cap E_{t_n}) = p(t_1)\,\text{pr}(E_{t_2-t_1} \cap E_{t_3-t_1} \cap ... \cap E_{t_n-t_1})$

for all positive integers n and $0 \leqslant t_1 \leqslant t_2 \leqslant ... \leqslant t_n$. Then $p(.)$ (with $p(0) = 1$) is called a *p-function*. We interpret this by saying that there is a *regenerative phenomenon* Φ which surely happens at $t = 0$, and that

(49) $\qquad\qquad\qquad \{\Phi \text{ happens at } t\} = E_t,$

the probability of either of these events being $p(t)$. The essence of a regenerative phenomenon is the amnesia induced by its own occurrence, and in this sense Kingman's theory is a natural generalization of renewal theory (though quite distinct from some other well-known generalizations of that theory). The behaviour of p-functions is simplest when they satisfy the condition

(50) $\qquad\qquad\qquad \lim_{t \to 0} p(t) = 1,$

and they are then called *standard*. We shall only be concerned here with the set of standard p-functions, which is always called $\mathscr{P}$.

As in the renewal case, a semigroup operation is present; in fact $\mathscr{P}$ is a commutative associative semigroup under the binary operation

(51) $\qquad\qquad\qquad (p_1 \otimes p_2)(t) = p_1(t)p_2(t);$

this is proved exactly as in the renewal case. If we give to $\mathscr{P}$ the topology of simple convergence, and set

(52) $\qquad\qquad\qquad \Delta(p) = \log(1/p(1)),$

then I showed [37] that $\mathscr{P}$ is a Delphic semigroup, I found all the infinitely divisible elements of $\mathscr{P}$, and I showed that $\mathscr{P}$ *must* contain primes because it is easy to produce a non-infinitely divisible p in $\mathscr{P}$ (and so primes have to exist in virtue of Theorem (A) of the Delphic theory). One can indicate the nature of the infinitely divisible elements by remarking that

(53) $\qquad\qquad\qquad p(t) = \exp(-\lambda \min(\tau, t))$

defines one such for any non-negative λ and any (perhaps infinite) non-negative τ, and that the general infinitely divisible p has a unique Choquet representation as a product-integral of p-functions like (53).

These results from [37] raised a number of difficult questions, both in the general theory of Delphic semigroups, and also in the individual theories of special examples. Some of the latter were seen to be of significance in (for example) operational research because it was shown in [37] that the infinitely divisible p-functions constitute the general class of solutions to a problem of pedestrian-and-road traffic.

Davidson in his three Delphic papers (reprinted here, 2.3, 2.4 and 2.5) gave sweeping though not perhaps exhaustive answers to these questions, and they still constitute the latest word on the subject. As they do not make very easy reading (for the analysis involved is very complicated), I will indicate briefly some of the holes in the subject which his papers fill.

First there is obviously the question of the analogue of the class I_0. In the classical (cf) case this is still unsolved. But for the Delphic semigroup $\mathscr{R}^+$ of *positive* renewal sequences Davidson identifies it completely; by a very difficult argument he shows that it consists exactly of the 'exponentials'

$$(54) \qquad\qquad u_n = \mathrm{e}^{-\lambda n} \quad (n = 0, 1, 2, \ldots),$$

a result which was and still is very surprising.

This immediately raises the question, what about the class I_0 for $\mathscr{P}$? Does it consist exactly of the exponentials $p(t) = \mathrm{e}^{-\lambda t}$? It is easy to show that these p-functions do have the I_0-property, and indeed no others have been found, but it is not certain that they do not exist. Davidson has found a large class of infinitely divisible ps which are not in I_0 (for example, it includes the function $p(t) = 1/(1+t)$, and the functions $p(t) = \exp(-at^b)$ for $0 < b < 1$). More curious are the results of his study of the p-functions defined above in equation (53). He shows that these do not belong to I_0 if $\lambda\tau > 1$ (λ and τ do not matter individually, but only through their product), and he gives a heuristic argument which *suggests* that possibly they *do* belong to I_0 if $\lambda\tau \leqslant 1$. It is still not known whether this conjecture of Davidson's is true. If it is, then there are elements of I_0 whose squares are not in I_0.

Next there is the question of the primes (or 'simples' as Davidson calls them) in $\mathscr{R}^+$ and in $\mathscr{P}$. In $\mathscr{R}^+$ he found a wide (dense) class of primes, one example of which is

$$(55) \qquad\qquad u_n = \frac{1-q}{1+q}(1+(-1)^{n-1}q^n) \quad (n > 0)$$

(with $u_0 = 1$, of course). He also showed that the set of all primes in $\mathcal{R}^+$ is *enormous*; in fact, it is 'residual' (the complement of a meagre set). So perhaps their complete identification is beyond our reach. Davidson also constructed some dense primes in $\mathcal{P}$, and here again showed that all the primes form a residual set.

By virtue of knowing class I_0 exactly for $\mathcal{R}^+$, Davidson was able to carry the general factorization theory for this example a great deal further; in fact he exhibits it as the direct product of two Delphic sub-semigroups.

In a subject so studded with Choquet theorems, one naturally checks every set in sight for convexity. Davidson shows that $\mathcal{R}^+$ is not additively convex, which is not perhaps surprising. It is trivial from the known analogue of the Lévy–Khinchin theorem for these Delphic semigroups that the infinitely divisible elements in both $\mathcal{R}^+$ and $\mathcal{P}$ form a *multiplicatively* convex set, for the logarithms have Choquet representations, but what is truly extraordinary is that for both semigroups the infinitely divisible elements *also* form an *additively* convex set. This again was discovered by Davidson, and he performed the indicated Choquet analysis, which is now very complicated. Moreover, it turns out that one is here in the situation in which the Choquet decomposition of the associated cone is *not* unique (as also happens, for example, in the Choquet analysis of doubly stochastic matrices). Perhaps this is 'why' the additive convexity is so puzzling. The proof (in both cases) depends on some algebra with Kaluza sequences which it is hard to interpret intuitively.

In Davidson's first Delphic paper he introduced a very useful variant of the definitions in which the topological conditions are replaced by parallel notions involving only sequential convergence. In his third Delphic paper he returned to the general theory and dealt with a query raised by Kingman: is DSG3 necessary, or can it be proved as a *theorem* by suitably strengthening DSG1 and DSG2? In Theorem 3 of this third paper Davidson gives a strengthening of the weaker Delphic axioms which enable him to *prove* a central limit *theorem* like DSG3 (in fact, its sequential analogue), so that Kingman's question is answered, and the abstract theory much enriched. The first section of that paper gives a good survey of the final position. The extra strength in what is assumed can satisfactorily be loosely described as an assumption that there are 'enough' continuous homomorphisms into the additive reals (or the circle group) to 'separate points'.

Davidson then derives the main examples afresh from this improved standpoint, and in particular gives a detailed Delphic study of the semigroup of cfs of laws on the line. The arithmetic properties of cfs of laws on the circle are also studied in this paper. Finally some ingenious 'spiral' semigroups are devised to destroy some otherwise plausible conjectures about I_0.

Before leaving Delphic semigroups we ought to mention that numerous other examples, of analytic rather than probabilistic origin, have been found by Lamperti and by Bingham (see e.g. [56], [5], [6], [7]). However, these authors have not been so much concerned with the deeper arithmetic problems.

(25) One aspect of Davidson's second Delphic paper has not been mentioned; this is less concerned with the theory of Delphic semigroups than with the analytic properties of $\mathscr{P}$ which have no obvious connection with its Delphic nature. For example, Kingman had shown [45] that a p-function $p(.)$ determines a cf for a law on the line by

$$\Phi(s) = p(|s|)$$

(so that in fact $\mathscr{P}$ is a sub-semigroup of the Lévy semigroup of all cfs), and therefore, in view of what we have said earlier, it will not be found surprising that the convergence $p_n \rightarrow p$ in the compact-open topology holds if and only if it holds componentwise (i.e. in the topology of simple convergence). In view of the Kendall–Lamperti result, one might expect that the c-o and s-c *topologies* on $\mathscr{P}$ would be distinct, but surprisingly this is not so: Davidson shows that they are *identical*. (It ought to be admitted that in this presentation the historical order has been reversed; Davidson's result came first and prompted our other enquiry.) He obtained this result from interesting inequalities about p-functions which have generated much further work. In this paper he also initiated and obtained interesting results for non-standard p-functions, but here a more definitive analysis has subsequently been given by Kingman [47].

One of the most tantalizing problems about (standard) p-functions is this: given $M = p(1)$, how small can $m(M; p) = \inf\{p(t): 0 \leqslant t \leqslant 1\}$ be? We can also ask, what is the value of

$$I(M) = \inf_{p \in \mathscr{P}} m(M; p)?$$

Note that the special value $t = 1$ in the definition of M is not significant, and can be adjusted to any other interesting value of t by a scale-change (this is associated with Davidson's apparently bizarre remark about 'letting $1 \rightarrow \infty$'). Davidson encountered these questions first in his study of the different topologies on $\mathscr{P}$, and in his thesis he showed that if $M > \frac{3}{4}$, then

$$I(M) \geqslant \tfrac{1}{2} + \sqrt{(M - \tfrac{3}{4})},$$

an inequality discovered independently by Blackwell and Freedman [8] in the more special context of Markovian p-functions. Let

$$c = \inf\{M : I(M) > 0\};$$

then evidently $M > c$ implies that $I(M) > 0$, and $I(M) > 0$ implies that $M \geqslant c$, while the quadratic inequality just quoted shows that $c \leqslant \frac{3}{4}$. Davidson showed by an example that $e^{-1} \leqslant c$, and he conjectured that $c = e^{-1}$, but this is still an open question. From one of Davidson's papers published for the first time in this book ('On Smith's Phenomenon', 3.3 of this book) it follows that in fact $c \leqslant \frac{2}{3}$. Progress with this problem is especially worth striving for, in view of the fact that the truth of even the relatively crude conjecture, $c \leqslant \frac{1}{2}$, would have important consequences.

To understand the situation a little more thoroughly it is helpful to introduce

$$d = \inf\{M : \text{there exists } \delta(M) > 0 \text{ such that each } p(t) \geqslant \delta \text{ on } [0, \delta]\}$$

(notice that δ is not to depend on $p \in \mathscr{P}$!). To know the value of d is to know something about the way in which $p(t) \to 1$ as $t \to 0$, and information about d is relevant to the problem of 'Markov groups'. What Davidson did in his last paper was to show that $d \leqslant \frac{2}{3}$, but Kingman, Reuter and Williams have pointed out that (trivially) $d \leqslant c$, and that (because each p-function is super-multiplicative) we also have $c \leqslant d$, so that $c = d$, and Davidson's second inequality actually yields $c \leqslant \frac{2}{3}$.

Bloomfield [9] (and also Davidson) showed that $M \leqslant 1 + m \log m$, and in this book (3.2) Bloomfield gives a charming new proof of this result using queueing theory. Davidson liked to illustrate the state of play on these problems by drawing an (M, m) diagram; Figure 1 shows his last version of this. Notice the notch at $M = 0.63 \ldots = 1 - e^{-1}$; this records his *belief* that $c \,(= d)$ could be no larger than this value, at any rate, in view of the 'peak height' of the Bloomfield–Davidson curve, but even this much is still an open question. The dotted curve appears to be $M = 1/(2 - m)$ and is evidently associated with an attack on the Markov group problem (because it cuts the M-axis at $M = \frac{1}{2}$). David S. Griffeath has just shown that this locus is associated in a natural way with Davidson's proof that $c \leqslant \frac{2}{3}$. Davidson had also initiated a computer-search for 'maximally oscillating' p-functions, and this will perhaps be continued at Cambridge. [As this work goes to press, it is pleasant to be able to report that Davidson's conjecture $c = d \leqslant 1 - e^{-1}$ has been shown to be true by A. G. Cornish.]

(26) There have been several historical inversions in this article, and now we have to admit to yet another. Long before any attention had been paid to its semigroup character, the class $\mathscr{P}$ of standard Kingman

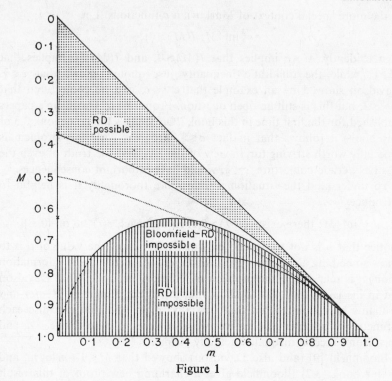

Figure 1

p-functions was seen to be of fundamental importance for the theory of countable-state temporally homogeneous Markov processes in continuous time. The reason for this is connected with the fact that if $p_{ij}(.)$ denotes a typical transition function for such a process, assumed to satisfy the usual continuity condition

$$(56) \qquad\qquad \lim_{t\to 0} p_{ij}(t) = \delta_{ij},$$

then each 'diagonal' transition function (e.g. $p_{11}(.)$) is an element of $\mathscr{P}$. The class of all such standard diagonal Markov transition functions is called $\mathscr{PM}$, and one of Kingman's first discoveries was that $\mathscr{PM}$ *is a proper subset of* $\mathscr{P}$. This is shocking and exciting because, in the analogous situation in discrete time, the corresponding statement is *false*; that is, if $p_{ij}^{(n)}$ is a typical element of the *n*th iterate of a stochastic matrix, then trivially each diagonal element like $p_{11}^{(n)}$ determines a renewal sequence by the formula

$$u_n = p_{11}^{(n)} \quad (n = 0, 1, 2, ...),$$

and also *every renewal sequence can be represented in this way.* So here we have an entirely unexpected and very puzzling distinction between the discrete and continuous time versions of Markovian theory.

Provoked by this discovery, Kingman commenced a search for a characterization of $\mathscr{PM}$ as a subset of $\mathscr{P}$, which took seven years, but which was triumphantly successful. The result, which is somewhat too complicated to describe here, was presented in [49] and is elaborated upon in [50]. Much else of great interest emerged during the chase; a theory of linked regenerative phenomena, of stationary regenerative phenomena, of non-standard regenerative phenomena, and so on. Here I want to pick out just two aspects of Kingman's work which will give some idea of its characteristic flavour.

(27) Most readers, or at least those of my generation, will remember the concept of 'depth' in number theory. The prime number theorem was considered to be deeper than, say, the unique decomposition theorem because no one could see how to prove the former without an apparently irrelevant excursion into complex function theory. Such a concept, if well founded, can be an invaluable guide in the construction of proofs. For example, it was the comment by R. Rado that 'there seem to be uncountably many independent necessary conditions' which led Jane Speakman [70] to a successful solution to the difficult problem of characterizing intrinsically all those ideals $\mathscr{I}$ of subsets A of a countably infinite set (say the positive integers) which admit an explicit representation of the form,

(57) $$A \in I \quad \text{if and only if} \quad \sum_{n \in A} a_n < \infty$$

($\sum a_n$ being a divergent series of positive terms). Unfortunately the concept of depth in number theory proved to be illusory; as we all know there now exists an 'elementary' (if very difficult) proof of the prime number theorem which is wholly phrased in the language of real-variable theory. These remarks are offered as a preface to the statement that in Markov chain theory Kingman has introduced a concept of depth which is guaranteed against such a failure, because of the precise way in which it is formulated. A theorem about Markov chains is called *deep* if it holds for $\mathscr{PM}$ but not for all of $\mathscr{P}$, and *shallow* if it holds for $\mathscr{P}$ as a whole. (Of course, in the way we are setting out the argument here, the concept applies only to theorems concerned with the diagonal elements of families of transition probabilities, but it could be extended beyond this.)

Here are some shallow theorems about elements of $\mathscr{PM}$: each 'skeleton' $u_n = p(nh)$ $(n = 0, 1, 2, ...)$ defines a renewal sequence u; p is strictly positive and uniformly continuous on $[0, \infty)$, it has a (perhaps infinite) right-hand derivative at $t = 0$, and is differentiable everywhere in $(0, \infty)$ save at most on a countable set (at all points of which the one-sided derivatives both exist); $\Phi(t) = p(|t|)\,(-\infty < t < \infty)$ defines a cf; $p(t)$ tends to a limit as $t \to \infty$; p satisfies an integral equation of the Volterra form

$$(58) \qquad p(t) + \int_0^t p(t-s)\,m(s)\,ds = 1 \quad (t \geqslant 0),$$

where the non-negative monotonic decreasing function m determines p uniquely; there is a unique non-negative totally finite measure μ on the Borel sets of $(0, \infty]$ such that

$$(59) \qquad s + \int_{(0,\infty]} \frac{1 - e^{-sx}}{1 - e^{-x}}\,\mu(dx) = 1/r(s),$$

where

$$(60) \qquad \int_0^\infty e^{-st} p(t)\,dt = r(s) \quad (0 < s < \infty);$$

there is a one-to-one correspondence between the measures μ which occur in equation (59) and the elements p in $\mathscr{P}$; the function m which occurs in equation (58) is given by

$$(61) \qquad m(s) = \int_{(s,\infty]} \frac{\mu(dx)}{1 - e^{-x}} \quad (s > 0).$$

All these results are true for $\mathscr{P}$, so are shallow theorems for $\mathscr{PM}$.

Here for the sake of the contrast are two deep theorems for $\mathscr{PM}$: p is continuously differentiable for $t > 0$; the measure μ in equation (59) is absolutely continuous with respect to Lebesgue measure and admits a strictly positive Radon–Nikodym derivative. These are false for $\mathscr{P}$, in general.

Because of the μ–p correspondence mentioned in connection with equations (59) and (60), it is clear that the problem of characterizing $\mathscr{PM}$ reduces to that of characterizing those measures μ which correspond to p-functions in $\mathscr{PM}$, and this is how Kingman's characterization of the latter class was constructed. As he has remarked, the crude division of Markovian theorems into shallow and deep can now be further refined. If Π_1 and Π_2 are two properties possessed by *all Markovian p*-functions, then by reason of the correspondence just mentioned Π_1 and Π_2 can be identified with the corresponding sets of measures μ, so that a p-function in $\mathscr{P}$ will have the property Π_j $(j = 1, 2)$ if and only if the corresponding measure μ belongs to the set of measures which (without risk of ambiguity) we can also call Π_j. We can now say that, for Markov p-functions, *the fact that they all have property* Π_1 *is deeper than the fact that they all have property* Π_2 *when, for the corresponding sets of measures,*

(62) $$\Pi_1 \subset \Pi_2, \quad \Pi_1 \neq \Pi_2.$$

We thus have a well-defined partial ordering for the theorems concerning diagonal Markovian transition functions. With respect to this partial ordering it turns out that continuous differentiability of p is less deep than the absolute continuity of μ (in fact, continuous differentiability of p is equivalent to non-atomicity for μ).

(28) This is a good place to mention another quite different shallow/deep classification for Markovian theorems, which has so far been very little explored. It is known (the best example was given by Feller [23]) that the Chapman–Kolmogorov relations

(63) $$p_{ij}(u+v) = \sum_k p_{ik}(u) p_{kj}(v) \quad (u, v \geqslant 0)$$

together with, say, continuity for the p_{ij}s, and $p_{ij}(0) = \delta_{ij}$, do *not* characterize Markov processes. Now a very large part of Markovian theory is based on equation (63) coupled with the relatively trivial requirements that each $p_{ij}(t) \geqslant 0$ and $\sum_j p_{ij}(t) = 1$. This is in fact that part of Markovian theory (for which see, e.g. Chung [10, 11], Dynkin [21], Freedman [25, 26], Reuter [68] and Williams [76]) which reduces to the study of transition semigroups of bounded operators acting on the Banach space l_1. Some quite difficult results have this character; for example the famous theorem of Lévy and Austin, known as the *Lévy dichotomy*:

> for fixed i and j, and $0 < t < \infty, p_{ij}(t)$ is either
> everywhere zero or everywhere strictly positive.

Now if any countable-state stochastic process has continuous versions of the transition probabilities

$$p_{ij}(t) = \mathrm{pr}\,(\{w\colon X_{t+\tau}(w) = j \mid X_\tau(w) = i\})$$

which do not depend on τ and which satisfy equation (63), then by the Daniell–Kolmogorov theorem these determine a name-class for *Markov* processes, and so any semigroup-type Markov theorem (e.g. any theorem which can be *stated* in terms of the transition probabilities alone) must be true for the parent stochastic process *whether that process was Markov or not*. Such theorems thus enjoy a new sort of shallowness; we may call them CK-theorems and it is evident that the Lévy dichotomy is one such. Unfortunately non-Markovian CK-processes have not yet turned up in any practical contexts, and therefore we really know very little about them except that they exist.

(29) We turn now to another aspect of Kingman's work. I have already hinted above (and it is true) that one can extend the notion of a p-function to matrices with components $p_{ij}(.)$ which generalize semigroups of stochastic matrices in the same way that the original concept of a p-function generalizes the property of being a diagonal element of a semigroup of stochastic matrices. All this has been worked out by Kingman and analogues of most of the $\mathscr{P}$-theorems have been given by him in this wider setting. Now we know that the Lévy dichotomy will not permit a Markovian $p_{ij}(.)$ to attain the value 0 (or 1) (still less to be constantly equal to 0 (or 1) over a subinterval) on the open half-line $(0, \infty)$, unless $p_{ij}(t) = 0$ (or 1) for that i and j, and *all* t. This fact suggested the following conjecture: if $0 < c < 1$, then a Markovian transition function $p_{ij}(.)$ cannot have the property

(64) $$p_{ij}(t) = c \quad \text{for } 0 \leqslant \tau_1 < t < \tau_2 < \infty.$$

Kingman (who has characterized Markovian transition functions of the form p_{ij} as well as those of the form p_{ii}) has used his characterization to show that the conjecture is *false*.

(30) Another conjecture, made at the same time [35], has a slightly different origin. Suppose that an otherwise unknown standard semigroup of Markov transition functions $p_{ij}(.)$ satisfies the following conditions:

(65)   *for each i and j,* $p_{ij}(t) = f_{ij}(t)$ *for* $0 \leqslant t \leqslant T_{ij}$.

Here the (strictly positive) Ts and the (continuous) fs are supposed given.

Conjecture. Then the p_{ij}s are uniquely determined for all $t \geqslant 0$.

If this is true, it could prove the starting point for a new approach to the theory of the generation of Markov processes, for we could proceed as follows:

Let T denote the system $\{T_{ij}: i, j = 1, 2, ...\}$, and let $(T; f)$ denote the system $\{(T_{ij}, f_{ij}): i, j = 1, 2, ...\}$, for any choice of positive T_{ij}s and continuous f_{ij}s. Then the Ts, if partly ordered by

(66) $$T \leqslant T' \quad \text{when } T_{ij} \leqslant T'_{ij} \quad \text{for all } i \text{ and } j,$$

form a directed system (directed 'downwards'), and relative to this the $(T; f)$s form a net which is an inductive spectrum in the sense of Dugundji ([20], p. 420), the mapping $\Phi_{T', T}$ being defined by

(67) $$\Phi_{T', T}(T'; f') = (T; f),$$

where f_{ij} is the restriction of f'_{ij} to $[0, T_{ij}]$. If we then form the inductive limit of this inductive spectrum, we obtain the system of *germs* at $t = 0+$.

The conjecture, if true, will tell us that if two standard Markov transition semigroups 'belong' to (equivalently *have*) the same germ, then they must be *identical*. We shall then have the situation that a given germ either determines *a unique*, or *no*, Markov transition semigroup; that is, we shall have Markovian germs and non-Markovian germs. We could then propose the problem of identifying the Markovian germs, and the further problem of generating a Markov transition semigroup when its germ is given. It would be natural to look for help with these problems to non-standard real-variable theory, and to hope, by defining $p_{ij}(t)$ over all non-negative 'infinitesimal' t, to repair the damage caused by the failure of the derivative matrix, with components

(68) $$q_{ij} = p'_{ij}(0+) \quad (i,j = 1, 2, ...),$$

to characterize a Markov transition semigroup uniquely. But all this is possibly vacuous speculation while the status of the conjecture remains in doubt. Perhaps this is the appropriate place to record the fact that we do not even know what matrices $Q = \{q_{ij} : i,j = 1, 2, ...\}$ can occur as at equation (68) in connection with Markov processes. For a recent contribution to this problem see Williams [77].

My second conjecture italicized above is still open. One way of approaching it is to formulate the stronger *conjecture*: if $p \in \mathscr{PM}$, then p $(= p_{11})$ is uniquely determined when its values over some non-degenerate interval $[0, T]$ are given. This, however, as Kingman has shown, is *false*, and in a sense almost maximally so. For he has proved that if the 'section' of p which is given is not exponential $(p(t) = \mathrm{e}^{-\lambda t})$, then there are *always* at least two distinct ways of continuing it (in $\mathscr{PM}$) to $[0, \infty)$.

This last result of Kingman destroys any hope one might have had that the class $\mathscr{PM}$ would prove to be quasi-analytic near the origin. Some quasi-analytic subclasses of $\mathscr{PM}$ have indeed been found [35], using a fascinating new theory of quasi-analyticity due to Neuberger [64, 65] which is based on successive differences and not (as classically) on successive derivatives. (It is this last feature of Neuberger's theory which makes it relevant, for it has been known for some time [79] that $\mathscr{PM}$-functions need not belong to C^∞.) Recent work (Beurling [3]; Kato [33]) seems to show that these quasi-analytic classes may in fact consist entirely of *analytic* $\mathscr{PM}$-functions, so perhaps *strictly* quasi-analytic classes of $\mathscr{PM}$-functions are still to be found.

(31) These remarks show some of the big strides which have been made in Markovian theory since 1950, when Kolmogorov [53] remarked that one still did not know of a non-analytic Markovian transition function. (One was found almost immediately after this by Yuskevic [79].)

(32) On taking leave of renewal sequences, *p*-functions and Delphic semigroups we may note that these ideas have contributed to the classical factorization theory of cfs, by providing one approach to the very remarkable Goldie–Steutel theorem on mixtures of exponential distributions mentioned above (just after formula (34)) (see [28] and [71]), and also by facilitating practical calculations in concrete queueing contexts (Kingman [48] and Tweedie [72]).

(33) This review has certainly not covered everything which might be included under the rather broad umbrella of stochastic analysis, and indeed it is rather heavily biased towards those aspects of the subject in which the United Kingdom workers have been most interested. Normally exception might well be taken to this, but on the present occasion I have taken as my brief, to provide a sketch of the mathematical activities of the group within which Davidson worked, in order to give some measure of unity to this book containing his own publications together with individual contributions by his friends, actual and virtual. Even so there are glaring gaps, and much of what has been omitted is of equal interest and of comparable importance.

REFERENCES

1. A. D. Barbour, "The principle of the diffusion of arbitrary constants", *J. Appl. Prob.*, **9** (1972), 519–541.
2. J. G. Basterfield, *Some problems in probability*, Ph.D. dissertation, Cambridge (1971).
3. A. Beurling, "On analytic extension of semi-groups of operators", *J. Functional Anal.* **6** (1970), 387–400.
4. P. Billingsley, *Convergence of Probability Measures*, Wiley, New York (1968).
5. N. H. Bingham, "Factorisation theorems and domains of attraction for generalised convolution algebras", *Proc. Lond. math. Soc.* (3) **23** (1971), 16–30.
6. ——, "Random walk on spheres", *Ztschr. Wahrsch'theorie & verw. Geb.* **22** (1972), 169–192.
7. ——, "Positive-definite functions on spheres", *Proc. Cam. Phil. Soc.*, to appear.
8. D. Blackwell and D. Freedman, "On the local behaviour of Markov transition probabilities", *Ann. Math. Stat.* **39** (1968), 2123–2127.
9. P. Bloomfield, "Lower bounds for renewal sequences and *p*-functions", *Ztschr. Wahrsch'theorie & verw. Geb.* **19** (1971), 271–273.
10. K. L. Chung, *Markov Chains*, 2nd ed., Springer, Berlin (1967).
11. ——, *Lectures on Boundary Theory for Markov Chains* (*Annals of Math. Studies* **65**), Princeton U.P., Princeton (1970).
12. P. J. Cohen, "Factorisation in group algebras", *Duke Math. J.* **26** (1959), 199–205.

13. D. J. Daley, "On a class of renewal functions", *Proc. Cam. Phil. Soc.* **61** (1965), 519–526.
14. P. J. Daniell, "Integrals in an infinite number of dimensions", *Ann. of Math.* (2) **20** (1918–9), 281–288.
15. ——, "Functions of limited variation in an infinite number of dimensions", *Ann. of Math.* (2) **21** (1919–20), 30–38.
16. H. E. Daniels, "Contribution to a discussion", *J. R. statist. Soc. B* **19** (1957), 224–225 and 230–231.
17. ——, "The distribution of the total size of an epidemic", *Proc. 5th Berkeley Symp.* **4** (1967), 281–293.
18. J. L. Doob, "Probability in function space", *Bull. Am. math. Soc.* **53** (1947), 15–30.
19. ——, *Stochastic Processes*, Wiley, New York (1953).
20. J. Dugundji, *Topology*, Allyn & Bacon, Boston (1966).
21. E. B. Dynkin, *Markov Processes* (2 vols.), Springer, Berlin (1965).
22. K. S. Fahady, M. P. Quine and D. Vere-Jones, "Heavy traffic approximations for the Galton–Watson process", *Adv. Appl. Prob.* **3** (1971), 282–300.
23. W. Feller, "Non-Markovian processes with the semigroup property", *Ann. Math. Statist.* **30** (1959), 1252–1253.
24. ——, *An Introduction to Probability Theory and its Applications* (vol. I, 3rd ed.; vol. II, 2nd ed.), Wiley, New York (1968–71).
25. D. Freedman, *Markov Chains*, Holden Day, San Francisco (1971).
26. ——, *Approximating Countable Markov Chains*, Holden Day, San Francisco (1971).
27. B. V. Gnedenko and I. N. Kovalenko, *Introduction to Queueing Theory*, Moscow (1966). (In English: Jerusalem (1968).)
28. C. Goldie, "A class of infinitely divisible random variables", *Proc. Cam. Phil. Soc.* **63** (1967), 1141–1143.
29. P. R. Halmos, *Measure Theory*, Van Nostrand, Toronto (1950).
30. D. L. Iglehart, "Diffusion approximations in applied probability", *Lectures in Applied Math.* (Am. Math. Soc.) **12** (1967), 235–254.
31. S. Johansen, "An application of extreme-point methods to the representation of infinitely divisible distributions", *Ztschr. Wahrsch'theorie & verw. Geb.* **5** (1966), 304–316.
32. T. Kaluza, "Über die Koeffizienten reziproker Potenzreihen", *Math. Zeit.* **28** (1928), 161–170.
33. T. Kato, "A characterisation of holomorphic semigroups", *Proc. Am. math. Soc.* **25** (1970), 495–498.
34. D. G. Kendall, "Extreme-point methods in stochastic analysis", *Ztschr. Wahrsch'theorie & verw. Geb.* **1** (1963), 295–300.
35. ——, "Some recent advances in the theory of denumerable Markov processes", *Trans. 4th Prague Conf. on Information Theory etc.* (1967), 11–17.
36. ——, "Renewal sequences and their arithmetic", *Symposium on Probability Methods in Analysis* (*Lecture Notes in Math.* **31**), Springer, Berlin (1967) (2.1 of this book).
37. ——, "Delphic semigroups, infinitely divisible regenerative phenomena, and the arithmetic of *p*-functions", *Ztschr. Wahrsch'theorie & verw. Geb.* **9** (1968), 163–195 (2.2 of this book).

38. D. G. Kendall, "Foundations of a theory of random sets", *Stochastic Geometry*, Wiley, London, Chapter 6.2.
39. ——, "Separability and measurability for stochastic processes: a survey" (5.4 of this book).
40. D. G. Kendall and J. Lamperti, "A remark on topologies for characteristic functions", *Proc. Cam. Phil. Soc.* **68** (1970), 703–705.
41. D. P. Kennedy, "The continuity of the single server queue", *J. Appl. Prob.* **9** (1972), 370–381.
42. A. Ya. Khinchin, "The arithmetic of distribution laws", *Bull. Univ. Etat Moscou, Sér. Int., Sect. A, Math. et Mécan.* **1** (fasc. 1) (1937), 6–17.
43. J. F. C. Kingman, "The single server queue in heavy traffic", *Proc. Cam. Phil. Soc.* **57** (1961), 902–904.
44. ——, "On queues in heavy traffic", *J. R. statist. Soc. B* **24** (1962), 383–392.
45. ——, "The stochastic theory of regenerative events", *Ztschr. Wahrsch' theorie & verw. Geb.* **2** (1964), 180–224.
46. ——, "An approach to the study of Markov processes", *J. R. statist. Soc. B* **28** (1968), 417–447.
47. ——, "On measurable *p*-functions", *Ztschr. Wahrsch'theorie & verw. Geb.* **11** (1968), 1–8.
48. ——, "An application of the theory of regenerative phenomena", *Proc. Cam. Phil. Soc.* **68** (1970), 697–701.
49. ——, "Markov transition probabilities, V", *Ztschr. Wahrsch'theorie & verw. Geb.* **17** (1971), 89–103.
50. ——, *Regenerative Phenomena*, Wiley, London (1972).
51. F. B. Knight, "On the random walk and Brownian motion", *Trans. Am. math. Soc.* **103** (1962), 218–228.
52. A. N. Kolmogorov, *Grundbegriffe der Wahrscheinlichkeitsrechnung*, Springer, Berlin (1933). (In English: Chelsea, New York (1950).)
53. ——, "On the differentiability of the transition probabilities in homogeneous Markov processes with a denumerable number of states", *Učenye Zapiski MGY* 148 (*Mat.*) **4** (1951), 53–59.
54. ——, "On the approximation of distributions of sums of independent summands by infinitely divisible distributions", *Sankhyā (A)* **25** (1963), 159–174.
55. J. Lamperti, "On the coefficients of reciprocal power-series", *Am. Math. Monthly* **65** (1958), 90–94.
56. ——, "The arithmetic of certain semigroups of positive operators", *Proc. Cam. Phil. Soc.* **64** (1968), 161–166.
57. P. Lévy, *Théorie de l'Addition des Variables Aléatoires*, 2nd ed., Gauthier Villars, Paris (1954).
58. T. Lewis, "The factorisation of the rectangular distribution", *J. Appl. Prob.* **4** (1967), 529–542.
59. Yu. V. Linnik, *Decomposition of Probability Distributions*, Oliver & Boyd, Edinburgh (1964).
60. E. Lukacs, *Characteristic Functions*, 2nd ed., Griffin, London (1970).
61. P. A. Meyer, *Probability and Potentials*, Blaisdell, Waltham (1966).
62. A. V. Nagaev and A. N. Starcev, "The asymptotic analysis of a stochastic model of an epidemic", *Theory of Probability* **15** (1970), 98–107.

63. E. Nelson, "Regular probability measures on function space", *Ann. of Math.* **69** (1959), 630–643.
64. J. W. Neuberger, "A quasi-analyticity condition in terms of finite differences", *Proc. Lond. math. Soc.* (3) **14** (1964), 245–259.
65. ——, "Analyticity and quasi-analyticity for one-parameter semigroups", *Proc. Am. math. Soc.* **25** (1970), 488–494.
66. J. Neveu, *Mathematical Foundations of the Calculus of Probability*, Holden Day, San Francisco (1965).
67. R. E. A. C. Paley and N. Wiener, *Fourier Transforms in the Complex Domain*, Amer. Math. Soc., New York (1934).
68. G. E. H. Reuter, "Denumerable Markov processes and the associated contraction semigroups on *l*", *Acta Math.* **97** (1957), 1–46. [See also *J. Lond. math. Soc.* **34** (1959), 81–91 and **37** (1962), 63–73.]
69. W. Rudin, "Factorisation in the group algebra of the real line", *Proc. Nat. Acad. Sci. U.S.A.* **43** (1957), 339–340.
70. Jane M. O. Speakman, "An algebraic characterisation of convergence ideals", *J. Lond. math. Soc.* **44** (1968), 26–30.
71. F. W. Steutel, "Note on the infinite divisibility of exponential mixtures", *Ann. Math. Statist.* **38** (1967), 1303–1305. [See also **39** (1968), 1153–1157 and **40** (1969), 1130–1131.]
72. R. L. Tweedie, "Queues with server breakdown", in preparation.
73. W. Whitt, "A guide to the application of limit theorems for sequences of stochastic processes", *Oper. Res.* **18** (1970), 1207–1213.
74. ——, "The continuity of queues", to appear.
75. D. V. Widder, *The Laplace Transform*, Princeton University Press, Princeton (1941).
76. D. Williams, "A new method of approximation in Markov chain theory", *Proc. Lond. math. Soc.* (3) **16** (1966), 213–240.
77. ——, "A note on the *Q*-matrices of Markov chains", *Ztschr. Wahrsch' theorie & verw. Geb.* **7** (1967), 116–121.
78. A. M. Yaglom, "Certain limit theorems of the theory of branching stochastic processes", *Dokl. Akad. Nauk U.S.S.R.* (N.S.) **56** (1947), 795–798.
79. A. A. Yuskevic, "On differentiability of transition probabilities of homogeneous Markov processes with a countable number of states", *Učenye Zapiski MGY* 186 (*Mat.*) **9** (1959), 141–159.

2
DELPHIC SEMIGROUPS, AND RELATED ARITHMETIC PROBLEMS

2.1

Renewal Sequences and their Arithmetic

D. G. KENDALL

[Reprinted from *Symposium in Probability Methods in Analysis* (*Lecture Notes in Math.* **31**) (1967) 147–175.]

2.1.1 THE KALUZA PROPERTY

In 1928 Th. Kaluza [8] noticed that if the formal power-series

$$(1) \qquad\qquad U(z) = \sum_{n \geqslant 0} u_n z^n$$

has real coefficients satisfying the conditions

$$(2) \qquad u_0 = 1, \quad u_n \geqslant 0, \quad \begin{vmatrix} u_{n-1} & u_n \\ u_n & u_{n+1} \end{vmatrix} \geqslant 0 \quad (n \geqslant 1),$$

then the formal power-series

$$(3) \qquad\qquad F(z) = \sum_{r \geqslant 1} f_r z^r$$

defined by the identity

$$(4) \qquad\qquad U(z) = 1 + U(z) F(z)$$

will also have real non-negative coefficients. Kaluza added the extra condition $u_1 > 0$ to the conditions (2), but this is unnecessary, for if $u_1 = 0$ then conditions (2) imply that $U(z) = 1$ and $F(z) = 0$, so that then $f_r = 0$ for each r. If this degenerate case is excluded then every $u_n > 0$, and the conditions (2) then merely assert that $u_0 = 1$ and that

$$u_1/u_0, u_2/u_1, u_3/u_2, \ldots$$

is an increasing sequence of strictly positive terms. There is no limit imposed on the rate of growth of this sequence and we could, for example, have $u_n = e^{n^2}$. The situation becomes more interesting if we confine our attention to *bounded* Kaluza sequences $\{u_n : n = 0, 1, 2, \ldots\}$. Because

$$u_n = \frac{u_n}{u_0} = \frac{u_n}{u_{n-1}} \frac{u_{n-1}}{u_{n-2}} \cdots \frac{u_1}{u_0},$$

and because the factors in the last product never decrease with increasing n, it is clear that the boundedness of $\{u_n : n = 0, 1, 2, \ldots\}$ implies that each ratio u_n/u_{n-1} must be less than or equal to unity, and so a Kaluza sequence is bounded if and only if it satisfies the condition

(5) $$1 = u_0 \geqslant u_1 \geqslant u_2 \geqslant u_3 \geqslant \ldots \geqslant 0.$$

(We have expressed the final statement here in such a way as to include the degenerate case $U(z) = 1$, $F(z) = 0$.) Each of the power-series $U(z)$ and $F(z)$ will now have a radius of convergence of at least unity, and we shall have $\sum f_r \leqslant 1$. Thus (Lamperti [12]) *every bounded Kaluza sequence is a renewal sequence.*

Plainly not all renewal sequences are bounded Kaluza sequences; the renewal sequence $(1, 0, 1, 0, 1, \ldots)$ illustrates this fact. On the other hand, the Kaluza theorem enables us to write down renewal sequences which are not otherwise easily recognized as such; for example,

(6) $$(1, 2^{-t}, 3^{-t}, 4^{-t}, \ldots),$$

where t is any non-negative real number. The question then arises: *Can we characterize probabilistically those renewal sequences which are bounded Kaluza sequences?* We shall give such a characterization in this paper, by first constructing a theory of *infinitely divisible renewal sequences.* Encouraged by this, we shall then investigate the arithmetic of renewal sequences in general.

2.1.2 RENEWAL SEQUENCES AND REGENERATIVE PHENOMENA

When we speak of a renewal sequence $\{u_n : n = 0, 1, 2, \ldots\}$ we have of course in mind the following situation; $(\Omega, \mathscr{A}, \mathrm{pr})$ is a probability-space on which $\{E_n : n = 0, 1, 2, \ldots\}$ is a sequence of events (elements of the σ-algebra $\mathscr{A}$) such that $E_0 = \Omega$, and

(7) $$\mathrm{pr}\left(\bigcap_{j=1}^{k} E_{n_j}\right) = \mathrm{pr}(E_{n_1}) \prod_{j=1}^{k-1} \mathrm{pr}(E_{n_{j+1}-n_j})$$

for all $k = 2, 3, \ldots$ and all increasing sequences $n_1, n_2, \ldots, n_k$. We interpret this situation by inventing a regenerative phenomenon† Φ which is said to 'happen' at just those times t for which the sample-point ω lies in the measurable subset E_t of the sample-space Ω. The requirement (7) can then

---

† It has until now been usual to call Φ a regenerative or recurrent *event*. We find it confusing to speak thus of Φ as if it were an element of $\mathscr{A}$.

be described informally by saying that the mechanism responsible for the occurrence of the phenomenon Φ suffers a complete loss of memory whenever the phenomenon Φ occurs.

The corresponding renewal sequence $\{u_n : n = 0, 1, 2, ...\}$ is then defined by

$$(8) \qquad u_n = \text{pr}(E_n) \quad (n = 0, 1, 2, ...).$$

Here it is clear that $u_0 = 1$ and that $0 \leqslant u_n \leqslant 1$, but renewal sequences also satisfy more complicated non-linear relations, of which

$$(9) \qquad u_{m+n} \geqslant u_m u_n,$$

proved by noting that

$$u_{m+n} = \text{pr}(E_{m+n}) \geqslant \text{pr}(E_m \cap E_{m+n}) = \text{pr}(E_m)\text{pr}(E_n),$$

is the simplest (and will frequently be useful to us later). If we write

$$f_r = \text{pr}(\Phi \text{ happens when } t = r, \text{ and not at any earlier } t > 0),$$

then we shall have

$$(10) \qquad f_r \geqslant 0 \quad (r = 1, 2, ...) \quad \text{and} \quad \Sigma f_r \leqslant 1,$$

and

$$(11) \qquad u_n = f_n + f_{n-1}u_1 + f_{n-2}u_2 + ... + f_1 u_{n-1} \quad (n \geqslant 1).$$

Conversely, if $\{f_r : r = 1, 2, ...\}$ is given and satisfies the conditions (10), and if $\{u_n : n = 0, 1, 2, ...\}$ is defined by equation (11) and by $u_0 = 1$, then this will be a renewal sequence; that is, we can set up a model of a regenerative phenomenon Φ by constructing $\{E_n : n = 0, 1, 2, ...\}$ on a probability-space $(\Omega, \mathscr{A}, \text{pr})$ in such a way that $E_0 = \Omega$ and equation (7) is satisfied, and equation (8) holds. On introducing the power-series (1) and (3), one finds that the recurrent relation (11) (with $u_0 = 1$) is equivalent to the identity (4). All this is of course entirely familiar; it is set out here merely in order to establish the notation.

It may also be helpful to the reader if we recall the very simple proof of Kaluza's result. From equation (4), when $n \geqslant 1$, we have

$$0 = u_n - \sum_{j=1}^{n} u_{n-j} f_j$$

and

$$f_{n+1} = u_{n+1} - \sum_{j=1}^{n} u_{n-j+1} f_j,$$

so that

$$u_n f_{n+1} = \sum_{j=1}^{n} (u_{n+1} u_{n-j} - u_n u_{n-j+1}) f_j.$$

Now we need only consider the case in which $u_1 > 0$, and then every $u_n > 0$, and the ratio u_{m+1}/u_m is non-decreasing for increasing m. The last identity together with $f_1 = u_1 > 0$ then shows recurrently that each $f_r \geqslant 0$.

2.1.3 INFINITELY DIVISIBLE RENEWAL SEQUENCES

We write $\mathscr{R}$ for the class of all renewal sequences $\{u_n : n = 0, 1, 2, ...\}$, and $\mathscr{R}_0$ for the class of *infinitely divisible* renewal sequences defined by the requirement that

(12) $(1, u_1^t, u_2^t, ...)$

is to belong to $\mathscr{R}$ for each real non-negative t. This is a very natural definition. In the first place, such sequences exist; the sequence (6) in fact provides us with the example

$$u_n = \frac{1}{n+1} \quad (n = 0, 1, 2, ...).$$

Next, it is known (Kingman [11]) that

(a) if $\{u_n' : n = 0, 1, 2, ...\}$ and $\{u_n'' : n = 0, 1, 2, ...\}$ are renewal sequences, then so is $\{u_n' u_n'' : n = 0, 1, 2, ...\}$;

(b) if $\{u_n(k) : n = 0, 1, 2, ...\}$ is a renewal sequence for each k, and if $u_n(k) \to v_n$ as $k \to \infty$, for each n, then $\{v_n : n = 0, 1, 2, ...\}$ is a renewal sequence.

This being so, it will be clear that if T denotes the set of non-negative real numbers t such that the sequence (12) is a renewal sequence, then T is a closed additive semigroup containing $t = 0$ and $t = 1$. The class $\mathscr{R}_0$ is that subclass of $\mathscr{R}$ defined by the requirement that the semigroup T is to be the half-line $[0, \infty)$.

From the closure of T it is clear that $T = [0, \infty)$ if and only if 0 is a cluster-point of T. In particular, a renewal sequence $\{u_n : n = 0, 1, 2, ...\}$ belongs to $\mathscr{R}_0$ if and only if the sequence of mth roots $\{u_n^{1/m} : n = 0, 1, 2, ...\}$ is a renewal sequence for every positive integer m.

We shall content ourselves here with characterizing $\mathscr{R}_0$, and we defer for the time being the wider question of just what closed semigroups T can be associated in this way with a general renewal sequence.

If $\{u_n: n = 0, 1, 2, ...\}$ is a bounded Kaluza sequence, then so is the sequence (12) for each $t \geqslant 0$. Thus Lamperti's observation shows that if $\mathcal{R}_1$ is the class of bounded Kaluza sequences, then

$$(13) \qquad \mathcal{R}_1 \subseteq \mathcal{R}_0 \subseteq \mathcal{R}.$$

This suggests that we may best reach an understanding of the place of $\mathcal{R}_1$ in $\mathcal{R}$ by considering it as a subclass of $\mathcal{R}_0$. Our principal result is that, apart from some technical details which we shall meet presently, $\mathcal{R}_1$ and $\mathcal{R}_0$ are essentially identical.

We conclude this section by recalling briefly the proofs of (*a*) and (*b*); (*b*) is most easily proved by use of equation (11), but (*a*) has an illuminating probabilistic proof which will provide a pattern for later calculations. Let Φ' and Φ'' be regenerative phenomena defined on two different probability-spaces, and let $\{u'_n: n = 0, 1, 2, ...\}$ and $\{u''_n: n = 0, 1, 2, ...\}$ be the associated renewal sequences. Construct the direct product of the two probability-spaces,

$$(\Omega' \times \Omega'', \mathcal{A}' \times \mathcal{A}'', \mathrm{pr}' \times \mathrm{pr}''),$$

and on this define a new regenerative phenomenon $\Phi' \cap \Phi''$ which 'happens' at time t if and only if Φ' and Φ'' *both* 'happen' at time t; i.e. if and only if $(\omega', \omega'') \in E'_t \times E''_t$. Then $\Phi' \cap \Phi''$ will have $\{u'_n u''_n: n = 0, 1, 2, ...\}$ as its renewal sequence.

2.1.4 THE REDUCTION OF $\mathcal{R}_0$ TO $\mathcal{R}'_0$

We find ourselves embarrassed by the zero components of the vector $\{u_n: n = 0, 1, 2, ...\}$, but within $\mathcal{R}_0$ there is a simple means of getting rid of them. Let us put

$$(14) \qquad v_n = \lim_{m \to \infty} u_n^{1/m} = \begin{cases} 1 & \text{if } u_n > 0, \\ 0 & \text{if } u_n = 0. \end{cases}$$

Then by (*b*), the sequence $\{v_n: n = 0, 1, 2, ...\}$ belongs to $\mathcal{R}$, and indeed to $\mathcal{R}_0$. We may have $u_n = 0$ for each $n \geqslant 1$, in which case $v_n = u_n$ for every n. If we leave this case on one side for the moment then we can write N for the first integer $n \geqslant 1$ for which $u_n > 0$, and then (from inequality (9)) we shall have $v_{kN} = 1$ for every k. From equation (7) we now obtain (for the Φ^* associated with $\{v_n: n = 0, 1, 2, ...\}$)

$$\mathrm{pr}(E_{kN+r}) = \mathrm{pr}(E_{kN} \cap E_{kN+r}) = \mathrm{pr}(E_{kN}) \mathrm{pr}(E_r) = 0,$$

if $0 < r < N$. Thus Φ^* recurs almost surely after N steps, and

$$u_n > 0 \quad \text{when } N \mid n,$$
$$u_n = 0 \quad \text{when } N \nmid n,$$

and so in $\{u'_n : n = 0, 1, 2, ...\}$ we have a zero-free member of $\mathcal{R}_0$ if we put

(15) $$u'_n = u_{Nn}.$$

We have used here the selection principle (Kingman [11])

(c) if $\{u_n : n = 0, 1, 2, ...\} \in \mathcal{R}$, then $\{u_{kn} : n = 0, 1, 2, ...\} \in \mathcal{R}$ for every $k \geqslant 0$.

Like (a), this property of $\mathcal{R}$ is best proved probabilistically; i.e. via equation (7).

We now see that $\mathcal{R}_0$ has been partitioned into

(i) the special element $(1, 0, 0, 0, ...)$;

(ii) equivalence classes of renewal sequences, each class being characterized by a unique zero-free infinitely divisible renewal sequence

$$\{u'_n : n = 0, 1, 2, ...\}$$

defined as at equation (15).

We write $\mathcal{R}'_0$ for the set of equivalence classes at (ii), identifying each member of $\mathcal{R}'_0$ with the associated zero-free infinitely divisible renewal sequence $\{u'_n : n = 0, 1, 2, ...\}$ This allows us to work with zero-free infinitely divisible renewal sequences only, from now on. (To obtain the general member of $\mathcal{R}_0$ other than $(1, 0, 0, 0, ...)$ we have only to choose a positive integer N and insert $(N-1)$-ads of zeros into a general member of $\mathcal{R}'_0$.)

If $\{u_n : n = 0, 1, 2, ...\} \in \mathcal{R}'_0$ then it is natural to write

(16) $$u_n = e^{-x_n} \quad (n = 0, 1, 2, ...),$$

where $x_0 = 0$ and $0 \leqslant x_n < \infty$ for all n. From (a) and the infinite divisibility it follows that the system $\mathscr{C}$ of all vectors $\mathbf{x} = \{x_n : n = 0, 1, 2, ...\}$ is a proper pointed convex cone, the vertex $(0, 0, 0, ...)$ corresponding to $u_n = 1$ (all n), i.e. to the almost sure regenerative phenomenon. We can interpret (b) in the present situation if we think of $\mathscr{C}$ as lying in the vector space R^∞ with the topology of simple convergence, for (b) then implies that $\mathscr{C}$ is closed.

As usual we obtain a base for the cone by slicing it by a closed hyperplane which avoids the vertex; the hyperplane defined by $x_1 = 1$ is convenient. The base $\mathscr{B}$ so obtained is convex, closed and metrizable (for R^∞ is such), and we shall now see that it is compact. For inequality (9) tells us that (on $\mathscr{C}$)

(17) $$x_{m+n} \leqslant x_m + x_n,$$

and so on the base $\mathscr{B}$ we must have

(18) $$0 \leqslant x_n \leqslant n x_1 = n.$$

Thus

$$\mathscr{B} \subseteq \underset{n \geqslant 0}{\bigtimes} [0, n],$$

and the compactness of $\mathscr{B}$ now follows from the theorem of Tychonov (the Cartesian product of compacts is compact). ($\bigtimes$ = sign for Cartesian product.)

Accordingly ([2]; for a simple proof see [1]) $\mathscr{B}$ possesses a set $\partial\mathscr{B}$ of extreme points which is non-empty and is a $\mathscr{G}_\delta$, and if $\mathbf{y}$ denote a typical extreme point then an arbitrary element $\mathbf{x}$ of this cone $\mathscr{C}$ has a Choquet representation of the form

(19) $$x_n = \int_{\partial\mathscr{B}} y_n \mu_{\mathbf{x}}(d\mathbf{y}) \quad (n = 0, 1, 2, \ldots),$$

where $\mu_{\mathbf{x}}$ is a totally finite measure on the Borel subsets of $\partial\mathscr{B}$.

We shall identify the elements of $\partial\mathscr{B}$ and so make equation (19) explicit, thus obtaining the explicit (and, as it turns out, faithful) parametrization

(20) $$u_n = \exp\left\{ -\int_{\partial\mathscr{B}} y_n \lambda_{\mathbf{u}}(d\mathbf{y}) \right\} \quad (n = 0, 1, 2, \ldots)$$

the measure $\lambda_{\mathbf{u}}$ being uniquely determined by the zero-free infinitely divisible renewal sequence $\{u_n : n = 0, 1, 2, \ldots\}$ in $\mathscr{R}'_0$.

2.1.5 THE RELATION BETWEEN $\mathscr{R}_1$ AND $\mathscr{R}_0$

We have already remarked that $(1, 0, 0, \ldots)$ is the only bounded Kaluza sequence having zero elements. Let $\mathscr{R}'_1$ denote $\mathscr{R}_1$ with this exceptional element removed. We now prove

Theorem 1. *The class $\mathscr{R}'_1$ of zero-free bounded Kaluza sequences is identical with the class $\mathscr{R}'_0$ of zero-free infinitely divisible renewal sequences.*

Proof. We have only to prove that $\mathscr{R}'_0 \subseteq \mathscr{R}_1$. Suppose then that $\{u_n : n = 0, 1, 2, \ldots\}$ belongs to $\mathscr{R}'_0$ and define $\mathbf{x}$ as at equation (16) above. Then for each non-negative real number t there will exist non-negative real numbers $f_r(t)$ $(r = 1, 2, \ldots)$ with sum not exceeding unity, such that

(21) $$\sum_{n=0}^{\infty} e^{-t x_n} z^n = \left\{ 1 - \sum_{r=1}^{\infty} f_r(t) z^r \right\}^{-1} \quad (|z| < 1).$$

Obviously $f_r(0) = \delta_{r1}$; also $e^{-t x_n} \to 1$ as $t \to 0$ for each n, and so, because

3

the quantities $f_r(t)$ are polynomials in the quantities $u_n(t) = e^{-tx_n}$, it follows that

$$\lim_{t \to 0} f_r(t) = f_r(0) = \delta_{r1}.$$

We now assert that

(22)
$$\lim_{t \to 0} \frac{1 - f_1(t)}{t} = x_1$$

and that

(23)
$$\lim_{t \to 0} \frac{f_r(t)}{t} = 2x_{r-1} - x_r - x_{r-2} \quad (r = 2, 3, \ldots).$$

For a proof of equations (22), and of (23) when $r = 2$, we have only to recall that

$$f_1(t) = e^{-tx_1}$$

and that

$$f_2(t) = e^{-tx_2} - e^{-2tx_1}.$$

The result (23) for general r is then easily established by an inductive argument based on the identity

$$f_n = (1 - f_1) + f_1(1 - u_{n-1}) - \sum_{j=2}^{n-1} f_j u_{n-j} - (1 - u_n) \quad (n \geq 3),$$

which is simply a rearrangement of equation (11).

Now $f_r(t) \geq 0$, and so equation (23) tells us that $\mathbf{x}$ is a *concave* vector, i.e. that

$$u_n u_{n-2} \geq u_{n-1}^2 \quad (n = 2, 3, \ldots),$$

so that the last condition in (2) is satisfied and $\{u_n : n = 0, 1, 2, \ldots\}$ belongs to $\mathscr{R}_1'$, as required. (That the quantities u_n are positive and bounded and that $u_0 = 1$ is of course obvious.)

2.1.6   THE EXTREME RAYS OF THE CONE $\mathscr{C}$

We now change our point of view and study the analytical object $\mathscr{R}_1$ in order to obtain a better understanding of the probabilistic object $\mathscr{R}_0$. We have seen that the base $\mathscr{B}$ of the cone $\mathscr{C}$ is just the set of vectors $\mathbf{x}$ specified by

(24)
$$x_0 = 0, \quad x_1 = 1, \quad x_n \geq 0, \quad \Delta^2 x_n \leq 0,$$

and we shall show (as is presumably well known) that the extreme points of the convex set defined by the relations (24) are precisely the vectors

$\mathbf{y}(k)$, where

(25)
$$y_n(k) = \begin{cases} n & (0 \leqslant n \leqslant k), \\ k & (k \leqslant n) \end{cases}$$

and $k = 1, 2, ..., \infty$. We note that the relations (24) imply that

$$x_{m+n} \leqslant x_m + x_n,$$

that $x_n \leqslant n$ and that x_n is non-decreasing for increasing n.

That the $\mathbf{y}(k)$ are extreme is easily shown; suppose that $\mathbf{y}$ is some extreme point of $\mathscr{B}$ different from all of these. Then for some (finite) positive integer k we shall have

(26)
$$y_n = n \quad \text{for } n = 0, 1, 2, ..., k \quad \text{and} \quad k < y_{k+1} < k+1.$$

Now put

$$\mathbf{x} = \mathbf{y} + qp^{-1}(\mathbf{y} - \mathbf{y}(k)),$$

where $0 < p < 1$ and $q = 1 - p$. Then $\mathbf{y} = p\mathbf{x} + q\mathbf{y}(k)$, and we shall have contradicted the extreme character of $\mathbf{y}$ unless $\mathbf{x}$ fails to satisfy the relations (24). (Note that $\mathbf{x}$ and $\mathbf{y}(k)$ differ at least in their $(k+1)$th components.) But $\mathbf{y} \geqslant \mathbf{y}(k)$, because these two vectors agree up to $n = k$ and $\mathbf{y}$ is non-decreasing. Thus $\mathbf{x}$ must fail to be concave and can only do so on the triplet $(k-1, k, k+1)$. We must therefore have

$$2(y_k - qy_k(k)) < (y_{k-1} - qy_{k-1}(k)) + (y_{k+1} - qy_{k+1}(k)),$$

so that $2y_k < y_{k-1} + y_{k+1} + q$ whenever $0 < q < 1$. Now let $q \to 0$, and recall that $\mathbf{y}$ itself is concave; we find that we must have $2y_k = y_{k-1} + y_{k+1}$, so that $y_{k+1} = k+1$, which contradicts equation (26). Thus no such $\mathbf{y}$ can exist.

The integral representation (19) now takes on the simple form

(27)
$$\mathbf{x} = \sum_{1 \leqslant j < \infty} \lambda_j \mathbf{y}(j) + \lambda_\infty \mathbf{y}(\infty),$$

where the coefficients λ are non-negative real numbers such that $\sum \lambda_j < \infty$. The λs depend on x, and are uniquely determined by $\mathbf{x}$, as the following evaluations show:

(28)
$$\lambda_j = -\Delta^2 x_{j-1} \quad (j = 1, 2, ...),$$

(29)
$$\lambda_\infty = x_1 - \sum_{1 \leqslant j < \infty} \lambda_j.$$

We can thus rewrite equation (27) in the completely explicit form

(30)
$$\mathbf{x} = x_1 \mathbf{y}(\infty) + \sum_{1 \leqslant j < \infty} \Delta^2 x_{j-1} \{\mathbf{y}(\infty) - \mathbf{y}(j)\}.$$

On assembling the preceding results we obtain

Theorem 2. *An infinitely divisible renewal sequence* $\{u_n : n = 0, 1, 2, ...\}$ *is either the degenerate sequence* $\{1, 0, 0, 0, ...\}$, *or it admits a unique representation of the following form: for some positive integer* h, *and for some real non-negative* λ*s satisfying* $\sum \lambda_j < \infty$, $u_n = 0$ *when* h *does not divide* n *and*

$$(31) \qquad u_n = \exp\left\{ -\sum_{j=1}^{n/h} j\lambda_j - \frac{n}{h}\left(\sum_{j > n/h} \lambda_j + \lambda_\infty \right) \right\}$$

when h *does divide* n.

If we put $v_n(j) = \exp(-y_n(j))$, for $j = 1, 2, ..., \infty$, then when $h = 1$ we can write this relation more concisely in the form

$$(32) \qquad \mathbf{u} = \prod_{j=1}^{j=\infty} \mathbf{v}(j)^{\lambda_j},$$

the powers and product being formed componentwise. The more general case in which $h > 1$ can then be obtained from this by interpolating $(h-1)$-ads of zeros.

When $h = 1$, there are elegant formulae corresponding to equations (28) and (29) which express the λs in terms of the us. The first of these,

$$(33) \qquad \lambda_j = \log \frac{u_{j-1} u_{j+1}}{u_j^2} \quad (j = 1, 2, ...)$$

brings out very clearly the connection between the Kaluza property and the fact that the λs are non-negative. Similarly the formula for λ_∞,

$$(34) \qquad \lambda_\infty = \log \left\{ \frac{1}{u_1} \prod_{j \geqslant 1} \frac{u_j^2}{u_{j-1} u_{j+1}} \right\}$$

$$= \lim_{j \to \infty} \log(u_j/u_{j+1}),$$

shows that the non-negative character of λ_∞ reflects the non-increasing character of an infinitely divisible renewal sequence. This last formula also shows that if $\lambda_\infty = 0$ then $u_j/u_{j+1} \to 1$, while if $\lambda_\infty > 0$ then

$$u_j/u_{j+1} \to e^{\lambda_\infty} > 1,$$

and then $u_n \to 0$ geometrically fast. Thus Φ is transient if $\lambda_\infty > 0$. It is, however, not the case that Φ is always persistent when $\lambda_\infty = 0$. Thus, if we put $\lambda_j = \alpha/j^2$ for $j = 1, 2, ...$ and $\lambda_\infty = 0$ then we find that $u_n \sim C(\alpha)/n^\alpha$, and so Φ is persistent in this case if and only if $\alpha \leqslant 1$.

As for nullity, it is easily verified that $u_n \to 0$ if and only if one at least of the following conditions holds:

$$(35) \qquad \sum j\lambda_j = \infty, \quad \lambda_\infty > 0.$$

2.1.7 INFINITELY DIVISIBLE REGENERATIVE PHENOMENA

Up to this point we have attached the adjective 'infinitely divisible' only to the renewal sequence itself, and not to the regenerative phenomenon Φ described by it. But if we identify regenerative phenomena on different probability spaces provided that they have the same renewal sequence, then we can say that Φ is infinitely divisible precisely when in this sense it is equivalent to

$$(36) \qquad \Phi_{m1} \cap \Phi_{m2} \cap \ldots \cap \Phi_{mm},$$

for each positive integer m. Here, for each m, Φ_{mk} $(k = 1, 2, \ldots, m)$ are to be independent copies of a regenerative phenomenon Φ_m, and the regenerative phenomenon (36) is to be understood to 'happen' when and only when *every one* of $\Phi_{m1}, \Phi_{m2}, \ldots, \Phi_{mm}$ 'happens'.

The variety of infinite divisibility envisaged here is quite different from that occurring in the classical theory of Khintchine, Lévy and Kolmogorov, and it is also quite different from the infinite divisibility of point processes studied by Goldman [5], Lee [13] and Matthes [15]. In fact it is clear (cf. Grenander [7]) that if $*$ denotes *any* commutative associative product for random variables of *any* kind, then we can say that such a random variable is infinitely divisible if it is distributionally equivalent to the $*$-product of m independent copies of some other random variable, for each $m = 2, 3, \ldots$. We obtain the classical theory when the random variables are real-valued, and $*$ denotes addition. We obtain the kind of infinite divisibility usually considered for point processes when $*$ denotes the formation of the set-theoretic *union* of the two realized sets of points. We obtain a dual notion of infinite divisibility for point-processes, not so far as I am aware studied by previous writers, when $*$ denotes the formation of the *intersection* of the two realized sets of points, and it is this interpretation (in the special context of regenerative phenemona viewed as 'indicators' of point-processes) that we are concerned with here.

2.1.8 THE REGENERATIVE PHENOMENA ASSOCIATED WITH THE EXTREME RAYS OF $\mathscr{C}$

The canonical decomposition (32) for zero-free infinitely divisible renewal sequences suggests that we should examine the character of the regenerative phenomena associated with the irreducible factors $\mathbf{v}(1), \mathbf{v}(2), \ldots, \mathbf{v}(\infty)$. We lose nothing of interest by restricting attention to the zero-free case, for the degenerate renewal sequence $(1, 0, 0, \ldots)$ corresponds to a Φ which almost surely never recurs, while an infinitely divisible renewal sequence for which $h>1$ corresponds to a Φ whose recurrence times are almost surely multiples of h.

The simplest case is that of the infinitely divisible renewal sequence $\mathbf{v}(\infty)$. Here and later it will make the notation more easily comprehensible if we introduce an appropriate positive finite λ (in this case λ_∞), and then put $e^{-\lambda} = \rho$. Thus we shall in the present case discuss the Φ for which the renewal sequence is

$$(v_n(\infty))^{-\log \rho} \quad (n = 0, 1, 2, ...).$$

But this is just $(1, \rho, \rho^2, ...)$, where $0 < \rho < 1$, and so Φ is fully described by saying that it recurs immediately after an occurrence with probability ρ, and that when it once stops 'happening' it stops forever. In the notation of Section 2.1.1, $U(z) = (1-\rho z)^{-1}$ and $F(z) = \rho z$, so that the (defective) recurrence-time distribution just has mass ρ at $t = 1$. The next simplest case is that associated with $\mathbf{v}(1)$. Here the renewal sequence to be considered is $(1, \rho, \rho, \rho, ...)$, where as before $0 < \rho < 1$. Here $U(z) = 1 + \rho z/(1-z)$ and $F(z) = \rho z/(1-(1-\rho)z)$, so that Φ is now persistent and has the recurrence-time distribution

$$\mathrm{pr}\,(T = r) = \rho(1 - \rho)^{r-1} \quad (r = 1, 2, ...).$$

Finally we must consider the more complicated case when the basic renewal sequence is $\mathbf{v}(j)$ with $1 < j < \infty$. This time we have to describe the Φ for which

$$u_n = \rho^n \quad (0 \leqslant n \leqslant j), \quad u_n = \rho^j \quad (j \leqslant n) \quad (0 < \rho < 1).$$

We shall then have

$$U(z) = \frac{1 - \rho^j z^j}{1 - \rho z} + \frac{\rho^j z^j}{1-z},$$

and now

(37)

$$F(z) = \rho z + (1-\rho)\,z\cdot \rho^j z^j \cdot \sum_{m=0}^{\infty} \left\{ (1-\rho^j)z \cdot \frac{1 + \rho z + \rho^2 z^2 + ... + \rho^{j-1} z^{j-1}}{1 + \rho + \rho^2 + ... + \rho^{j-1}} \right\}^m.$$

In this case also, therefore Φ is persistent, but the recurrence-time T for Φ now has a rather complicated distribution which we can describe as follows.

(*i*) With probability $\rho, T = 1$.

(*ii*) If $T > 1$, then

$$T = 1 + j + \sum_{i=1}^{m} \tau_i,$$

where $\mathrm{pr}\,(m = s\,|\,T > 1) = \rho^j(1 - \rho^j)^s$ $(s = 0, 1, 2, ...)$, and where (given that $T > 1$ and that $m = s$) the random variables $\tau_1, \tau_2, ..., \tau_m$ are conditionally independent and distributed over the possibilities $1, 2, ..., j$ with probabilities proportional to $1, \rho, ..., \rho^{j-1}$.

To make a model for the regenerative phenomenon with a general zero-free infinitely divisible renewal sequence as at equation (32) we have only to construct the direct product of a countable sequence of probability-spaces carrying regenerative phenomena of the types just described (with appropriate ρs in each case), and then to identify Φ with the *simultaneous* occurrence of each of the component phenomena.

2.1.9 THE CENTRAL LIMIT THEOREM FOR RENEWAL SEQUENCES

We are now going to construct an analogue of the central limit theorem (in its general (triangular array) form [6]) for renewal sequences. First we need some notation. We shall write **u** for a renewal sequence $\{u_n : n = 0, 1, 2, ...\}$, $\mathbf{u}^t$ for the corresponding sequence of tth powers, and $\lim \mathbf{u}(r)$ for the termwise limit of a sequence of renewal sequences $\mathbf{u}(r)$ $(r = 1, 2, ...)$ as $r \to \infty$ (by (b) of Section 2.1.3, this will again be a renewal sequence). We shall write $\mathbf{u} * \mathbf{v}$ for the renewal sequence whose nth term is $u_n v_n$, and similarly for any larger number of factors. We note that *any* sequence of renewal sequences $\mathbf{u}(r)$ contains a convergent subsequence; this is proved by an obvious compactness argument depending on the fact that $0 \leqslant u_n(r) \leqslant 1$ for all r and n. By a triangular array we shall mean an array

$$\mathbf{u}(1, 1),$$

$$\mathbf{u}(2, 1), \quad \mathbf{u}(2, 2),$$

$$\mathbf{u}(3, 1), \quad \mathbf{u}(3, 2), \quad \mathbf{u}(3, 3),$$

$$\cdots \qquad \cdots \qquad \cdots \qquad \cdots$$

of renewal sequences, there being i entries in the ith row $(i = 1, 2, ...)$. We shall call this a *null* array when

$$(38) \qquad \lim_{i \to \infty} u_1(i, j) = 1,$$

the limit being uniform with respect to j $(1 \leqslant j \leqslant i)$. By the ith marginal product of a triangular array we shall mean the renewal sequence

$$(39) \qquad \mathbf{v}(i) = \mathbf{u}(i, 1) * \mathbf{u}(i, 2) * ... * \mathbf{u}(i, i),$$

and we shall say that a triangular array is convergent (necessarily to a renewal sequence) when $\lim \mathbf{v}(i)$ exists. We are now going to prove the basic

Theorem 3. *If a null triangular array of renewal sequences is convergent then the limit is infinitely divisible, and either the limit is* $(1, 0, 0, 0, \ldots)$ *or each of the terms is positive. Also, every positive infinitely divisible renewal sequence can be obtained in this way.*

Proof. Let us prove the second (easier) half of the theorem first. If $\mathbf{v}$ is a positive infinitely divisible renewal sequence, then we can exhibit it as the limit of a null triangular array by putting

$$u_n(i,j) = v_n^{1/i} \quad (j = 1, 2, \ldots, i: i \geqslant 1);$$

similarly if $\mathbf{v}$ is $(1, 0, 0, 0, \ldots)$ then we can put

$$u_n(i, j) = \exp\left(-\frac{n}{\sqrt{i}}\right),$$

and we again obtain a null triangular array with $\mathbf{v}$ as limit.

We now turn to the first half of the theorem, and suppose that we are presented with a convergent null triangular array, with limit $\mathbf{v}$. The first step is to prove that either (*i*) the limit is $(1, 0, 0, 0, \ldots)$ or (*ii*) every $v_n > 0$. It will suffice to deduce from $v_1 = 0$ that $v_{m+1} = 0$ for all $m \geqslant 1$, because we know that if $v_1 > 0$ then $v_{m+1} \geqslant v_1^{m+1} > 0$. We make use here of an inequality † noted by Kingman [11]:

(40) $$u_{r+s} \leqslant 1 + u_r u_s - \max(u_r, u_s).$$

From this is follows that

$$u_{m+1} \leqslant 1 + u_m u_1 - u_m = 1 - (1 - u_1) u_m \leqslant 1 - (1 - u_1) u_1^m.$$

Now

$$1 - (1 - \tau) \tau^m \leqslant \sqrt{\tau}$$

if $1 - \delta \leqslant \tau \leqslant 1$, provided that the positive number δ is small enough, and so since we shall have

$$1 - \delta < u_1(i, j) \leqslant 1$$

for all $j \leqslant i$, provided that $i \geqslant I(\delta)$, we can conclude that

(41) $$0 \leqslant u_{m+1}(i, j) \leqslant \sqrt{(u_1(i, j))}$$

for all $j \leqslant i$ provided that $i \geqslant I_m$, for any fixed m. But this implies that $0 \leqslant v_{m+1} \leqslant \sqrt{v_1}$ and so the desired conclusion follows. From this point

---

† This is most easily proved by noticing that $u_r(1 - u_s) \leqslant 1 - u_{r+s}$ (an easy consequence of equation (7)) and then appealing to symmetry.

onwards we can without loss of generality assume that each v_n is positive, and it only remains for us to show that v is infinitely divisible.

In the first place it is clear that if n is fixed then $u_m(i,j)>0$ for $m=1,2,\dots,n$ and $1\leqslant j\leqslant i$, provided that i exceeds some i_0 (which we henceforth suppose to be the case). Accordingly (using equation (11)) we can write $v_n(i)v_{n-2}(i)/v_{n-1}(i)^2$ in the form

$$(42) \qquad \prod_{j=1}^{i}\left\{\frac{1+A(i,j)+f_n(i,j)/u_{n-1}(i,j)}{1+B(i,j)}\right\},$$

where

$$A(i,j) = \sum_{h=2}^{n-1} f_h(i,j)\,u_{n-h}(i,j)/u_{n-1}(i,j)-(1-f_1(i,j))$$

and

$$B(i,j) = \sum_{h=2}^{n-1} f_h(i,j)\,u_{n-h-1}(i,j)/u_{n-2}(i,j)-(1-f_1(i,j)),$$

if $n\geqslant 3$. From equation (38) and the inequality $1\geqslant u_m\geqslant u_1^m$ we know that $u_m(i,j)\to 1$ as $i\to\infty$, uniformly for $1\leqslant m\leqslant n$ and $1\leqslant j\leqslant i$, while from $\sum f_r\leqslant 1$ and $f_1=u_1$ we know that

$$1-f_1(i,j) \quad \text{and} \quad f_r(i,j)$$

each tend to zero as $i\to\infty$, uniformly for $2\leqslant r\leqslant n$ and $1\leqslant j\leqslant i$. This being so, the numerator and denominator of the jth factor in the product (42) will be of the form $1+z$, with $|z|\leqslant\frac{1}{2}$, for all sufficiently large i, and so can be written in the form $\exp(z+\rho z^2)$, where $|\rho|\leqslant 1$. It follows that

$$(43) \qquad \log\left\{\frac{v_n(i)v_{n-2}(i)}{v_{n-1}(i)^2}\right\}$$

$$\geqslant \sum_{j=1}^{i}\left\{\frac{f_n(i,j)}{u_{n-1}(i,j)}+(A(i,j)-B(i,j))-(A(i,j)+f_n(i,j)/u_{n-1}(i,j))^2-B(i,j)^2\right\}$$

$$\geqslant \sum_{j=1}^{i}\left\{\frac{f_n(i,j)}{u_{n-1}(i,j)}-M(1-u_1(i,j))^2\right\}$$

for all sufficiently large i, where M depends on n only. Now $0\leqslant u_1\leqslant 1$ and so $u_1\leqslant\exp\{-(1-u_1)\}$. As

$$\prod_{j=1}^{i} u_1(i,j) = v_1(i)\to v_1>0,$$

it follows that

$$\sum_{j=1}^{i}(1-u_1(i,j))$$

is bounded as $i\to\infty$, and therefore (using equation (38) again) that

$$\sum_{j=1}^{i}(1-u_1(i,j))^2\to 0 \quad \text{as } i\to\infty.$$

Accordingly the inequality (43) implies that

$$\log\frac{v_n v_{n-2}}{v_{n-1}^2}$$

is non-negative; i.e. that $v_n v_{n-2}-v_{n-1}^2\geqslant 0$ for $n=3,4,5,\ldots$. Now this inequality must in any case hold when $n=2$, in view of the identity $u_2=f_2+f_1^2\geqslant f_1^2=u_1^2$, which holds for all renewal sequences. Thus $\{v_n: n=0,1,2,\ldots\}$ is a positive bounded Kaluza sequence, and so is infinitely divisible by Theorem 1.

2.1.10 THE ARITHMETIC OF RENEWAL SEQUENCES†

The convolution semigroup of probability-laws on R_1 (or, equivalently, the multiplication semigroup of their characteristic functions) has exceedingly interesting arithmetic properties which have been unravelled by Lévy, Khintchine and others in a classical series of investigations. (See, for example, Chapter 5 of Linnik's book [14].) We shall now make some general remarks which suggest that there may exist a similar 'arithmetic of renewal sequences', and then we shall proceed to construct it.

We shall confine ourselves in this discussion to *aperiodic* renewal sequences; that is, we shall suppose that the set of values of n for which $u_n>0$ has unity as its greatest common divisor. This will have the effect, in particular, of eliminating the sequence $(1,0,0,0,\ldots)$ from consideration (it obviously has every other renewal sequence as a factor) and of restricting the infinitely divisible renewal sequences to the class of those which are zero-free. We lose very little by this limitation, because if $\{u_n: n=0,1,2,3,\ldots\}$ is periodic with period $h=\text{g.c.d.}\{n: u_n>0\}\geqslant 2$, and

† In my Loutraki Lectures, Section 2.1.10 of this paper was replaced by a sketch of a similar 'arithmetic' for general 'Delphic' semigroups. An account of this will be presented in [10].

if $\mathbf{u} = \mathbf{v} * \mathbf{w}$ is a factorization into two renewal-sequence factors, then we can confine our attention to those values of n which are divisible by h, and then in virtue of property (c) of Section 2.1.4 we shall obtain a corresponding factorization of the aperiodic renewal sequence

$$\{u_{nh}: \quad n = 0, 1, 2, \ldots\}$$

into the renewal sequences $\{v_{nh}: n = 0, 1, 2, \ldots\}$ and $\{w_{nh}: n = 0, 1, 2, \ldots\}$, both now necessarily also aperiodic.

It is known (Chung [3]) that every renewal sequence $\mathbf{u}$ can be identified with the sequence of 'top-left-hand-corner' elements $\{p_{11}^{(n)}: n = 0, 1, 2, \ldots\}$ for a Markov chain in which we can take the state 1 to be aperiodic when the renewal sequence is such, and then (Kendall [9]) we know that we can write

$$(44) \qquad u_n = \int_C e^{in\theta} \mu(d\theta) \quad (n = 0, 1, 2, \ldots),$$

where μ is a probability measure on the Borel subsets of the circumference C of the unit circle which is symmetric about the diameter from $\theta = 0$ to $\theta = \pi$, and which (apart from a possible atom at $\theta = 0$) is absolutely continuous with respect to Lebesgue measure. It follows that we can identify the double-sequence

$$(\ldots, u_2, u_1, u_0, u_1, u_2, \ldots)$$

with the characteristic 'function' of the probability law μ in such a way that the $*$-multiplication we have been considering for renewal sequences corresponds to the operation of convolution for the associated measures μ; in other words, renewal sequences with the operation of $*$-multiplication form a system which is isomorphic with a sub-semigroup of the convolution semigroup of probability laws on C. (This assertion is true even if we relax the requirement of aperiodicity; the only effect of periodicity is to permit further atoms to μ, located at the hth roots of unity.)

This fact makes it worth while for us to look for a factorization theory of renewal sequences, but it does not guarantee the existence of one, and while it may suggest true theorems it does not enable us to dispense with the necessity for proving them. This is because *we do not know how to characterize the sub-semigroup of measures μ which correspond by way of formula (44) to renewal sequences $\{u_n: n = 0, 1, 2, \ldots\}$*. Some necessary conditions on μ have already been mentioned, and while they can be added to (Feller [4] has shown that in the aperiodic case μ has a density which is *continuous* away from $\theta = 0$), even these augmented conditions are, as we shall see, not sufficient. For let $0 < r < 1$, and consider the continuous positive probability density

$$\lambda'(\theta) = (2\pi)^{-1} \frac{1-r^2}{1+2r\cos\theta+r^2} \quad (0 \leqslant \theta < 2\pi),$$

which has the required symmetry. We can rewrite this as

$$(2\pi)^{-1}\left\{1 + \sum_{n=1}^{\infty} (-r)^n(e^{in\theta} + e^{-in\theta})\right\},$$

and so

$$v_n = \int_C e^{in\theta}\,\lambda(d\theta) = \int_C e^{in\theta}\,\lambda'(\theta)\,d\theta = (-r)^n \quad (n = 0, 1, 2, \ldots).$$

Obviously this is not a renewal sequence, but $\mathbf{u} = \mathbf{v} * \mathbf{v}$ *is* a renewal sequence (indeed, is an infinitely divisible one). Thus the factorization $\mu = \lambda * \lambda$ (where μ corresponds to $\mathbf{u}$) is meaningful for the factorization theory of probability laws but is irrelevant to the theory we wish to construct. Accordingly, while we shall imitate the classical proofs wherever possible, we shall frequently have to support them by arguments specially designed for the present situation.

We start by showing that there exist *indecomposable* aperiodic renewal sequences; that is, *aperiodic renewal* sequences $\mathbf{u}$ with the following two properties:

(*i*) $\mathbf{u} \neq (1, 1, 1, \ldots)$;

(*ii*) if $\mathbf{u} = \mathbf{v} * \mathbf{w}$, then $\mathbf{v} = (1, 1, 1, \ldots)$ and $\mathbf{w} = \mathbf{u}$, or vice versa. With the notation of equations (1) and (3), let

$$f_1 > 0, \quad f_2 = 1 - f_1 > 0, \quad f_r = 0 \quad (r = 3, 4, \ldots),$$

so that

$$u_1 = f_1 \quad \text{and} \quad u_2 = f_1^2 + f_2.$$

Then Φ is persistent and aperiodic, and its recurrence-time distribution is limited to the two values $T = 1$ and $T = 2$, each of which carries positive probability. Suppose if possible that there is a *proper* decomposition $\mathbf{u} = \mathbf{v} * \mathbf{w}$, so that each of $\mathbf{v}$ and $\mathbf{w}$ is an aperiodic renewal sequence distinct from $(1, 1, 1, \ldots)$. Let Φ' and Φ'' denote the two associated regenerative phenomena. Then, as previously explained, we can identify Φ with $\Phi' \cap \Phi''$. This shows that each of Φ' and Φ'' must be persistent, so that their recurrence-time distributions are not defective; let us write $\{g_r : r = 1, 2, \ldots\}$ and $\{h_r : r = 1, 2, \ldots\}$ for these. Now $f_1 = g_1 h_1$, so that each of g_1 and h_1 is positive; also each of g_r and h_r must vanish for $r > 2$

because we cannot have $T > 2$. Finally g_2 and h_2 must be positive, because $g_2 = 0$ implies that $g_1 = 1$ and so that $v_n = 1$ for all n, and this is excluded. From

$$f_1 = u_1 = v_1 w_1 = g_1 h_1$$

and

$$f_1^2 + f_2 = u_2 = v_2 w_2 = g_1^2 h_1^2 + g_1^2 h_2 + g_2 h_1^2 + g_2 h_2$$

we obtain

$$g_1 h_1 + g_2 h_2 + g_1^2 h_2 + g_2 h_1^2 = f_1 + f_2 = 1.$$

However we also know that

$$(g_1 + g_2)(h_1 + h_2) = 1,$$

and so $g_1 h_2(1 - g_1) + g_2 h_1(1 - h_1) = 0$, and yet this expression is equal to $g_1 g_2 h_2 + h_1 h_2 g_2 > 0$. Thus *the coefficients in the expansion of*

(45) $$(1 - f_1 z - f_2 z^2)^{-1}$$

form an indecomposable aperiodic renewal sequence when $f_1 + f_2 = 1$ and f_1 and f_2 are positive.

This example is by no means exceptional. In fact the next theorem will show (for example) that if $\{u_n : n = 0, 1, 2, \ldots\}$ is an aperiodic renewal sequence such that $u_n < u_{n+1}$ for at least one n, then either **u** is itself indecomposable or it has an indecomposable proper factor.

Theorem 4. *If the aperiodic renewal sequence $\{u_n : n = 0, 1, 2, \ldots\}$ is not infinitely divisible, then either it is indecomposable or it has an indecomposable proper factor.*

Proof. Let **u** be an aperiodic renewal sequence which is not indecomposable and suppose further that in every proper decomposition neither factor is indecomposable. We must prove that **u** is infinitely divisible. We first prove the elementary

Lemma. *If in a renewal sequence $u_n = 1$ for some positive n, say for the first time when $n = m$, then $u_n = 1$ whenever m divides n, and otherwise $u_n = 0$.*

This follows from Kingman's inequality (40), which with $r + s = m$ shows that either $u_r = u_s = 0$ or $u_r = u_s = 1$. The second possibility can only occur if $r = 0$ and $s = m$, or vice versa, and so either $m = 1$ or $m > 1$ and $u_n = 0$ for $1 \leqslant n < m$. It is then clear that the recurrence-time T is almost surely equal to m, and the truth of the lemma is evident.

On applying the lemma in the present context we see that we can assume that u_n is less than unity for every positive n, for we have supposed **u** to be aperiodic, and $\mathbf{u} = (1, 1, 1, ...)$ is trivially infinitely divisible. We can recall that by hypothesis $\mathbf{u}$ is aperiodic and so is not $(1, 0, 0, 0, ...)$.

Now let $m \geqslant 1$ be such that $0 < u_m < 1$, and let m be fixed until further notice. We define the functional $X(\mathbf{u})$ by

$$(46) \qquad\qquad u_m = \exp\{-X(\mathbf{u})\},$$

so that

$$(47) \qquad\qquad 0 < X(\mathbf{u}) < \infty,$$

and we observe that $X(\mathbf{u} * \mathbf{v}) = X(\mathbf{u}) + X(\mathbf{v})$. It is relevant here that the functional X is positive and finite on proper components $\mathbf{v}$ of $\mathbf{u}$, for obviously they must be such that $v_m > 0$, and as they must be aperiodic like $\mathbf{u}$, and are not allowed to be $(1, 1, ...)$, the possibility $v_m = 1$ is excluded by the lemma. Accordingly every proper decomposition of $\mathbf{u}$ implies a division of $X(\mathbf{u})$ into a finite number of finite positive parts.

We shall employ $X(\mathbf{u})$ in place of what Linnik [14] calls Khintchine's functional. We first assert that each proper component $\mathbf{v}$ of $\mathbf{u}$ has a proper subcomponent $\mathbf{v}'$ for which $X(\mathbf{v}')$ is arbitrarily small. Consider for a fixed proper component $\mathbf{v}$ of $\mathbf{u}$ the infimum of $X(\mathbf{w})$, where $\mathbf{w}$ ranges through all components of $\mathbf{v}$. Suppose if possible that this infimum is positive. Then we can construct a sequence of proper components of $\mathbf{v}$ on which the functional converges to its infimum, and from this sequence, using compactness and (b) of Section 2.1.3, we can pick out a subsequence which converges to a component $\mathbf{w}'$ of $\mathbf{v}$ such that $X(\mathbf{w}')$ is equal to the infimum. But $\mathbf{w}'$ is clearly a proper component of $\mathbf{v}$, and so of $\mathbf{u}$, so admits a proper decomposition, and on exploiting this fact we can contradict the infimum character of $X(\mathbf{w}')$.

We now consider all possible decompositions

$$(48) \qquad\qquad \mathbf{u} = \mathbf{v}(1) * \mathbf{v}(2) * ... * \mathbf{v}(k)$$

into exactly k proper components. We may as well assume that the components at equation (48) have been arranged in an order of increasing X-values, and in relation to the set of all such decompositions we write α (obviously positive and finite) for the supremum of $X(\mathbf{v}(1))$. Compactness ensures that this bound is attained, and we suppose that it is attained at equation (48). We assert that $X(\mathbf{v}(j)) = \alpha$ for all $j = 1, 2, ..., k$. For if this were not so we should have

$$\alpha = X(\mathbf{v}(1)) = ... = X(\mathbf{v}(i)) < X(\mathbf{v}(i+1)) \leqslant ... \leqslant X(\mathbf{v}(k))$$

for some i such that $1 \leqslant i < k$. From $\mathbf{v}(i+1)$ we can peel off i components for each of which X is arbitrarily small (but positive), and if we transfer each of these in turn to $\mathbf{v}(1), \mathbf{v}(2), ..., \mathbf{v}(i)$ then we can retain the original ordering but increase $X(\mathbf{v}(1))$ above α, which gives a contradiction. Thus a k-fold decomposition can be found in which, for each of the k components, $X(\mathbf{v}) = X(\mathbf{u})/k$.

We next observe that the preceding arguments can be imitated exactly when $m' > m'' \geqslant 1$, and both $0 < u_{m'} < 1$ and $0 < u_{m''} < 1$, if we use instead of the functional $X(\mathbf{u})$ the modified functional

$$Y(\mathbf{u}) = \log \frac{1}{u_{m'} u_{m''}}.$$

If X' and X'' are the X-functionals based on the choices $m = m'$, and m'', then obviously $Y = X' + X''$, and so we are able to conclude that if $u_{m'}$ and $u_{m''}$ both lie in the open interval $(0, 1)$ then for each k it is possible to find a proper decomposition of $\mathbf{u}$ into k factors $\mathbf{u}(k, j)$ $(j = 1, 2, ..., k)$ in such a way that

$$\max\{X'(\mathbf{u}(k, j)), X''(\mathbf{u}(k, j))\} \leqslant Y(\mathbf{u}(k, j)) = Y(\mathbf{u})/k$$

whenever $j \leqslant k$.

We now consider the triangular array

$$\mathbf{u},$$
$$\mathbf{u}(2, 1), \quad \mathbf{u}(2, 2),$$
$$\mathbf{u}(3, 1), \quad \mathbf{u}(3, 2), \quad \mathbf{u}(3, 3),$$
$$\ldots \qquad \ldots \qquad \ldots \qquad \ldots$$

and note that the marginal products

$$\mathbf{u}(k, 1) * \mathbf{u}(k, 2) * \ldots * \mathbf{u}(k, k)$$

are all equal to $\mathbf{u}$, so that the array converges to $\mathbf{u}$. If then we can show that the array is a null array, it will follow from Theorem 3 that the limit $\mathbf{u}$ must be infinitely divisible, and the proof of Theorem 4 will be complete.

That the array is a null array is a consequence of the fact that $\mathbf{u}$ has been assumed to be aperiodic. We recall that for any renewal sequence the set $\{n : u_n > 0\}$ is an addition semigroup, and from the assumed aperiodicity of $\mathbf{u}$ we know that in the present instance this semigroup has unity as its greatest common divisor. There will therefore exist in the semigroup a finite set $s_1, s_2, ..., s_t$ of positive integers without a common prime factor, and as all sufficiently large positive integers can be written in the form

$$a_1 s_1 + a_2 s_2 + \ldots + a_t s_t,$$

where the as are non-negative integers, it follows that all sufficiently large positive integers belong to the semigroup. In particular, therefore, we can find positive integers $m'' = m$ and $m' = m + 1$ such that $u_{m'}$ and $u_{m''}$ are both positive, and as before we can assume that each of these quantities is less than unity because otherwise the infinite divisibility of **u** would be immediate. We may therefore suppose the triangular array to have been constructed with such a choice of m' and m'', and accordingly it is a feature of the array that

$$u_{m'}(k, j) \to 1 \quad \text{and} \quad u_{m''}(k, j) \to 1$$

as $k \to \infty$, uniformly for $j = 1, 2, ..., k$, where the essential new feature of the situation is that $m' = m'' + 1$.

We now appeal once again to Kingman's inequality (40), using it to obtain

$$0 \leqslant u_r(1 - u_1) \leqslant 1 - u_{r+1}$$

as a universal inequality for renewal sequences. From this we deduce that

(49) $$1 - u_1(k, j) \leqslant \frac{1 - u_{m'}(k, j)}{u_{m''}(k, j)},$$

and hence that $u_1(k,j)$ tends to unity j-uniformly as $k \to \infty$, i.e. that we have a null array. Theorem 4 has therefore been established.

We now prove a theorem which asserts that every aperiodic renewal sequence can be built up out of infinitely divisible and indecomposable factors in at least one way (we make no assertion about unicity).

Theorem 5. *If* **u** *is any aperiodic renewal sequence, then we can always write*

(50) $$\mathbf{u} = \mathbf{v}(1) * \mathbf{v}(2) * ... * \mathbf{w},$$

where each **v** *is indecomposable,* **w** *is infinitely divisible, there are not more than countably many* **v***s, and the* **v***s or* **w** *may be absent.*

Remark. It will appear from the proof of Theorem 5 that a factorization (50) always exists in which **w** has no indecomposable factor.

Proof. If **u** is infinitely divisible we put **w** = **u** and omit the **v**s. If **u** is indecomposable we put **v**(1) = **u** and omit **w** and the rest of the **v**s. We can therefore exclude these two cases. Then, by Theorem 4, **u** has an indecomposable proper factor. The complementary factor may be infinitely divisible, or indecomposable, in either of which cases we arrive at the desired decomposition. If this is not so, then the complementary factor has an indecomposable proper subfactor. Rather than say 'and so on', we now use Zorn's lemma. We consider the system of objects S specified as follows:

(*i*) each S is a mapping from the set of all indecomposable aperiodic renewal sequences **v** to the set $(0, 1, 2, ...)$;

(*ii*) for any positive integer k, for any choice of the k distinct indecomposable aperiodic renewal sequences $\mathbf{v}(1), \mathbf{v}(2), ..., \mathbf{v}(k)$, and for any S, the renewal sequence

$$\mathbf{v}(1)^{S(\mathbf{v}(1))} * \mathbf{v}(2)^{S(\mathbf{v}(2))} * ... * \mathbf{v}(k)^{S(\mathbf{v}(k))}$$

is a factor of $\mathbf{u}$.

Now for any indecomposable aperiodic renewal sequence $\mathbf{v}$, we must have $v_n < 1$ for all $n \geq 1$; this implies† that $S(\mathbf{v})$ has a finite (attained) least upper bound for each fixed $\mathbf{v}$. We partly order the system of objects S by asserting that $S' \leq S''$ if and only if $S'(\mathbf{v}) \leq S''(\mathbf{v})$ for each indecomposable aperiodic $\mathbf{v}$, and we than observe that the system of objects S has the Zorn property: each chain of objects S has an upper bound in the system. This being so, there exists a maximal S; let us choose one and fix it throughout the following argument.

Because we may without loss of generality suppose that $\mathbf{u}$ is not infinitely divisible, we can take it that $0 < u_m < 1$ for some $m \geq 1$; and with this value of m we construct the X-functional as before. $X(\mathbf{v})$ must be finite and positive for each $\mathbf{v}$ for which $S(\mathbf{v}) > 0$, and so there can be only countable many such $\mathbf{v}$s. Let us label them as $\mathbf{v}(1), \mathbf{v}(2), ...$ Each product

$$\mathbf{v}(1)^{S(\mathbf{v}(1))} * \mathbf{v}(2)^{S(\mathbf{v}(2))} * ... * \mathbf{v}(k)^{S(\mathbf{v}(k))}$$

is a factor of $\mathbf{u}$ and may be identical with $\mathbf{u}$. We have nothing left to prove if we can construct $\mathbf{u}$ in this way, and so we may as well suppose that there is always a complementary proper factor, $\mathbf{w}(k)$, say. We cannot be sure that $\lim \mathbf{w}(k)$ exists, but it must do so (by compactness) if $k \to \infty$ in some suitable sequence $(k_1, k_2, ...)$, where $k_r \uparrow \infty$ as $r \to \infty$. In this way we can be sure that there exists a renewal sequence $\mathbf{w}$ such that

$$\mathbf{u} = \mathbf{v}(1) * \mathbf{v}(1) * ... * \mathbf{v}(1) * \mathbf{v}(2) * ... * \mathbf{w},$$

where $\mathbf{v}(k)$ occurs $S(\mathbf{v}(k))$ times. Because $\mathbf{w}$ is a factor of $\mathbf{u}$ it must be aperiodic, and if it were indecomposable or possessed an indecomposable proper factor then S could be extended and would not be maximal. Thus, by Theorem 4, $\mathbf{w}$ is infinitely divisible, and the desired decomposition of $\mathbf{u}$ has been found. Of course it may turn out that $\mathbf{w}$ is $(1, 1, 1, ...)$, in which case it could be omitted.

We have now completed the construction of the promised factorization theory for aperiodic renewal sequences. Our results imply that the sub-semigroup of probability measures μ associated as at equation (44) with

† Because as usual we understand aperiodicity to exclude the possibility
$$\mathbf{u} = (1, 0, 0, ...).$$

aperiodic renewal sequences **u** must have a convolution-arithmetic which is very similar to the convolution-arithmetic of all probability measures on *C*. This suggests that in seeking to characterize such special probability measures μ one might usefully look for properties of the full convolution semigroup which are not shared by the sub-semigroup. As all the positive characteristics we have looked at are shared by both, it may be more profitable to look at negative ones, and to ask how far the pathologies of the classical convolution-arithmetic occur in the present context; we shall not, however, discuss such questions further here.

2.1.11 POSTSCRIPT: THE DELAY TO PEDESTRIANS CROSSING A ROAD

Consider a pedestrian wishing to cross a road carrying a single lane of traffic. We suppose that it takes a (finite) time *c* to cross the road when it is free of traffic, and that the vehicles (whose size will be neglected and whose speed is conventionally taken to be unity) form a Poisson stream with intensity α. If we are given that it is possible to cross at time zero, then the chance that it will be possible to cross at a later time *t* is readily seen to be

$$(51) \qquad p(t) = \exp\{-\alpha \min(c,t)\} \quad (0 \leqslant t < \infty);$$

here $0 < c < \infty$. The possibility of crossing is evidently a regenerative phenomenon in continuous time, and the function (51) is the continuous-time analogue of the renewal sequence. Regenerative phenomena Φ in continuous time, and their associated *p*-functions, have been intensively investigated by Kingman (see [11] for a summary of his work). The factorization problems treated in the present paper are being separately studied in the *p*-function context, and an account of this work will be published elsewhere (Kendall [10]). The role of the particular *p*-function (51), for $0 < c \leqslant \infty$, corresponds exactly to that of the renewal sequence

$$(52) \qquad u_n = \rho^{\min(j,n)} \quad (n = 0, 1, 2, ...)$$

(where $j = 1, 2, ..., \infty$), in that it is infinitely divisible and corresponds to an extreme ray of the associated Choquet cone, so that the most general infinitely divisible *p*-function can be canonically expressed as an integral-product involving the functions (51). Obviously we can provide a similar traffic-theoretic interpretation for equation (52) when *j* is finite, if we allow pedestrian and vehicles to move discontinuously in discrete time, and this interpretation may be found of some interest in view of the fact that the recurrence-time analysis of Section 2.1.8 proved to be rather complicated. It may be remarked parenthetically that the recurrence-time analysis associated with equation (51) is equally complicated. For a study of this see Tanner [17].

We can interpret the infinite divisibility of equation (51) or of equation (52) either by dissecting α (i.e. sorting out the cars according to, say, colour or first name of driver), or by considering the problem of crossing several equally wide lanes of traffic without waiting between them, the time- and space-zero for each lane being adjusted appropriately for the effect of the crossing-time c. (When pedestrians are allowed to cross one lane and then wait on an 'island' until they can cross the next, the problem is more complicated (Mayne, [16]).) The multi-lane traffic problem is of especial interest because it supplies a concrete realization of the composition of point-processes by intersection rather than by union.

REFERENCES

1. F. F. Bonsall, "On the representation of points of a convex set", *J. Lond. math. soc.* **38** (1963), 332–334.
2. G. Choquet, "Les cônes convexes faiblement complets", *Proc. Intern. Congr. Math.* (*Stockholm*, 1962), 317–330.
3. K. L. Chung, *Markov Chains with Stationary Transition Probabilities*, 2nd edn., Springer, Berlin (1967).
4. W. Feller, "On the Fourier representation for Markov chains and the strong ratio theorem", *J. Math. Mech.* **15** (1966), 273–283.
5. J. R. Goldman, *Stochastic point processes: limit theorems and infinite divisibility*, Thesis, Princeton University (1965).
6. B. V. Gnedenko and A. N. Kolmogorov, *Limit Distributions for Sums of Independent Random Variables*, English translation, Addison-Wesley, Cambridge, Mass. (1954).
7. U. Grenander, *Probabilities on Algebraic Structures*, New York (1963).
8. Th. Kaluza, "Über die Koeffizienten reziproker Potenzreihen", *Math. Zeit.* **28** (1928), 161–170.
9. D. G. Kendall, "Unitary dilations of Markov transition operators", in *Surveys in Probability and Statistics*, ed. U. Grenander, Stockholm (1959).
10. ——, "Delphic semigroups, infinitely divisible regenerative phenomena, and the arithmetic of p-functions", *Ztschr. Wahrsch'theorie & verw. Geb.* **9** (1968), 163–195. (2.2 of this book).
11. J. F. C. Kingman, "An approach to the study of Markov processes", *J. R. statist. Soc. B*, **28** (1966), 417–447.
12. J. Lamperti, "On the coefficients of reciprocal power-series", *Am. Math. Monthly* **65** (1958), 90–94.
13. P. M. Lee, *Infinitely divisible stochastic processes*, Ph.D. dissertation, Cambridge University (1966).
14. Yu. V. Linnik, *Decomposition of Probability Distributions*, English translation, Edinburgh (1964).

15. K. Matthes, "Unbeschränkt teilbare Verteilungsgesetze stationärer zufäl-
linger Punktfolgen", *Wiss. Zeit. Hochschule für Electrotechnik Ilmenau*, **9**
(1963), 235–238.

16. A. J. Mayne, "Some further results in the theory of pedestrians and road
traffic", *Biometrika*, **41** (1954), 375–389.

17. J. C. Tanner, "The delay to pedestrians crossing a road", *Biometrika*, **38**
(1951), 383–392.

2.2

Delphic Semigroups, Infinitely Divisible Regenerative Phenomena, and the Arithmetic of p-functions

D. G. KENDALL

ἀπὸ δὲ τοῦ Κωρυκίου
χαλεπὸν ἤδη καὶ ἀνδρὶ εὐζώνῳ
πρὸς τὰ ἄκρα ἀφικέσθαι τοῦ Παρνασσοῦ

PAUSANIAS

[Reprinted from Z. *Wahrsch'theorie & verw. Geb.* **9** (1968), 163–195.]

2.2.1 DELPHIC SEMIGROUPS

2.2.1.1 *Introduction*

In the first part of this paper we discuss what we shall call Delphic semigroups; these are (very roughly) commutative topological semigroups which obey the central limit theorem for triangular arrays, and we shall prove that such a semigroup necessarily has an arithmetic very similar to the convolution arithmetic of distribution functions on R.

The present investigation arose from a study of the arithmetic of the multiplicative semigroup of renewal sequences, an account of which will be found in the published version [9] of my contributions to the Loutraki Symposium on Probability Methods in Analysis. A brief sketch of the general theory of Delphic semigroups was also given at Loutraki, but is not included in [9], and reference to that paper will be unnecessary unless the reader wishes to see the ultimate details of a simple concrete example. Another (but more sophisticated) example will be presented here in Section 2.2.2. The principal results of Section 2.2.1 of the present paper were announced without proofs in [10].

73

2.2.1.2 *Definitions*

Let $\mathscr{G}$ be a commutative semigroup with a (unique) neutral element e, and let $\mathscr{G}$ carry a topology such that the mapping $(u, v) \rightarrow uv$ is continuous and $\mathscr{G}$ itself is Hausdorff. Suppose further that $\mathscr{G}$ is provided with a continuous homomorphism Δ to the additive semigroup of non-negative real numbers, so that

(1) $$\Delta(uv) = \Delta(u) + \Delta(v)$$

and

(2) $$0 \leqslant \Delta(u) < \infty.$$

Obviously $\Delta(e)$ must be zero, and while in general Δ might vanish on other elements of $\mathscr{G}$, we shall exclude this possibility by making the assumption

(a) $\Delta(u) = 0$ *if and only if u is the neutral element e.*

It is to be noted that Δ need not be an isomorphism. The topological and algebraic structures on $\mathscr{G}$ will be further related by the assumption

(b) *for each u in $\mathscr{G}$, the factors of u form a compact set.*

By a *triangular array* we shall understand a system

$$u(1, 1),$$
$$u(2, 1), \quad u(2, 2),$$
$$u(3, 1), \quad u(3, 2), \quad u(3, 3),$$
$$\dots \qquad \dots \qquad \dots \qquad \dots$$

of elements of $\mathscr{G}$, and we shall call

$$u(i, 1)\, u(i, 2) \dots u(i, i)$$

the ith *marginal product* for the array. We shall call an array: *convergent* when the marginal products converge to some u in $\mathscr{G}$; *null* when $\Delta(u(i, j))$ converges to zero as i tends to infinity, uniformly for $1 \leqslant j \leqslant i$.

We shall say that u in $\mathscr{G}$ is *indecomposable* when it has both the properties

(i) $u \neq e$,

and

(ii) $u = vw$ implies that one of v and w is e (the other factor therefore being equal to u),

and we shall say that u is *infinitely divisible* when it possesses a kth root in $\mathscr{G}$, for every $k = 2, 3, \dots$. The neutral element is necessarily infinitely divisible (and because of the assumptions about Δ it has no factor distinct from itself); there may be no indecomposable elements. If u in $\mathscr{G}$ is not indecomposable then we shall call it *decomposable*; thus e has this property, and so has every infinitely divisible u.

Finally we make the assumption

(c) *if a null triangular array is convergent, then the limit is infinitely divisible.*

If $\mathscr{G}$ can be associated with a homomorphism Δ in such a way that the pair $(\mathscr{G}, \Delta)$ has all these properties, then $\mathscr{G}$ will be called a *Delphic semigroup*. Assumption (c) makes precise the loose statement in Section 2.2.1.1 that our semigroups will be required to obey the central limit theorem. We shall now show that such semigroups possess all the general properties discovered by Khintchine in his study [11] of the arithmetic of distribution functions on R.

In order to show that delphic semigroups do exist we mention the

Example. *Take $\mathscr{G}$ to be the half-open interval $(0, 1]$ with multiplication as the binary operation and the usual topology. Take $\Delta(u) = -\log u$. Then (a) holds trivially, and $\{u' : u' | u\}$ is the compact interval $[u, 1]$, so that (b) holds. Finally, every element is infinitely divisible (and so none is indecomposable), so that (c) also holds trivially. Thus our axioms are consistent.*

2.2.1.3 *The analogues of Khintchine's theorems*

First we have the trivial

Theorem 1. *An infinitely divisible element u of $\mathscr{G}$ can always be represented as the limit of a convergent null triangular array.*

Proof. For each i let $u^{1/i}$ denote some fixed ith root of u in $\mathscr{G}$ and put $u(i, j) = u^{1/i}$ for $j = 1, 2, \dots, i$. This gives us a triangular array in which each marginal product is equal to u, and so the array converges to u. It is obviously null, because $\Delta(u(i, j)) = \Delta(u)/i$.

Next we have the classification theorem:

Theorem 2. *The elements u of $\mathscr{G}$ can be classified as follows. Either*

(i) *u is indecomposable, or*

(ii) *u is decomposable (possibly infinitely divisible) and has an indecomposable factor, or*

(iii) *u is infinitely divisible (possibly neutral) and has no indecomposable factors.*

Proof. The classes defined by (*i*), (*ii*) and (*iii*) are obviously mutually exclusive and we have to prove that they are exhaustive. Suppose then that u has neither of the properties (*i*) and (*ii*), and that u is not the neutral element (which we know to have property (*iii*)). Then u can be decomposed into two non-neutral factors, and each one of its non-neutral factors can itself be decomposed into two non-neutral factors. We must prove that u is infinitely divisible. We note that $\Delta(u) > 0$.

As a first step we prove that if w runs through the non-neutral factors of v, where v is a given non-neutral factor of u, then $\inf \Delta(w) = 0$. To see this we recall that $\{v': v' | v\}$ is compact, and that Δ maps this compact set continuously into $[0, \infty)$; it attains its lower bound (zero) at $v' = e$. We shall show that e is a cluster point of $\{v': v' | v\}$, and this will suffice, for it will imply the existence of a sequence $w_1, w_2, \ldots$ of elements of $\mathscr{G}$, all of which are non-neutral factors of v, such that $\Delta(w_n) \rightarrow \Delta(e) = 0$. The neutral element e must be a cluster-point of $\{v': v' | v\}$, because otherwise $\{w: e \neq w | v\}$ (which is not empty) would also be compact, and Δ would attain its lower bound on this non-empty compact set, say at $w = w_0$. Now w_0, being a non-neutral factor of v, and so of u, could by hypothesis be split into non-neutral factors w_1 and w_2, each themselves non-neutral factors of v, and from the resulting inequality $\Delta(w_1) < \Delta(w_0)$ we should be able to contradict the minimal character of $\Delta(w_0)$.

Now consider all k-fold decompositions

(3) $$u = v(1) v(2) \ldots v(k)$$

of u; we shall prove that among these there is one decomposition for which $\Delta(v(i))$ is the same for $i = 1, 2, \ldots, k$. Each (ordered) decomposition (3) determines a unique point $[v(1), v(2), \ldots, v(k)]$ in the Cartesian product $\mathscr{G}^k$, and the non-empty set K of points $\mathscr{G}^k$ determined in this way is closed and therefore compact (for it is covered by the compact Cartesian product $\{v: v | u\}^k$). The function on $\mathscr{G}^k$ evaluated by

$$\min \{\Delta(v(1)), \Delta(v(2)), \ldots, \Delta(v(k))\}$$

is continuous on K and so is bounded there and attains its upper bound M; let equation (3) correspond to a point at which this upper bound is attained. Without losing any generality we can suppose that the factors in equation (3) have been rearranged so that $\Delta(v(i))$ is non-decreasing for increasing i. We should like $\Delta(v(i))$ to be independent of i and so we suppose that it is not. Then, for some r in $1 \leqslant r < k$, we shall have

$$M = \Delta(v(1)) = \Delta(v(2)) = \ldots = \Delta(v(r)) < \Delta(v(r+1)) \leqslant \ldots \leqslant \Delta(v(k)),$$

and because a decomposition like equation (3) into k non-neutral factors is certainly possible we know that $M > 0$. Thus each $v(i)$ is non-neutral;

in particular $v(r+1)$ is a non-neutral factor of u. It follows that there exist non-neutral factors w of $v(r+1)$ with arbitrarily small Δ-values. We can therefore express $v(r+1)$ in the form

$$v(r+1) = w(1)\, w(2) \ldots w(r)\, v^*(r+1),$$

where none of $\Delta(w(1)), \Delta(w(2)), \ldots, \Delta(w(r))$ exceeds an arbitrarily assigned positive ε, and all are positive. We now replace equation (3) by

$$u = v^*(1)\, v^*(2) \ldots v^*(k),$$

where

$$v^*(i) = v(i)\, w(i) \quad \text{for } 1 \leqslant i \leqslant r,$$

$$v^*(i) = v^*(r+1) \quad \text{for } i = r+1,$$

$$v^*(i) = v(i) \qquad \text{for } r+2 \leqslant i \leqslant k.$$

By choice of ε we can ensure that

$$\max_{1 \leqslant i \leqslant r} \Delta(v^*(i)) < \Delta(v^*(r+1)) < \Delta(v^*(r+2)) = \min_{r+1 < s \leqslant k} \Delta(v^*(i)),$$

so that

$$\min \Delta(v^*(i)) = \min_{1 \leqslant i \leqslant r} \Delta(v^*(i)) > \Delta(v(1)) = M,$$

and this contradicts the maximal character of M. Thus the chosen decomposition (3) must be such that

(4) $$\Delta(v(i)) = \Delta(u)/k \quad \text{for } i = 1, 2, \ldots, k.$$

We now set up a triangular array having u for each marginal product and such that $\Delta(u(i,j)) = \Delta(u)/i$; this will be convergent and null, and so its limit u must be infinitely divisible, as required.

Theorem 2 has the following immediate consequences.

Corollary 1. *If u is not indecomposable and if none of its factors is indecomposable, then u is infinitely divisible.*

Corollary 2. *One and only one of the following alternatives must hold:*

(a) *every element of $\mathscr{G}$ is infinitely divisible;*

(b) *$\mathscr{G}$ contains an indecomposable element.*

Finally we have the representation theorem:

Theorem 3. *For each u in $\mathscr{G}$ we can write*

(5) $$u = v(1)^{m_1} v(2)^{m_2} \ldots w$$

in at least one way. Here the ms are positive integers, the vs are distinct and indecomposable (there is at most a countable infinity of vs in the representation and there may be none) and the factor w is infinitely divisible and has no indecomposable subfactor (it may be the neutral element).

Proof. We fix u in $\mathscr{G}$, and we consider the system of all mappings φ (from the set $\mathscr{V}$ of all indecomposable elements v in $\mathscr{G}$ to the non-negative integers) which are such that, for each $k \geqslant 1$ and for each set of k distinct indecomposable elements $v_1, v_2, \ldots, v_k$,

(6) $$v_1^{\varphi(v_1)} v_2^{\varphi(v_2)} \ldots v_k^{\varphi(v_k)}$$

is a factor of u. This definition breaks down if $\mathscr{V}$ is vacuous, but then (by Corollary 2) u must be infinitely divisible and have no indecomposable factors, and so equation (5) holds trivially in the form $u = w$. If $\mathscr{V}$ is non-vacuous then such mappings certainly exist; thus we could take $\varphi(v) = 0$ for each v in $\mathscr{V}$.

It is clear that we must have

(7) $$\psi(v) \equiv \sup_{\varphi} \varphi(v) < \infty$$

for each v in $\mathscr{V}$, for if this were false for some such v then v^n would be a factor of u for every positive integer n and yet $\Delta(v) > 0$ and $\Delta(u) < \infty$, so that the inequality $n\Delta(v) \leqslant \Delta(u)$ would give a contradiction. It is important to notice that ψ need not itself be a φ.

We impose a partial ordering on the set of all φs by writing $\varphi' \leqslant \varphi''$ whenever $\varphi'(v) \leqslant \varphi''(v)$ for every v in $\mathscr{V}$. We claim that this partly-ordered set has the Zorn property. For let $\{\varphi_\alpha : \alpha \in A\}$ be a chain of φs, and put

(8) $$\varphi^*(v) \equiv \sup_{\varphi} \varphi_\alpha(v) \leqslant \psi(v) < \infty.$$

This defines a finite integer-valued function φ^* which dominates every φ_α, but we must show that it has the defining property of a φ-mapping associated with the word (6). Suppose then that the positive integer k and distinct indecomposables $v_1, v_2, \ldots, v_k$ have been given. For each v the supremum at equation (8) is attained, and so $\alpha_1, \alpha_2, \ldots, \alpha_k$ exist such that

$$\varphi^*(v_j) = \varphi_{\alpha_j}(v_j) \quad (j = 1, 2, \ldots, k),$$

and because the φ_αs form a chain we shall have $\varphi_{\alpha_j} \leqslant \varphi_\beta$ $(j = 1, 2, \ldots, k)$ where β is some one of $(\alpha_1, \alpha_2, \ldots, \alpha_k)$. By hypothesis

(9) $$v_1^{\varphi_\beta(v_1)} v_2^{\varphi_\beta(v_2)} \ldots v_k^{\varphi_\beta(v_k)}$$

is a factor of u, and as the word (6) with $\varphi = \varphi^*$ is a factor of the word (9), it too must be a factor of u. Thus our constructed φ^* is a valid φ, and the Zorn property holds.

It follows that the system of mappings φ has at least one maximal element. Choose one, and let φ denote this henceforth. If $\varphi \equiv 0$, then (by Corollary 1) u must be infinitely divisible and have no indecomposable factor and equation (5) holds in the form $u = w$. If $\{v : v \in \mathscr{V}$ and $\varphi(v) > 0\}$ is not empty then it is at any rate countable, since for any finite subset of it we shall have

$$\sum \Delta(v_j) \leqslant \sum \varphi(v_j)\Delta(v_j) \leqslant \Delta(u) < \infty.$$

We therefore enumerate $\{v : v \in \mathscr{V}$ and $\varphi(v) > 0\}$ as $v(1), v(2), \ldots$, where this sequence may terminate, and we write $m_r = \varphi(v(r))$. We now have factorizations as follows, for so long as the $v(r)$s last out:

$$u = v(1)^{m_1} z_1,$$
$$u = v(1)^{m_1} v(2)^{m_2} z_2,$$
$$\ldots \quad \ldots \quad \ldots \quad \ldots$$

If the sequence of $v(r)$s terminates with $v(n)$ then by Corollary 1 and maximality it follows that z_n must have the properties desired for w, and equation (5) holds.

Suppose then that the sequence does not terminate. We invoke

Lemma 1. *If u_r $(r = 1, 2, \ldots)$ and u are elements of $\mathscr{G}$ such that $u_r | u_s | u$ whenever $r \leqslant s$, then $u_\infty = \lim u_r$ exists and $u_r | u_\infty | u$ for all $r \geqslant 1$.*

If we put

$$u_r = v(1)^{m_1} v(2)^{m_2} \ldots v(r)^{m_r}$$

then we are in the situation of the Lemma, and so the limit

$$u_\infty = \prod_{j \geqslant 1} v(j)^{m_j}$$

exists in the $\mathscr{G}$-topology, has each u_r as a factor, and is itself a factor of u. We can therefore write $u_\infty = u_r w_r$ and $u = u_\infty w$. These last two formulae give $u = u_r w_r w$ $(r = 1, 2, \ldots)$, and if w were an indecomposable v (or had an indecomposable v as a factor) then the maximality of φ would be contradicted. Thus (by Corollary I) w is infinitely divisible and has no indecomposable factor, and $u = u_\infty w$ is the desired representation (5).

We now turn to the proof of Lemma 1. We shall obtain this as a consequence of

Lemma 2. *If a and b are elements of $\mathscr{G}$ such that $a | b$, then the set*

$$\{u : a | u | b\}$$

is compact.

Proof of Lemma 2. Let the functions f and g with domain $\mathscr{G}^2$ and range in $\mathscr{G}$ be defined as follows:

$$f(u', u'') = au',$$
$$g(u', u'') = au'u'';$$

they are continuous. Now

$$\{u: a\,|\,u\,|\,b\} = \{au': au'\,|\,b\}$$
$$= \{au': au'u'' = b \text{ for some } u''\}$$
$$= f(g^{-1}(\{b\})).$$

Also $g^{-1}(\{b\})$ is closed, and is a subset of the compact set

$$\{(u', u''): u'\,|\,b, \text{ and } u''\,|\,b\}$$

(this being the Cartesian square of the compact set $\{u: u\,|\,b\}$). Thus $g^{-1}(\{b\})$ is compact, and so its continuous image $f(g^{-1}(\{b\}))$ is compact.

Proof of Lemma 1. The track of the sequence $\{u_r: r = 1, 2, ...\}$ lies in the compact set $\{u': u'\,|\,u\}$, and so the sequence has at least one cluster-point u^* which is a factor of u, and it will be convergent if and only if this cluster-point is unique. Suppose then that u^* and u^{**} are cluster-points of the sequence, necessarily both factors of u; we must show that $u^* = u^{**}$. For any fixed r, u^{**} is also a cluster-point of the sequence

$$\{u_s: s = r, r+1, ...\},$$

and so of a sequence whose track lies in the compact set $\{u': u_r\,|\,u'\,|\,u\}$; thus u^{**} must lie in that compact set, and so $u_r\,|\,u^{**}$ for every r. But this means that the whole sequence $\{u_r: r = 1, 2, ...\}$ has its track in the compact set $\{u': u'\,|\,u^{**}\}$, and so now we can conclude that $u^*\,|\,u^{**}$, and (by symmetry) that $u^{**}\,|\,u^*$.

Suppose therefore that $u^{**} = u^*c$ and that $u^* = u^{**}d$; then

$$\Delta(u^{**}) = \Delta(u^*) + \Delta(c),$$

and

$$\Delta(u^*) = \Delta(u^{**}) + \Delta(d),$$

and from the non-negative and finite character of Δ we deduce that $\Delta(c) = \Delta(d) = 0$, and so that $c = d = e$, whence $u^* = u^{**}$, as required. Lemma 1 (and so also Theorem 3) has now been established.

It is of some interest to note that the infinite product in equation (5) is unconditionally convergent (the order in which the factors are taken does not matter). This is readily shown by a further use of Lemma 1.

It will be observed that the proofs of Theorems 1 to 3, like the theorems themselves, closely follow the pattern evolved in the original work on the convolution semigroup of distributions on R, and in fact the only substantial variations are those which have been forced on us by the absence of a metric topology (which prevents us from using sequential compactness arguments).

It seems likely that the appeal to Zorn's lemma cannot be avoided; one would doubt its necessity only if the representation (5) could be shown to be unique, and we know that this is not so, even for the convolution semigroup of distributions on R.

2.2.1.4. *Semigroups of aperiodic renewal sequences*

A detailed study of the arithmetic of renewal sequences was presented in [9], using direct methods. Here we shall merely show how the relevant results in [9] can be reformulated so as to become special cases of the theorems proved in the last section.

Let S denote an arbitrary additive semigroup of non-negative integers such that $0 \in S$ and g. c. d. $\{n: n \in S\} = 1$. Such a semigroup (which cannot be $\{0\}$) is finitely generated, so we choose a minimal set $\{g_j: j = 1, 2, ..., k)$ of generators for it. Then each $g_j > 0$, and the g_js have no common prime factor; it follows that every sufficiently large positive integer can be expressed linearly in terms of the g_js with non-negative coefficients.

Now let $\mathscr{G}(S)$ denote the set of all renewal sequences

$$u = \{u_n: n = 0, 1, 2, ...\}$$

which are such that

(10) $$u_{g_1} > 0, \quad u_{g_2} > 0, ..., u_{g_k} > 0.$$

This is of course just the set of renewal sequences for which $u_n > 0$ whenever $n \in S$. We shall give $\mathscr{G}(S)$ the metric topology of simple convergence which it acquires as a subset of R^∞, and we make it into a semigroup by writing

$$\{u'_n: n = 0, 1, 2, ...\}\{u''_n: n = 0, 1, 2, ...\} = \{u'_n u''_n: n = 0, 1, 2, ...\};$$

that this associative commutative binary operation yields a renewal sequence is well known. The neutral element will be $e = \{1, 1, 1, ...\}$. Clearly the semigroup operation is continuous.

We shall now prove

Theorem 4. *If S is an additive semigroup of non-negative integers* $0, s_1, s_2, ...$ *without a common prime factor, then $\mathscr{G}(S)$ is a Delphic semigroup.*

Proof. Let $\mathscr{R}$ denote the full semigroup of all renewal sequences. Clearly $\mathscr{G}(S)$ is a sub-semigroup of $\mathscr{R}$, and moreover all $\mathscr{R}$-factorizations of an element of $\mathscr{G}(S)$ are actually $\mathscr{G}(S)$-factorizations. Thus u in $\mathscr{G}(S)$ is infinitely divisible (indecomposable) if and only if it has this property as an element of $\mathscr{R}$. We shall use this fact in the proof of Theorem 4, but we note in parenthesis that it will also be needed as a link between Theorem 4 and the results of [9].

We now define the mapping Δ by

$$(11) \qquad \Delta(u) \equiv \sum_{j=1}^{k} \log\left(1/u_{g_j}\right),$$

and so obtain a continuous homomorphism from $\mathscr{G}(S)$ into the additive semigroup of non-negative reals. We must show that the pair $(\mathscr{G}(S), \Delta)$ satisfies (*a*), (*b*) and (*c*).

In order to prove that (*a*) holds we suppose that $\Delta(u) = 0$. Then $u_{g_j} = 1$ for $j = 1, 2, ..., k$, and from the lemma in Section 10 of [9] it follows that for some positive integer m

$$u_n = \begin{cases} 1 & \text{when } m \mid n, \\ 0 & \text{otherwise.} \end{cases}$$

But the g_js have no common prime factor and so m must be unity, and then $u = e$, as required.

Let us give $\mathscr{R}$ (like $\mathscr{G}(S)$) the metric topology induced from R^∞. It is known (see [9] and the references given there) that $\mathscr{R}$ is sequentially closed, and so it is closed as a subset of R^∞; also it is a subset of a countable Cartesian product of unit segments, and so it is compact. The topological semigroup $\mathscr{G}(S)$, however, may not be compact. But if u is an element of $\mathscr{G}(S)$ then

$$(12) \quad \{u' : u' \in \mathscr{G}(S), \text{ and } u' \mid u \text{ in } \mathscr{G}(S)\} = \{u' : u' \in \mathscr{R}, \text{ and } u' \mid u \text{ in } \mathscr{R}\},$$

and so in verifying (*b*) for $\mathscr{G}(S)$ it will suffice for us to show that the set on the right-hand side of equation (12) is a closed subset of $\mathscr{R}$ (hence compact), and in testing for closure we can do so sequentially, because the topology is metric. Suppose then that $u(r)$ is an $\mathscr{R}$-factor of $u \in \mathscr{G}(S)$ for $r = 1, 2, ...,$ and suppose that $u(r) \to u(\infty)$, so that $u(\infty) \in \mathscr{R}$. We shall have

$$u = u(r) u^*(r)$$

where $u^*(r) \in \mathscr{R}$, and in some subsequence $\lim u^*(r) = u^*(\infty)$ will exist and will belong to $\mathscr{R}$. Clearly then $u(\infty) \mid u$ in $\mathscr{R}$, the set at (12) is closed (so compact), and (*b*) holds.

Finally we must prove (*c*) for $\mathscr{G}(S)$. Suppose then that we are given a null triangular array in $\mathscr{G}(S)$ which converges to a limit u *which is in* $\mathscr{G}(S)$.

We must prove that u is infinitely divisible in $\mathscr{G}(S)$, and in accordance with a remark above we need only show that it is infinitely divisible in $\mathscr{R}$.

From nullity we know that

$$u_{g_s}(i,j) \to 1 \quad \text{as} \quad i \to \infty, \text{ uniformly for } 1 \leqslant j \leqslant i,$$

for $s = 1, 2, ..., k$. If m is large enough we also know that

$$m = \sum_{s=1}^{k} p_s g_s \quad \text{and} \quad m+1 = \sum_{s=1}^{k} q_s g_s,$$

where the ps and qs are non-negative integers. By one of the basic renewal inequalities it follows from this that

$$1 \geqslant u_m(i,j) \geqslant \prod_{s=1}^{k} u_{g_s}(i,j)^{p_s} \to 1$$

and

$$1 \geqslant u_{m+1}(i,j) \geqslant \prod_{s=1}^{k} u_{g_s}(i,j)^{q_s} \to 1$$

as $i \to \infty$, uniformly for $j = 1, 2, ..., i$. But from the inequality of Kingman used at (49) in [9] we know that

$$0 \leqslant 1 - u_1(i,j) \leqslant (1 - u_{m+1}(i,j))/u_m(i,j),$$

and so $u_1(i,j) \to 1$ as $i \to \infty$, uniformly for $j = 1, 2, ..., i$. We can now invoke the triangular central limit theorem for renewal sequences (proved in Section 9 of [9]) in order to deduce that u is infinitely divisible in $\mathscr{R}$, and so (c) holds and $\mathscr{G}(S)$ is Delphic.

We can now deduce most of the content of Section 10 of [9] from the theorems of Section 2.2.1.3 of the present paper. If u is an aperiodic renewal sequence then it belongs to $\mathscr{G}(S)$ for some† S, and its $\mathscr{R}$-factors are its $\mathscr{G}(S)$-factors; also the arithmetical character of u and its factors is the same whether we think of them as elements of $\mathscr{R}$ or of $\mathscr{G}(S)$. Thus the conclusions of our general theorems in this particular case translate into the apparently stronger assertions given in [9].

From the remarks at the beginning of Section 10 of [9] it will be clear that $\mathscr{G}(S)$ for each S is isomorphic to a sub-semigroup of the usual convolution semigroup of probability measures on the circle, and of course our arithmetical theorems can be transferred to these sub-semigroups. Their identification and detailed study now becomes a matter of some interest.

† For example, we can take $S = (n: u_n > 0)$.

2.2.2 INFINITELY DIVISIBLE REGENERATIVE PHENOMENA, AND THE ARITHMETIC OF p-FUNCTIONS

2.2.2.1. $\mathscr{P}$ as a Delphic semigroup

We suppose that the reader is already familiar with at least the elements of the theory of p-functions due to J. F. C. Kingman [13, 14, 15]. The expository accounts in [16] and [7] may be found useful to those who do not choose to enter into all the details. We shall write $\mathscr{P}$ for the class of all 'standard' p-functions; i.e. those which satisfy the condition, $p(t) \rightarrow 1$ as $t \downarrow 0$. The main result of this section will be

Theorem 5. *The Kingman semigroup $\mathscr{P}$ is a Delphic semigroup.*

Proof. First we observe that $\mathscr{P}$ is a commutative semigroup with the law of composition

$$(13) \qquad (p_1 p_2)(t) = p_1(t) p_2(t) \quad (0 \leqslant t < \infty).$$

As in the parallel situation for renewal sequences, the probabilistic proof of this is almost trivial (set up models† for the regenerative phenomena Φ_1 and Φ_2 on distinct probability spaces $(\Omega_1, \mathscr{F}_1, \mathrm{pr}_1)$ and $(\Omega_2, \mathscr{F}_2, \mathrm{pr}_2)$, and on the direct product of these spaces define a new regenerative phenomenon Φ by saying that 'Φ happens at time t if and only if *both* Φ_1 and Φ_2 happen at time t'), while an analytical verification would be a somewhat formidable task. (Notice though that the probabilistic argument we have just sketched merely shows that $p_1 p_2$ is a p-function, and we must also show that it is standard; this however is obvious from (13).) Accordingly we have a commutative semigroup (which we propose to call the Kingman semigroup) with neutral element e defined by $e(t) = 1$ (all t).

We now define the functional Δ, by

$$(14) \qquad \Delta(p) \equiv \log(1/p(1)) \quad (p \in \mathscr{P}).$$

Then $\Delta(p)$ will be a non-negative real number because $0 < p(t) \leqslant 1$, and obviously, from the properties of the logarithm, Δ provides a homomorphism. It vanishes when $p = e$, and in order to verify (*a*) we must prove the converse. Suppose then that $\Delta(p) = 0$, so that $p(1) = 1$. Then from the Kingman inequalities

$$(15) \qquad p(s)p(t) \leqslant p(s+t) \leqslant 1 + p(s)p(t) - \max\{p(s), p(t)\}$$

---

† This Φ-formulation of the theory of p-functions is an exact analogue of the Φ-formulation of the theory of renewal sequences given in Section 2 of [9]. See also Section 3 of [7].

we see that $p(n) = 1$ for every non-negative integer n, and then that

$$1 = p(n) = p(n-t+t) \leqslant 1 + p(n-t)p(t) - p(n-t) \quad (0 \leqslant t \leqslant n),$$

so that (because $p(n-t) > 0$) we can conclude that $p(t) \geqslant 1$, and thus that $p = e$.

Before we can proceed any further we must commit ourselves to a topology for $\mathscr{P}$. We shall use the topology induced on $\mathscr{P}$ by regarding it as a subset of the Cartesian product $C = [0, 1]^{[0,\infty)}$; i.e. the topology of simple convergence. Then obviously $\mathscr{P}$ is Hausdorff, and the semigroup composition is continuous. Also Δ can now be seen to be a continuous homomorphism (from the fact that $p(1)$ never vanishes).

In order to verify (*b*) we take a fixed element q in $\mathscr{P}$ and consider the set $\{p: p|q\}$. If p belongs to this set, then $q = pp_1$ for some p_1 in $\mathscr{P}$ and so we shall have

(16) $$q(t) \leqslant p(t) \leqslant 1 \quad (0 \leqslant t < \infty).$$

Now suppose that we have a net $\{p_\alpha: \alpha \in A\}$ in $\mathscr{P}$, where p_α is a factor of q for each α in the directed set A. The net has a cluster-point p_∞ in the compact space C, and a suitably chosen subnet will converge to p_∞. From Theorem 1 (and its corollary) in [13] it follows that p_∞ is a p-function, and equation (16) shows that it is standard. But we also know that q/p_α is a p-function in $\{p: p|q\}$, and the corresponding subnet will converge to q/p_∞, which is a p-function, is standard and satisfies $q = p_\infty \cdot q/p_\infty$. Thus p_∞ belongs to $\{p: p|q\}$; that is, the net $\{p_\alpha: \alpha \in A\}$ has a cluster-point in $\{p: p|q\}$, and so this latter set is compact, as was required.

Finally we have to verify (*c*). Suppose then that we are given a triangular array (with (i, j)th element p_{ij}) which is null and converges to $p_\infty \in \mathscr{P}$. We must show that p_∞ is infinitely divisible. Take any $\tau > 0$ and write

(17) $$u_n(i, j) \equiv p_{ij}(n\tau) \quad (n = 0, 1, 2, \ldots).$$

Then we know that $u(i, j)$ is a renewal sequence for each (i, j), and that $u_n(i, j) > 0$ for all n. Obviously in $\{u(i, j): i = 1, 2, \ldots; j = 1, 2, \ldots, i\}$ we have a triangular array of renewal sequences in $\mathscr{G}(S)$ where S is the semigroup of *all* non-negative integers. The marginal products of this new array converge to $u^\infty \in \mathscr{G}(S)$, where

(18) $$u_n^\infty = p_\infty(n\tau) \quad (n = 0, 1, 2, \ldots).$$

We claim that the new array, like the original one, is null. Because S is generated by the single element 1 we can take the Delphic homomorphism for $\mathscr{G}(S)$ to be

$$\Delta_\tau(v) = \log(1/v_1),$$

so that

$$\Delta_\tau(u(i, j)) = \log(1/p_{ij}(\tau)).$$

We therefore have to show that $p_{ij}(\tau) \to 1$ as $i \to \infty$ uniformly for $1 \leqslant j \leqslant i$, and we have to establish this for each $\tau > 0$. We know that it is true when $\tau = 1$, from the nullity of the original triangular array, and so from the left-hand inequality of (15) it is true for $\tau = n$, where n is any positive integer. Let p_i denote the ith marginal product of the original array. For any $\tau > 0$ choose a positive integer $n > \tau$ and choose I so that

$$p_i(n - \tau) > \tfrac{1}{2} p_\infty(n - \tau) > 0 \quad \text{for } i \geqslant I.$$

Now the right-hand inequality at (15) shows that

$$1 - p_{ij}(\tau) \leqslant \frac{1 - p_{ij}(n)}{p_{ij}(n - \tau)},$$

and we also know that

$$p_{ij}(n - \tau) \geqslant p_i(n - \tau) > \tfrac{1}{2} p_\infty(n - \tau) > 0 \quad \text{for } i \geqslant I;$$

it follows that the desired uniform-convergence statement can be transferred from the array at $t = n$ to the array at $t = \tau$, as desired.

It follows that $\{u(i, j)\}$ is a null triangular array in $\mathscr{G}(S)$ which converges to u^∞ in $\mathscr{G}(S)$, and so by the results of Section 2.2.1.4. of the present paper we can conclude that u^∞ is infinitely divisible in $\mathscr{G}(S)$. This, however, implies that

$$p_\infty(n\tau)^{1/k} \quad (n = 0, 1, 2, \ldots)$$

is a renewal sequence for each $\tau > 0$, for every $k \geqslant 1$, and thus (by Proposition 6 of [13]) it follows that $p_\infty^{1/k}$ belongs to $\mathscr{P}$. We have therefore shown that Axiom (c) is satisfied, and the theorem is proved.

2.2.2.2. *Consequences of Theorem 5 for the Kingman semigroup*

The first interesting consequence of Theorem 5 is that (by Theorem 2) *the elements of $\mathscr{P}$ can be partitioned into the three subsets*

(i) *the indecomposable elements of $\mathscr{P}$,*

(ii) *the decomposable elements of $\mathscr{P}$ which have at least one indecomposable factor,*

(iii) *those of the infinitely divisible elements of $\mathscr{P}$ which have no indecomposable factors.*

We shall now show that *each of these three subsets is non-empty*.

It is convenient to deal with class (*iii*) first. This of necessity contains *e*, but we shall find a non-trivial member. This class is the analogue of what Linnik, when writing about the convolution arithmetic of probability distributions on *R*, calls 'the class I_0'. In that arithmetic many remarkable theorems have been proved about the class I_0 but it has never been completely identified, although some members are known (for example, the normal distributions). What might be called 'the I_0-problem' for the Delphic semigroup of renewal sequences $\mathscr{G}(S)$ when *S* is the semigroup of all non-negative integers has recently been completely solved by Davidson [3], who has made the surprising discovery that in this case I_0 is generated by a single element; in fact he shows that $u \in I_0$ if and only if

$$u_n = \rho^n$$

for some ρ satisfying $0 < \rho \leqslant 1$. For the Delphic semigroup $\mathscr{P}$ we shall show that $I_0\backslash\{e\}$ is non-empty by proving that

$$p(t) = e^{-\alpha t}$$

defines an element of I_0 whenever $0 \leqslant \alpha < \infty$. It is known that each such *p* is in $\mathscr{P}$.

For any *q* in $\mathscr{P}$ let us define a function *y* by

(19) $$q(t) = e^{-y(t)} \quad (0 \leqslant t < \infty),$$

so that $y(t)$ is non-negative, finite, continuous, and vanishes at $t = 0$. Also (from the left-hand inequality of (15)) *y* is subadditive:

(20) $$y(s+t) \leqslant y(s) + y(t).$$

Now suppose that *q* and *r* are members of $\mathscr{P}$ such that $q(t)r(t) = e^{-\alpha t}$ for all *t*. Let *q* and *r* transform to *y* and *z* as at equation (19); then we shall have

$$y(s+t) \leqslant y(s) + y(t),$$
$$z(s+t) \leqslant z(s) + z(t),$$

and

$$\alpha(s+t) = \alpha s + \alpha t,$$

so that (since $y(\tau) + z(\tau) = \alpha\tau$ for all τ) equality must hold throughout; that is, *y* and *z* must be additive and not merely subadditive. But each of *y* and *z* is continuous, and so $y(t) = \beta t, z(t) = \gamma t$, and therefore *every factor of p is a power of p*; moreover it is obvious that *p* is infinitely divisible. This shows that $p(t) = e^{-\alpha t}$ enjoys in $\mathscr{P}$ the same sort of stability which Cramér proved for the normal distribution in the convolution semigroup of probability distributions on R, and in particular it shows that *such p-functions belong to I_0*.

In view of Davidson's theorem concerning $\mathcal{G}(S)$ it is now natural to ask whether there are any other members of I_0. We do not yet know the answer to this question.

We have shown that class (*iii*) is non-empty. If we can find an element p of $\mathcal{P}$ which is not infinitely divisible then it cannot belong to class (*iii*), so must belong to class (*i*) or to class (*ii*). If p belongs to class (*i*) then p^2 will belong to class (*ii*), while if p belongs to class (*ii*) then it must possess at least one indecomposable factor, which will belong to class (*i*). Thus we shall then know that all three classes are non-empty.

It is easy to find such a non-infinitely-divisible p; for example we may take p to be the p-function constructed by Kingman at the end of Section 6 of his paper [13]; for this

$$p(t) = e^{-t} \quad (0 \leqslant t \leqslant 1)$$

and

$$p(t) = e^{-t} + (t-1)e^{-(t-1)} \quad (1 \leqslant t \leqslant 2).$$

Now if p were infinitely divisible in the arithmetic of $\mathcal{P}$ then $u(\tau)$, where

$$u(\tau)_n = p(n\tau) \quad (n = 0, 1, 2, \ldots)$$

and $\tau > 0$, would be infinitely divisible in the arithmetic of $\mathcal{G}(S)$ and so (by the results of [9]) $p(n\tau)$ would be a non-increasing function of n for each fixed $\tau > 0$. But

$$p(1+s) = e^{-1}\{1 + (e-1)s + o(s)\}$$

as $s \downarrow 0$, and so $p(1+s) > p(1)$ for all sufficiently small positive values of s. With a suitably small choice of τ, therefore, $u(\tau)$ will not be a non-increasing sequence, and so p is not infinitely divisible.

From Theorem 3 of Section 2.2.1.3 and Theorem 5 of the present section we learn that *an arbitrary member p of $\mathcal{P}$ can be represented in at least one way in the form*

(21) $$p(t) = q_1(t)^{m_1} q_2(t)^{m_2} \ldots r(t) \quad (0 \leqslant t < \infty),$$

where the ms are positive integers, the qs (of which there may be none, a finite number, or a countable infinity) are distinct and indecomposable, and where r (which may be e) belongs to I_0.

In our verification of (c) for $\mathcal{P}$ we have of course already shown that $\mathcal{P}$ *obeys the triangular form of the central limit theorem.*

Kingman (Theorem 5 of [13]) has shown that each p in $\mathcal{P}$ is the characteristic function of a probability measure on R, so that the arithmetic of $\mathcal{P}$

can be identified with the convolution arithmetic of the particular sub-semigroup of the full semigroup of probability measures on R for which the characteristic functions of the constituent measures belong to $\mathscr{P}$. The properties of $\mathscr{P}$ which we have just established resemble those found by Khintchine for the full semigroup of probability measures on R, but are *not* consequences of the Khintchine theorems. The situation is in fact similar to that discussed in Section 10 of [9]. (See Section 2.2.2.5 below for some further remarks on this topic.)

2.2.2.3. *Infinitely divisible p-functions*

We now turn our attention to the infinitely divisible elements of the Kingman semigroup $\mathscr{P}$. If p is infinitely divisible then so is $p^{m/n}$ for any positive integers m and n, and hence (by the corollary to Theorem 1 of [13]) so is p^u for all $u \geqslant 0$, whether rational or not. From this it follows that if as before we put

$$p(t) = \exp(-x(t)) \quad (0 \leqslant t < \infty)$$

then for infinitely divisible p the functions x constitute a positive convex cone whose vertex $x \equiv 0$ ($p \equiv 1$) corresponds to the neutral element of $\mathscr{P}$. We therefore embark on a Choquet analysis of this cone (for Choquet's theorem see [20]).

If p is infinitely divisible in $\mathscr{P}$ then the sequence $u(\tau)$ defined by $u(\tau)_n = p(n\tau)$ ($n = 0, 1, 2, \ldots$), where $\tau > 0$, will be a strictly positive renewal sequence and will be an infinitely divisible element of the semigroup $\mathscr{G}(S)$ defined in Section 2.2.1.4, where now S is the semigroup of *all* non-negative integers. Thus, by Theorem 1 of [9], $u(\tau)$ is a Kaluza sequence and so

$$p(n\tau + \tau)\, p(n\tau - \tau) \geqslant p(n\tau)^2$$

and

$$x(n\tau + \tau) - 2x(n\tau) + x(n\tau - \tau) \leqslant 0,$$

from which it follows that

$$x(n\tau + k\tau) - 2x(n\tau) + x(n\tau - k\tau) \leqslant 0$$

whenever $\tau \geqslant 0$, $k \geqslant 0$ and $n \geqslant k$. But (Theorem 2 of [13]) p is uniformly continuous on $[0, \infty)$, and so

$$x(\alpha + \beta) - 2x(\alpha) + x(\alpha - \beta) \leqslant 0$$

when $0 \leqslant \beta \leqslant \alpha < \infty$; that is, x is a concave function (and we already know that x is continuous and non-negative, and that $x(0) = 0$).

On the other hand, let x be any non-negative concave function such that $x(0) = x(0+) = 0$. Then x is continuous on $[0, \infty)$, and if we put

$$p(t) \equiv \exp(-x(t)) \quad (0 \leqslant t < \infty)$$

then we obtain a function p which is continuous and positive, and is bounded above by unity, and which assumes the value 1 at $t = 0$. Also $u(\tau)_n \equiv p(n\tau)$ will define a bounded Kaluza sequence and so will yield a renewal sequence for each $\tau > 0$ (by Lamperti's result quoted in Section 1 of [9]). From Proposition 6 of [13] it then follows that p must belong to $\mathscr{P}$, and since all the assumptions we have made about x are equally well satisfied by every positive multiple of x we can conclude that $p^u \in \mathscr{P}$ for each $u \geqslant 0$. Accordingly we have proved

Theorem 6. *The relation* $p(t) = \exp(-x(t))$ $(0 \leqslant t < \infty)$ *sets up a one-one correspondence between the infinitely divisible p of $\mathscr{P}$, and the non-negative concave functions x on $[0, \infty)$ such that* $x(0) = x(0+) = 0$.

Kingman in Section 3.4 of [16] has already noticed that $p = e^{-x}$ yields a p-function whenever x has the properties set out in the theorem, and it is immediately clear that such p-functions will be infinitely divisible. What is new in Theorem 6 is the assertion that *all* the infinitely divisible elements of $\mathscr{P}$ can be obtained in this way.

In order to apply Choquet's theorem we must augment the cone of x-functions specified in Theorem 6 by deleting the condition $x(0+) = x(0)$. The added functions (which we must now *require* to be finite) will not correspond to standard p-functions (they will not be continuous at $t = 0$); they will be discarded later, and are here added merely for compactification purposes. The augmented system of x-functions thus consists of *all finite non-negative concave functions which vanish at $t = 0$*; they are necessarily continuous for *positive t*, and necessarily non-decreasing. We must now determine the extreme rays of this cone, which we shall call $\mathscr{K}$.

A ray through $x \in \mathscr{K}$ (where $x \neq 0$) is said to be extreme when the data

$$x = x' + x'', \quad x' \in \mathscr{K}, \quad x'' \in \mathscr{K},$$

imply that each of x' and x'' is a multiple of x (so lies on the ray). Suppose first that $x(0+) > 0$; then we get such a decomposition if we write

$$x'(t) = \begin{cases} \varepsilon & (t > 0), \\ 0 & (t = 0), \end{cases} \quad x''(t) = x(t) - x'(t),$$

provided that $0 \leqslant \varepsilon \leqslant x(0+)$, and thus x is minimal (lies on an extreme ray) only if $x(t)$ is constant for $t > 0$. On the other hand the minimality of all such xs is obvious, for in any decomposition x' and x'' would each have to be linear, so constant, for $t > 0$.

With this case out of the way we can confine attention to those minimal xs which are continuous even at $t = 0$. Because each x is concave and non-decreasing it will have a finite right-hand and a finite left-hand derivative for each $t > 0$, and a (perhaps infinite) right-hand derivative at $t = 0$. Let the right-hand derivative at $t = 0$ be g_0, and suppose if possible that for some $t = t_1 > 0$ the right-hand derivative g_1 satisfies the inequality $0 < g_1 < g_0$. Then we write

$$x'(t) = g_1 t \quad (0 \leqslant t \leqslant t_1), \quad x'(t) = g_1 t_1 + x(t) - x(t_1) \quad (t_1 \leqslant t < \infty),$$

$$x''(t) = x(t) - g_1 t \quad (0 \leqslant t \leqslant t_1), \quad x''(t) = x(t_1) - g_1 t_1 \quad (t_1 \leqslant t < \infty).$$

With these definitions, $x = x' + x''$ is a decomposition within $\mathcal{H}$, and neither of x' and x'' is the zero function. For x to be minimal, therefore, it would have to be linear in $[0, t_1]$ and constant in $[t_1, \infty)$, but this would contradict the assumption $g_1 > 0$. Accordingly the minimality of x must imply that for $t > 0$ the right-hand derivative of x can have no values other than g_0 and zero.

It is now clear that the possibilities $g_0 = 0$ and $g_0 = \infty$ can be eliminated, for if $g_0 = 0$ then $x = 0$ and so x does not determine a ray, while if $g_0 = \infty$ then $x(t)$ must be constant for $t > 0$ and discontinuous at $t = 0$, contrary to our current assumptions.

This means that among the xs which are continuous at $t = 0$ the only candidates for minimality are

$$x(t) = gt \quad (0 < g < \infty),$$

and

$$x(t) = g \min (t, t_1) \quad (0 < g < \infty; \, 0 < t_1 < \infty),$$

and the actual minimality of each of these follows from the remark that every decomposition $x = x' + x''$ within $\mathcal{H}$ must be such that each of x' and x'' is piecewise linear.

Accordingly the complete set of minimal elements of $\mathcal{H}$ is given by

$$(22) \qquad x(t) = \min (\alpha t, \beta) \quad (0 \leqslant t < \infty),$$

where $0 < \alpha \leqslant \infty$ and $0 < \beta \leqslant \infty$, and α and β are not both infinite.

We must now slice through $\mathcal{H}$ by a hyperplane which avoids the origin; the hyperplane $\{x: x(1) = 1\}$ will be convenient. Let this hyperplane cut $\mathcal{H}$ in the set B (the *base* of the cone). As usual we embed the cone $\mathcal{H}$ in the Cartesian product $R^{[0,\infty)}$, which is a Hausdorff space, and we examine the induced topology on the (closed) base B. This is identical with the topology induced from $R^{\text{non-negative rationals}}$ in virtue of the continuous and non-decreasing character of the x-functions, for $t > 0$, and so B is *metrizable*. On the other hand the inequalities

$$t \leqslant x(t) \leqslant 1 \quad (0 \leqslant t \leqslant 1),$$

$$1 \leqslant x(t) \leqslant t \quad (1 \leqslant t < \infty),$$

which hold for every $x \in B$, show that B is *compact*. These facts enable us to apply Choquet's theorem, and we can do so as soon as we have provided the extreme rays of $\mathcal{K}$ with a parametrization from a suitable homeomorph of the Choquet boundary $\partial \mathcal{K}$ (which we can take to be the set of points ∂B in which the extreme rays meet B).

On referring back to equation (22) we see that ∂B consists precisely of the points x_λ, where

(23) $$x_\lambda(t) = \min(t, \lambda) \quad (0 \leqslant t < \infty)(1 \leqslant \lambda \leqslant \infty)$$

and

(24) $$x_\lambda(t) = \min(t/\lambda, 1) \quad (0 \leqslant t < \infty)(0 \leqslant \lambda \leqslant 1);$$

also the usual topology on $\{\lambda : 0 \leqslant \lambda \leqslant \infty\}$ makes $\{x_\lambda : 0 \leqslant \lambda \leqslant \infty\}$ a homeomorphic image of ∂B, as required. The graphs of the functions x_λ are shown in Figure 1.

We shall therefore have

(25) $$x(t) = \int_{[0,\infty]} x_\lambda(t) \, \sigma(d\lambda) \quad (0 \leqslant t < \infty),$$

where σ is a totally finite measure on the Borel sets of $[0, \infty]$, in the sense

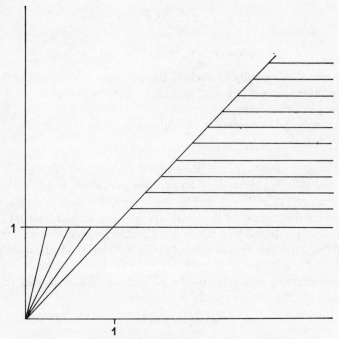

Figure 1. The extreme points x_λ of B

that every such σ defines in this way an element x of $\mathscr{K}$, and every x in $\mathscr{K}$ can be expressed in the form (25) for at least one choice of the measure σ. We can rewrite equation (25) in the form

$$(26) \qquad x(t) = x_0(t)\, \sigma(\{0\}) + \int_{(0,\infty]} \frac{\min(t,\lambda)}{\min(\lambda,1)}\, \sigma(d\lambda).$$

Now for $0 \leqslant t \leqslant 1$ the integrand of the last integral is non-negative and bounded above by unity for all λ in $(0,\infty]$, and for each such fixed λ it tends to zero as $t \to 0$. It follows that the integral is continuous at $t = 0$, and so those xs in $\mathscr{K}$ which are not continuous at $t = 0$ arise when and only when σ has an atom at $\lambda = 0$. If therefore we exclude the possibility of an atom at $\lambda = 0$ we shall get a similar representation for the set of xs which correspond to infinitely divisible p-functions.

On noticing that the ratio of the functions $\min(\lambda,1)$ and $1 - e^{-\lambda}$ is bounded away from zero and from infinity we then see that the xs associated with infinitely divisible elements of $\mathscr{P}$ are precisely the functions of the form

$$(27) \qquad x(t) = \int_{(0,\infty]} \frac{\min(t,\lambda)}{1 - e^{-\lambda}}\, \sigma(d\lambda) \quad (0 \leqslant t < \infty),$$

where again σ is an arbitrary totally finite measure on the Borel sets of $(0,\infty]$.

Now if we take the Cartesian product of σ on $\{\lambda : 0 < \lambda \leqslant \infty\}$ and Lebesgue measure on $\{u : 0 \leqslant u < \infty\}$, then the right-hand side of equation (27) is the integral of the function $(1 - e^{-\lambda})^{-1}$ over the set

$$\{(\lambda, u) : 0 \leqslant u \leqslant \min(t,\lambda),\ 0 < \lambda \leqslant \infty\},$$

and if we apply Fubini's theorem and evaluate the integral the other way round we find that

$$(28) \qquad x(t) = \int_0^t du \int_{[u,\infty]} \frac{\sigma(d\lambda)}{1 - e^{-\lambda}}.$$

It follows that $x(t)$ has a left-hand derivative (for $t > 0$) equal to

$$(29) \qquad \int_{[t,\infty]} \frac{\sigma(d\lambda)}{1 - e^{-\lambda}}$$

and a right-hand derivative (for $t \geqslant 0$) equal to

$$(30) \qquad \int_{(t,\infty]} \frac{\sigma(d\lambda)}{1 - e^{-\lambda}}$$

Thus σ is uniquely determined when x is known. We can now collect our various results to obtain

Theorem 7. *The relation*

$$(31) \qquad p(t) = \exp\left\{-\int_{(0,\infty]} \frac{\min(t,\lambda)}{1-e^{-\lambda}}\,\sigma(d\lambda)\right\}$$

sets up a one-one correspondence between the totally finite measures σ on the Borel sets of $(0,\infty]$ and the infinitely divisible elements p of the Kingman semigroup $\mathscr{P}$.

This result was announced without proof in the author's note [8]. The unicity of the representation implies that the set of infinitely divisible elements of $\mathscr{P}$ can be given a lattice structure. Here $p_1 \prec p_2$ (where p_1 and p_2 are each infinitely divisible) will mean $p_1 | p_2$ (the cofactor being *required* to be infinitely divisible), and then

$$p_1 \prec p_2 \rightleftarrows \sigma_1 \leqslant \sigma_2.$$

It is very important to observe that it is possible for two infinitely divisible elements p_1 and p_2 of $\mathscr{P}$ to be such that $p_1 | p_2$ even when it is not the case that $p_1 \prec p_2$; when this happens the ratio p_2/p_1 will be in $\mathscr{P}$ but will not be infinitely divisible, and $p_2(t)/p_1(t)$ will still be given by equation (31) but the measure σ will no longer be non-negative. In such a case p_2/p_1 must be indecomposable or possess an indecomposable factor, and so p_2 will belong to $I(\mathscr{P})\backslash I_0(\mathscr{P})$ (the class of infinitely divisible elements which do possess indecomposable factors). An example of this state of affairs has been constructed by Davidson [3].

We conclude this section with a few examples. Let us first of all look at the infinitely divisible p-functions which correspond to minimal xs. These are given by

$$(32) \qquad p_{s,\lambda}(t) = e^{-s\min(t,\lambda)} \quad (0 \leqslant t < \infty),$$

where $0 < s < \infty$ and $0 < \lambda \leqslant \infty$.

Other (not minimal) infinitely divisible elements of $\mathscr{P}$ are easily constructed with the aid of Theorem 6. For example,

$$(33) \qquad p(t) = \exp\left(-(1-e^{-t})\right),$$

and

$$(34) \qquad p(t) = \exp\left(-t^{\alpha}\right),$$

where $0 < \alpha < 1$. In cases like this it is quite easy to find the representing measures. Thus when p is given by equation (33) then $x(t) = 1 - e^{-t}$, and as x is differentiable we know that σ is completely non-atomic (there is not even an atom at $\lambda = \infty$ because $x'(t) \to 0$ as $t \to \infty$). Because x' is here absolutely continuous we can solve equations (29) and (30) to find that

$$(35) \qquad \sigma(d\lambda) = e^{-\lambda}(1 - e^{-\lambda})\, d\lambda \quad (0 < \lambda < \infty).$$

The situation with equation (34) is similar, but now

$$(36) \qquad \sigma(d\lambda) = \alpha(1 - \alpha)(1 - e^{-\lambda})\frac{d\lambda}{\lambda^{2-\alpha}} \quad (0 < \lambda < \infty).$$

Two quantities of special interest in connection with any p-function are

$$q = -p'(0+) \quad \text{and} \quad \omega = \lim_{t \to \infty} p(t).$$

For infinitely divisible p-functions it is easy to express these in terms of the measure σ; in fact

$$(37) \qquad q = \int_{(0,\infty]} \frac{\sigma(d\lambda)}{1 - e^{-\lambda}}$$

and

$$(38) \qquad \omega = \exp\left\{-\int_{(0,\infty]} \frac{\lambda\sigma(d\lambda)}{1 - e^{-\lambda}}\right\}.$$

From equation (37) it follows that p will describe what Kingman has called 'an event of type B' when and only when $\int \sigma(d\lambda)/\lambda$ is convergent at $\lambda = 0$. When this integral is divergent, p is unstable and so describes 'an event of type KR'.

2.2.2.4. *The operator T_c, and an interpretation for infinitely divisible p-functions*

An element p of $\mathscr{P}$, whether infinitely divisible or not, is called *stable* when $q = -p'(0+)$ is finite. Kingman has shown (Theorems 3 and 4 of [13]) that every element p of $\mathscr{P}$ (whether stable or not) has a Laplace transform

$$(39) \qquad r(\theta) = \int_0^\infty p(t)e^{-\theta t}\, dt \quad (0 < \theta < \infty)$$

which can be expressed in the form

$$(40) \qquad r(\theta) = \left\{\theta + \int_{(0,\infty]} (1 - e^{-\theta x})\,\mu(dx)\right\}^{-1},$$

where μ is a σ-finite measure on the Borel sets of $(0,\infty]$ such that

$$(41) \qquad \int_{(0,\infty]} (1 - e^{-x})\,\mu(dx) < \infty,$$

and conversely that each μ satisfying inequality (41) determines through equation (40) a function r which is the Laplace transform as at equation (39) of some $p \in \mathscr{P}$, the relation between p and μ being one–one. Those measures μ which correspond to stable ps are easily recognized, for they are precisely the μs which are totally finite, and in fact we always have

$$(42) \qquad\qquad q = \mu\{(0, \infty]\}.$$

Thus, when we are concerned only with stable ps, we can put them into one-one correspondence with totally finite μs, because if μ is totally finite then the condition (41) will be automatically satisfied.

Suppose then that p is a stable element of $\mathscr{P}$. We define a new function $T_c p$ by

$$(43) \qquad \begin{aligned} (T_c p)(t) &= e^{-qt} & (0 \leqslant t \leqslant c), \\ (T_c p)(t) &= p(t-c) e^{-qc} & (c \leqslant t < \infty). \end{aligned}$$

Here c is any positive real number. We now have

Theorem 8. *If p is a stable element of $\mathscr{P}$, and $q = -p'(0+)$, then $T_c p$ is also a stable element of $\mathscr{P}$, and it is infinitely divisible if and only if p has that property.*

Proof. First we note that $T_c p$ is continuous and that it has $-q$ as right-hand derivative at the origin, so that the theorem will follow if we can show that

$$r_c(\theta) = \int_0^\infty e^{-\theta t} (T_c p)(t)\, dt$$

can be expressed in the form (40) with the measure totally finite. If for brevity we write

$$q + \theta = Q, \quad e^{-(q+\theta)c} = E \quad \text{and} \quad \int_{(0,\infty]} e^{-\theta x} \mu(dx) = M,$$

so that

$$r_c(\theta) = \frac{1-E}{Q} + \frac{E}{Q-M},$$

then

$$1/r_c(\theta) = \theta + \int_{(0,\infty]} (1 - e^{-\theta x}) \mu_c(dx),$$

where μ_c is the measure with total mass q defined by

$$\int_{(0,\infty]} e^{-\theta x} \mu_c(dx) = EM \bigg/ \left(1 - \frac{1-E}{Q} M\right) = EM \sum_n \left(\frac{1-E}{\geqslant 0 \, Q} M\right)^n.$$

The only detail of the argument which is not immediately evident is that μ_c assigns no mass to the point $x = 0$. This however follows from the fact that μ_c is a countable sum of finite convolution-products of measures no one of which has an atom at $x = 0$. Thus p_c is an element of $\mathscr{P}$ with representing measure μ_c, and it is stable since μ_c is totally finite.

In order to check the statement about infinite divisibility we write as usual $p = e^{-x}$ and then note that

(44)
$$x_c(t) = qt \qquad (0 \leqslant t \leqslant c),$$
$$x_c(t) = qc + x(t-c) \quad (c \leqslant t < \infty).$$

From these formulae we see that x_c (necessarily continuous and non-negative, and vanishing at $t = 0$) is concave if and only if x is so, and this completes the proof of the theorem.

Now p and $T_c p$, both stable, correspond to 'events of type B' in the sense of Kingman. How are these two 'events of type B' related to one another? We shall find that the answer to this question links in an unexpected way the Choquet analysis of the last section with some problems in the statistical theory of road traffic.

As in Section 6 of Kingman's paper [13] we can associate p with a stable regenerative phenomenon Φ in the following way. Let X_n $(n = 1, 2, \ldots)$ be independent positive random variables, where all the Xs with an odd suffix have the distributiom

$$e^{-qx} q \, dx \quad (0 < x < \infty),$$

and all the Xs with an even suffix have the arbitrary distribution function F (here $F(0+) = 0$, but $F(\infty)$ may be less than unity). Let $S(\omega)$ denote the random set

$$[0, X_1) \cup (X_1 + X_2, X_1 + X_2 + X_3)$$
$$\cup (X_1 + X_2 + X_3 + X_4, X_1 + X_2 + X_3 + X_4 + X_5) \cup \ldots,$$

and write

$$E(t) \equiv (\omega : t \in S(\omega)).$$

Let us say 'Φ happens at time t' when $t \in S(\omega)$, i.e. when $\omega \in E(t)$ (so that Φ certainly happens at $t = 0$, since $E(0) = \Omega$ (the whole of the sample space)). Then Φ is a regenerative phenomenon, and its p-function is given by

$$p(t) = \text{pr}(E(t)) = \text{pr}(t \in S(\omega)),$$

and

(45)
$$1/r(\theta) = \theta + q - q \int_0^\infty e^{-\theta x} \, dF(x).$$

Let us mention two familiar practical situations in which such a phenomenon can be recognized. In the first we consider a one-way flow of traffic on a motorway. It is well known ([1, 21, 19]) that there is a tendency for the vehicles to collect into groups called 'platoons', and we can take the even-labelled Xs to represent the lengths of road occupied by individual platoons, while the odd-labelled Xs represent the unoccupied stretches of road between the platoons.

In the second example we consider any sort of queueing system with a Poissonian input, and identify the even-labelled Xs with the 'busy' periods during which at least one server is occupied, while the odd-labelled Xs are to be identified with the 'free' periods during which all servers are idle.

Now consider a pedestrian wishing to cross the road in the traffic example. We suppose that the traffic is flowing uniformly at constant speed V, and that the pedestrian requires a closed time interval of length T in which to cross the road, so that he needs a closed *spatial* traffic gap equal to $c = VT$. Let us say that the phenomenon Φ_c occurs at just those moments at which it is safe for him to start his crossing, $t = 0$ being such a moment. Evidently this phenomenon is regenerative, and its p-function will be found to be $T_c p$.

In the second example (with say a single server) we may imagine that the server is required to carry out some ancillary task which will take a time c, and then Φ_c happens at just those moments at which it is possible for him to be detached for a (closed) period of length c without interfering with his main duties. (To make this example realistic we must suppose that advance information about the input is available to the supervisor.)

In the traffic example it is interesting to observe what happens when the platoons become very small (i.e. the case of (mini-) cars travelling independently, so that each vehicle is a platoon on its own). We must now require the distribution function F to take up all its (unit) growth very near to $x = 0$. By exploiting this idea we can with a little ingenuity construct a sequence of elements p_n of $\mathscr{P}$ for all of which $p_n'(0+) = q$, and yet such that $\lim p_n(t) = 1$ for all $t \geqslant 0$. On applying this limiting process to equation (43) we obtain as the limit of $T_c p_n$ the p-function defined by

$$(46) \qquad \begin{aligned} p(t) &= \mathrm{e}^{-qt} \quad (0 \leqslant t \leqslant c), \\ p(t) &= \mathrm{e}^{-qc} \quad (c \leqslant t < \infty). \end{aligned}$$

This, however, is a typical member of the set of minimal infinitely divisible p-functions specified at equation (32). In other words *the minimal infinitely divisible element $p_{s,\lambda}$ of $\mathscr{P}$ corresponds to the phenomenon of safe road-crossing, where the (small, independent) vehicles travel with speed V, a time*

T is needed in order to cross the road, $\lambda = VT$, *and s is the spatial traffic density.*

Inasmuch as the most general infinitely divisible element of $\mathscr{P}$ has been canonically expressed at equation (31) as a product-integral of such minimal infinitely divisible *p*-functions, we can say that *the most general infinitely divisible p-function corresponds to the problem of crossing a one-way lane of traffic in which the vehicles are moving independently with a velocity distribution governed by the measure σ.* This interpretation is not quite perfect, however, because it does not allow for the possibility of σ having an atom at $\lambda = \infty$. We can incorporate this, if we allow for the possibility of the arbitrary erection of a permanent sign saying *No Crossing*, at an epoch later than $t = 0$ by the amount of an exponential delay. The situation (with no such atom) is illustrated in Figure 2.

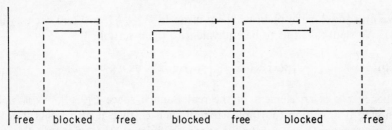

| free | blocked | free | blocked | free | blocked | free |

Figure 2. An infinitely divisible regenerative phenomenon, Φ. (Φ occurs when it is safe to cross the road, on which there is a one-way stream of traffic of two different speeds. The dotted lines separate the free and blocked segments of the spatial axis. This corresponds to an infinitely divisible member of the Kingman semigroup for which the representing measure σ consists of just two atoms)

2.2.2.5. *Convergence and compactness theorems for* $\mathscr{P}$

For the time being $\mathscr{P}$ will retain the topology of simple convergence given to it in Section 2.2.2.1. When $p \in \mathscr{P}$, r will always be used to denote its Laplace transform as at equation (39), but it will now be convenient to write the Kingman representation equation (40) in a slightly different form. To this end we let Λ denote the set of *all totally finite measures* λ *on the Borel subsets of* $[0, \infty]$ *which have no atom at* $x = 0$. The measure μ in Kingman's formula defines such a λ by way of the relation

$$(47) \qquad \lambda(A) = \int_{A \cap (0, \infty]} (1 - e^{-x}) \mu(dx),$$

and conversely equation (47) associates with each $\lambda \in \Lambda$ a unique σ-finite measure μ satisfying condition (41). Accordingly we can set up a series of one–one correspondences $p \sim r \sim \mu \sim \lambda$ in such a way that $p \sim \lambda$ associates the whole of $\mathscr{P}$ in a one–one manner with the whole of Λ, and the new version of the Kingman representation is then

$$(48) \qquad r(\theta) = \left\{ \theta + \int \frac{1 - e^{-\theta x}}{1 - e^{-x}} \lambda(dx) \right\}^{-1} \quad (0 < \theta < \infty).$$

The main results of the present section will be (*i*) a 'continuity theorem' identifying convergence in $\mathscr{P}$ with convergence in Λ, and (*ii*) a 'compactness principle' for $\mathscr{P} \cong \Lambda$. The correspondences $p \sim r \sim \mu \sim \lambda$ will be taken as canonical from henceforth and used freely without comment.

Because $\mathscr{P}$ still has the topology of simple convergence, the assertion $p_n \to p$ (for p and each p_n in $\mathscr{P}$) means that

$$(49) \qquad p_n(t) \to p(t) \quad \text{as } n \to \infty \text{ for } 0 \leqslant t < \infty.$$

We must next make plain what we mean by $\lambda_n \to \lambda$ (for λ and each λ_n in Λ). We shall take this to be equivalent to the assertion,

$$(50) \qquad \int \varphi(x) \lambda_n(dx) \to \int \varphi(x) \lambda(dx) \quad \text{as } n \to \infty$$

whenever φ is an element of the real Banach space $C[0, \infty]$. As is well known, if condition (50) holds for all such φ, then it also holds if φ is continuous for $0 \leqslant t \leqslant \infty$ save for at most a finite number of simple discontinuities at positive finite values of x, provided that the limit measure λ has no atoms at these values of x.

We shall now establish the following two theorems.

Theorem 9. *If* p_n $(n = 1, 2, ...)$ *and* p *all belong to* $\mathscr{P}$, *and if*

$$\lambda_n \ (n = 1, 2, ...)$$

and λ *are the corresponding members of* Λ, *then*

$$p_n \to p \quad \text{if and only if} \quad \lambda_n \to \lambda.$$

Theorem 10. *If* p_n $(n = 1, 2, ...)$ *and* p *all belong to* $\mathscr{P}$, *and if* $p_n \to p$, *then*

(*i*)   *the set of functions* $\{p_n : n = 1, 2, ...\}$ *is equi-uniformly continuous on* $[0, \infty)$;

(*ii*) $p_n(t) \to p(t)$ *uniformly on every compact* t-*set*;

(*iii*)   *if* $t_n \to t$ (*finite*), *then* $p_n(t_n) \to p(t)$.

Proof of Theorem 9. (*a*) First suppose that $p_n \to p$. Then by bounded convergence $r_n \to r$, and from equation (48) (by differencing) it then follows that

$$\int e^{-\theta x} \lambda_n(dx) \to \int e^{-\theta x} \lambda(dx)$$

for all real θ greater than *or equal to* 0. (This last detail follows from equation (48) on putting $\theta = 1$.) Now the exponential polynomials $\alpha + \sum \beta_j e^{-\theta_j x}$ (with all $\theta s > 0$) are dense in $C[0, \infty]$, and this and positivity therefore tells us that the condition (50) holds, so that $\lambda_n \to \lambda$ as required.

(*b*) Now suppose conversely that $\lambda_n \to \lambda$. For $0 < \theta < \infty$ the integrand in the integral at equation (48) belongs to $C[0, \infty]$, and so certainly

$$r_n(\theta) \to r(\theta) \text{ for } 0 < \theta < \infty.$$

We now make use of the inequality

(51) $$0 \leqslant 1 - p(t) \leqslant \int \frac{\min(x, t)}{1 - e^{-x}} \lambda(dx),$$

which is a simple consequence of the integral equation

(52) $$1 - p(t) = \int_0^t p(t - s) \mu(s, \infty] \, ds$$

(proved as Proposition 7 in Kingman [13]) because the integral cannot exceed

$$\int_0^t ds \int_{(s, \infty]} \mu(dx);$$

on inverting the integrations we obtain the right-hand side of inequality (51). Accordingly if $0 \leqslant t \leqslant \delta$ (finite) we can write $|1 - p_n(t)| \leqslant U_n + V_n$, where

$$U_n = \int_{[0, \delta]} \frac{x}{1 - e^{-x}} \lambda_n(dx),$$

$$V_n = t \int_{[\delta, \infty]} \frac{1}{1 - e^{-x}} \lambda_n(dx).$$

Let us define U and V similarly, the measure λ_n being replaced by λ. We know that λ has no atom at $x = 0$, and that it has at most a countable set of atoms altogether, and so we can arrange both that $x = \delta > 0$ is not an atom of λ, and that

$$U = \int_{[0, \delta]} \frac{x}{1 - e^{-x}} \lambda(dx) < \varepsilon,$$

where ε is to be positive but is otherwise at our choice. If we think of U as an integral over $[0, \infty]$ then the integrand belongs to $C[0, \infty]$ save for a simple discontinuity at $x = \delta$, and so $U_n \to U$ as $n \to \infty$. It follows that $U_n < \varepsilon$ for all $n \geqslant$ some $N(\varepsilon)$.

In much the same way we see that $t^{-1} V_n$ converges to a finite limit as $n \to \infty$, and so $V_n \leqslant Mt$ (where M depends on ε), and therefore (when $0 \leqslant t \leqslant \delta$)

$$|1 - p_n(t)| < \varepsilon + Mt + \max_{m < N} |1 - p_m(t)| \quad (n = 1, 2, \ldots);$$

this, however, enables us to deduce that

$$\lim_{t \to 0} p_n(t) = 1 \quad \text{uniformly for } n = 1, 2, \ldots.$$

On recalling the inequality

(53) $$|p(t') - p(t'')| \leqslant 1 - p(|t' - t''|)$$

(a consequence of inequality (15)) we see that the set of functions

$$\{p_n : n = 1, 2, \ldots\}$$

is actually equi-uniformly continuous over the whole of $[0, \infty)$. Note that, in proving this, we have merely assumed that $\lambda_n \to \lambda$, where λ and all λ_n are in Λ. All values of p-functions lie in the compact interval $[0, 1]$, and so by selecting a subsequence of $\{1, 2, \ldots\}$ we can arrange that $p_n(t)$ has a limit for all rational $t \geqslant 0$, when $n \to \infty$ in the subsequence. Equi-uniform continuity will now ensure that $p_n(t) \to p^*(t)$ for *all* $t \geqslant 0$, where p^* is continuous on $[0, \infty)$, so that (from the corollary on p. 185 of [13]) the function p^* must be an element of $\mathscr{P}$. So far as we yet know, however, p^* depends on the subsequence selected above.

From $p_n \to p^*$ (n in the subsequence) and part (*a*) of the present proof we can conclude that $r_n \to r^*$ (n in the subsequence). But also $r_n \to r$, and so $r^* = r$ and $p^* = p$. The freedom of choice which we had in selecting the subsequence now guarantees that $p_n \to p$ when n tends to infinity through the whole sequence of positive integers, and the proof of Theorem 9 is now complete.

Proof of Theorem 10. We are given that $p_n \to p$ and so from Theorem 9 we know that $\lambda_n \to \lambda$. Part (*b*) of the proof of Theorem 9 then tells us that $\{p_n : n = 1, 2, \ldots\}$ is an equi-uniformly continuous set of functions on $[0, \infty)$, as required by (*i*). Parts (*ii*) and (*iii*) of the theorem then follow by standard arguments.

If p and p' are any two members of $\mathscr{P}$, let us define a metric ρ on $\mathscr{P}$ by

(54) $$\rho(p, p') \equiv \sum_1^\infty 2^{-m} \max_{[0, m]} |p(t) - p'(t)|.$$

This converts $\mathscr{P}$ into a complete metric space, and Theorem 10 (*ii*) shows that $p_n \to p \in \mathscr{P}$ if and only if p_n converges to p in the sense of the metric ρ. We can describe this metric topology for $\mathscr{P}$ in another way; if

$$S(K, G) \equiv \{f : f(t) \in G \text{ when } t \in K\},$$

where K is any compact and G any open subset of the real line, then the sets $S(K, G)$ form a sub-base for what is called the compact-open topology on R^∞, and this induces on $\mathscr{P}$ a metric topology which is exactly that realized by the metric (54). *We may therefore associate our sequential convergence concept* (49) *either with the topology of simple convergence or, if we prefer, with the compact-open topology.* In some circumstances the second choice is preferable, because in metric topologies it is unnecessary to distinguish between the different sorts of compactness. Notice however that while they yield the same *sequential* convergence, the two topologies ('simple convergence' and 'compact-open') on $\mathscr{P}$ are *not* necessarily identical, and it is conjectured that they are in fact distinct. [This conjecture has now been disproved by Davidson.]†

We now prove

Theorem 11. *If we give $\mathscr{P}$ the compact-open topology, then the conditionally compact sets will be precisely the sets of functions p in $\mathscr{P}$ which are equi-continuous at $t = 0$.*

Proof. We first note that, because the inequality (53) holds for all elements of $\mathscr{P}$, there is no need to distinguish between equi-uniform continuity and the local version of this property at $t = 0$. Next, because the compact-open topology is metric, we can test for compactness sequentially whenever this is convenient. Now let E be an equi-uniformly continuous subset of $\mathscr{P}$, and let p_n ($n = 1, 2, ...$) be a sequence of elements of E. As usual we can pick out a subsequence of values of n for which $p_n(t)$ converges to some $q(t)$ for every rational t, and then because of the equi-uniform continuity we know that in fact $p_n(t)$ will converge to $q(t)$ for every t, with $q \in \mathscr{P}$, provided that we keep to the chosen subsequence. Thus E is conditionally compact, because from dominated convergence and Theorem 10 (*ii*) it will follow that $\rho(p_n, q) \to 0$ when $n \to \infty$ in the subsequence.

On the other hand let E be conditionally compact, and suppose if possible that it is not an equi-continuous family of functions at $t = 0$. Then for some $\varepsilon > 0$ there exists a sequence $\{t_n : n = 1, 2, ...\}$ of real numbers such that $t_n \downarrow 0$, and a sequence $\{p_n : n = 1, 2, ...\}$ of elements of E, such that

$$1 - p_n(t_n) > \varepsilon \quad (n = 1, 2, ...).$$

† See 2.4 of this book.

From the conditional compactness of E there must exist a subsequence of $\{1, 2, \ldots\}$ in which $\rho(p_n, p) \to 0$, where $p \in \mathscr{P}$, and then in this subsequence Theorem 10 shows that we must have $p_n(t_n) \to p(0) = 1$, which gives a contradiction.

Theorems 10 and 11 recall known properties of characteristic functions of probability-laws, and might in fact have been derived in this way by making use of a second representation theorem of Kingman (Theorem 5 of [13]). This asserts that any $p \in \mathscr{P}$ determines the characteristic function $\psi(t) = p(|t|) \, (-\infty < t < \infty)$ of a probability law

$$(55) \qquad\qquad c\delta(du) + f(u) \, du \quad (-\infty < u < \infty),$$

where the atom at $u = 0$ has mass

$$(56) \qquad\qquad c = 1 \Big/ \Big\{ 1 + \int_{(0,\infty]} x \mu(dx) \Big\}$$

and where the density f is symmetric and is defined for $u > 0$ by

$$(57) \qquad f(u) = \frac{1}{\pi} \mathrm{Re} \left[1 \Big/ iu + \int_{(0,\infty)} (1 - e^{-iux}) \mu(dx) + \mu(\{\infty\}) \right].$$

Here μ is the measure canonically associated with p by equation (40). It is one of the assertions of Kingman's theorem that f is non-negative and that $c + \int f(u) \, du = 1$. If desired, μ can be transformed into λ as at equation (47) above, so that the formulae (55 − 57) provide another version of the canonical one–one association of $\mathscr{P}$ with Λ. A precursor of this result (in the special case in which p is a diagonal Markov transition function, and without the explicit formula for c and f) was given by the author [6] using unnecessarily sophisticated Hilbert space techniques; his arguments were subsequently greatly simplified by Feller [4]. Kingman established his more general result by inverting the Laplace transformation involved in his earlier representation (40). In general it seems to be agreed that there is something accidental and unnatural about the Fourier representation, and certainly equation (40) is much the more useful. We must however record it here, because one or two of the preceding results could have been obtained very rapidly in this way by noting that, if $p \in \mathscr{P}$, then

$$(58) \qquad\qquad \psi(t) = p(|t|) \quad (-\infty < t < \infty)$$

is a characteristic function; thus (for example) we can identify

$$p_n \to p, \quad t_n \to t (\text{finite}) \Rightarrow p_n(t_n) \to p(t)$$

with a special case of a known result about characteristic functions (Loève [18], p. 192).

It is this Fourier representation to which we referred at the end of Section 2.2.2.2 of this paper, and the laws (55) (with c and f expressed in terms of μ (or λ) as at equations (56) and (57)) constitute the sub-semigroup of the full convolution semigroup of probability-laws on R of which $\mathscr{P}$ is the Fourier image. The convolution of two such laws must therefore be a law of the same form, although this is far from being analytically evident, and it seems at present to be a hopeless task to try to construct the associative binary operation on the measures μ (or λ) which corresponds to convolution for the corresponding laws (55). The formulae (56) and (57) give an explicit and faithful parametrization of the laws (55), but progress at this level is gravely impeded by the fact that we do not know any *intrinsic* characterization of the pairs (c, f) for which the laws (55) have characteristic functions belonging to $\mathscr{P}$. In fact the current fashion [5, 17] for replacing characteristic-function arguments by 'direct' ones formulated in terms of the distribution function itself is maximally inappropriate in the present situation. For us the laws (55) are an artificial construct, and it is the system of the p-functions which is the primary object of interest.

2.2.2.6. *The multiplicatively stable p-functions, and their 'domains of attraction'*

Let us now ask which elements p of $\mathscr{P}$ have the property that, for every positive c_1 and c_2, there exists a positive c_3 such that

$$(59) \qquad p(c_1 t)p(c_2 t) = p(c_3 t) \quad (0 \leqslant t < \infty);$$

we shall call such p-functions *multiplicatively stable*. Many of our results could be lifted directly from the theory of characteristic functions, using the Fourier representation associated with formulae (55)—(57), but we shall argue directly, partly because it seems to us more natural to do so, and partly for the sake of the methodology which this will oblige us to develop (for which see Section 2.2.2.7 below).

It is obvious that if equation (59) holds then for each $n = 1, 2, \ldots$ there must exist a finite positive b_n such that

$$(60) \qquad p(t)^n = p(b_n t) \quad (0 \leqslant t < \infty),$$

and this rather than equation (59) will be our starting point. It will turn out that all solutions to equation (60) automatically satisfy the apparently stronger requirement (59). We shall always assume that p is not the neutral element e, because e trivially has both properties (59) and (60).

From equation (60) it follows at once that we must have

$$p(b_{rs} t) = p(b_r b_s t),$$

and so $p(t) = p(c^k t)$, where $c = b_{rs}/b_r b_s$, for all t and all $k = 1, 2, \ldots$. We know [13] that $p(\infty)$ exists, and on letting $k \to \infty$ we find (if $c \neq 1$) that either $p(t) = 1$ for all t or that $p(t) = p(\infty)$ for all t (and then necessarily $p(\infty) = 1$); in each case we obtain a contradiction to our assumption that $p \neq e$. All this can be avoided if $c = 1$, and only then. Accordingly we have learnt that the sequence $\{b_n\}$ is *multiplicative*: $b_{rs} = b_r b_s$. It follows immediately from this that $b_1 = 1$.

It also follows from equation (60) that

$$p((b_n/b_{n+1})^k t) = p(t)^{(n/(n+1))^k},$$

and from this (on letting $k \to \infty$) we see that either $p(\infty) = 1$ or $b_n \leqslant b_{n+1}$. But the first of these alternatives can be eliminated, for if $p(\infty) = 1$ then Theorem 6 of [13] would force us to conclude that $\mu = 0$ and $p = e$. Thus the sequence $\{b_n\}$ is non-decreasing as well as multiplicative, and a standard number-theoretical argument then show that $b_n = n^c$ for some finite positive c ($b_n = 1$ is excluded because this would lead to $p = e$).

Now write as usual $p = e^{-x}$, so that $nx(t) = x(n^c t)$, and let $y(u) = x(u^c)$; then $ny(u) = y(nu)$ and so by continuity $y(u) = \alpha u$ ($\alpha > 0$) and

(61) $p(t) = \exp(-\alpha t^{1/c}) \quad (0 \leqslant t < \infty).$

If this is to be a p-function then clearly it will have to be an infinitely divisible one, and so $\alpha t^{1/c}$ must be concave; this shows that $1 \leqslant c < \infty$. On the other hand if c satisfies these inequalities then (by Theorem 6) equation (61) will define an infinitely divisible p-function which is obviously multiplicatively stable. Thus we have proved

Theorem 12. *The multiplicatively stable elements of $\mathscr{P}$ (defined as at equation (59) or (60)) are precisely the functions*

$$p(t) = \exp(-\alpha t^{1/c}) \quad (0 \leqslant t < \infty),$$

where $\alpha \geqslant 0$ and $1 \leqslant c < \infty$. Such a p-function satisfies equation (60) with $b_n = n^c$.

These special infinitely divisible p-functions were included among the examples at the end of Section 2.2.2.4, and their Choquet representations were worked out there.

It is clear that all we have done so far is to rediscover certain of the symmetrical stable laws on R. We turn now to the question of their 'domains of attraction'.

We therefore seek to characterize those triples $(p; p^*; a_n)$, where p and p^* are non-neutral elements of $\mathscr{P}$ and $\{a_n\}$ is a sequence of finite positive real numbers, such that

(62) $p(t/a_n)^n \to p^*(t) \quad as \quad n \to \infty \quad (0 \leqslant t < \infty).$

For the time being we do not assume that p^* satisfies the conditions of Theorem 12, but it will turn out in the course of the argument that it must do so; by the 'domain of attraction' of p^* we shall then mean the set of non-neutral ps in $\mathscr{P}$ which can be associated with a suitable sequence $\{a_n\}$ in such a way that $(p; p^*; a_n)$ satisfies condition (62).

Suppose first that $(p; p^*; a'_n)$ and $(p; p^{**}; a''_n)$ each satisfy condition (62), where no one of p, p^*, p^{**} is e. Let us write $p_n(t) = p(t/a'_n)^n$, so that

$$p_n(t) \to p^*(t) \quad \text{and} \quad p_n((a'_n/a''_n)t) \to p^{**}(t)$$

as $n \to \infty$. Then from (*iii*) of Theorem 10 it follows that the ratio a'_n/a''_n is bounded away from 0 and from ∞, because otherwise we could prove that $p^{**}(t) = p^*(0) = e$ or that $p^*(t) = p^{**}(0) = e$. If β is any finite non-zero cluster point of the sequence of numbers a'_n/a''_n then we shall have

$$p^*(\beta t) = p^{**}(t) \quad \text{and} \quad 0 < \beta < \infty.$$

Such a β must be unique (for otherwise we could once again force p^* and p^{**} to be the neutral element e), and so we have proved that *any two solutions* $(p; p^*; a'_n)$ *and* $(p; p^{**}; a''_n)$ *to condition* (62) *are necessarily related as follows*:

$$a'_n \sim \beta a''_n, \quad p^*(\beta t) = p^{**}(t) \quad (n = 1, 2, \ldots; 0 \leqslant t < \infty);$$

here $0 < \beta < \infty$.

This is the appropriate analogue to what is sometimes called the 'convergence of types' lemma (cf. [18], p. 203).

Next we show that *if condition* (62) *is true, then* $a_n \to \infty$ *and* $a_{n+1}/a_n \to 1$. For we must have $p(t/a_n) \to 1$ as $n \to \infty$, because $p^*(t)$ cannot be zero, and as $p(t) < 1$ for $0 < t \leqslant \infty$ (in virtue of inequality (15) and the fact that $p \neq e$) it is clear that t/a_n must tend to zero when $n \to \infty$; this shows that $a_n \to \infty$. Now $(p; p^*; a_{n+1})$ must also satisfy condition (62), because we know that $p(t/a_{n+1}) \to 1$ when $n \to \infty$, and therefore we can appeal to the 'convergence of types' lemma to deduce that $a_{n+1} \sim a_n$, as required. (As usual, β must be equal to unity because otherwise $p^* = e$.)

Now suppose that condition (62) holds, and let us write $p_k(t) = p(t)^k$, so that p_k is a non-neutral element of $\mathscr{P}$, and

$$p_k(t/a_n)^n \to p^*(t)^k, \quad p_k(t/a_{nk})^n \to p^*(t).$$

From the 'convergence of types' lemma we deduce that $p^*(t)^k = p^*(\beta_k t)$ where $0 < \beta_k < \infty$, *whence* p^* *has the form* (61) *and* $\beta_k = k^c$ *with* $1 \leqslant c < \infty$; *also* $a_{nk}/a_n \to k^c$ *as* $n \to \infty$, *so that the sequence* $\{a_n\}$ *has regular variation with exponent c*.

We are left with the following problem; let p^* be defined as at equation (61) (with $1 \leqslant c < \infty$, and $\alpha > 0$, so that p^* is not e), and let $\{a_n\}$ always be a sequence of positive real numbers such that

(i) $a_n \to \infty$;

(ii) $a_{n+1}/a_n \to 1$;

(iii) $\{a_n\}$ has regular variation with exponent c.

What elements p of $\mathscr{P}$ then satisfy condition (62) with the given p^*, for some suitable choice of $\{a_n\}$? In order to deal with this question we shall need some new technical aids which it will be convenient to develop systematically. We shall therefore do this in Section 2.2.2.7, and return to the above question in Section 2.2.2.8.

2.2.2.7. *Some 'Croftian' theorems*

Let $\mathscr{F}$ be a class of functions (say of a positive real variable t) and let P be a property which the values $f(t)$ of $f \in \mathscr{F}$ may or may not enjoy when $t \to \infty$ in any specified manner. Then by a 'Croftian' theorem we shall mean a result of the following type: 'If $f \in \mathscr{F}$ has property P whenever $t \to \infty$ through any one of a certain set of sequences of t-values, then f has property P when $t \to \infty$ unrestrictedly'.

The prototype of all such results is a famous theorem of H. T. Croft [2], subsequently extended by Kingman [12a, 12b]. In Croft's theorem the special sequences were arithmetic progressions $\{nh: n = 1, 2, ...\}$. Among the innovations introduced by Kingman was the consideration of special sequences of the form $\{a_n h: n = 1, 2, ...\}$, and in the case when $\{a_n\}$ increases strictly to ∞ as limit he showed that the condition

$$\limsup (a_{n+1}/a_n) = 1$$

is necessary and sufficient for the Croftian inferences to hold. For our special purposes here it would be undesirable to impose a restriction of monotony on $\{a_n\}$, and so we shall require Theorems 13–15 below. The apparently greater generality is deceptive, because Theorems 13–15 are in fact corollaries to those of Kingman; we shall, however, give the proofs as these are so short. We shall then prove a new Croftian theorem (Theorem 16) which is the objective of the present digression.

Theorem 13. *Let* $\{G_k\}$ *be a countable family of unbounded open subsets of* $(0, \infty)$ *and let* $\{a_n\}$ *be a sequence of positive real numbers such that*

$$(63) \qquad \limsup_{n \to \infty} a_n = \infty \quad and \quad \limsup_{n \to \infty} (a_{n+1}/a_n) = 1.$$

Then there is a dense set H in $(0, \infty)$ such that, for each $h \in H$ and each $k \geqslant 1$, $a_n h \in G_k$ for infinitely many n.

Proof. Let $I = (\alpha, \beta)$ where $0 < \alpha < \beta < \infty$. Then

$$\bigcup_{n \geqslant m} a_n I$$

is unbounded and in fact contains all sufficiently large positive numbers, so that this set meets each G_k. Thus

$$H = \bigcap_{k \geqslant 1} \bigcap_{m \geqslant 1} \left\{ \bigcup_{n \geqslant m} a_n^{-1} G_k \right\}$$

is dense in $[0, \infty)$, because each of the (open) unions is dense and Baire's theorem applies. This completes the proof. Note that we only need $\limsup (a_{n+1}/a_n) \leqslant 1$, but that then equality must hold, for otherwise $\{a_n\}$ would converge to zero.

Theorem 14. *Let f be a continuous function from $(0, \infty)$ to R and let the positive sequence $\{a_n\}$ satisfy condition (63). Let J be an open subinterval of $(0, \infty)$ and let $\{F_k\}$ be a sequence of closed sets in R such that, for each h in J, there is a value of k depending on h for which $f(a_n h) \in F_k$ for all sufficiently large n. Then for some k, $f(t) \in F_k$ for all sufficiently large t.*

Proof. Suppose if possible that the hypotheses hold and that the conclusion does not. Then for each k there exists a t-sequence converging to ∞ on which $f(t)$ lies outside F_k, and so lies inside the open set $R \backslash F_k$. Thus $G_k = f^{-1}(R \backslash F_k)$ is an unbounded open set in $(0, \infty)$, and we are told that, for each h in J, the point $a_n h$ lies outside a particular one of these sets, $G_{k(h)}$ say, for all $n \geqslant N(h)$. Now take $h \in J \cap H$ to obtain a contradiction (in virtue of the preceding theorem).

Theorem 15. *Let f be a continuous function from $(0, \infty)$ to R and let the positive sequence $\{a_n\}$ satisfy condition (63).*

(i) *If $\{f(a_n h): n = 1, 2, \ldots\}$ is bounded for each separate h in some open interval of $(0, \infty)$, then f is bounded on $[1, \infty)$.*

(ii) *If $f(a_n h) \to L(h)$ as $n \to \infty$, for each h in some open interval of $(0, \infty)$, then*

$$\lim_{t \to \infty} f(t) = L,$$

where $L = L(h)$ for some h.

Proof of (i). Take $F_k = [-k, k]$, and apply the preceding theorem.

Proof of (ii). We may assume that $L(h)$ is finite; the general case in which $-\infty \leqslant L(h) \leqslant \infty$ can then be obtained from the finite case by a preliminary mapping of $[-\infty, \infty]$ onto a compact segment. Now $g_n(h) = f(a_n h)$ defines a function g_n which is continuous on the open interval mentioned in the theorem and the sequence $\{g_n\}$ converges pointwise there to the function $L(\cdot)$. Such a function $L(\cdot)$ must have a dense set of points of continuity; let $t = a$ be one such, and contract the open interval about $t = a$ until the range of $L(\cdot)$ lies with $(L(a) - \varepsilon, L(a) + \varepsilon)$, where $\varepsilon > 0$. If we now take

$$F_1 = F_2 = \ldots = [L(a) - \varepsilon, L(a) + \varepsilon],$$

then the preceding theorem tells us that $|f(t) - L(a)| \leqslant \varepsilon$ for all sufficiently large t. As the value of ε is at our choice, the proof is complete; evidently the limit L is the common value $L(h)$ of the function $L(\cdot)$ throughout the open interval on which it is defined.

We can now prove a new Croftian theorem which will enable us to recognize a continuous function of regular growth. Note that in the following theorem *we do not assume f to be monotone* (this is the detail— essential for our purposes—which distinguishes the present result from a known one to be found in [5] (Lemma 2, p. 270)).

Theorem 16. *Let f be a continuous positive real-valued function on $(0, \infty)$, and let the positive sequence $\{a_n\}$ satisfy equations (63). For all t in an open interval T of $(0, \infty)$ let*

$$f(a_n t) \sim b_n C(t) \quad as \ n \to \infty,$$

where b_n and $C(t)$ are finite and positive for $t \in T$ and $n \geqslant 1$. Then, for some finite real number c,

$$f(st)/f(t) \to s^c \quad as \ t \to \infty \quad (0 < s < \infty).$$

Proof. If each of h and sh lie in T then

$$f(a_n h) \sim b_n C(h) \quad and \quad f(a_n sh) \sim b_n C(sh)$$

as $n \to \infty$, and so if T and sT overlap and if we write $g_s(t) = f(st)/f(t)$, then we can assert that

$$g_s(a_n h) \to C(sh)/C(h) \quad as \ n \to \infty$$

for every h in the non-vacuous open interval $T \cap (s^{-1}T)$. Now g_s is continuous and so Theorem 15 *(ii)* applies and tells us that

$$g_s(t) \to D(s) \quad as \ t \to \infty,$$

where $D(s)$ is finite and positive, provided that s is chosen so that sT meets T.

Now let S denote the set of positive real numbers s such that $g_s(t)$ converges to a finite positive limit as $t \to \infty$. Clearly $1 \in S$, $S^{-1} \subseteq S$, and $S^2 \subseteq S$, so that S is a subgroup of the multiplicative group $(0, \infty)$. Also S contains each s for which T meets sT; in particular, S contains a neighbourhood of $s = 1$. It follows that $S = (0, \infty)$, and we can now take $D(s)$ to be defined over $(0, \infty)$.

Evidently $D(s_1 s_2) = D(s_1) D(s_2)$, and $D(\cdot)$ is finite, positive, and measurable, so that $D(s) = s^c$ for some real c. The proof of the theorem is now complete.

2.2.2.8. *The 'domain of attraction' problem*

We now return to the problem formulated at the end of Section 2.2.2.6. Suppose that p is a non-neutral member of $\mathscr{P}$ such that $(p; p^*; a_n)$ satisfies condition (62), where p^* and $\{a_n\}$ have the properties stated in the last paragraph of Section 2.2.2.6. We write as usual $p = e^{-x}, p^* = e^{-x^*}$, and so we know that

$$nx(t/a_n) \to x^*(t) \quad \text{as } n \to \infty \text{ for } 0 \leqslant t < \infty,$$

and thus that

$$y(a_n u) \sim \alpha n^{-1} u^{-1/c} \quad \text{as } n \to \infty,$$

where $y(u) = x(1/u)$ and $0 < u < \infty$. The function y is continuous and positive on $(0, \infty)$ (because y cannot vanish for such a value of u unless $p = e$), and thus Theorem 16 applies and we find that $y(vu)/y(u) \to v^{-1/c}$ as $u \to \infty$ for $0 < v < \infty$. It follows that

(64) $$x(st)/x(t) \to s^{1/c} \quad \text{as } t \to 0, \text{ for } 0 < s < \infty;$$

that is, *x is of regular variation as its argument tends to zero, the exponent of regularity being* $1/c$.

Conversely let us suppose that a non-neutral p has this property, p^* being given as before. Then $x(t) \to 0$ as $t \to 0$, and so we can find a sequence $\{a_n\}$ of positive real numbers such that $a_n \to \infty$ and $x(1/a_n) \sim \alpha/n$ (indeed, we could take equality here), and then

$$nx(1/a_n) \to \alpha \quad \text{as } n \to \infty.$$

Because x has regular variation it will follow from this that

$$nx(t/a_n) \to \alpha t^{1/c} \quad \text{as } n \to \infty \text{ for } 0 \leqslant t < \infty,$$

and so that $(p; p^*; a_n)$ satisfies condition (62). This shows that *the set of non-neutral elements p of $\mathscr{P}$ for which $\log(1/p)$ has regular variation at $t = 0$, with index* $1/c$, *is the exact domain of attraction of the multiplicatively stable p-function* (61). Notice that we do not assert that $(p; p^*; a_n)$ will satisfy condition (62) for *any* sequence $\{a_n\}$ satisfying (*i*)–(*iii*), but only that at least one such sequence will exist.

On collecting together the above results we are able to state

Theorem 17. *Let p and p^* belong to $\mathscr{P}\backslash e$ and let $\{a_n\}$ be a sequence of positive real numbers, and suppose that*

$$(65) \qquad \lim_{n-\infty} p(t/a_n)^n = p^*(t) \quad (0 \leqslant t < \infty).$$

Then

(a) $p^*(t) = \exp(-\alpha t^{1/c}) \quad (0 \leqslant t < \infty),$

(b) $p(t) = 1 - t^{1/c} L(t)$, where $L(\cdot)$ varies slowly as $t \to 0$,

and

(c) $a_n \to \infty$, $a_{n+1}/a_n \to 1$, and $a_{mn}/a_n \to m^c \, (m \geqslant 1)$ as $n \to \infty$, for some c in the range $1 \leqslant c < \infty$, and some $\alpha > 0$.

If p and p^* satisfy (a) and (b) then a sequence $\{a_n\}$ can be found for which the limit-relation (65) holds, so that (b) describes the exact domain of attraction of the multiplicatively stable p-function at (a).

If $p \in \mathscr{P}\backslash e$ and if p satisfies (b), then the associated p^* and $\{a_n\}$ for which (65) holds form an equivalence class under the equivalence relation defined by

$$(66) \qquad (p^*; a'_n) \equiv (p^{**}; a''_n) \quad \text{when } a'_n \sim \beta a''_n \text{ and } p^*(\beta t) = p^{**}(t),$$

where $0 < \beta < \infty$.

Proof. Almost all of the assertions of the theorem have been proved. The statement at (*b*) is equivalent to the regular variation of $x = \log(1/p)$ at $t = 0$, with exponent $1/c$. The fact that $(p; p^{**}; a''_n)$ satisfies equation (65) whenever $(p; p^*; a'_n)$ does so and condition (66) holds, may be proved as follows. Write $p_n(t) = p(t/a'_n)^n$, so that $p_n \in \mathscr{P}\backslash e$, and $p_n \to p^*$. Now $t(a'_n/a''_n) \to \beta t$ (finite), and so by (*iii*) of Theorem 10 we can conclude that

$$p(t/a''_n)^n = p_n(t(a'_n/a''_n)) \to p^*(\beta t) = p^{**}(t),$$

as required.

It is obvious that if p^* is the function at (*a*) in the last theorem and if $p = p^*$, then condition (65) holds when $a_n = n^c$. If we restrict the sequence $\{a_n\}$ in this way, then by analogy with the definitions usually adopted in the theory of characteristic functions we can suitably call the set of ps satisfying condition (65) (for the given p^*) the domain of 'normal' attraction for that multiplicatively stable p-function. We then have

Theorem 18. *If p^* denotes the multiplicatively stable p-function*

$$p^*(t) = \exp(-\alpha t^{1/c}) \quad (0 \leqslant t < \infty),$$

where $\alpha > 0$ *and* $1 \leqslant c < \infty$, *then its domain of normal attraction consists of those p-functions for which*

$$(67) \qquad\qquad p(t) = 1 - \alpha t^{1/c} \, (1 + o(1))$$

near $t = 0$.

Proof. It is obvious that all such ps do lie in the domain of normal attraction, so that we need only prove the converse assertion. Now if p lies in the domain of normal attraction then

$$nx(t/n^c) \to \alpha t^{1/c} \quad \text{as } n \to \infty \text{ for } 0 \leqslant t < \infty.$$

Write $y(u) = u^{1/c} x(1/u) \, (0 < u < \infty)$. Then

$$y(t^{-1} n^c) = nt^{-1/c} x(t/n^c) \to \alpha$$

as $n \to \infty$, for every positive finite t. We can now use part (*ii*) of the Croftian Theorem 15 to conclude that $y(u) \to \alpha$ as $u \to \infty$. This, however, is equivalent to equation (67).

In conclusion I wish to thank the numerous friends with whom I have been privileged to discuss this investigation, and especially Professor D. A. Kappos and his colleagues in the University of Athens, who also made possible the pilgrimage to the Corycian cave with which it all began.

REFERENCES

1. W. D. Ashton, *The Theory of Road Traffic Flow*, Methuen, London (1966).
2. H. T. Croft, "A question of limits", *Eureka*, **20** (1957), 11–13.
3. R. Davidson, private communication, to be published. (See 2.3 of this book.)
4. W. Feller, On the Fourier representation for Markov chains and the strong ratio theorem", *J. Math. Mech.* **15** (1966), 273–283.
5. ——, *An Introduction to Probability Theory and its Applications*, vol. II, Wiley, New York (1966).
6. D. G. Kendall, "Unitary dilations of one-parameter semigroups", *Proc. Lond. math. Soc.*, III Ser. **9** (1959), 417–431.
7. ——, "Recent advances in the theory of denumerable Markov processes", *Trans. 4th Prague Conf. Information Theory, statist. Decision Functions, Random Processes*, Academia (for Czech. Acad. Sci.), Prague, pp. 11–27 (1967).
8. ——, Comments on J. F. C. Kingman's paper, (published together with [16]).
9. ——, "Renewal sequences and their arithmetic". *Symposium on Probability* (*Lecture Notes in Math.* **31**), Springer, Berlin, Heidelberg and New York (1967). (2.1 of this book.)
10. ——, "Delphic semi-groups", *Bull. Am. math. Soc.* **73** (1967), 120–121.
11. A. Ya. Khintchine, "The arithmetic of distribution laws", *Bull. Univ. Etat Moscou, Sér. Int., Sect. A: Math et Mécan.* **1** (1937) Fasc. 1, 6–17.

12(a) J. F. C. Kingman, "Ergodic properties of continuous-time Markov processes and their discrete skeletons", *Proc. Lond. math. Soc.* III Ser. **13** (1963), 593–604.

12(b) ——, "A note on limits of continuous functions", *Quart. J. Math., Oxford* II Ser. **15** (1964), 279–282.

13. ——, "The stochastic theory of regenerative events", Z. *Wahrsch'theorie & verw. Geb.* **2** (1964), 180–224.

14. ——, "Linked systems of regenerative events", *Proc. Lond. math. Soc.* III Ser. **15** (1965), 125–150.

15. ——, "Some further analytical results in the theory of regenerative events", *J. math. Analysis. Appl.* **11** (1965), 422–433.

16. ——, "An approach to the study of Markov processes", *J. R. statist. Soc.* B, **28** (1966), 417–447.

17. A. N. Kolmogorov, "On the approximation of distributions of sums of independent summands by infinitely divisible distributions", *Sankhya* (A) **25** (1963), 159–174.

18. M. Loève, *Probability Theory*, 3rd ed., van Nostrand, Princeton (1963).

19. A. J. Mayne, "Some further results in the theory of pedestrians and road traffic", *Biometrika*, **41** (1954), 375–389 and **45** (1958), 291.

20. R. R. Phelps, *Lectures on Choquet's Theorem*, van Nostrand, Princeton, (1966).

21. J. C. Tanner, "The delay to pedestrians crossing a road", *Biometrika*, **38** (1951), 383–392.

2.3

Arithmetic and Other Properties of Certain Delphic Semigroups: I

R. DAVIDSON

[Reprinted from *Z. Wahrsch'theorie & verw. Geb.* **10** (1968), 120–145.]

2.3.1 SUMMARY

Some aspects of Delphic semigroups in general—in particular, the idea of an hereditary sub-semigroup, which has many uses in connection with Delphic semigroups—are first treated. After that, attention is directed to the arithmetic of $\mathscr{R}^+$, the semigroup of positive renewal sequences. In a Delphic semigroup the aboriginal elements are the 'simples' and the members of 'I_0'; a class of simples of $\mathscr{R}^+$ is constructed and the simples are shown to be residual. I_0 is explicitly identified, and this leads to a canonical factorization of $\mathscr{R}^+$. The properties of division in $\mathscr{R}^+$ are discussed.

2.3.2 INTRODUCTION

Professor D. G. Kendall promulgated Delphic semigroups in [6] and developed their theory fully in [7]. An adequate account will be found in our Section 2.3.3, if in it we substitute topological for sequential assumptions (in particular, compactness for sequential compactness). The subject had its origin in the work of Khintchine on the convolution semigroup $\mathscr{W}$ of probability laws on the line. He derived for this semigroup the three basic theorems listed in Section 2.3.3. In 1966 Kendall [5] proved a triangular central limit theorem for the semigroup $\mathscr{R}$ of aperiodic renewal sequences, and showed that the three basic theorems hold for $\mathscr{R}$. He was then led to conceive the notion of a Delphic semigroup in the abstract, of which $\mathscr{W}$ and $\mathscr{R}$ are concrete examples—almost: they are not quite proper Delphic semigroups. (It is convenient to make the Delphic hypotheses

115

rather restrictive, and then for a given semigroup to show that one can modify it in an unimportant way to satisfy these hypotheses.)

Kendall himself showed (in [7]) that $\mathscr{P}$, the semigroup of standard Kingman p-functions, was (properly) Delphic. Mr. J. G. Basterfield has shown that $\mathscr{W}$ is almost Delphic, and one can show that $\mathscr{L}$, the semigroup of probability laws on the circle, is also almost Delphic. Professor J. Lamperti has shown that certain semigroups of positive operators associated with the ultraspherical polynomials are Delphic, and Mr. N. H. Bingham that certain semigroups dual to these are almost Delphic. So there are many almost Delphic semigroups; from the arithmetical point of view the most interesting ones are $\mathscr{W}$, $\mathscr{P}$, $\mathscr{R}$ and $\mathscr{L}$ (in that order), because for these there is a strong interaction between the analytical theory of the semigroup and the probability theory of the objects which compose it. The arithmetic of $\mathscr{W}$ has been studied for the last thirty years; here we direct our attention to $\mathscr{P}$ and $\mathscr{R}^+$ (the Delphic modification of $\mathscr{R}$).

Sections 2.3.3 and 2.3.4 are concerned with Delphic semigroups in general. In Section 2.3.3 we derive the three basic theorems from sequential assumptions; the value of this is explained in that section. In Section 2.3.4 we introduce the idea of an hereditary sub-semigroup, and show how useful this idea is when we want to prove some particular semigroup almost (or properly) Delphic.

Arithmetic is largely the science of division, so in any Delphic semigroup there are three classes of its elements which specially concern us:

the 'simples', which have no divisors bar themselves and the identity;

the 'infinitely divisible' or 'i.d.' elements, which have nth roots for every positive integer n;

'I_0', the set of i.d. elements all of whose divisors are i.d.

One of Kendall's basic Delphic theorems says that each member of a Delphic semigroup can be built up as a product of simples and a member of I_0, so a knowledge of I_0 and the simples is fundamental to a study of the arithmetic of the semigroup.

Sections 2.3.5–2.3.10 of this paper are devoted to the arithmetic of $\mathscr{R}^+$; a second paper (to follow) describes some non-arithmetical properties of $\mathscr{R}^+$ and $\mathscr{P}$ and the arithmetic of $\mathscr{P}$. For each semigroup we find a dense class of simples; this, together with a result that follows from the hereditariness of $\mathscr{R}^+$ and $\mathscr{P}$ in certain larger semigroups, shows that the simples are residual in $\mathscr{R}^+$ and $\mathscr{P}$. So we can say that certain subsets of $\mathscr{R}^+$ and $\mathscr{P}$ contain simples despite not being able to produce any such simples explicitly. An interesting unsolved problem is whether the class $\mathcal{PM}$ of diagonal Markov functions contains any of the simples of $\mathcal{P}$.

We find I_0 for $\mathcal{R}^+$ precisely: it is a sub-semigroup of $\mathcal{R}^+$; and we can factor $\mathcal{R}^+$ into the direct product of its I_0 with another of its hereditary sub-semigroups, which we call $\mathcal{R}^+/\mathcal{D}$ and which is generated by the simples of $\mathcal{R}^+$. Coupled with the work of Kendall [5] these results provide a very good view of the structure of $\mathcal{R}^+$. For $\mathcal{P}$ we have but a partial characterization of I_0: we have two results which exclude large classes of i.d. elements from I_0, but there is much still to be done in this direction.

Turning to the non-arithmetical properties of $\mathcal{R}^+$ and $\mathcal{P}$, we show that the sets of i.d. elements of $\mathcal{P}$ and $\mathcal{R}^+$ are convex under addition; and do a Choquet analysis of both sets, identifying the extreme elements and showing that in neither case is the Choquet representation unique. Indeed, the set of extremals is residual among the i.d. elements of $\mathcal{P}$. A probabilistic interpretation of the additive convexity is given.

From the Kingman inequalities (see [8]) for the elements of $\mathcal{P}$ we deduce an inequality which—among other things—is enough to show that neither $\mathcal{P}$ nor $\mathcal{R}^+$ can be convex under addition. But the inequality's principal use is this: there are two interesting topologies on $\mathcal{P}$—'uniform convergence on compacta' and 'pointwise convergence'. The former is metrizable, but it is usually more natural to use the latter. Kendall [7] shows that they agree about convergence of sequences; from this and my inequality we show that they are the same topology, so that the topology of pointwise convergence on $\mathcal{P}$ is metrizable. This result is of great use in the analysis of $\mathcal{P}$.

2.3.3 SEQUENTIALLY DELPHIC SEMIGROUPS

There are three main theorems of the subject of Delphic semigroups:

Theorem 1. *Every i.d. element can be represented as the limit of a triangular array (for the definition of which see (h) below).*

Theorem 2. *In every Delphic semigroup there are three exclusive and exhaustive classes:*

    (a) *the simples*
    (b) *those elements which, themselves not simple, possess a simple factor*
    (c) I_0.

Theorem 3. *For each u in the semigroup there is at least one representation* $u = w . \prod v_i$, *where w is in I_0 and each v_i is simple. The product is at most countable and may be finite or void.*

5

We shall derive these theorems in a sequential way.

G is said to be a sequentially Delphic semigroup if G is commutative and has an identity e, and satisfies the following conditions:

(*a*) There exists a set

$$M \subset \prod_{i=1}^{\infty} G$$

and a map L of M into G. (M is the set of sequences of elements of G that possess limits, and L is the operation of taking the limit. Our definition implies that G is 'Hausdorff'.)

(*b*) If the sequence (u_n) is in M and (n') is a subsequence of (n), then $(u_{n'})$ is also in M and $L(u_{n'}) = L(u_n)$.

(*c*) Let u be arbitrary in G. If (u_n) is such that for every subsequence (n') of (n) there is a sub-subsequence (n'') of (n') with $(u_{n''})$ in M and $L(u_{n''}) = u$, then (u_n) is itself in M and $L(u_n) = u$.

(*d*) If (u_n) and (v_n) are in M, then so is $(u_n \cdot v_n)$ and

$$L(u_n \cdot v_n) = L(u_n) \cdot L(v_n).$$

(*e*) There is a homomorphism D of G into the non-negative reals under addition, such that if (u_n) is in M then the limit

$$\lim_{n \to \infty} D(u_n)$$

exists and equals $D(L(u_n))$.

(*f*) $D(u) = 0$ if and only if $u = e$.

(*g*) Let the symbol $|$ mean 'divides'. If $u_n | u$ for every n, then there is a subsequence (n') of (n) such that $(u_{n'})$ is in M and $L(u_{n'}) | u$.

(*h*) Suppose that we have a triangular array $(u_{ni}) (1 \leqslant i \leqslant n < \infty)$, and that

$$\max_{1 \leqslant i \leqslant n} D(u_{ni}) \to 0 \quad \text{as} \quad n \to \infty.$$

Let

$$u_n = \prod_{i=1}^{n} u_{ni}$$

for each n. If the sequence (u_n) of row-products is in M, then $L(u_n)$ is i.d.

Lemma 1. *If $u_n = u$ for each n, then (u_n) is in M and $L(u_n) = u$.*

Proof. By (g), there is a subsequence (n') of (n) such that $(u_{n'})$ is in M. Put $v = L(u_{n'})$; then $v|u$. But the terms of (u_n) are the same as those of $(u_{n'})$, being all of them equal to u. So by (a) (u_n) is in M and $L(u_n) = v$. By (e) we have $D(u) = D(v)$; and since $v|u$, by (f) and the non-negative nature of D we have $v = u$.

Lemma 2. *If (u_n) and u are such that $u_r|u_s|u$ whenever $r \leqslant s$, then (u_n) is in M and $u_r|L(u_n)|u$ for every r.*

Proof. Let $(n'), (n'')$ be arbitrary subsequences of (n); then by (g) there are sub-subsequences (m') of (n'), (m'') of (n'') such that $(u_{m'})$ and $(u_{m''})$ are in M. Let $L(u_{m'}) = u'$ and $L(u_{m''}) = u''$. Fix r' in (m') and consider the sub-sequence (s') of (m') given by $s' \geqslant r'$. Define $v_{s'}$ by $v_{s'}.u_{r'} = u_{s'}$; then by the hypothesis of the lemma $v_{s'}$ is in G and $v_{s'}|u$ for each s'. So by (g) there is a subsequence (t') of (s') such that $(v_{t'})$ is in M and $L(v_{t'}) = w$ say. But by (b) $(u_{t'})$ is in M and $L(u_{t'}) = u'$; so by (d) and Lemma 1 we have $u' = u_{r'}.w$ and hence $u_{r'}|u'$. By the hypothesis of the lemma we have at once that $u_{r''}|u'$ for all r'' in (m''); using (a), (b) and (g) we can pass to the limit to obtain $u''|u'$ and by symmetry $u'|u''$. By (e) and (f) then $u' = u'' = v$ say. Thus each choice of subsequence (n^*) of (n) leads to a sub-subsequence (m^*) with (u_{m^*}) in M and $L(u_{m^*}) = v$. By (c), then, (u_n) is in M and $L(u_n) = v$ also. In the same way that we showed above that $u_{r'}|u'$ for each r' we can show that $u_r|v$ for each r; and by (g) and (b) $v|u$.

Theorem 1 is now trivial. For the nth roots of the i.d. element u taken n times form a triangular array satisfying (h); the convergence of the row products to u follows from Lemma 1.

Theorem 2 can be restated: *Let u be not simple and let no factor of u be simple, and suppose that $u \neq e$. Then u is i.d.* The proof of this will follow Kendall's [7] proof closely.

Proof of Theorem 2. Clearly $D(u) \neq 0$. Let $v|u$ and $v \neq e$, and consider

$$V = (w: w|v \text{ and } w \neq e).$$

We show that $\inf(D(w): w \in V) = 0$. For suppose not: call the non-zero infimum k. There is a sequence (w_i) of members of V with

$$\lim_{i \to \infty} D(w_i) = k;$$

by (g) there is a subsequence (i') of (i) with $(w_{i'})$ in M. Let $L(w_{i'}) = w^*$: then $w^*|v$, and by (e) $D(w^*) = k$. But k is the infimum of the D-values of all the non-trivial factors of v, so by (f) w^* must be simple—but w^* is a factor of u, a contradiction.

Consider now all k-fold decompositions

$$Q: u = \prod_{i=1}^{k} v(i).$$

Define

$$m(Q) = \min_{1 \leqslant i \leqslant k} D(v(i)),$$

so that $0 \leqslant m(Q) \leqslant (1/k) . D(u)$; and let

$$h = \sup_{Q} m(Q).$$

Then there is a sequence (Q_r) with $m(Q_r) \to h$ as $r \to \infty$. By (g) we can find a subsequence (r') of (r) such that

(1)   for each i, $(v_{r'}(i))$ is in M and $L(v_{r'}(i)) = v'(i)$ say;

(2) $$u = \prod_{i=1}^{k} v'(i);$$

(3) $$m(Q_{r'}) \to h \text{ as } r' \to \infty.$$

Let the limiting decomposition be Q'. Then $m(Q') = h$. Suppose that the $v'(i)$ have been ordered so that $D(v'(i))$ increases with i; then $h = D(v'(1))$. Now either all the Ds are equal or there is a least l with $1 \leqslant l < k$ and $D(v'(l)) < D(v'(l+1))$. By the first paragraph of this proof we can write

$$v'(l+1) = w'(1) \ldots w'(l) . v''(l+1),$$

where $D(w'(1)), \ldots, D(w'(l))$ are all arbitrarily small but non-zero. If we choose them so that $D(v''(l+1)) > D(v'(l))$, we shall obtain a decomposition

$$Q'': u = \overline{v'(1) . w'(1)} \ldots \overline{v'(l) . w'(l)} . v''(l+1) . v'(l+2) \ldots v'(k)$$

which has $m(Q'') > h$, a contradiction. So in Q' all the Ds must be equal, entailing $D(v'(i)) = (1/k) . D(u)$ for each i $(1 \leqslant i \leqslant k)$.

We have a Q' for each k now: call it $Q'(k)$. The $Q'(k)$ form a triangular array whose row products are all equal to u, so by Lemma 1 the array converges to u. Further,

$$\max_{1 \leqslant i \leqslant k} D(v'_k(i)) = (1/k) . D(u) \to 0 \quad \text{as } k \to \infty.$$

Thus the conditions of (h) are verified and so u is infinitely divisible.

Theorem 3 follows from Lemma 2 without modification of Kendall's [7] proof.

A Delphic semigroup whose topology is first countable is also sequentially Delphic with the limiting operation induced by the topology; if the topology of the semigroup is metrizable, the semigroup is Delphic if and only if it is sequentially so. The exercise of getting the three main theorems under sequential assumptions has the following use:

Our condition (e) says effectively that the homomorphism D is continuous. In the special semigroups at which we shall look, D is defined as the value of a real function at a point; and the evaluation map is continuous in the product topology. Suppose, however, that we have to define D as the value of an integral. Except in special circumstances (see e.g. Bourbaki [1]) the theorem of dominated convergence only works for sequences, so unless we can find a metrizable topology for the semigroup—as we can in the case of the semigroup of probability laws on the line—we are constrained to adopt the sequentially Delphic formulation.

2.3.4 HEREDITARY SUB-SEMIGROUPS

Let G be a semigroup and let H be a non-empty sub-semigroup of G. H is said to be hereditary in G if for every u in H, the set $F(u)$ of G-factors of u lies entirely in H. It is of course standard that if G is commutative, so is H; and if G is a topological semigroup and we give H the subspace topology, H is a topological semigroup. It is also clear that if G has an identity then H contains the identity. The notion of hereditariness has many applications in the arithmetic of semigroups. We first note a sufficient condition for the fulfilment of Kendall's [6] Axiom B for Delphic semigroups.

Proposition 1. *If H is an hereditary subsemigroup (with the subspace topology) of a compact Hausdorff semigroup G, then for every u in H the set $F_H(u)$ of H-factors of u is compact in H.*

Proof. Let $F_G(u)$ be the set of G-factors of u; then $F_G(u)$ is closed in G. For let (v_a) be a net of members of $F_G(u)$ converging to v, and consider the cofactors w_a. Since G is compact there is a subnet (w_b) of (w_a) converging to some element, w say, of G. Since G is topological we have $v.w = u$; so v is a G-factor of u. Since G is compact and $F_G(u)$ is now seen to be closed, $F_G(u)$ is compact in G. But H is hereditary in G, so $F_H(u) = F_G(u)$; accordingly $F_H(u)$ is compact in H.

Lemma 3. *Let H be hereditary in G, and let u be an element of H. Then u is simple in H if and only if u is simple in G, and u is i.d. in H if and only if* *u* is i.d. in *G*. (It is clear that the terms 'simple' and 'i.d.', though originally defined for Delphic semigroups, extend immediately to all semigroups.)

The proof is trivial.

Proposition 2. *If G is a Delphic semigroup with homomorphism* D_G *and H is hereditary in G, then if we put* $D_H(u) = D_G(u)$ *when u is in H and give H the subspace topology, H is Delphic.*

The proof—verification of Kendall's [6] Delphic axioms—is immediate from Lemma 3 and the Delphic nature of *G*.

Proposition 3. *Let G be a compact commutative Hausdorff topological semigroup with identity. Let H be an hereditary subsemigroup of G, and give H the subspace topology. If H is Delphic (with homomorphism D say) then the set of simples in H—SH, say—is a* G_δ *in H.*

Proof. (We follow Parthasarathy *et al.* [10].) Define for all $c > 0$

$$E(c) = \left\{ u : \begin{array}{l} u \text{ is in } H, \text{ and there are } v, w \text{ in } H \\ \text{such that } D(v) \geqslant c, \ D(w) \geqslant c \text{ and } u = v.w \end{array} \right\}.$$

We show that $E(c)$ is closed in *H*. Let (u_a) be a net of members of $E(c)$ converging to u^* in *H*, and let v_a, w_a be chosen to correspond to u_a as in the definition of $E(c)$. Since *G* is compact we can find a subnet (b) of (a) such that (v_b) and (w_b) converge to v^*, w^* (say) in *G*. By continuity of multiplication we have $v^* . w^* = u^*$; so since u^* is in *H* and *H* is hereditary in G, v^* and w^* are in *H*. It follows that $D(v_b)$ and $D(w_b)$ converge to $D(v^*)$ and $D(w^*)$, so $D(v^*) \geqslant c$ and $D(w^*) \geqslant c$. Thus $u^* = v^* . w^*$ is in $E(c)$, implying that $E(c)$ is indeed closed.

Define

$$Q = \bigcup_{r=1}^{\infty} E(1/r).$$

If *u* is in *SH* then $u = v.w$ implies that $D(v) = 0$ or $D(w) = 0$, so that *u* is not in *Q*. On the other hand, if *u* in *H* is not simple it is in $E(c)$ for some $c > 0$; so we have that $Q = H \backslash SH$. Since $E(c)$ is closed in *H* for each $c > 0$, *Q* is an F_σ in *H*; so *SH* must be a G_δ in *H*.

Corollary. *If H is open in G, then SH is a* G_δ *in G also.*

Proposition 4. *If semigroups A, B, C are such that A is hereditary in B and B is hereditary in C, then A is hereditary in C.*

The proof is trivial.

Example 1. Let $\mathscr{Y}(\mathscr{R},\mathscr{R}^+)$ be the semigroup of all (aperiodic, positive) renewal sequences. Then $\mathscr{R}$ is hereditary in $\mathscr{Y}$ and $\mathscr{R}^+$ hereditary in $\mathscr{R}$.

Example 2. Let $\mathscr{G}(\mathscr{P},\mathscr{Q})$ be the semigroup of all (standard, stable) p-functions of Kingman [8]. Then $\mathscr{P}$ is hereditary in $\mathscr{G}$ and $\mathscr{Q}$ in $\mathscr{P}$.

The verification of these examples is easy.

Proposition 5. *Let G be a commutative Hausdorff semigroup, and let* (G_a) *(a in some index set A) be a collection of hereditary subsemigroups of G that together cover G. If each* G_a *is Delphic then Theorems 1, 2 and 3 hold for G.*

The proof is trivial.

The use of this last result is to get round a difficulty posed sometimes by the absence of a continuous homomorphism from G to the non-negative reals under addition. For example, there is no non-zero continuous homomorphism from the set $\mathscr{R}$ of aperiodic renewal sequences to the reals under addition (this non-arithmetic fact about $\mathscr{R}$ is proved in Section 2.3.11). But if $k(l) = (k_1, ..., k_l)$ is any set of positive integers with h.c.f. $(k(l)) = 1$, the set $\mathscr{R}_{k(l)}$ of renewal sequences u such that $u_{k_i} \neq 0$ for $1 \leqslant i \leqslant l$ is a Delphic semigroup (see Kendall [7]); and

$$D_{k(l)}(u) = - \sum_{i=1}^{l} \log(u_{k_i})$$

is a suitable Delphic homomorphism for $\mathscr{R}_{k(l)}$. It is clear that $\mathscr{R}_{k(l)}$ is hereditary in $\mathscr{R}$ and that the $\mathscr{R}_{k(l)}$, for all possible choices of $k(l)$, cover $\mathscr{R}$. So we can deduce the three main theorems for $\mathscr{R}$, first proved by Kendall [5].

In this case it happens that each $\mathscr{R}_{k(l)}$ is open in $\mathscr{R}$; and it will frequently happen that each G_a is open in G, since the usual way in which we get an hereditary sub-semigroup G_a of G is by demanding that some continuous functional on G be non-zero on G_a. Let D_a be the Delphic homomorphism on G_a in Proposition 5; if u in G is not in G_a, put $D_a(u) = +\infty$.

Proposition 6. *Under the conditions of Proposition 5 suppose that for all a in A,* G_a *is open in G. Let* $(u(i, j))$ $(1 \leqslant j \leqslant i < \infty)$ *be a triangular array of elements of G whose row products*

$$u(i) = \prod_{j=1}^{i} u(i, j)$$

converge to u in G as $i \to \infty$. *Let* G_b *be a covering Delphic semigroup in which u lies. If*

$$\lim_{i \to \infty} \max_{1 \leqslant j \leqslant i} D_b(u(i, j)) = 0$$

then u is i.d. in G.

Since G_b is open and was assumed to be Delphic, the proof is trivial.

2.3.5 THE ARITHMETIC OF $\mathscr{R}^+$: DEFINITIONS AND BASIC PROPERTIES

The sequence $u = (u_n)$ $(0 \leqslant n < \infty)$ of real numbers bounded between zero and unity is said to be a renewal sequence if and only if there is a sequence $f = (f_r)$ $(1 \leqslant r < \infty)$ of real numbers such that $f_r \geqslant 0$ for all r and

$$\sum_{r=1}^{\infty} f_r \leqslant 1$$

related to u by the equation

$$\left(1 - \sum_{r=1}^{\infty} f_r z^r\right) \cdot \sum_{n=0}^{\infty} u_n z^n = 1$$

for all complex z with $|z| < 1$.

If u is a renewal sequence, there exists a sequence of random variables $X(n)$, indicators of events $F(n)$, such that

$$\text{pr}\left(\bigcap_{i=1}^{k} \{X(n_i) = 1\}\right) = u_{n_1} \cdot \prod_{i=2}^{k} u_{n_i - n_{i-1}}$$

for all positive integers k and all choices of $0 \leqslant n_1 \leqslant n_2 \leqslant \ldots \leqslant n_k < \infty$. The events $F(n)$ are said to be manifestations of the regenerative phenomenon F. Suppose we have two renewal sequences v and w, with corresponding phenomena G and H and sequences of random variables $Y(n)$ and $Z(n)$; we can take G and H to be independent. If we define a new sequence of random variables $X(n)$ by $X(n) = Y(n).Z(n)$, then the $X(n)$ are indicators of events $F(n)$ that are manifestations of a regenerative phenomenon F say; and u, defined by $u_n = v_n.w_n$, is a renewal sequence (Kendall [5]). By the definition of the $X(n)$, F manifests itself at time n if and only if both G and H do so.

$\mathscr{R}^+$ is the set of renewal sequences all of whose terms are positive. Since $u_n \geqslant f_1^n$ for all n, and $u_1 = f_1$, $\mathscr{R}^+$ is precisely the set of all renewal sequences u with $u_1 > 0$. $\mathscr{R}^+$ is a Delphic semigroup (Kendall [7]). The multiplication operation on $\mathscr{R}^+$ is term-by-term multiplication of the sequences; the topology is that of term-by-term convergence. The identity e is given by $e_n = 1$ for all n; and the homomorphism D is given by $D(u) = -\log(u_1)$.

2.3.6 CONSTRUCTION OF SIMPLES IN $\mathscr{R}^+$

Let $\mathscr{S}\mathscr{R}^+$ be the set of simples in $\mathscr{R}^+$. Let $\mathscr{R}^*$ be the set of all u in $\mathscr{R}^+$ whose f-sequences f satisfy the conditions

$$(4) \qquad \sum_{r=1}^{\infty} f_r = 1 \quad \text{and} \quad f_1 < 1.$$

Proposition 7. *Let u in $\mathscr{R}^*$ be not simple and have f-sequence f. Then there is an integer $l > 1$ such that*

$$(5) \qquad f_{lr} > 0 \quad \text{for all integers } r > 0.$$

Proof. First we observe that for any renewal sequence u with f-sequence f

$$(6) \qquad u_1 = f_1, \quad \text{and} \quad f_1 = 1 \quad \text{if and only if } u = e.$$

By hypothesis there exist v and w, neither of them equal to e, such that $u = v.w$. Let g, h be the f-sequences of v, w; then by conditions (4) and (6) we have, since $u_1 = v_1.w_1$,

$$(7) \qquad 0 < g_1 < 1 \quad \text{and} \quad 0 < h_1 < 1.$$

Let F, G, H be the regenerative phenomena corresponding, as in the last section, to u, v, w; and let $F(n), G(n), H(n)$ be the corresponding sequences of events. Then $F(n) = G(n) \cap H(n)$, and so

$$(8) \qquad F(n) \subset G(n) \quad \text{for all } n.$$

By standard renewal theory,

$$\sum_{r=1}^{\infty} f_r = \text{pr}\left(\bigcup_{n=1}^{\infty} F(n)\right).$$

Hence by conditions (4) and (8)

$$1 = \sum_{r=1}^{\infty} f_r = \text{pr}\left(\bigcup_{n=1}^{\infty} F(n)\right) \leqslant \text{pr}\left(\bigcup_{n=1}^{\infty} G(n)\right) = \sum_{r=1}^{\infty} g_r \leqslant 1.$$

Therefore

$$(9) \qquad \sum_{r=1}^{\infty} g_r = 1 \quad \text{and similarly} \quad \sum_{r=1}^{\infty} h_r = 1.$$

From this and the inequalities (7) it follows that

(10)   there are integers $l > 1$ and $m > 1$ such that $g_l > 0$ and $h_m > 0$;

without loss of generality we have $l \geqslant m$.

Let now r be an arbitrary positive integer, and define S_F, S_G, S_H to be the sets of times n of occurrence of $F(n), G(n), H(n)$ for $0 \leqslant n \leqslant rl$. We note that if S is any set of non-negative integers not exceeding rl then the statement '$S_F = S$' is an F-measureable event, and similarly for S_G and S_H.

Let S_g be the set of integers

$$\{0, l, 2l, 3l, \ldots, (r-1)l, rl\}$$

(where ... denotes arithmetic progression); let S_h be the set of integers

$$\left\{ \begin{array}{c} 0;\ 1, 2, \ldots, l-m+1;\ l+1, l+2, \ldots, 2l-m+1;\ \ldots; \\ (r-2)l+1, (r-2)l+2, \ldots, (r-1)l-m+1;\ (r-1)l+1, \\ (r-1)l+2, \ldots, rl-1, rl \end{array} \right\}.$$

By conditions (7) and (10) and the regenerative natures of G and H we have

$$\operatorname{pr}(S_G = S_g) > 0 \quad \text{and} \quad \operatorname{pr}(S_H = S_h) > 0;$$

so since G and H are independent and $F(n) = G(n) \cap H(n)$ for all n,

(11) $\operatorname{pr}(S_F = S_g \cap S_h) > 0.$

But by inspection $S_g \cap S_h = \{0\} \cup \{rl\}$, so inequality (11) is the same statement as

(12) $\operatorname{pr}\left(F(0) \cap F(rl) \cap \bigcap_{s=1}^{rl-1} F'(s)\right) > 0$

(' here denotes complementation). The left-hand member of inequality (12), however, is precisely f_{rl} (this is standard); so, r being arbitrary, we have $f_{rl} > 0$ for all $r > 0$ and some $l > 1$.

Proposition 8. *Let u in $\mathscr{R}^*$ have f-sequence f. If u satisfies either*
   (C)   *There is an integer N such that $f_n = 0$ for all $n > N$, or*
   (C')   *For all r such that $f_r > 0$ there is an s such that s is a multiple of r and $f_s = 0$,*
then u lies in $\mathscr{S}\mathscr{R}^+$.

Proof. It is clear that if u satisfies (C) it also satisfies (C'); and (C') denies the conclusion of Proposition 7.

Example 3. *Define u in* $\mathscr{R}^*$ *by its f-sequence f:*

$$f_r = 0 \quad if\ r > 2; \quad f_1 = p, \quad f_2 = q \quad (0 < p = 1 - q < 1).$$

Then u satisfies (C) *and so is in* $\mathscr{S}\mathscr{R}^+$. *(This is Kendall's* [5] *example of a simple u.) We evaluate explicitly* $u_n = (1 + (-1)^n q^{n+1})/(1+q)$.

This example is very important because it is one of the few u in $\mathscr{S}\mathscr{R}^+$ known explicitly. Here is another:

Example 4. *Define u in* $\mathscr{R}^*$ *by its f-sequence f:*

$$f_{2r} = 0 \quad for\ all\ r > 0;$$

$$f_{2r+1} = p \cdot q^r \quad (0 < p = 1 - q < 1) \quad for\ all\ r \geqslant 0.$$

Then u satisfies (C') *and so is in* $\mathscr{S}\mathscr{R}^+$. *The explicit form of u is*

$$u_0 = 1; \quad u_n = (1-q)(1 + (-1)^{n-1} q^n)/(1+q) \quad for\ all\ n \geqslant 1.$$

Example 5. *It might be conjectured that whenever u in* $\mathscr{R}^*$ *has f-sequence f such that* $f_k = 0$ *for some integer* $k > 0$, *u is in* $\mathscr{S}\mathscr{R}^+$. *But this does not hold. For define v, w in* $\mathscr{R}^*$ *by their f-sequences g, h:*

$$\left.\begin{array}{l} g_1 = g_5 = \tfrac{1}{2} \\ h_1 = h_2 = \tfrac{1}{2} \end{array}\right\} \quad all\ other\ g_r, h_s\ vanishing.$$

Put $u = v \cdot w$, *and let u have f-sequence f. By direct computation we find* $f_3 = 0$; *by arguments of the sort that we used in Proposition 7, one can show that* f_r *vanishes whenever there is an integer* $s \geqslant 0$ *such that* $r = 5s + 3$ *or* $5s + 4$. *And* $u = v \cdot w$ *is in* $\mathscr{R}^*$ *and is not simple.*

Example 6. *There is a u in* $\mathscr{S}\mathscr{R}^+$ *whose f-sequence f satisfies*

$$\sum_{r=1}^{\infty} f_r < 1.$$

For let u be given by $f_{2r} = 0$ $(r \geqslant 1)$; $f_{2r-1} = \tfrac{1}{2} \cdot r^{-2}$ $(r \geqslant 1)$. *Then u is in* $\mathscr{R}^+$, *and*

$$\sum_{r=1}^{\infty} f_r = \frac{\pi^2}{12} < \frac{10}{12} < 1.$$

Suppose that u has a factorization $u = v \cdot w$, *where v, w have f-sequences g, h. Then* $g_1 > 0$ *and* $h_1 > 0$. *Suppose that* g_1 *is the only* g_r *that does not vanish. Immediately we have*

$$h_r = f_r \cdot (g_1)^{-r} \quad for\ all\ r \geqslant 1;$$

so we must have

$$\sum_{r=1}^{\infty} h_r = \sum_{r=1}^{\infty} f_r \cdot (g_1)^{-r} \leqslant 1.$$

Inspection of the f_r shows at once that for this to be true we need $g_1 = 1$, that is, $v = e$. Hence for the factorization to be non-trivial there must be integers $l > 1$ and $m > 1$ such that $g_l > 0$ and $h_m > 0$.

Now we pick up the argument of Proposition 7—at its formula (10)—to show that if there is a non-trivial factorization of u, then there is a positive integer n such that $f_{rn} > 0$ for all $r \geqslant 1$. But this contradicts the definition of u, so u must be in $\mathscr{S}\mathscr{R}^+$.

2.3.7 TOPOLOGICAL QUESTIONS AND THE CATEGORY OF $\mathscr{S}\mathscr{R}^+$

Let $\mathscr{X}$ be the space of sequences $x = (x_n)$ $(0 \leqslant n < \infty)$ of real numbers x_n with $x_0 = 1$ and $0 \leqslant x_n \leqslant 1$ for all n. It is standard that the topology T induced by the metric

$$d(x, y) = \sum_{n=0}^{\infty} 2^{-n} \cdot |x_n - y_n|$$

is the topology of term-by-term convergence. By Tychonov's theorem $(\mathscr{X}, T)$ is compact. Let $\mathscr{Y}$ be the set of all renewal sequences u: it is standard that $\mathscr{Y}$ is closed in $(\mathscr{X}, T)$ and so is compact since $(\mathscr{X}, T)$ is compact. So $\mathscr{Y}$, under the subspace topology T, is a complete (compact) metric space. Define a new topology U on $\mathscr{Y}$ to be the topology of term-by-term convergence of the f-sequences.

Lemma 4. *On $\mathscr{Y}$, $T = U$.*

Proof. For all $r \geqslant 1$ there are polynomials P_r and Q_r of r variables $x_1, \ldots, x_r$ such that for every u in $\mathscr{Y}$

(13) $u_r = P_r(f_1, \ldots, f_r)$ and $f_r = Q_r(u_1, \ldots, u_r),$

where f is the f-sequence of u. Now U is metrizable, a suitable metric being

$$d^*(u, v) = \sum_{r=1}^{\infty} 2^{-r} \cdot |f_r - g_r|$$

(where g is the f-sequence of v). Since T is metric, it and U will be equivalent topologies for $\mathscr{Y}$ if and only if they agree on convergence of sequences—that is, if and only if

(14) $(u(n)) \to u$ term-by-term $\Leftrightarrow (f(n)) \to f$ term-by-term

where $f(n)$ for each n is the f-sequence of $u(n)$. But condition (14) is obvious in view of equations (13).

Proposition 9. $\mathscr{S}\mathscr{R}^+$ *is dense in* $\mathscr{Y}$.

Proof. Let v be any member of $\mathscr{Y}$, and let its f-sequence be g. We shall construct a sequence $(u(n))$ of members of $\mathscr{S}\mathscr{R}^+$ T-converging to v as $n \to \infty$; $u(n)$ will be given by its f-sequence $f(n)$. There are four cases:

(a) $g_r = 0$ for all r. For all $n \geqslant 1$ put

$$f_1(n) = 2^{-n}, \quad f_n(n) = 1 - 2^{-n} \quad \text{and} \quad f_r(n) = 0 \quad \text{if } r \neq 1 \text{ or } n.$$

(b) $g_1 = 1$, so that $g_r = 0$ for all $r > 1$. For all $n \geqslant 1$ put

$$f_1(n) = 1 - 2^{-n}, \quad f_2(n) = 2^{-n} \quad \text{and} \quad f_r(n) = 0 \quad \text{for all } r \geqslant 3.$$

(c) $g_1 = 0$ but there is an integer s with $g_s > 0$. Let t be the least such s. For all $n > t+1$ put

$$f_1(n) = 2^{-n} \cdot g_t, \quad f_r(n) = 0 \quad \text{for } 1 < r < t, \quad f_t(n) = (1 - 2^{-n}) \cdot g_t,$$

$$f_r(n) = g_r \quad \text{for } t < r \leqslant n, \quad f_{n+1}(n) = 1 - \sum_{i=1}^{n} f_i(n)$$

and

$$f_r(n) = 0 \quad \text{for all } r > n+1.$$

(d) $0 < g_1 < 1$. For all $n \geqslant 1$ put

$$f_r(n) = g_r \quad \text{for } 1 \leqslant r \leqslant n, \quad f_{n+1}(n) = 1 - \sum_{i=1}^{n} f_i(n)$$

and

$$f_r(n) = 0 \quad \text{for all } r > n+1.$$

In each case we see that $u(n)$ is in $\mathscr{R}^*$ for all relevant n and satisfies condition (C) of Proposition 8. So each $u(n)$ is in $\mathscr{S}\mathscr{R}^+$; and it is clear that in each case the sequence $(u(n))$ U-converges to v as $n \to \infty$. By Lemma 4 this is equivalent to T-convergence.

Corollary. $\mathscr{R}^+ \backslash \mathscr{S}\mathscr{R}^+$ *is dense in* $\mathscr{Y}$.

Proof. For each integer $r \geqslant 2$ define $u(r)$ by $u_n(r) = (1 - (1/r))^n$ for all n. Then $u(r)$ is in $\mathscr{R}^+ \backslash \mathscr{S}\mathscr{R}^+$, being, indeed, in $I_0(\mathscr{R}^+)$ (see Proposition 11). As $r \to \infty$, the sequence $(u(r))$ T-converges to e. So if v is any member of $\mathscr{R}^+$, the sequence $(u(r).v)$ of members of $\mathscr{R}^+ \backslash \mathscr{S}\mathscr{R}^+$ T-converges to v; whence $\mathscr{R}^+ \backslash \mathscr{S}\mathscr{R}^+$ is dense in $\mathscr{R}^+$. But *a fortiori* from the proposition $\mathscr{R}^+$ is dense in $\mathscr{Y}$, so $\mathscr{R}^+ \backslash \mathscr{S}\mathscr{R}^+$ must be dense in $\mathscr{Y}$ also.

Lemma 5. $\mathscr{R}^+$ *is open in* $\mathscr{Y}$.

Proof. u is in $\mathscr{Y}\backslash\mathscr{R}^+$ if and only if $u_1 = 0$. So $\mathscr{Y}\backslash\mathscr{R}^+$ is closed in $\mathscr{Y}$.

A subset S of a topological space (X, T) is said to be residual in X if S is non-meagre, and $X\backslash S$ is meagre, in X.

Proposition 10. (*i*) $\mathscr{S}\mathscr{R}^+$ *is residual in* $\mathscr{Y}$. (*ii*) $\mathscr{S}\mathscr{R}^+$ *is residual in* $\mathscr{R}^+$.

Proof. Ad (*i*): $\mathscr{R}^+$ is hereditary in $\mathscr{Y}$ by Example 1, and $\mathscr{Y}$ is a compact metric semigroup. So by Proposition 3 $\mathscr{S}\mathscr{R}^+$ is a G_δ in $\mathscr{R}^+$; and is a G_δ in $\mathscr{Y}$ also, for $\mathscr{R}^+$ is open in $\mathscr{Y}$. But from Proposition 9 $\mathscr{S}\mathscr{R}^+$ is dense in $\mathscr{Y}$; so $\mathscr{Y}\backslash\mathscr{S}\mathscr{R}^+$ is a countable union of nowhere-dense sets, i.e. is meagre in $\mathscr{Y}$. But $\mathscr{Y}$ is a complete metric space, so (by Baire's category theorem) is non-meagre in itself. $\mathscr{S}\mathscr{R}^+$ must therefore be residual in $\mathscr{Y}$.

Ad (*ii*): It is a standard topological fact (see e.g. Čech [2]) that if (A, T) is a topological space and $B \subset C \subset A$ with C open, then B is meagre in C if and only if B is meagre in A. Now by Lemma 5 $\mathscr{R}^+$ is open in $\mathscr{Y}$; so we can apply this, with $(A, T) = (\mathscr{Y}, T), B = \mathscr{S}\mathscr{R}^+$ and $C = \mathscr{R}^+$, to (*i*). (*ii*) is then immediate.

Corollary. *There is a u in* $\mathscr{S}\mathscr{R}^+$ *with f-sequence f such that* $f_r > 0$ *for all r.*

For the set $\mathscr{Y}_k$ of u in $\mathscr{Y}$ such that $f_k = 0$ is closed and has void interior; so $\mathscr{Z}$ is meagre in $\mathscr{Y}$, where

$$\mathscr{Z} = \bigcup_{k=1}^{\infty} \mathscr{Y}_k$$

—that is, $\mathscr{Z}$ is the set of u in $\mathscr{Y}$ which have $f_k = 0$ for some k. A fortiori, then, $\mathscr{S}\mathscr{R}^+ \cap \mathscr{Z}$ is meagre in $\mathscr{Y}$. The corollary then follows from the first part of the Proposition.

2.3.8 I_0 FOR $\mathscr{R}^+$

Recall that I_0 is the set of all i.d. elements of $\mathscr{R}^+$ all of whose factors are i.d.

For all real p and integers n such that $1 \leqslant n \leqslant \infty$ and $0 < p \leqslant 1$ define $v(n, p)$ by $v_r(n, p) = p^r$ for $1 \leqslant r \leqslant n$; $v_r(n, p) = p^n$ for all $r > n$. Then Kendall [5]) $v(n, p)$ is i.d. in $\mathscr{R}^+$, and any i.d. u in $\mathscr{R}^+$ can be written

$$(15) \qquad u = \prod_{1 \leqslant n \leqslant \infty} v(n, p_n)$$

for some sequence (p_n) such that $0 < p_n \leqslant 1$ for each n and

$$\prod_{1 \leqslant n \leqslant \infty} p_n > 0.$$

Proposition 11. $v(\infty, p)$ *is in* I_0 *for all* p *such that* $0 < p \leqslant 1$.

Proof. $v(\infty, p)$ has f-sequence $f(p)$ given by

(16) $$f_1(p) = p; \quad f_r(p) = 0 \quad \text{for all } r \geqslant 2.$$

Suppose that w, with f-sequence h, is a factor of $v(\infty, p)$. Then either w is of the form $v(\infty, p')$ for some p' or there is an integer $s \geqslant 2$ such that $h_s > 0$. But since the cofactor of w in $v(\infty, p)$ is in $\mathscr{R}^+$ we see at once that $h_s > 0$ entails $f_s(p) > 0$, a contradiction of equation (16). So w is of the form $v(\infty, p')$ for some $p' \geqslant p > 0$ and therefore is i.d. in $\mathscr{R}^+$. Thus $v(\infty, p)$ has only i.d. factors.

Theorem 4. *If* $1 \leqslant n < \infty$ *and* $0 < p < 1$, *then* $v(n, p)$ *is not in* I_0.

Proof. For any positive integer n and any p in $]0, 1[$ define the sequence $u(n)$—which need not be in $\mathscr{R}^+$—by

(17) $$u_r(n) = v_r(n, p)/v_r(n+1, p') \quad \text{for all } r \geqslant 0,$$

where $p' = (1/q)$ and q, a function of n and p, is given by

(18) $$q = 1/(2\sqrt{p} - p) \quad \text{when } n = 1;$$

(19) $$q = (\sqrt{(4p - 3p^2)} - p)/(2p(1-p)) \quad \text{for all } n \geqslant 2.$$

We first show that $q > 1$ for all n and p. If $n = 1$, the result is obvious from equation (18). If $n \geqslant 2$, we note that equation (19) is the same statement as

(20) q is the positive root of the equation $Y(q) = q^2 p(1-p) + qp - 1 = 0$.

It is clear that $Y(q) \to \infty$ as $q \to \pm\infty$ and $Y(1) < 0$, so the equation has a root exceeding unity. Since $Y(0) = -1 < 0$, this is the only positive root; so we do indeed have $q > 1$. This implies that the denominator of the right-hand member of equation (17) makes sense, for now $p' = (1/q) < 1$. We next show that

(21) $$0 \leqslant u_r(n) \leqslant 1 \quad \text{for all } r \text{ and } n.$$

If $n = 1$, equation (17) shows that condition (21) will hold if and only if $pq^2 \leqslant 1$. But this is easily verified from equation (18) and the fact that $0 < p < 1$. If $n \geqslant 2$, equation (17) shows that condition (21) holds if and only if

(22) $$p^n q^{n+1} \leqslant 1 \quad \text{for all } n \geqslant 2.$$

For this it suffices that

(23) $$p^2 q^3 \leqslant 1 \quad \text{and} \quad pq \leqslant 1.$$

Let, then, $y = p^2 q^3$. In view of condition (20), y satisfies the equation

(24) $$Z(y) = y^{\frac{2}{3}} p^{-\frac{1}{3}}(1-p) + y^{\frac{1}{3}} p^{\frac{1}{3}} - 1 = 0;$$

and $y > 0$ by definition. Now $Z(1) = (1 + p^{\frac{2}{3}})(p^{-\frac{1}{3}} - 1) > 0$ and $Z(0) = -1 < 0$, so equation (24) has a root in the range $0 < y < 1$. Since Z is quadratic in $y^{\frac{1}{3}}$, equation (24) has but one other root; and this is easily seen to be negative. So $0 < p^2 q^3 < 1$. Since $q > 1$ we have *a fortiori* $pq < 1$; so condition (23) holds as required, yielding condition (21) for all r and n.

Because condition (21) holds, the function $U_n(z)$ defined formally by the power series

(25) $$U_n(z) = \sum_{r=0}^{\infty} u_r(n) . z^r$$

is regular and equal to the sum of the series for all z such that $|z| < 1$. It is clear from equation (17) that $U_n(0) = u_0(n) = 1$; so if we put

(26) $$F_n(z) = 1 - (1/U_n(z)),$$

F_n is defined and regular in some neighbourhood of $z = 0$. So there is a $d > 0$ such that when $|z| < d$ we have the Taylor development

(27) $$F_n(z) = \sum_{r=1}^{\infty} f_r(n) . z^r.$$

We want to show that for all n, $u(n)$ lies in $\mathscr{R}^+$. For this it is necessary and sufficient that for all $n \geqslant 1$

(28) $$f_1(n) > 0,$$

(29) $$f_r(n) \geqslant 0 \quad \text{for all } r > 1,$$

and

(30) $$\sum_{r=1}^{\infty} f_r(n) \leqslant 1.$$

It is clear from equation (17) that the inequality (28) is always satisfied.

Define $t = pq$. From condition (23) we have

(31) $$0 < t < 1.$$

From equation (17) we have directly

(32) $$U_n(z) = \frac{1 - t^{n+1} z^{n+1}}{1 - tz} + \frac{t^n q z^{n+1}}{1-z}.$$

If we define

(33) $$H_n(z) = 1 + t^{n-1}(q-1) z^n - (1-t) . \sum_{r=1}^{n-1} t^{r-1} z^r,$$

then equations (32) and (26) give us, after some reduction,

(34) $$F_n(z) = tz \cdot H_n(z)/H_{n+1}(z).$$

Let $n = 1$. Then

$$F_1(z) = tz(1+(q-1)z)/(1-(1-t)z+t(q-1)z^2) = F(z), \quad \text{say.}$$

(This last formula is due to Kendall.) Consider the equation

$$X(x) = x^2 - (1-t)x + t(q-1) = 0.$$

Because of equation (18) and the definition of t the roots of this are equal; let their common value be a. Since $a+a = 1-t$, inequality (31) shows that $0 < a < 1$; and we have

(35) $$F(z) = z(t+a^2 z)/(1-az)^2.$$

We see at once that condition (29) is satisfied for $n = 1$; and since $|a| < 1$ the radius of convergence of the Taylor series of F exceeds unity, so that we can put $z = 1$ in equation (35) to obtain

$$\sum_{r=1}^{\infty} f_r(1) = F(1) = (t+a^2)/(1-a)^2 = 4tq/(1+t)^2 = 1,$$

by equation (18) and the definition of t. Hence condition (30) as well as conditions (28) and (29) is satisfied for $n = 1$, so that

(36) $$u(1) \text{ is in } \mathscr{R}^+.$$

We turn to the case $n \geqslant 2$. Define

(37) $$h = (1-t)/t \quad \text{and} \quad k = (q-1)/t,$$

so that by inequality (31) $h > 0$. If we put also

(38) $$w = tz,$$

we find from equation (33) that

$$H_n(z) = 1 - h \cdot \sum_{r=1}^{n-1} w^r + k \cdot w^n.$$

But equation (19) shows that $k = h^2$, so

(39) $$H_n(z) = 1 - h \cdot \sum_{r=1}^{n-1} w^r + h^2 \cdot w^n;$$

and by equation (34),

$$F_n(z) = w \cdot H_n(z)/H_{n+1}(z).$$

By formulae (38) and (31) we see that condition (29) will be satisfied if and only if the Taylor coefficients at $w = 0$ of $F_n(z)/w$ are all non-negative.

The classical form for the coefficients of the quotient of two power-series has been given by Hagen [4] and Markushevich [9]. They show that if we have two power-series

$$A(w) = \sum_{r=0}^{\infty} a_r w^r, \quad B(w) = \sum_{r=0}^{\infty} b_r w^r \quad (b_0 \neq 0)$$

with radii of convergence R_a, R_b, then provided $|w| < \min(R_a, R_b, r_b)$ (where r_b is the least of the moduli of the zeros of $B(w)$),

(40) $$\frac{A(w)}{B(w)} = \sum_{r=0}^{\infty} (-1)^r \cdot D_r \cdot (b_0)^{-(r+1)} \cdot w^r.$$

Here D_r is the $(r+1)$-rowed principal minor of the matrix (c_{ij}), where

$$c_{i0} = a_i, \quad c_{ij} = 0 \quad \text{if } j \geq i+2 \quad \text{and} \quad c_{ij} = b_{i-j+1} \quad \text{if } 0 < j \leq i+1,$$

so that

(41) $$(c_{ij}) = \begin{bmatrix} a_0 & b_0 & 0 & 0 & 0 & 0 & \cdots \\ a_1 & b_1 & b_0 & 0 & 0 & 0 & \cdots \\ a_2 & b_2 & b_1 & b_0 & 0 & 0 & \cdots \\ \cdot & \cdot & \cdot & \cdot & \cdot & \cdot & \\ \cdot & \cdot & \cdot & \cdot & \cdot & \cdot & \\ \cdot & \cdot & \cdot & \cdot & \cdot & \cdot & \end{bmatrix}.$$

In our case put

(42) $$A(w) = H_n(z), \quad B(w) = H_{n+1}(z).$$

Then in the matrix (41), $a_0 = b_0 = 1$;

$$a_i = b_j = -h \quad \text{if } 1 \leq i \leq n-1, \ 1 \leq j \leq n;$$

$$a_n = b_{n+1} = h^2; \quad \text{and} \quad a_i = b_j = 0 \quad \text{if } i \geq n+1, \ j \geq n+2.$$

(c_{ij}) will from now on be supposed to have these a- and b-values. Now we do a sequence of row-operations on (c_{ij}). Let its rows be $R_i \ (i \geq 0)$. There exists a set $(x_{ij}) \ (0 \leq i < j < \infty)$ of real numbers x_{ij} such that if we take in sequence

$$x_{0j} R_0 \text{ from } R_j \text{ for each } j > 0, \quad \text{then}$$

$$x_{1j} R_1 \text{ from } R_j \text{ for each } j > 1, \quad \text{then}$$

$$x_{2j} R_2 \text{ from } R_j \text{ for each } j > 2 \quad \text{and so on,}$$

we end up with the matrix

(43)

$$
\begin{bmatrix}
A_0 & 1 & 0 & 0 & 0 & \cdots \\
A_1 & 0 & 1 & 0 & 0 & \cdots \\
A_2 & 0 & 0 & 1 & 0 & \cdots \\
\cdot & \cdot & \cdot & \cdot & \cdot & \cdots \\
\cdot & \cdot & \cdot & \cdot & \cdot \\
\cdot & \cdot & \cdot & \cdot & \cdot
\end{bmatrix}
$$

which has numbers A_r $(r \geqslant 0)$ in the first column, 1s in the first super-diagonal, and zeros elsewhere. Since in the matrix (41) we had $b_0 = 1$, we have

$$(44) \qquad D_r = (-1)^r A_r \quad \text{for all } r \geqslant 0,$$

where the D_r are as in equation (40). The coefficients in the Taylor development of $F_n(z)/w$ will all be non-negative, then, if and only if $A_r \geqslant 0$ for all $r \geqslant 0$.

Since $A(w)$ and $B(w)$ are polynomials, we have—by looking at the row operations we used to get the matrix (43)—a recurrence relation:

$$(45) \qquad A_{n+r+1} = h . \sum_{s=1}^{n} A_{r+s} - h^2 . A_r \quad \text{for all } r \geqslant 0;$$

equivalently,

$$A_{n+r+1} - h . A_{n+r} = h . \sum_{s=2}^{n-1} A_{r+s} + h(A_{r+1} - h . A_r).$$

We have now to treat the case $n = 2$ by itself. From equation (41) and our row-operations it is easy to see that in this case $A_0 = 1$, $A_1 = 0$ and $A_2 = h + h^2$. These values and equation (45) then yield

$$(46) \qquad
\begin{cases}
\text{if } h = 1, & \begin{cases} A_r = \tfrac{1}{2}r + 1 & \text{if } r \text{ is even;} \\ A_r = \tfrac{1}{2}(r-1) & \text{if } r \text{ is odd;} \end{cases} \\[2ex]
\text{if } h \neq 1, & \begin{cases} A_r = c^r(1 - c^{r+2})/(1 - c^2) & \text{if } r \text{ is even;} \\ A_r = c^{r+3}(1 - c^{r-1})/(1 - c^2) & \text{if } r \text{ is odd.} \end{cases} \quad (c = \sqrt{h})
\end{cases}
$$

Now by equations (44), (42), (40), (38) and (34) we have

$$(47) \qquad f_{r+1} = t^{r+1} . A_r \quad \text{for all } r \geqslant 0,$$

and by inequality (31) $t > 0$. Since equations (46) imply that $A_r \geqslant 0$ for all $r \geqslant 0$, we get condition (29) for $n = 2$.

We return to the general case, $n \geqslant 3$. Let P_r for each $r \geqslant 1$ be the proposition that $A_r - h \cdot A_{r-1} \geqslant 0$, and let Q_r for each $r \geqslant 0$ be the proposition that $A_r \geqslant 0$. We want to demonstrate Q_r for all $r \geqslant 0$. We have from equation (45) that

$$\text{for all } r \geqslant 1, \quad P_r \cap \bigcap_{s=1}^{n-1} Q_{r+s} \Rightarrow R_{r+n},$$

where $R_r = P_r \cap Q_r$. *A fortiori*, then,

(48) $$\text{for all } r \geqslant 1, \quad \bigcap_{s=0}^{n-1} R_{r+s} \Rightarrow R_{r+n}.$$

By direct computation in equation (41) we find

(49) $A_0 = 1, \quad A_r = 0 \quad \text{for } 1 \leqslant r \leqslant n-1, \quad A_n = h + h^2 \quad \text{and} \quad A_{n+1} = h^3$.

From equations (49) and (45) then we have $A_{n+r+1} \geqslant h \cdot A_{n+r}$ for $1 \leqslant r \leqslant n-1$, for in this range of r-values the $-h^2 \cdot A_r$ term in equation (45) vanishes. So

(50) $$Q_r \text{ holds for } \quad 0 \leqslant r \leqslant n+1$$

and

(51) $$R_r \text{ holds for } \quad n+2 \leqslant r \leqslant 2n.$$

We now show that R_{2n+1} holds. For each $s \geqslant 2$ define

$$B_s(h) = h^2((1+h)^s - h(1+h)^{s+2}).$$

From equations (49) and (45) we find

(52) $$A_r = B_{r-n}(h) \quad \text{for } n+2 \leqslant r \leqslant 2n;$$

and again by equations (49) and (45)

(53) $$A_{2n+1} = B_{n+1}(h) - h^2(1+h)^2.$$

But now from the definition of the B_s we have for all $n \geqslant 3$

$$B_{n+1}(h) - h^2(1+h)^2 \geqslant h \cdot B_n(h),$$

so by equations (52) and (53)

$$A_{2n+1} - h \cdot A_{2n} \geqslant 0 \quad \text{for all } n \geqslant 3;$$

that is, P_{2n+1} holds .Condition (51), however, tells us that R_{2n}—and so Q_{2n}— holds; so we get Q_{2n+1} as well as P_{2n+1}. Thus R_{2n+1} holds; this and the inductive mechanism (48) entail R_r for all $r \geqslant n+2$, and condition (50) then shows that Q_r holds for all $r \geqslant 0$ and $n \geqslant 3$. From equation (47) and the definition of Q_r, then, we have condition (29) for all $n \geqslant 3$ (and, in view of equation (46), for $n = 2$ also).

We now have to check condition (30) for $n \geq 2$. Since $A_r \geq 0$ for all $n \geq 2$ and all $r \geq 0$, equation (45) shows that

$$A_{n+r+1} \leq h . \sum_{s=1}^{n} A_{r+s} \quad \text{for all } r \geq 0;$$

so

$$0 \leq A_{r+1} \leq h . \sum_{s=0}^{r} A_s \quad \text{for all } r \geq n+2.$$

This shows at once that there is a non-negative constant K such that

$$0 \leq A_r \leq K . (1+h)^r \quad \text{for all } r \geq n+3.$$

But by equation (37) $h = (1-t)/t$, and by equation (47) $f_r = t^r . A_{r-1}$; so for all $r \geq n+4$

$$0 \leq f_r \leq t^r . K . t^{-(r-1)} = K/t.$$

Since by inequality (31) $t > 0$, this implies

$$\limsup_{r \to \infty} |f_r|^{(1/r)} \leq 1,$$

so the radius d of convergence of the Taylor series of $F_n(z)$ is at least unity. Now F_n has a singularity at the point $z = d$, for all the Taylor coefficients of F_n are non-negative (Titchmarsh [11]). Since by equation (34) F_n is a rational function this singularity must be a pole, so that

$$\lim_{z \to d} |F_n(z)| = \infty.$$

By inspection, however,

$$\lim_{z \to 1} |F_n(z)| = 1,$$

so we must in fact have $d > 1$. So we can put $z = 1$ in $F_n(z)$ to get

$$\sum_{r=1}^{\infty} f_r(n) = F_n(1) = 1.$$

Thus condition (30) is satisfied for all $n \geq 2$, so for all $n \geq 2$ $u(n)$ is in $\mathscr{R}^+$; $u(1)$ is in $\mathscr{R}^+$ by condition (36).

By the definition (17) of $u(n)$, $v(n,p)$ has $u(n)$ as a factor, and we have just shown that $u(n)$ is in $\mathscr{R}^+$; the cofactor $v(n+1, p')$ is of course (being i.d.) in $\mathscr{R}^+$ also. But $u(n)$ is not a Kaluza-sequence (in the sense of Kendall [5]), because

$$u_{n+2}(n) . u_n(n) = p^n q^{n+1} . p^n q^n < (p^n q^{n+1})^2 = (u_{n+1}(n))^2.$$

So (Kendall [5]) $u(n)$ is not i.d. in $\mathscr{R}^+$; whence $v(n,p)$, having $u(n)$ as a factor in an $\mathscr{R}^+$-factorization, cannot be in I_0.

Corollary. I_0 *consists precisely of the elements* $v(\infty, p)$ *for* $0 < p \leqslant 1$.

For Proposition 11 shows that I_0 contains these elements. On the other hand, let u, i.d. in $\mathscr{R}^+$, be not of the form $v(\infty, p)$ for any p. Then u has $v(n, p')$ as a factor for some finite n and some p' such that $0 < p' < 1$, by the representation (15). So by Theorem 4, any i.d. u that is not of the form $v(\infty, p)$ has a factor that is not i.d. and so cannot be in I_0.

2.3.9 AN EXAMPLE ON SEMIGROUPS OF POWERS OF ELEMENTS OF $\mathscr{R}^+$

For any u in $\mathscr{R}^+$ define $S(u)$ to be the additive semigroup of strictly positive real numbers x such that $(u)^x$ is in $\mathscr{R}^+$. $S(u)$ clearly always contains the positive integers; if u is i.d., $S(u)$ is the entire positive half-line. Dugué [3] has given examples of more interesting $S(u)$ for the semigroup of probability laws on the line; our example here—a by-product of the proof of Theorem 4—has

$$S(u) = [1, \infty[.$$

Example 7. *Let u be given by $u_0 = 1$, $u_1 = \frac{1}{3}$ and $u_n = \frac{4}{9}$ for all $n \geqslant 2$. Then $(u)^x$ is in $\mathscr{R}^+$ if and only if $(x = 0$ or$)x \geqslant 1$.*

Proof. We have, in the notation of the last section, $v(1, \frac{1}{4}) = v(2, \frac{3}{4}) \cdot u$. By (18) and (36) of the last section u is in $\mathscr{R}^+$. Define in the usual way

$$U_x(z) = \sum_{r=0}^{\infty} u_r^x \cdot z^r \quad \text{and} \quad F_x(z) = 1 - (1/U_x(z));$$

then by equations (33) and (34)

$$F_x(z) = t^x z (1 + (q^x - 1)z)/(1 - (1 - t^x)z + t^x(q^x - 1)z^2),$$

where $q = \frac{4}{3}$ and $t = \frac{1}{3}$. It is easy to check by taking partial fractions that if $x \geqslant 1$ all the Taylor coefficients of F_x are non-negative, whereas if $x < 1$ they oscillate in sign. If $x \geqslant 1$ we see that the zeros of the denominator of F_x lie without the unit circle, so that we can put $z = 1$ in $F_x(z)$ to see that the sum of its Taylor coefficients does not exceed unity. Thus if $x \geqslant 1$ conditions (28), (29) and (30) are verified for $(u)^x$, whereas if $x < 1$ condition (29) is violated. Accordingly $S(u) = [1, \infty[$, as stated.

It is an interesting unsolved problem, whether the u of this example is simple. If it were, light would be thrown on the factorization theory of $\mathscr{R}^+$, for $(u)^3$ would have more than one (essentially different) factorization.

2.3.10 FACTORIZATION THEORY OF $\mathscr{R}^+$

We defined 'simples' as those members of a semigroup that had no non-trivial factors. The question arises of how far it is justifiable, in a given semigroup—here $\mathscr{R}^+$—to replace the word 'simple' by 'prime'.

One very strong property of the primes among the integers is

(Q): If p is prime and p divides $a.b$, then p divides a or b or both.

Proposition 12. *(Q) does not hold in any Delphic semigroup in which I_0 does not exhaust the i.d. elements. (So (Q) does not hold for $\mathscr{R}^+$.)*

Proof. Let u be an i.d. element of the semigroup not lying in I_0. Then u has a simple factor, v say. Since u is i.d., the element $u^{(1/n)}$ is in the semigroup for all positive integers n. If D is the Delphic homomorphism, we have $D(u^{(1/n)}) = (1/n).D(u)$. Since $D(v)$ is fixed, there is an m such that $D(u^{(1/m)}) < D(v)$, so that v cannot be a factor of $u^{(1/m)}$. Since $u = (u^{(1/m)})^m$, (Q) is contradicted.

We propose the following four possible axioms of factorization for a Delphic semigroup G. Let a, b be elements of G.

$(Q1)$ If every simple dividing a divides b, then a divides b;

$(Q2)$ If every integral simple-power dividing a divides b, then a divides b;

$(Q3)$ If every finite product of simples dividing a divides b, then a divides b;

$(Q4)$ If every product of simples dividing a divides b, then a divides b.

It is clear that if (Qi) holds in G then (Qj) also holds if $j > i$.

Proposition 13. *Not even (Q4) holds for $G = \mathscr{R}^+$.*

Proof by example. Let a and q be fixed so that $0 < a < 1$ and $0 < q < 1$. Let u and v in $\mathscr{R}^+$ be given by

$$u_n = (1 + (-1)^n q^{n+1})/(1+q) \quad \text{and} \quad v_n = a^n \quad \text{for all } n.$$

By Example 3 u is simple in $\mathscr{R}^+$; and $v(= v(\infty, a)$ of Section 2.3.8) is in $\mathscr{R}^+$ also. Let $u' = u.v$, and let the f-sequences of u, u' be f, f'. By looking at the generating functions of f, f' we find

(54) $\qquad f_1 = 1 - q, \quad f_2 = q \quad \text{and} \quad f_r = 0 \quad \text{for all } r > 2;$

(55) $\qquad f'_1 = a(1-q), \quad f'_2 = a^2 q \quad \text{and} \quad f'_r = 0 \quad \text{for all } r > 2.$

Since we assumed $a < 1$, and $u' = u.v$, it is clear that u' is not a factor of u. On the other hand, we shall show that every product of simples dividing u' also divides u.

Suppose we have a factorization of u': $u' = x.y$, the f-sequences of x and y being g, h. Since $g_1 > 0$ and $h_1 > 0$, $f'_r = 0$ for all $r > 2$ implies

(56) $$g_r = h_r = 0 \quad \text{for all } r > 2.$$

Also by direct computation

$$\begin{aligned}
f'_3 &= u'_3 - 2u'_1 u'_2 + u'^3_1 = x_3 y_3 - 2x_1 y_1 x_2 y_2 + x_1^3 y_1^3 \\
&= g_3 h_3 + 2g_1 g_2 h_3 + 2h_1 h_2 g_3 + g_1^3 h_3 + h_1^3 g_3 + 2g_1 g_2 h_1 h_2 \\
&= 2g_1 g_2 h_1 h_2 \quad \text{by equation (56)} \\
&= 0 \quad \text{by equation (55),}
\end{aligned}$$

so that at least one of g_2 and h_2 vanishes. They cannot both vanish, for then $x.y = u'$ would be of the form $v(\infty, p)$; so let $h_2 = 0$ and $g_2 \neq 0$. Put $b = h_1$ and $c = a/b$. Since $h_2 = 0$ we obtain at once from equation (55)

(57) $$g_1 h_1 = f'_1 = a(1-q) \quad \text{and} \quad g_2 h_1^2 = f'_2 = a^2 q,$$

so

$$g_1 = c(1-q) \quad \text{and} \quad g_2 = c^2 q.$$

Since g is the f-sequence of an element of $\mathcal{R}^+$, we have $g_1 + g_2 \leqslant 1$, which yields $c \leqslant 1$ (and so $b \geqslant a$). So y is of the form $v(\infty, b)$ for some $b \geqslant a$; and from equation (57) $x = u.v(\infty, c)$ for some $c \leqslant 1$. Now y is trivially not simple, and x is simple if and only if $c = 1$. So the unique simple factor of u' is u.

Hence, every product of simples s dividing u' must be an integral power of u: $s = u^r$ say. Let the f-sequence of u^r be $f(r)$. An elementary calculation shows that unless $r = 1$, $f_3(r) > 0$. Therefore by equation (56) if u^r divides u' then $r = 1$. Consequently the only product of simples s that divides u' is $s = u$, which implies that every product of simples dividing u' also divides u. But, as we noted earlier, u' does not itself divide u.

We can get round this difficulty, at any rate partially, by introducing a new Delphic semigroup that inherits most of the arithmetic of $\mathcal{R}^+$.

Let u be in $\mathcal{R}^+$ with f-sequence f. Define $q(u)$ by

(58) $$q = \sup(t : t \geqslant 1 \text{ and } \sum_{r=1}^{\infty} f_r t^r \leqslant 1).$$

It is standard from complex variable theory that

(59) $$\sum_{r=1}^{\infty} f_r q^r \leqslant 1,$$

(60) $$F(z) = \sum_{r=1}^{\infty} f_r z^r \quad \text{is regular for } |z| < q,$$

(61) $$|F(z)| < 1 \quad \text{for } |z| < q,$$

and either

$$(62) \qquad\qquad \sum_{r=1}^{\infty} f_r q^r = 1$$

or F has an essential singularity at the point $z = q$ (or possibly both).
  By formulae (60), (61) and (62),

$$U(z) = \sum_{n=0}^{\infty} u_n z^n = 1/(1 - F(z))$$

is regular for $|z| < q$ and has a singularity at the point $z = q$. Accordingly

$$(63) \qquad\qquad \liminf_{n \to \infty} (u_n)^{-(1/n)} = q.$$

But since $u_{m+n} \geqslant u_m . u_n$ for all $m, n \geqslant 0$, the sequence $(-\log(u_n))$ is sub-additive, from which it follows that the limit inferior in equation (63) is a genuine limit. For $|z| < q$, then, we have

$$\sum_{n=0}^{\infty} u_n z^n = 1 \Big/ \Big(1 - \sum_{r=1}^{\infty} f_r z^r \Big);$$

so if we put $w = z/q$, we have for $|w| < 1$

$$\sum_{n=0}^{\infty} (u_n q^n) . w^n = 1 \Big/ \Big(1 - \sum_{r=1}^{\infty} (f_r q^r) . w^r \Big).$$

In view of inequality (59) it follows that the sequence u^* given by $u_n^* = u_n q^n$ for all n is in $\mathscr{R}^+$, and by equation (58) this last statement holds for no larger q. So for each u in $\mathscr{R}^+$ there is a largest $q(u)$ such that u^* given by $u_n^* = u_n q^n$ is in $\mathscr{R}^+$, and

$$q(u) = \lim_{n \to \infty} (u_n)^{-(1/n)}.$$

Now we can put an equivalence relation $\mathscr{D}$ on $\mathscr{R}^+$ by saying $u\mathscr{D}v$ if and only if $u^* = v^*$. We can identify the quotient space with the set $\mathscr{R}^+/\mathscr{D}$ of those u in $\mathscr{R}^+$ with $q(u) = 1$; the topology of $\mathscr{R}^+/\mathscr{D}$ is to be that induced on it as a *subspace* of $\mathscr{R}^+$.

Proposition 14. *$\mathscr{R}^+/\mathscr{D}$ is an hereditary sub-semigroup of $\mathscr{R}^+$.*

Proof. If u and v lie in $\mathscr{R}^+/\mathscr{D}$, then

$$\lim_{n \to \infty} (u_n v_n)^{-(1/n)} = \lim_{n \to \infty} (u_n)^{-(1/n)} . \lim_{n \to \infty} (v_n)^{-(1/n)} = 1.1 = 1,$$

so $\mathscr{R}^+/\mathscr{D}$ is a sub-semigroup. Again, let u be in $\mathscr{R}^+/\mathscr{D}$ and let v be an

$\mathscr{R}^+$-factor of u. Then since $v_n \geqslant u_n$ for all n,

$$1 \leqslant \lim_{n \to \infty} (v_n)^{-(1/n)} \leqslant \lim_{n \to \infty} (u_n)^{-(1/n)} = 1,$$

so v is in $\mathscr{R}^+/\mathscr{D}$; it follows that $\mathscr{R}^+/\mathscr{D}$ is hereditary in $\mathscr{R}^+$.

Corollary. $\mathscr{R}^+/\mathscr{D}$, *with the subspace topology and the induced homomorphism, is Delphic.*

This follows from Proposition 2.

We call u in $\mathscr{R}^+$ geometric if u is of the form $v(\infty,p)$ with $0 < p \leqslant 1$, and Geometric if it is of the form $v(\infty,p)$ with $0 < p < 1$. Since $\mathscr{R}^+$ is Delphic every u in $\mathscr{R}^+$ has a representation $u = w . \prod u(r)$, where each $u(r)$ is simple and w is in I_0, the product being at most countable and possibly finite or void. By Theorem 4 w is geometric, so we may rewrite the representation as

$$(64) \qquad u = w . \prod_{r=1}^{\infty} u(r),$$

where w is geometric and each $u(r)$ is either simple or equal to the identity e.

Proposition 15. *u in $\mathscr{R}^+$ is in $\mathscr{R}^+/\mathscr{D}$ if and only if it has a representation* (64) *in which* $w = e$.

Proof. If u in $\mathscr{R}^+$ has a representation (64) in which $w \neq e$, then w is Geometric: $w_n = p^n$ for all n and some p such that $0 < p < 1$. Then $q(u) \geqslant 1/p > 1$ and so u is not in $\mathscr{R}^+/\mathscr{D}$. This proves the 'only if'.

Contrariwise, suppose we have

$$u = \prod_{r=1}^{\infty} u(r),$$

where each $u(r)$ is either simple or equal to e. Define for each $n \geqslant 1$

$$y_n = -(1/n) . \log(u_n),$$

so that $y_n \geqslant 0$ for all $n \geqslant 1$, and

$$\lim_{n \to \infty} y_n = \log(q(u))$$

exists and is finite. Define also for each n and $r \geqslant 1$, $y_n(r) = -(1/n) . \log(u_n(r))$. Again $y_n(r) \geqslant 0$ for all n and $r \geqslant 1$; so since for all $n \geqslant 1$, $u_n(r) \geqslant (u_1(r))^n$,

$$(65) \qquad 0 \leqslant y_n(r) \leqslant y_1(r) \quad \text{for all } n \text{ and } r \geqslant 1.$$

Since $u(r)$ for each r is either simple or equal to e, no representation (64) of $u(r)$ can have a Geometric component; whence

$$(66) \qquad \lim_{n \to \infty} y_n(r) = \log(q(u(r))) = 0.$$

Now

$$\sum_{r=1}^{\infty} y_1(r) = y_1 < \infty,$$

so by inequality (65) the functions $y_n(r)$ of r are dominated by the summable function $y_1(r)$ of r. Therefore by dominated convergence

$$\lim_{n\to\infty} y_n = \lim_{n\to\infty} \sum_{r=1}^{\infty} y_n(r) = \sum_{r=1}^{\infty} \lim_{n\to\infty} y_n(r) = 0$$

by equation (66), whence

$$q(u) = \exp\left(\lim_{n\to\infty} y_n\right) = 1$$

and so u is in $\mathscr{R}^+/\mathscr{D}$.

Corollary. *Suppose we have two representations* (64) *of an element u of $\mathscr{R}^+$:* $u = w.v = w'.v'$, *where v, v' are products of simples and w, w' are geometric. Then $w = w'$ and $v = v'$.*

Proof. The proposition tells us that $q(v) = q(v') = 1$, so since the q-functional is multiplicative we must have $q(w) = q(w')$. We know that there are p, p' in $[0, 1]$ such that $w = v(\infty, p)$ and $w' = v(\infty, p')$; and $q(w) = 1/p, q(w') = 1/p'$. So $q(w) = q(w')$ implies that $w = w'$ and so also $v = v'$.

This corollary tells us that $\mathscr{R}^+$ is the direct product of its two (hereditary, Delphic) sub-semigroups $\mathscr{R}^+/\mathscr{D}$ and $I_0(\mathscr{R}^+)$.

Since every member of $\mathscr{R}^+/\mathscr{D}$ is a (possibly empty) product of simples, the axiom $(Q4)$ defined at the start of this section holds trivially in $\mathscr{R}^+/\mathscr{D}$. Because the topology of $\mathscr{R}^+/\mathscr{D}$ is Hausdorff, the set of factors of any element of $\mathscr{R}^+/\mathscr{D}$ is closed, because compact. It follows at once that $(Q3)$ holds for $\mathscr{R}^+/\mathscr{D}$, again since every element of $\mathscr{R}^+/\mathscr{D}$ is a product of simples; so the factorization theory of $\mathscr{R}^+/\mathscr{D}$ is indeed better than that of $\mathscr{R}^+$. We do not know, however, whether $(Q1)$ or $(Q2)$ hold for $\mathscr{R}^+/\mathscr{D}$; but $(Q1)$ is unlikely to hold—for if it did, every square of a simple would have an alternative decomposition.

$\mathscr{R}^+$ has a cancellation law, since each u in $\mathscr{R}^+$ has all its terms positive; but larger sub-semigroups of $\mathscr{Y}$ (the semigroup of all renewal sequences) in general do not. Take for example $\mathscr{R}$, the semigroup of all aperiodic renewal sequences. It is easy to see, by the arguments of Proposition 7, that the u in $\mathscr{R}$ given by its f-sequence f,

$$f_1 = 0, \quad f_2 = f_3 = \tfrac{1}{2}; \quad f_r = 0 \quad \text{for all } r > 3,$$

is simple in $\mathscr{R}$. For suppose $u = v.w$; let the f-sequences of v, w be g, h. Since $f_1 = 0$ we have $g_1 = 0$ or $h_1 = 0$ or both. Suppose both. Then since $f_2 > 0$ and $f_3 > 0$ we must have all of g_2, h_2, g_3, h_3 positive. But then $f_5 > 0$, a contradiction. So suppose $g_1 = 0$ but $h_1 > 0$. Since $f_2 > 0$ and $f_3 > 0$ we get $g_2 > 0$ and $g_3 > 0$. For a non-trivial factorization there must be an $s > 1$ such that $h_s > 0$. If s is even, since $g_2 > 0$ we get $f_s > 0$; if s is odd, since $g_2 > 0$ and $h_1 > 0$ we get $f_{s+1} > 0$. So by assumptions on f we must have $s = 2$. But then we find $f_4 > 0$, a contradiction. So the only factorizations are the trivial ones $u = u.e = e.u$ and u must be simple.

Define $v(x)$ by $v_1(x) = x$; $v_n(x) = 1/(n+1)$ if $0 \leqslant n < \infty$ and $n \neq 1$. Then provided $\frac{4}{9} \leqslant x \leqslant 3\sqrt{3}/9$, $v(x)$ is a Kaluza-sequence and so is an i.d. element of $\mathscr{R}$. Further we have for all x, x' in the range given, $u.v(x) = u.v(x')$, so $\mathscr{R}$ has not a cancellation law.

2.3.11 A NON-EXISTENCE THEOREM FOR $\mathscr{R}$

Proposition 16. *If D is a continuous homomorphism from $\mathscr{R}$, the set of all aperiodic renewal sequences, to the reals under addition, then D is trivial: that is, $D(u) = 0$ for all u in $\mathscr{R}$.*

Proof. Of course $v(n, p)$ is in $\mathscr{R}$ for all positive integers n and all p such that $0 < p \leqslant 1$. So for each $n \geqslant 1$ there is a real constant a_n such that

$$(67) \qquad\qquad D(v(n, p)) = a_n . \log(p).$$

In this, let $p = \frac{1}{2}$ and let $n \to \infty$. Then the sequence $(v(n, \frac{1}{2}))$ converges to $v(\infty, \frac{1}{2})$; and so by the continuity of D at $v(\infty, \frac{1}{2})$

$$D(v(\infty, \tfrac{1}{2})) = \lim_{n \to \infty} D(v(n, \tfrac{1}{2})) = \log(\tfrac{1}{2}) . \lim_{n \to \infty} a_n,$$

so that

$$(68) \qquad\qquad \lim_{n \to \infty} a_n \text{ exists and is finite.}$$

From equation (67), if u is i.d. in $\mathscr{R}^+/\mathscr{D}$ we have

$$(69) \qquad D(u) = \sum_{n=1}^{\infty} a_n . \log(p_n), \quad \text{where} \quad u = \prod_{n=1}^{\infty} v(n, p_n).$$

Let us define $b_1 = a_2 - 2a_1$ and

$$b_r = a_{r+1} - 2a_r + a_{r-1} \quad \text{for all } r \geqslant 2.$$

Then if u is i.d. in $\mathscr{R}^+/\mathscr{D}$ we have, for all $n \geqslant 2$

$$(70) \quad -\sum_{r=1}^{n} a_r . \log(p_r) = \sum_{r=1}^{n} b_r . \log(u_r) + a_n . \log(u_{n+1}) - a_{n+1} . \log(u_n).$$

Let $\mathscr{Q}$ be the set of i.d. u in $\mathscr{R}^+/\mathscr{D}$ such that

$$\lim_{n \to \infty} u_n > 0.$$

If u is in $\mathscr{Q}$ we can let $n \to \infty$ in equation (70) to obtain from formulae (69) and (68)

(71) $$D(u) = -\sum_{n=1}^{\infty} b_n . \log(u_n).$$

Now $v(1, \frac{1}{2})$ is trivially in $\mathscr{Q}$; so evaluation of the sum (71) at $u = v(1, \frac{1}{2})$ shows that

$$\sum_{n=1}^{\infty} b_n$$

converges, perhaps conditionally.

Let $\mathscr{Q}/\mathscr{Q}$ be the set of u in $\mathscr{R}^+$ which are quotients of two members of $\mathscr{Q}$. Then u in $\mathscr{R}^+$ is in $\mathscr{Q}/\mathscr{Q}$ if

(72) $$\sum_{n=1}^{\infty} n . |\log(u_{n+1}) - 2 . \log(u_n) + \log(u_{n-1})| < \infty;$$

for let us write (for each $n \geqslant 1$) $q_n = \log(u_{n+1}) - 2 . \log(u_n) + \log(u_{n-1})$, and

$$q_n^+ = \max(0, q_n), \quad q_n^- = \max(0, -q_n).$$

Then we know that

$$\sum_{n=1}^{\infty} n . q_n^+ < \infty \quad \text{and} \quad \sum_{n=1}^{\infty} n . q_n^- < \infty;$$

and $q_n^+ - q_n^- = q_n$.

Define u^+ and u^- by $u_0^+ = u_0^- = 1$; when $n \geqslant 1$,

$$-\log(u_n^+) = \sum_{r=1}^{n} r . q_r^+ + n . \sum_{r=n+1}^{\infty} q_r^+$$

and similarly for u_n^-. Direct evaluation and the definitions of q_n^+, q_n^- show that u^+, u^- are in $\mathscr{Q}$ and u is their quotient.

We obviously have formula (71) for each u in $\mathscr{Q}/\mathscr{Q}$. Further it is clear that if u is in $\mathscr{R}^+$ and

$$L = \lim_{n \to \infty} u_n > 0$$

and there exist constants $K > 0$ and $d > 1$ such that, for all $n, |u_n - L| \leqslant K/d^n$, (72) is satisfied, and so u is in $\mathscr{Q}/\mathscr{Q}$. In particular let $\mathscr{S}^*\mathscr{R}^+$ be the set of u in $\mathscr{R}$ whose f-sequences f are such that

(a) $0 < f_1 < 1$,

(b) there is an N such that $f_n = 0$ for all $n > N$, and

(c) $\sum_{r=1}^{N} f_r = 1$.

Then by what we have just done, $\mathscr{S} * \mathscr{R}^+ \subset \mathscr{Q}/\mathscr{Q}$. But we saw in the proof of Proposition 9 that $\mathscr{S} * \mathscr{R}^+$ was dense in $\mathscr{Y}$ (and so also dense in $\mathscr{R}$), so $\mathscr{Q}/\mathscr{Q}$ is dense in $\mathscr{R}$. Thus if D vanishes on $\mathscr{Q}/\mathscr{Q}$ it is trivial.

If $b_n = 0$ for all n, D vanishes on $\mathscr{Q}/\mathscr{Q}$. So if D is non-trivial, there exists a least positive integer k such that $b_k \neq 0$; without loss of generality $b_k > 0$. We have of course $b_n = 0$ whenever $1 \leqslant n \leqslant k-1$.

Consider u in $\mathscr{R}$ defined by its f-sequence f:

$$f_r = 0 \quad \text{if } r \leqslant k \text{ or } r > 2k+1;$$

$$f_r = 1/(k+1) \quad \text{if } k+1 \leqslant r \leqslant 2k+1.$$

Then

$$L = \lim_{r \to \infty} u_r = 2/(3k+2) > 0,$$

and $u_n > 0$ for all $n \geqslant k+1$.

For each positive integer n define $u(n)$ in $\mathscr{R}^+/\mathscr{D}$ by its f-sequence $f(n)$:

$$f_1(n) = 2^{-n}; \quad f_r(n) = 0 \quad \text{if } 1 < r \leqslant k \text{ or } r > 2k+1;$$

$$f_r(n) = (1 - 2^{-n})/(k+1) \quad \text{if } k+1 \leqslant r \leqslant 2k+1.$$

Then

$$L_n = \lim_{r \to \infty} u_r(n) = 2/(3k(1 - 2^{-n}) + 2) > 0,$$

so that

(73) $4/(3k+4) \geqslant L_n \downarrow L = 2/(3k+2) \quad \text{as } n \to \infty.$

Let us write

$$u_r = L + v_r,$$

so that $v_r \to 0$ as $r \to \infty$ and

$$u_r(n) = L_n + v_r(n),$$

so that $v_r(n) \to 0$ as $r \to \infty$ for each n.

Put

$$F_n(z) = \sum_{r=1}^{2k+1} f_r(n) . z^r \quad \text{and} \quad F(z) = \sum_{r=1}^{2k+1} f_r . z^r,$$

and let

$$1 - F_n(z) = (1-z) . G_n(z) \quad \text{and} \quad 1 - F(z) = (1-z) . G(z).$$

Then G_n and G are polynomials of degree $2k$. By inspection of $F(z)$, we see that $G(z)$ does not vanish in or on the unit circle; so there are constants $m > 0$ and $d > 1$ such that

$$|G(z)| \geqslant m \quad \text{for all } z \text{ such that } |z| \leqslant d.$$

Define for all polynomials

$$A(z) = \sum_{r=0}^{2k} a_r \cdot z^r$$

of degree $2k$ with real coefficients

$$M(A) = \inf \left\{ \left| \sum_{r=0}^{2k} a_r \cdot z^r \right| : |z| \leqslant d \right\}.$$

Then M is a continuous function on the Euclidean space $(a_0, \ldots, a_{2k})$ of $(2k+1)$ dimensions; and $M(G) \geqslant m > 0$. So there is an $N \geqslant 1$ such that for all $n \geqslant N$, $M(G_n) \geqslant \frac{1}{2}m > 0$, since G_n converges to G in the Euclidean space as $n \to \infty$. Hence

(74)   There exist $d > 1$ and $M, m > 0$ and $N \geqslant 1$ such that
        for all $n \geqslant N$ and all $r \geqslant 1$, $\quad |v_r(n)| \leqslant (2M/m) \cdot d^{-r}$

by Cauchy's inequalities; for

$$\sum_{r=0}^{\infty} u_r(n) \cdot z^r = 1/(1 - F_n(z)) = L_n/(1-z) + H_n(z)/G_n(z),$$

where H_n is a polynomial of degree not greater than $2k-1$, and $H_n \to H$ (defined similarly for u) as $n \to \infty$.

Now provided $|z| \leqslant \frac{1}{2}$, $\log(1+z) = z \cdot g(z)$ where $|g(z)| < 2$. So whenever $v_r(n) \leqslant 1/(3k+2)$, formula (73) shows us that

(75) $$\log(u_r(n)) = \log(L_n) + (1/L_n) v_r(n) \cdot g_{r,n},$$

where $|g_{r,n}| < 2$.

In formula (74) now put $K = 2M/m$; and define the positive integer R by

$$R = \max(k+1, r_0),$$

where r_0 is the least integer r such that $K \cdot d^{-r} \leqslant 1/(3k+2)$. Then from formula (74) we have that there are $K > 0, d > 1, N \geqslant 1$ and $R \geqslant k+1$ such that for all $n \geqslant N$ and $r \geqslant R$

(76) $$|v_r(n)| \leqslant \min(1/(3k+2), K \cdot d^{-r}).$$

From formulae (76) and (75) we see that inequality (72) holds for $u(n)$ for all $n \geqslant N$, so the representation (71) holds for $u(n)$ for all $n \geqslant N$. Now consider

$$\lim_{n\to\infty} D(u(n)) = -\lim_{n\to\infty} \sum_{r=1}^{\infty} b_r . \log(u_r(n)).$$

We have

(77) $b_r = 0$ for $1 \leqslant r < k$;

(78) $b_k > 0$. So since $u_k(n) \to 0$, $-b_k . \log(u_k(n)) \to \infty$ as $n \to \infty$.

(79) When $k < r \leqslant R-1$, each $u_r(n)$ converges to the finite positive limit u_r, so the part of the sum from $k+1$ to $R-1$ converges.

When $r \geqslant R$, we can use formulae (76) and (75) to get

(80) $$\lim_{n\to\infty} \sum_{r=R}^{\infty} b_r \log(u_r(n)) = \lim_{n\to\infty} \log(L_n) \sum_{r=R}^{\infty} b_r + \lim_{n\to\infty} \sum_{r=R}^{\infty} b_r(1/L_n) v_r(n) g_{r,n}.$$

The first expression on the right-hand side of this converges to the finite limit

$$\log(L) . \sum_{r=R}^{\infty} b_r.$$

In the second expression, each individual term $b_r(1/L_n) v_r(n) g_{r,n}$ is dominated (in n) by

$$\tfrac{1}{2} . |b_r| . 2 . K . d^{-r} . (3k+2) = a_r, \quad \text{say;}$$

and

$$\sum_{r=R}^{\infty} a_r$$

converges (absolutely) since $d > 1$ and

$$\sum_{r=R}^{\infty} b_r$$

converges. Further, the term $b_r(1/L_n) v_r(n) g_{r,n}$ converges as $n \to \infty$ to the limit $b_r(\log(u_r) - \log(L))$, by equation (75) and the definitions of $u(n)$ and u; and this limit is finite, because $L > 0$ and $u_r > 0$ for $r \geqslant R (\geqslant k+1)$. So, by the theorem of dominated convergence, the second expression on the right-hand side of equation (80) converges, like the first, to a finite limit. From this, and formulae (77) and (79) we have that

$$\lim_{n\to\infty} \sum_{\substack{r=1 \\ r \neq k}}^{\infty} b_r . \log(u_r(n)) \quad \text{exists and is finite.}$$

But from formula (78)

$$\lim_{n \to \infty} b_k . \log(u_k(n)) = -\infty;$$

so since the sequence $(u(n))$ converges to u in $\mathscr{R}$ as $n \to \infty$, we see that D is discontinuous at u. Since the non-trivial homomorphism D was arbitrary, the proposition is proved.

REFERENCES

1. N. Bourbaki, *Éléments de Mathématique: Intégration*, Hermann, Paris, (1956).
2. E. Čech, *Topological Spaces*, Academia, Prague (1966).
3. D. Dugué, "Arithmetique des lois de probabilités", *Mémorial des Sciences Mathématiques* **137**, Gauthier-Villars, Paris (1957).
4. Revd. J. Hagen, "On division of series", *Am. J. Math.* **5** (1882), 236–237.
5. D. G. Kendall, "Renewal sequences and their arithmetic", *Symposium on Probability Methods in Analysis* (*Lecture Notes in Mathematics* **31**), Springer, Berlin, Heidelberg and New York (1967). (2.1 of this book.)
6. ——, "Delphic semigroups", *Bull. Am. math. Soc.* **73** (1) (1967), 120–121.
7. ——, "Delphic semigroups, infinitely divisible regenerative phenomena, and the arithmetic of p-functions", *Z. Wahrsch'theorie & verw. Geb.* **9** (1968), 163–195 (2.2 of this book.)
8. J. F. C. Kingman, "The stochastic theory of regenerative events", *Z. Wahrsch'theorie & verw. Geb.* **2** (1964), 180–224.
9. A. I. Markushevich, *Theory of Functions of a Complex Variable*, vol. 1, Prentice-Hall, Englewood Cliffs, N.J. (1965).
10. K. R. Parthasarathy, R. Ranga Rao, and S. R. S. Varadhan, "On the category of indecomposable distributions on topological groups", *Trans. Am. math. Soc.* **102** (1962), 200–217.
11. E. C. Titchmarsh, *The Theory of Functions*, Oxford University Press (1932).

2.4

Arithmetic and Other Properties of Certain Delphic Semigroups: II

R. DAVIDSON

[Reprinted from Z. Wahrsch'theorie & verw. Geb. **10** (1968), 146–172.]

2.4.1 SUMMARY

We show that the sets of infinitely divisible elements of the Delphic semigroups $\mathscr{R}^+$ (of positive renewal sequences) and $\mathscr{P}$ (of standard p-functions) are additively convex, and do a Choquet analysis in each case. We draw up the '(M, m) diagram' for members of $\mathscr{P}$, and deduce from it that the product topology on $\mathscr{P}$ is metrizable. Finally we look at the arithmetic of $\mathscr{P}$, showing that the simples are residual in it, and partially identifying 'I_0', the set of infinitely divisible elements without simple factors. Many examples are given.

2.4.2 INTRODUCTION

This paper is a continuation of our former paper [2]. In it we discuss some properties of the semigroups $\mathscr{R}^+$ of positive renewal sequences and $\mathscr{P}$ of the standard p-functions of Kingman [5]. A general introduction will be found in [2].

2.4.3 ADDITIVE CONVEXITY AND $\mathscr{R}^+$

We ask: Is it true that whenever u and v are in $\mathscr{R}^+$, so is w, where $w = a.u + b.v$ and $0 < a = 1 - b < 1$? The answer is No. For let u in $\mathscr{R}^+$ be given by

$$u_n = (1 + (-1)^n q^{n+1})/(1+q)$$

for all n and some fixed q such that $0 < q < 1$. Let $v = e$, the identity of $\mathscr{R}^+$,

150

so that $v_n = 1$ for all n. Put $w = \frac{1}{2}.u + \frac{1}{2}.v$; and let h be the f-sequence of w. Then $h_1 + h_2 = 1 + \frac{1}{4}q^2 > 1$, so w is not in $\mathcal{R}^+$. However, we have some positive results.

Proposition 1. *Let K be the set of i.d. elements of $\mathcal{R}^+$ together with the sequence z: $z_0 = 1$, $z_n = 0$ for all $n \geqslant 1$. Then K is additively convex.*

Proof. Let Q be the set of sequences u with $u_0 = 1$ and $0 < u_n \leqslant 1$ for all $n \geqslant 0$. For u in Q and all $n \geqslant 0$ set $r_n = u_n/u_{n+1}$, and let s_n, t_n be similarly defined for v, w in Q. Then (Kendall [3]) u in Q is i.d. in $\mathcal{R}^+$ if and only if $r_n \geqslant r_{n+1}$ for all n.

Now let v, w be in K, and let $u = a.v + b.w$, where $0 < a = 1 - b < 1$. If $v = w = z$ then $u = z \in K$. So suppose $w \neq z$; then u is i.d. in $\mathcal{R}^+$ if and only if for all $n \geqslant 0$

(1) $\qquad a^2(v_n v_{n+2} - v_{n+1}^2) + b^2(w_n w_{n+2} - w_{n+1}^2)$

$$+ ab(v_n w_{n+2} + w_n v_{n+2} - 2.v_{n+1} w_{n+1}) \geqslant 0.$$

If $v = z$ then the negative term in the third bracket of this vanishes; and the other two brackets are always non-negative, so inequality (1) holds and u is i.d. in $\mathcal{R}^+$. In the remaining case, $v \neq z$ and $w \neq z$, so that inequality (1) is equivalent to

(2) $\qquad a^2(s_n/s_{n+1} - 1) + b^2(t_n/t_{n+1} - 1) + ab(s_n/t_{n+1} + t_n/s_{n+1} - 2) \geqslant 0;$

this also to hold for all $n \geqslant 0$. Now the first two brackets of inequality (2) are non-negative since the sequences (s_n), (t_n) decrease; also

$$s_n/t_{n+1} - 1 \geqslant s_{n+1}/t_n - 1 \geqslant 1 - t_n/s_{n+1},$$

so that the third bracket is non-negative. So inequality (2) holds for all $n \geqslant 0$ and u is indeed in K.

The set K is multiplicatively, as well as additively, convex: that is, if u and v are in K then so is $u^a.v^b$ if $0 \leqslant a = 1 - b \leqslant 1$. Indeed, this was the analogy that prompted the present inquiry. Now u in $\mathcal{R}^+$ is i.d. if and only if $u^a.e^b$ is in $\mathcal{R}^+$ for all a and b such that $0 \leqslant a = 1 - b \leqslant 1$. Analogy suggests the proposition that 'u in $\mathcal{R}^+$ is i.d. if and only if $a.u + b.e$ is in $\mathcal{R}^+$ for all a and b such that $0 \leqslant a = 1 - b \leqslant 1$'. But (though by Proposition 1 the 'only if' statement holds) the 'if' statement does not hold, as the following example shows.

Example 1. *Let v and w be given by*

$$v_r = 2^{-r} \text{ if } r < 3; \quad v_r = 2^{-3} \text{ if } r \geqslant 3;$$

$$w_0 = 1; \quad w_1 = \tfrac{1}{3}; \quad w_r = \tfrac{4}{9} \text{ if } r \geqslant 2.$$

Then $v(= v(3, \frac{1}{2}))$ in the notation of Section 2.3.8 of [2]) is (i.d.) in $\mathscr{R}^+$, and we showed in Section 2.3.9 of [2] that w is also in $\mathscr{R}^+$. Define $u = v.w$. Then

$$u_0 = 1, \quad u_1 = \tfrac{1}{6}, \quad u_2 = \tfrac{1}{9} \quad and \quad u_r = \tfrac{1}{18} \quad for \ all \ r \geqslant 3.$$

So u, in $\mathscr{R}^+$, is not i.d., since

$$u_1.u_3 = \tfrac{1}{108} < \tfrac{1}{81} = u_2^2.$$

For all a and b such that $0 \leqslant a = 1 - b \leqslant 1$ define $u_a = a.e + b.u$; and put $c = b/18$, so that

$$u_a = (1, a + 3c, a + 2c, a + c, a + c, a + c, \ldots).$$

Define constants p, q and r by

$$p = (a + c)/(a + 2c), \quad q = (a + 3c)(a + 2c)/(a + c)$$

and

$$r = (a + 2c)^2/((a + c)(a + 3c)).$$

Then provided $0 \leqslant c \leqslant \tfrac{1}{18}$ (which we have since $0 \leqslant b \leqslant 1$), we have $p \leqslant 1$, $q \leqslant 1$ and $r \geqslant 1$. Further, $u_a = v(3, p).w_a$, where

$$w_a = (1, q, qr, qr, qr, qr, \ldots).$$

Of course $v(3, p)$ is in $\mathscr{R}^+$; by the methods of Section 2.3.9 of [2], w_a is in $\mathscr{R}^+$ if and only if $2\sqrt{(qr)} \leqslant 1 + q$, that is, if and only if

$$(18c - 1)(50c - 3) \geqslant 0.$$

But this last is satisfied because $0 \leqslant c \leqslant \tfrac{1}{18}$, so w_a—and hence u_a also—is in $\mathscr{R}^+$. We have therefore found a u such that $a.e + b.u$ is in $\mathscr{R}^+$ for all a and b such that $0 \leqslant a = 1 - b \leqslant 1$ although u is not i.d.

Proposition 1 showed that K was convex under addition. Now K is a closed subset of a compact metric space (the space of all real sequences bounded by zero and unity with the product topology) and is thus compact and metrizable. So K is a compact convex metrizable subset of the locally convex topological vector space of all real sequences under addition and the topology of pointwise convergence. According to Choquet's theorem, then, each u in K can be represented by a probability measure on K which assigns mass zero to the non-extremal elements of K. Interest thus centres on the extreme points of K.

Theorem 1. *(i) z is extremal.*

(ii) Let $u \neq z$. Define $r_{-1} = \infty$. Then u is extremal if and only if condition (3) below holds for all $n \geqslant 0$:

$$(3) \qquad\qquad r_n > r_{n+1} \ implies \ r_{n-1} = r_n \quad and \quad r_{n+1} = r_{n+2}.$$

Proof. (i) is trivial, since for every u in K, $u_n \geqslant 0$ for all $n \geqslant 0$. So we turn to (ii). Suppose condition (3) holds for all $n \geqslant 0$, and that $u = a.v + b.w$, with $0 < a = 1 - b < 1$. First suppose that exactly one of v, w (v, say) is in fact z. Then we have

$$u_0 = w_0 = 1; \quad u_n = b.w_n < w_n \quad \text{for all } n \geqslant 1.$$

Thus

$$r_{-1} = \infty > r_0 = b^{-1}.t_0 > t_0 \geqslant t_1 = r_1,$$

which contradicts condition (3) for $n = 0$.

So we can assume that $v \neq z$ and $w \neq z$. Then all r_n, s_n, t_n are strictly positive and finite for $n \geqslant 0$, so that u is in K if and only if inequality (2) holds for each $n \geqslant 0$. Now if for some n it should happen that $r_n = r_{n+1} = r$ say, formula (2) holds with equality (for that n) and we easily find that

$$s_n = s_{n+1} = t_n = t_{n+1} = r \quad \text{also.}$$

But condition (3) implies that $r_n = r_{n+1}$ often enough (as n runs through the non-negative integers) to ensure, by what we have just done, that $s_m = t_m = r_m$ for every non-negative integer m. Since $v_0 = w_0 = u_0 = 1$ we have at once

$$v_n = w_n = u_n \quad \text{for all } n \geqslant 0.$$

So every decomposition of u is trivial and u is therefore extremal.

Conversely, suppose that condition (3) does not hold for some $n \geqslant 0$. Suppose then that $n = N$ is the least $n \geqslant -1$ such that $r_n > r_{n+1} > r_{n+2}$. We have two cases:

(a) $N = -1$. Define w by

$$w_0 = 1, \quad w_1 = 1/r_1 \quad \text{and} \quad w_{n+1} = w_n/r_n \quad \text{for all } n \geqslant 1.$$

Then $t_0 = r_1$, and $t_n = r_n$ for all $n \geqslant 1$. So the sequence (t_n) decreases, and each $t_n \geqslant 1$, since the same is true of (r_n). Accordingly w is i.d. in R^+. It is easy to check that

$$(1 - r_1/r_0).z + (r_1/r_0).w = u \quad \text{and} \quad 0 < r_1/r_0 < 1.$$

Now $u \neq z$, so this is a non-trivial decomposition and u is not extremal.

(b) $N \geqslant 0$. Define v and w by

$$v_n = w_n = u_n \quad \text{for } n \leqslant N+1.$$

When $n > N+1$, let

$$v_n = x.u_n \quad \text{and} \quad w_n = y.u_n,$$

where x and y are defined by

$$r_{N+1} = x.r_N = y.r_{N+2}.$$

Clearly $0 < x < 1 < y < \infty$. If we put $a = (y-1)/(y-x)$ and $b = (1-x)/(y-x)$, then $0 < a = 1-b < 1$ and $u = a.v + b.w$; and the decomposition is non-trivial since $x < 1 < y$. All we have to do now is check that v and w are in K. We have from the definitions of v and w that

$$s_n = t_n = r_n \quad \text{for all } n \leqslant N \text{ and all } n \geqslant N+2;$$

$$s_{N+1} = r_N \quad \text{and} \quad t_{N+1} = r_{N+2}.$$

Because the sequence (r_n) is decreasing and all the r_n are not less than unity, the same is true of (s_n) and (t_n), so v and w are in K.

We have in this proof instituted a well-defined system of decomposition of the non-extremal members of K: the system removes one point of non-satisfaction of condition (3) at a time. But we have

Proposition 2. *The Choquet representation of u in K is not in general unique.*

Proof by example. Let $0 < a < 1, 0 < b < 1$ and $a \neq b$. Define u in K by

$$u_n = \tfrac{1}{2}(a^n + b^n) \quad \text{for } n \leqslant 2; \quad u_n = \tfrac{1}{2}(a^2 + b^2) \quad \text{for } n > 2.$$

Define the member $u(x)\,(0 \leqslant x \leqslant 1)$ of K by

$$u_n(x) = x^n \quad \text{for } n \leqslant 2; \quad u_n(x) = x^2 \quad \text{for } n > 2.$$

Then we have

$$u = \tfrac{1}{2}.u(a) + \tfrac{1}{2}.u(b)$$
$$= \tfrac{1}{2}((a+b)^2/(a^2+b^2)).u((a^2+b^2)/(a+b)) + \tfrac{1}{2}((a-b)^2/(a^2+b^2)).u(0).$$

All the sequences on the right-hand side of this are different; and $u(x)$ is extremal in K when $0 \leqslant x \leqslant 1$, from the theorem. So the proposition is proved.

2.4.4 ADDITIVE CONVEXITY AND THE KINGMAN SEMIGROUP $\mathscr{P}$

Let $\mathscr{G}$ be the set of all p-functions (see Kingman [5]), and let $\mathscr{P}$ be the set of standard ones—that is, the set of those continuous at the point $t = 0$ (and so everywhere). $\mathfrak{R}^+$ will mean the non-negative half-line.

p, q, r etc. will be members of $\mathscr{P}$ or $\mathscr{G}$ as the context demands; their x-functions x, y, z etc. are defined by $x(t) = -\log(p(t))$ for all t in $\mathfrak{R}^+$ (and similarly for y and z). If p is in $\mathscr{P}$ then x will be finite-valued, but if p is in $\mathscr{G}$ then x may take the value $+\infty$. $\mathscr{G}$ will be regarded here as a subset of the topological space (X, T), where X is the set of all real-valued functions on $\mathfrak{R}^+$ and T is the topology of pointwise convergence. (X, T) is of course

Hausdorff and $\mathscr{G}$ is a closed subset of the set Y of all functions on $\mathfrak{R}^+$ bounded by zero and unity (Kingman [5]). By Tychonov's theorem Y, and hence $\mathscr{G}$, is compact in (X, T).

Let K be the set of all i.d. elements of $\mathscr{P}$ (which is a semigroup under pointwise multiplication of the p-functions, and indeed Delphic (Kendall [4])). Then p in $\mathscr{P}$ lies in K if and only if its x-function x satisfies (Kendall [4])

(4)
$$\begin{cases} (i) \ \ 0 \leqslant x(t) < \infty \quad \text{for all } t \geqslant 0; \quad x(0) = 0; \\ (ii) \ \ x \text{ is continuous and concave (and so increasing).} \end{cases}$$

We want to know what is the closure, H say, of K in (X, T). So suppose that we have a net (p_a) with p_a in K for each a in the directed set A, and that this net T-converges to some p in X. At once, p is in $\mathscr{G}$, and we have

(5)
$$\begin{cases} (i') \ \ p_a(0) = 1 \quad \text{and} \quad 0 < p_a(t) \leqslant 1 \quad \text{for all } t \text{ in } \mathscr{R}^+; \\ (ii') \ \ p_a \text{ is continuous and convex (and so decreasing).} \end{cases}$$

(These conditions are derived from conditions (4) but are not equivalent to them, in the sense that there are p in $\mathscr{P}$ satisfying conditions (5) but not conditions (4): see Example 3.)

It follows from conditions (5) that the limit p of our net has

$$p(0) = 1, \quad 0 \leqslant p(t) \leqslant 1$$

for all t in $\mathfrak{R}^+$, and $p(s) \geqslant p(t)$ if $s \leqslant t$. Also p is convex; so since p is measurable (because decreasing) it must be continuous save perhaps at $t = 0$. Also if $p(s) = 0$ for some s, because p is decreasing it vanishes for $t \geqslant s$. Thus either $p(t) = 0$ for all $t > 0$, or $p(t) > 0$ for all $t \geqslant 0$, or $p(t) = 0$ if and only if $t \geqslant s > 0$. We show that the last is impossible. Define $w = p(\frac{1}{2}s)$; then $w > 0$, and we know that $p(0) = 1$ and $p(s) = 0$. For all a in A and all positive integers n define $q_{a,n}$ in K by $p_a(t) = (q_{a,n}(t))^n$ for all t in $\mathfrak{R}^+$; and let $p(t) = (q_n(t))^n$ similarly define q_n. Then the net $(q_{a\,n})$ T-converges to q, which is accordingly convex as a function of t. So for all positive integers n we have

$$w^{1/n} \leqslant \tfrac{1}{2}(1 + 0) = \tfrac{1}{2},$$

which as $w > 0$ is nonsense. So it is impossible that $p(t) = 0$ if and only if $t \geqslant s > 0$. On the other hand, if we put $p_a(t) = e^{-at}$ and let $a \to \infty$, we have a net (p_a) converging to p^*, where $p^*(t) = 0$ for all $t > 0$; $p^*(0) = 1$. So p^* is in H.

If we have $p(t) > 0$ for all $t \geqslant 0$, the x-function x of p is finite for all $t \geqslant 0$, so that the net (x_a) of the x-functions of (p_a) converges to x pointwise. So from conditions (4) x is concave, and continuous save perhaps at $t = 0$;

$x(0) = 0$ and x is non-negative on $\Re^+$. Hence

$$x(0) = 0; \quad x(t) = k + x'(t) \quad \text{for } t > 0,$$

where the constant k is non-negative, and $p'(= e^{-x})$ is in K. If $k = 0$, p is itself in K. If $k > 0$,

$$p(t) = e^{-x(t)} = 1 \quad \text{if } t = 0;$$

$$= e^{-k}.p'(t) \quad \text{if } t > 0.$$

Thus whether $k = 0$ or $k > 0$, we have

(*) $$p = (1 - e^{-k}).p^* + e^{-k}.p',$$

where p' is in K and p^* was defined above. The above form is thus necessary for p to lie in H. It also sufficient; for if we put

$$p_a(t) = \max(p(t), e^{-at}),$$

where p is of the form (*), and let $a \to \infty$, we have a net (p_a) of members of K, T-converging to p. Hence

$$H = (a.K + b.p^*), \quad 0 \leqslant a = 1 - b \leqslant 1.$$

(X, T) is a locally convex topological vector space under addition (Robertson and Robertson [9]). We can now state

Proposition 3. *H is compact convex and metrizable in* (X, T).

Proof. Compactness is evident since H is a closed subset of the compact Hausdorff space $\mathscr{G}$. On H, the topology T can be identified with the topology R of convergence on the rationals; R is a topology, and T is identical to it, since all the elements of H are monotone and continuous save perhaps at $t = 0$. But (H, R) is a subset of a countable product of unit cubes, so (H, T) is metrizable. As to convexity, a function p on $\Re^+$ is in K if and only if p is continuous and each discrete skeleton of p is a Kaluza-sequence (Kendall [4]). (By a Kaluza-sequence we mean a sequence (u_r) with $u_0 = 1, 0 \leqslant u_r \leqslant 1$ for all $r > 0$, and

$$u_r.u_{r+2} \geqslant u_{r+1}^2$$

for all $r \geqslant 0$.) Obviously an additive convex combination of two continuous functions is continuous, and (Proposition 1) we know that the set of all Kaluza-sequences is additively convex. So K is convex under addition. But H is the closure of K in the topological vector space (X, T), so H must itself be additively convex.

For any convex set Z let $ex Z$ be the set of extremals of Z. From what we have just done and Choquet's theorem, we have for every p in H a representation

$$p(t) = \int p_w(t) . dm(w),$$

where m is a probability measure on H with $m(H - \text{ex } H) = 0$.

Now that part of ex H which does not lie in K is precisely p^*: $p^*(0) = 1$, $p^*(t) = 0$ for all $t > 0$. Clearly p^* is in ex H, for all p-functions are non-negative; p^* is not in K since it is discontinuous at $t = 0$. But any p in H determines p' in K and c in $[0, 1]$ such that $p = c.p^* + (1 - c).p'$ (by our characterization of H in terms of K). So there can be no more extremals of H not lying in K.

Any Choquet representation of p in K in terms of ex H must concentrate all its mass on that part of ex H lying in K, else p would be discontinuous at $t = 0$. Now consider ex K; clearly ex K contains that part of ex H lying in K. But no element p of K can be a convex combination of two elements of H one or both of which is in $H - K$, again by the continuity of p at $t = 0$; so every element of ex K is also in ex H. So ex $K = $ ex $H - \{p^*\}$, and every element p of K has a Choquet representation in terms of elements of ex K:

$$p(t) = \int p_w(t) . dm(w),$$

where m is a probability measure on K with $m(K - \text{ex } K) = 0$.

For the rest of this section we shall be exclusively concerned with K. We want to find the set ex K, and shall be considering non-trivial decompositions of an element p of K: $p = a.q + b.r$, with $0 < a = 1 - b < 1$.

Because the x-function x of p is continuous concave and non-negative, it is absolutely continuous on $\mathfrak{R}^+$ and is thus the indefinite integral of its a.e. existing derivative x'. The right-hand derivative x'_+ of x exists everywhere (it may take the value $+\infty$ at $t = 0$): let us write $f = x'_+$ (g, h similarly defined for q, r). Then x is the indefinite integral of f.

A function f on $\mathfrak{R}^+$ is the f-function of a p in K if and only if

(6) $\begin{cases} f(t) \geqslant 0 \text{ for all } t \geqslant 0; f \text{ is decreasing and right-} \\ \text{continuous in } t; f \text{ is integrable over } [0, 1]. \end{cases}$

These conditions clearly ensure that f is finite save perhaps at $t = 0$. If the conditions hold, we see that x defined by

$$x(t) = \int_0^t f(s) \, ds$$

satisfies our conditions (4) for the x-function of a p in K; conversely the conditions (6) are easily verified for any f corresponding to a p in K.

p, q, r are of course differentiable from the right and

$$p'_+(t) = -f(t) \cdot p(t),$$

and similarly for q'_+, r'_+. Since $x = -\log(a \cdot q + b \cdot r)$ we have everywhere

$$f = (aqg + brh)/(aq + br).$$

In this expression only a and b are constants. Now f, g, h are a.e. differentiable with non-positive derivative (see Natanson [7]), and their derivatives are measurable. Except then on some exceptional set E of measure zero we have

$$(7) \quad \begin{cases} f'_+ \leqslant 0, g'_+ \leqslant 0, h'_+ \leqslant 0 \quad \text{and} \\ p^2 \cdot f'_+ = a^2 q^2 \cdot g'_+ + b^2 r^2 \cdot h'_+ + abqr(g'_+ + h'_+ - (g-h)^2). \end{cases}$$

Let S be a measurable set included in some interval I of the real line whose length mI may be finite or not. S is said to be absolutely dense (in I) if $S - N$ is dense in I whatever be the set N of measure zero. Equivalently, $(I - S) \cup N$ contains no interval for any set N of measure zero. (Kendall remarks that S is absolutely dense in I if and only if S meets every non-trivial subinterval of I in a set of positive measure.) Obviously if $I - S$ is of measure zero then S is absolutely dense in I; but the definition is non-trivial, for one can construct a pseudo-Cantor set of measure $\frac{1}{2}$ in $[0, 1]$ that is absolutely dense in $[0, 1]$.

For p in K define for all positive integers n

$$S_n = (t: f'_+(t) \text{ exists and does not exceed } -1/n).$$

There are precisely two possibilities:

(a) For all n, $\Re^+ - S_n$ is absolutely dense in $\Re^+$; or

(b) there is an integer n and a set N of measure zero such that $S_n \cup N$ contains a non-trivial interval.

Theorem 2. *In case* (a), p *is in* ex K; *in case* (b), *it is not.*

Proof. Suppose (a) holds. Put $T_n = \Re^+ - S_n$, so that T_n is absolutely dense in $\Re^+$ for every n. By definition of S_n we have for each n

$$(8) \quad \begin{cases} -1/n < f'_+ \text{ a.e. on } T_n, \text{ and so by condition (7)} \\ -1/n < p^{-2}(a^2 q^2 g'_+ + b^2 r^2 h'_+ + abqr(g'_+ + h'_+ - (g-h)^2)); \\ g'_+ \text{ and } h'_+ \text{ exist and are non-positive a.e. on } T_n. \end{cases}$$

Since T_n is absolutely dense on $\Re^+$, condition (8) holds, for every n, on a set T_n^* dense in $\Re^+$. For each positive integer k let I_k be the interval

$[0, k]$. On I_k, p, q and r are all bounded away from zero; it follows from condition (8) that there is a constant $A(k)$ such that for all n

(9) $(g-h)^2 \leqslant A(k)/n$ for all t in a set $T_n^* \cap I_k$ dense in I_k.

But g and h are decreasing and continuous from the right, so the satisfaction of condition (9) for all n requires that they be equal everywhere on I_k for each k. So $g(t) = h(t)$ for all t in $\mathfrak{R}^+$. Therefore $q = r$; their common value must be p; so every decomposition of p is trivial and p must be in ex K.

Conversely, suppose (*b*) holds. Then there is a constant $c > 0$ and an interval $[A, B] (A < B)$ such that a.e. on $[A, B]$ f_+' exists and does not exceed $-c$. We can take $0 < A < B < \infty$ without loss of generality.

Now we can write $f = f_1 + f_2$, where f_1 and f_2 are the *f*-functions of some p_1, p_2 in K, f_1 is absolutely continuous in every interval $[s, \infty] (s > 0)$, and f_2' vanishes a.e.: this is just the splitting of a measure into its absolutely continuous and non-absolutely-continuous parts. In terms of the *p*-functions involved we have $p = p_1 . p_2$; so if we show that p_1 is not extremal we shall have that p is not extremal either. In this decomposition of f we have $f_1' = f_2'$ a.e., so $f_1' \leqslant -c$ a.e. on $[A, B]$ and f_1 is the integral of its derivative. Again, we can write $f_1 = f_3 + f_4$, where f_3 and f_4 are absolutely continuous *f*-functions of p_3, p_4 in K, $f_4' \leqslant 0$ a.e., and

$$f_3'(t) = -c \quad \text{if } t \text{ lies in } [A, B]; \quad f_3'(t) = 0 \quad \text{otherwise.}$$

We have now $p = p_3 . p_4 . p_2$, and we shall show that p_3 has a non-trivial decomposition. p_3 will have $f_3(\infty)$ (this limit exists since f_3 is non-negative and decreasing) non-negative. Define p_5 by

$$p_3(t) = p_5(t) . \exp(-t . f_3(\infty)),$$

so that p_5 is in K (this is easy to check through the *f*-functions) and $f_5(\infty) = 0$. Also $f_5'(t) = -c$ if $A \leqslant t \leqslant B$; $f_5'(t) = 0$ otherwise. Let us now drop the subscript $_5$, so that we have

(10) $\begin{cases} p(t) = \exp(-x(t)), \quad \text{where } x(0) = 0 \quad \text{and} \\ x''(t) = -c \quad \text{if } A \leqslant t < B; \quad x''(t) = 0 \quad \text{if } t < A \text{ or } t \geqslant B; \\ x'(t) = c(B-A) \quad \text{if } t < A; \quad x'(t) = c(B-t) \quad \text{if } A \leqslant t < B, \quad \text{and} \\ x'(t) = 0 \quad \text{if } t \geqslant B. \end{cases}$

We now abandon all the notation of f, g, h, q, r, y, z so as to be able to bring in adequate new notation.

We want to prove that there are q and r, functions on $\mathfrak{R}^+$, such that

(11) $p = \frac{1}{2} . pq + \frac{1}{2} . pr$, pq and pr are in K, and

          neither of pq, pr is p itself.

Let y be that twice differentiable function on $\Re^+$ given by

$$y(0) = y'(0) = 0;$$

(12)
$$\begin{cases} y''(t) = y'(t) = 0 & \text{if } t < A \text{ or } t \geqslant B, \\ y''(t) = d \quad \text{and} \quad y'(t) = d(t-A) & \text{if } A \leqslant t < \tfrac{1}{2}(A+B), \\ y''(t) = -d \quad \text{and} \quad y'(t) = d(B-t) & \text{if } \tfrac{1}{2}(A+B) \leqslant t < B, \end{cases}$$

where d is the positive root of the equation

(13)
$$\tfrac{1}{2} . d^2 (B-A)^2 + d = c.$$

Define the function z by $z(t) = -\log(2 - e^{-y(t)})$ and the functions q and r by $q(t) = \exp(-y(t))$, $r(t) = \exp(-z(t))$. It is clear that the first and third conditions of (11) are satisfied, so all we have to do is check that $x+y$ and $x+z$ satisfy the conditions (4) for x-functions of members of K.

First, $x+y$ and $x+z$ are trivially continuous; and it is obvious from the definitions of x, y, z that they all vanish at $t = 0$, so that $x+y$ and $x+z$ vanish there also. That $x+y$ and $x+z$ are non-negative for all $t \geqslant 0$ will follow if we show that their derivatives are non-negative for all $t \geqslant 0$. By condition (10), x'; and by condition (12), y', is non-negative, so their sum is non-negative also. Now

$$z'(t) = -y'(t) . \exp(-y(t)) / (2 - \exp(-y(t))),$$

so since $\exp(-y(t)) \leqslant 1$ for all $t \geqslant 0$ we have

(14)
$$|z'(t)| \leqslant |y'(t)| \quad \text{for all } t \text{ in } \Re^+.$$

From equation (13), $d < c$, so by conditions (10) and (12) $|y'(t)| \leqslant |x'(t)|$ for all t in $\Re^+$; this, inequality (14) and $x'(t) \geqslant 0$ imply $(x' + z')(t) \geqslant 0$ for all t in $\Re^+$.

It remains only to show that $x+y$ and $x+z$ are concave. Now formulae (10), (12) and (13) show that $(x'' + y'')(t) \leqslant 0$ for all t in $\Re^+$, so $x+y$ is concave. On the other hand, we have for all t in $\Re^+$

(15)
$$z''(t) = \frac{2 . y'^2(t) . e^{-y(t)}}{(2 - e^{-y(t)})^2} - \frac{y''(t) . e^{-y(t)}}{2 - e^{-y(t)}} .$$

This and formula (12) show that $z''(t) = 0$ if $t < A$ or $B \leqslant t$. If $A \leqslant t < B$, since $\exp(-y(t)) \leqslant 1$ for all t in $\Re^+$ we have by equation (15)

$$z''(t) \leqslant 2 . \max_{[A,B]} (y'^2) + \max_{[A,B]} (y'')$$

$$\leqslant c, \quad \text{by equations (12) and (13).}$$

Using condition (10) we see now that $(x'' + z'')(t) \leqslant 0$ for all t in $\Re^+$, so

$x+z$ is concave. By the definitions of q and r in terms of y and z we see that the second condition of (11) is satisfied as well as the first and third; so p is not in ex K.

Proposition 4. *The Choquet representation is not in general unique.*

Proof by example. Define p_3, p_4 in K by their x-functions x_3, x_4:

$$x_3(0) = x_4(0) = x_3'(\infty) = x_4'(\infty) = 0;$$

$$x_3''(t) = -e^t/(e^t+1)^2 + e^3/(e^3+1)^2 \quad \text{if } 1 \leqslant t < 3$$

$$= -e^t/(e^t+1)^2 \quad \text{if } t<1 \text{ or } t \geqslant 3;$$

$$x_4''(t) = -e^3/(e^3+1)^2 \quad \text{if } 1 \leqslant t < 3$$

$$= 0 \quad \text{if } t<1 \text{ or } t \geqslant 3.$$

It is easily checked that these candidate x-functions satisfy the conditions (4), so that p_3 and p_4 really are in K. Define p in K by $p = p_3 \cdot p_4$. Then it is easy to verify that

$$(16) \qquad p(t) = \tfrac{1}{2}.1 + \tfrac{1}{2}.e^{-t} \quad \text{for all } t \geqslant 0.$$

But the p_* and p_{**} given by

$$(17) \qquad p_*(t) = 1 \quad \text{and} \quad p_{**}(t) = e^{-t} \quad \text{for all } t \geqslant 0$$

are extremal, so equation (16) is a Choquet representation of p.

We look now at p_4. We can use the method of the theorem to produce a decomposition $p_4 = \tfrac{1}{2}.p_1 + \tfrac{1}{2}.p_2$, where (as we see from conditions (10) and (12) applied to p_4 and $[A, B] = [1, 3]$)

$$(18) \qquad x_1'', \text{ existing for all } t \geqslant 0, \text{ is discontinuous at } t = 2.$$

We have, from the definitions of p_i $(1 \leqslant i \leqslant 4)$, that

$$(19) \qquad p = p_3(\tfrac{1}{2}.p_1 + \tfrac{1}{2}.p_2) = \tfrac{1}{2}.p_1 p_3 + \tfrac{1}{2}.p_3 p_2.$$

Let us put $q = p_1 \cdot p_3$. Then from formula (18) and the definition of p_3, q is everywhere twice differentiable and

$$q'' = p_1'' \cdot p_3 + 2 p_1' \cdot p_3' + p_1 \cdot p_3''.$$

The last two terms of this are continuous at $t = 2$, again by formula (18) and the definition of x_3'', p_3 is also continuous, and non-zero at $t = 2$. But p_1'' has a discontinuity at $t = 2$ by formula (18), so q'' must be discontinuous at $t = 2$. Thus q cannot be completely monotone—for if it were, it would be infinitely differentiable—and so, *a fortiori*, q has no Choquet representation in terms of p_* and p_{**} alone. Equation (19) then shows that p has a Choquet representation which is not equivalent to equation (16).

Proposition 5. ex H *is dense in* H.

Proof. Let p in K have x-function x, and define p_n for each positive integer n by its x-function x_n, as follows

For every non-negative integer k, x_n is to be linear on the segment

$$[k.2^{-n}, (k+1).2^{-n}]$$

and is to be equal to x at the end-points of the segment. By the theorem, p_n is in ex K for every n; and we showed earlier that ex $K \subset$ ex H, so p_n is in ex H for every n. Since x is continuous on $\Re^+$, the definition of x_n shows that $x_n(t) \rightarrow x(t)$ for each $t \geqslant 0$. Taking exponentials we see that (p_n) T-converges to p. So ex H is dense in K. But H was the T-closure of K, so ex H must be dense in H.

Corollary. ex H *is residual in* H.

For by Proposition 3, H is a compact metrizable space: and ex H is a G_δ in it (see Phelps [8]). The present proposition and Baire's category theorem yield the corollary.

All this additive convexity has a probabilistic interpretation: if p is in H, then $p(t)$ $(t \geqslant 0)$ is the probability that a certain stationary point-process on the line will put no points in a (given) interval of length t (this is due to Kendall). Taking additive convex combinations of the p then corresponds to mixing the processes. However, the process is not in general determined by the p-function.

2.4.5 THE (M, m) DIAGRAM AND SOME APPLICATIONS OF IT

For each p in $\mathscr{P}$ put $M(p) = p(1)$ and $m(p) = \inf\{p(t): 0 \leqslant t \leqslant 1\}$. In this section we investigate the possible pairs of values (M, m).

Proposition 6. *If* $M(p) > \frac{3}{4}$, *then* $m(p) \geqslant \frac{1}{2}(1 + \sqrt{(4M(p) - 3)})$.

Proof. (We omit explicit mention of p in M and m.) We have the Kingman inequality [5]: for all s and $t \geqslant 0$

$$(20) \qquad\qquad p(s+t) \leqslant 1 + p(s)p(t) - p(s).$$

Let $s + t = 1$ and put first $s = u$ and then $s = 1 - u$ in the inequality. Eliminating terms in $p(1 - u)$ from the resulting pair of inequalities we find

$$(21) \qquad\qquad 1 - M \geqslant p(u).(1 - p(u))$$

for all u such that $0 \leqslant u \leqslant 1$. But now $M > \frac{3}{4}$, and p is continuous in u, so since $p(0) = 1$ we must have

(22) $$p(u) \geqslant \tfrac{1}{2}(1 + \sqrt{(4M-3)})$$

for all u such that $0 \leqslant u \leqslant 1$.

This is a statement about what a p in $\mathscr{P}$ cannot do. It is satisfying to know something about what may happen.

Example 2. *Let* $e^{-1} < x \leqslant 1$. *Then there is a p in $\mathscr{P}$ with $M = x$ and $m = y$, where* $y = 1 + \log(x)$.

First, it is clear that if $x = 1$, $p(t) = 1$ will suffice for the example, so we can assume from now on that $e^{-1} < x < 1$. Define then

(23) $$z = -\log(y), \quad \text{so that} \quad x = \exp(-(1 - e^{-z}));$$

we observe that there is a biunique correspondence between z in $]0, \infty[$ and x in $]e^{-1}, 1[$. Define

(24) $$a = 1 + z - e^{-z},$$

so that $a > 0$; and define

(25) $$b = z/a = z/(1 + z - e^{-z}),$$

so that

(26) $$\tfrac{1}{2} < b < 1.$$

Define the regenerative event $E(t)$ thus. A particle oscillates between two states 0 and 1, the sojourn times in state 0 being exponential random variables with mean $1/a$, and the sojourn times in state 1 being a.s. fixed, equal to b. All sojourn times are to be independent, and at time 0 the particle is to be in state 0. Let $E(t)$ be the event that at time t the particle is in state 0. Then $E(.)$ is an 'event of type B' (Kingman [5]); and so if we define $p_a(t) = \mathrm{pr}\,(E(t))$, p_a is in $\mathscr{P}$. Further, we have by direct computation

(27) $$p_a(t) = \sum_{k=0}^{[t/b]} e^{-a(t-kb)} \cdot a^k \cdot (t-kb)^k/k!,$$

where as usual $[t]$ means the largest integer not exceeding t. Evidently

$$p_a(b) = e^{-a.b} = e^{-z} = y;$$

and by inequality (26)

$$p_a(1) = e^{-a.1} + a(1-b) \cdot e^{-a(1-b)} = \exp(-(1 - e^{-z})) = x,$$

by equations (23), (24) and (25). By formulae (24), (25), (26) and (27) and differentiation we check that p_a attains its minimum m in $[0, 1]$ at the point $t = b$, completing the demonstration of the example.

We shall say that the pair (M,m) is accessible if and only if there is a p in $\mathscr{P}$ with $M(p) = M$ and $m(p) = m$. We have the easy

Proposition 7. *If (M,m) is accessible, then so is (M',m) for all M' such that $m \leqslant M' \leqslant M$.*

Proof. Let p in $\mathscr{P}$ have $M(p) = M$ and $m(p) = m$. Let $u = \sup(t: 0 \leqslant t \leqslant 1$ and $p(t) \leqslant M')$. Since $M' \geqslant m$ and p is continuous, $p(u) = M'$. Define p' in $\mathscr{P}$ by $p'(t) = p(t/u)$ for all $t \geqslant 0$; then $M(p') = M'$ and $m(p') = m$, so (M',m) is accessible.

Example 2 and Propositions 6 and 7 can be summed up thus: If $0 < m \leqslant 1$ then there is an $m^*(m)$ such that when $M > m^*$ then (M,m) is inaccessible and when $m \leqslant M < m^*$ then (M,m) is accessible; and

$$\exp(-(1-m)) \leqslant m^* \leqslant 1 - m(1-m) \quad \text{if } m > \tfrac{1}{2},$$

whereas

$$\exp(-(1-m)) \leqslant m^* \leqslant \tfrac{3}{4} \quad \text{if } m \leqslant \tfrac{1}{2}.$$

Proposition 8. *$\mathscr{P}$ is not additively convex.*

Proof by example. Let p in $\mathscr{P}$ be given by formula (27) above with $z = 1$ therein and a and b given by equations (24) and (25), so that $m(p) = \mathrm{e}^{-1}$ and $M(p) = \exp(-(1-\mathrm{e}^{-1}))$. For all c such that $0 < c < 1$ define p_c by

$$p_c(t) = (1-c).p(t) + c.1 \quad \text{for all } t \geqslant 0.$$

Since $p_*(t) = 1$ is in $\mathscr{P}$, for $\mathscr{P}$ to be additively convex we must have p_c in $\mathscr{P}$ for all c in $]0,1[$. Let $M_c = M(p_c)$ and $m_c = m(p_c)$; then we have

$$M_c = (1-c).M(p) + c.1 \quad \text{and} \quad m_c = (1-c).m(p) + c.1.$$

Elementary calculation shows that there exists a $d < 1$ such that when $d < c < 1$, the point (M_c, m_c) is inaccessible, by Proposition 6, so that for such c, p_c cannot lie in $\mathscr{P}$.

Corollary. *The set of all positive renewal sequences is not additively convex.*

For if it were, by skeletal arguments we could show that $\mathscr{P}$ was additively convex too.

Example 3. *There is a p in $\mathscr{P}$ that satisfies the conditions (5) but is not i.d. in $\mathscr{P}$. For let p be defined as in the proof of Proposition 8 above. For all positive v define p_v by $p_v(t) = \mathrm{e}^{-vt}.p(t)$ for all $t \geqslant 0$. Now the function e^{-vt} is (i.d.) in $\mathscr{P}$, and so p_v, being the product of two members of $\mathscr{P}$, is itself in $\mathscr{P}$. p_v is not i.d. for any choice of v, for its x-function x_v is the sum of the linear function vt of t and the x-function x of p. This last cannot be concave, for p is not monotone. So x_v cannot be concave, and the conditions (4) now tell us* *that p_v is not i.d. Clearly all the requirements in conditions (5) are satisfied by p_v for all positive v, with the possible exception of the convexity of p_v. However, by use of the Volterra equation of Kingman [5], we can show that p_v is convex for all large enough v.*

Example 4. *In example 2 let $z \to \infty$. Then we find from equation (27) that*

$$p_a(t) \to e^{-n} . n^n/n! \quad if \ t = n \ for \ some \ integer \ n \geqslant 0,$$
$$\to 0 \quad otherwise:$$

an interesting example of pointwise convergence of a sequence of standard p-functions to a non-standard p-function.

2.4.6 TOPOLOGIES FOR $\mathscr{P}$

The elements of $\mathscr{P}$ are functions continuous on $[0, \infty]$ and bounded between zero and unity. There are accordingly several reasonable choices of topology:

(a) U, the topology of uniform convergence on the entire half-line;

(b) C, the topology of uniform convergence on compacta, alias the compact–open topology;

(c) T, the topology of pointwise convergence.

Clearly $U \supset C \supset T$. Now $U \neq C$; for let p_n be given by $p_n(t) = e^{-t/n}$ for all $t \geqslant 0$ and each positive integer n. Then the sequence (p_n) C-converges to p_* in $\mathscr{P}$ (where $p_*(t) = 1$); but the sequence does not converge under U. On the other hand, we shall show that $C = T$.

Define a new topology for $\mathscr{P}$: R, the topology of convergence on the rationals. That this is a Hausdorff topology for $\mathscr{P}$ follows from the continuity of the elements of $\mathscr{P}$. Also R is metrizable, for under R, $\mathscr{P}$ is a subset of a countable product of unit intervals. Obviously $R \subset T$, and so $R \subset C$ also.

Proposition 9. *Let p_n be in $\mathscr{P}$ for all positive integers n, and suppose that the sequence (p_n) R-converges to p in $\mathscr{P}$. Then (p_n) also converges to p under T.*

Proof. We first prove the statement

(*)   For all c such that $0 < c < \frac{1}{2}$ there exist $d > 0$ and an integer N such that whenever $0 \leqslant t \leqslant d$ and $n \geqslant N$, $p_n(t) \geqslant 1 - c$.

For define $a = c(1-c)$; then $0 < a < \frac{1}{4}$. Because p is in $\mathscr{P}$, p is continuous at $t = 0$; so there is a rational $d > 0$ such that

(28) $p(d) \geqslant 1 - \tfrac{1}{2}a.$

By the hypothesis of the proposition, since d is rational there is an integer N such that for all $n \geqslant N$

(29) $$p_n(d) \geqslant p(d) - \tfrac{1}{2}a \geqslant 1 - a,$$

by inequality (28). Now $0 < a < \tfrac{1}{4}$, so that $1 - a > \tfrac{3}{4}$. If we put $p_n^*(t) = p_n(t/d)$, Proposition 6 applies to p_n^*, so

$$p_n^*(t) \geqslant 1 - c \quad \text{if } 0 \leqslant t \leqslant 1;$$

whence

$$p_n(t) \geqslant 1 - c \quad \text{if } 0 \leqslant t \leqslant d,$$

for all $n \geqslant N$, proving (*).

Now for each $t \geqslant 0$ define

$$U(t) = \limsup_{n-\infty} p_n(t) \quad \text{and} \quad L(t) = \liminf_{n-\infty} p_n(t).$$

U and L are equal on the rationals by hypothesis; we have to show that they are equal everywhere. We have of course for all $t \geqslant 0$

(30) $$0 \leqslant L(t) \leqslant U(t) \leqslant 1.$$

Kingman's [5] inequalities

(31) $$p_n(s) \cdot p_n(t) \leqslant p_n(s + t) \leqslant 1 + p_n(s) \cdot p_n(t) - p_n(t),$$

together with (*), show that for all $t \leqslant d(c)$ and all $n \geqslant N(c)$

$$(1 - c) \cdot p_n(s) \leqslant p_n(s + t) \leqslant p_n(s) + c \quad \text{for all } s \geqslant 0.$$

Hence for all $t \leqslant d(c)$ and all $s \geqslant 0$

$$(1 - c) \cdot U(s) \leqslant U(s + t) \leqslant U(s) + c$$

and

$$(1 - c) \cdot L(s) \leqslant L(s + t) \leqslant L(s) + c.$$

Therefore by formula (30) we have for all $t \leqslant d(c)$ and all $s \geqslant 0$

$$|U(s + t) - U(s)| \leqslant c \quad \text{and} \quad |L(s + t) - L(s)| \leqslant c.$$

It follows that U and L are continuous. Their equality for all rational t then implies their equality everywhere.

Corollary. *On $\mathscr{P}$, $T = C$.*

Proof. Kendall [4] shows that

(32) $\mathscr{P}$ is a complete metric space under C; and

(33) if a sequence (p_n) of elements of $\mathscr{P}$ T-converges to an element p of $\mathscr{P}$, it converges to p under C also.

Suppose we have a sequence (p_n) of elements of $\mathscr{P}$ R-converging to an element p of $\mathscr{P}$. According to the Proposition the sequence converges to p

under T; so by condition (33) (p_n) C-converges to p. By condition (32), $(\mathscr{P}, C)$ is metrizable; and we noted in the definition of R that $(\mathscr{P}, R)$ was metrizable. So what we have just done implies that $R \supset C$. Since however we already have $C \supset T \supset R$, this implies $(R =) T = C$.

While C and R are equivalent topologies on $\mathscr{P}$, they are not equivalent as metrics. For according to condition (32) $(\mathscr{P}, C)$ is complete. But Example 4 shows that we can find a sequence (p_n) of elements of $\mathscr{P}$ which is Cauchy-convergent under R but which R-converges to a non-standard p-function, so cannot converge to an element of $\mathscr{P}$. So $(\mathscr{P}, R)$ is not complete and C and R cannot be equivalent metrics on $\mathscr{P}$. From now on, we shall use the symbol T for the topology on $\mathscr{P}$.

2.4.7 p-FUNCTIONS RIGHT-CONTINUOUS ON $]0, \infty[$

Let $\mathscr{PR}$ be the set of members of $\mathscr{G}$ that are right-continuous in $t > 0$. Let p^* be given by $p^*(0) = 1$, $p^*(t) = 0$ $(t > 0)$. Then $\mathscr{PR}$ contains all functions of the form

(§) $\qquad\qquad a . p + (1 - a) . p^* \quad (p \in \mathscr{P}, 0 \leqslant a \leqslant 1).$

For such a function is the product of p in $\mathscr{P}$ with the p-function of the regenerative event (E_t): the events E_t $(t > 0)$ are independent and each has probability a.

Theorem 3. *All members of $\mathscr{PR}$ are of the form* (§).[†]
This will follow from Propositions 10 and 11.

In this section, if $s, s+1, \ldots, s'$ is any set of consecutive integers, we write $t_{s \ldots s'}$ for the vector $(t_s, t_{s+1}, \ldots, t_{s'})$.

Proposition 10. *If $p \in \mathscr{PR}$, then*

$$\lim_{t \downarrow 0} p(t)$$

exists, equal to L say; and $p(t) \leqslant L$ for all $t > 0$.

Proof. Define for all positive integers n and all choices of

$$0 = t_0 < t_1 < t_2 < \ldots < t_n < \infty$$

(34) $\qquad H_n(t_{1 \ldots n}; p) = 1 - \sum_{i=1}^{n} p(t_i) + \sum_{1 \leqslant i < j \leqslant n} \sum p(t_i) . p(t_j - t_i) - \ldots$

$$+ (-1)^n . \prod_{i=1}^{n} p(t_i - t_{i-1}).$$

[†] Since this work was completed, the author has learnt that Professor Kingman has shown, independently, that if q in $\mathscr{G}$ is measurable and not a.e. zero, then q is of the form (§).

Then (Kingman [5]) $p(.)$ is in $\mathscr{G}$ if and only if

(35) $$H_n(t_{1\ldots n};p) \geqslant H_{n+1}(t_{1\ldots n+1};p) \geqslant 0$$

for all positive integers n and choice of $0 < t_1 < t_2 < \ldots < t_n < t_{n+1} < \infty$. Put

$$L_* = \liminf_{t \downarrow 0} p(t).$$

Let (s_r) be a sequence such that $s_r \downarrow 0$ and $p(s_r) \to L_*$ as $r \to \infty$.
 In equation (34), let

$$t_n \downarrow t_{n-1} \quad \text{by} \quad t_n = t_{n-1} + (s_r),$$

then

$$t_{n-1} \downarrow t_{n-2} \quad \text{by} \quad t_{n-1} = t_{n-2} + (s_r),$$

and so on until

$$t_2 \downarrow t_1 \quad \text{by} \quad t_2 = t_1 + (s_r).$$

Then from formula (35) we have for all $t > 0$ and $n \geqslant 1$

$$\binom{n}{0} - p(t) . \sum_{r=1}^{n} \binom{n}{r} (-1)^{r-1} L_*^{r-1} \geqslant 0.$$

In this, let $n \to \infty$. We see at once that

(36) $$p(t) \leqslant L_* \quad \text{for all } t > 0.$$

It follows at once that

$$\lim_{t \downarrow 0} p(t)$$

exists equal to L_*; this and inequality (36) complete the proof of the proposition.

Proposition 11. Let $p \in \mathscr{PR}$ and let

$$L = \lim_{t \downarrow 0} p(t).$$

If $L = 0$, then $p = p^*$. If $L > 0$, define q by $q(0) = 1$, $q(t) = p(t)/L$ for all $t > 0$; then $q \in \mathscr{P}$.

Proof. The first statement follows from inequality (36) above. So let $L > 0$; if $L = 1$, then $q = p$ is in $\mathscr{P}$ and there is nothing to prove, so let $L < 1$.

Let r be any function on $\mathfrak{R}^+$ such that $r(0) = 1$ and r is bounded by zero and unity. Define $F_0(r) = 1$, and $F_n(t_{1\ldots n};r)$ $(t_1 > 0, \ldots, t_n > 0)$ inductively by

(37) $$F_n(t_{1\ldots n};r) = F_{n-1}(t_1 + t_2, t_{3\ldots n};r) - r(t_1) . F_{n-1}(t_{2\ldots n};r).$$

Define further for each $n \geqslant 1$

(38) $$G_n(t_{1...n};r) = F_{n-1}(t_{1...n-1};r) - F_n(t_{1...n};r).$$

Then we check easily that the relation (37) is satisfied (for $n \geqslant 2$) by the Gs as well as by the Fs. Also we observe that

(39) $$F_n(t_{1...n};r) = H_n(t_1, t_1+t_2, ..., t_1+t_2+...+t_n;r),$$

where H_n is given by equation (34), so that r is in $\mathscr{G}$ if and only if

$$F_n(t_{1...n};r) \geqslant 0 \quad \text{and} \quad G_n(t_{1...n};r) \geqslant 0$$

for all positive integers n and choice of $t_{1...n} > 0$.

From equations (34), (37), (38) and (39) we see that the arguments of the function $r(\cdot)$ in F_n and G_n are

(40) $$\sum_{i=k}^{m} t_i \quad (1 \leqslant k \leqslant m \leqslant n).$$

For $0 \leqslant r \leqslant n < \infty$ define

(41) $F_{n,r}(t_{1...n})$ to be $F_n(t_{1...n};p)$, save that for all $m \leqslant r$ in the sum (40) we replace $p(\cdot)$ by $q(\cdot)$.

Define similarly $G_{n,r}(t_{1...n})$ for $1 \leqslant n < \infty$ and $0 \leqslant r \leqslant n$. Then we have $F_{n,0} \geqslant 0$ for all $n \geqslant 0$ and $G_{n,0} \geqslant 0$ for all $n \geqslant 1$, and we have to show that $F_{n,n} \geqslant 0$ for all $n \geqslant 0$ and $G_{n,n} \geqslant 0$ for all $n \geqslant 1$.

By formulae (37) and (41) and induction (on r, for each fixed $n-r$) we have

(42) $$F_{n,r}(t_{1...n}) = F_{n-1,r-1}(t_1+t_2, t_{3...n}) - q(t_1) \cdot F_{n-1,r-1}(t_{2...n})$$

for $1 \leqslant r \leqslant n < \infty$. From equation (37) we have, if $m \geqslant n+1$,

(43) $$F_{m,0}(t_1, 0^+, ..., 0^+, t_{2...n}) = F_{m-1,0}(t_1, 0^+, ..., 0^+, t_{2...n})$$
$$-p(t_1) \cdot F_{m-1,0}(0^+, ..., 0^+, t_{2...n}).$$

Let us henceforth write $\Omega = (0^+, ..., 0^+)$: a vector whose length will be obvious from the context. Then again from equation (37) if $m \geqslant n+1$

$$F_{m-1,0}(\Omega, t_{2...n}) = (1-L) \cdot F_{m-2,0}(\Omega, t_{2...n}),$$

so that

$$F_{m-1,0}(\Omega, t_{2...n}) = (1-L)^{m-n} \cdot F_{n-1,0}(t_{2...n}) \quad \text{for all } m \geqslant n.$$

From this and equation (43) we get for all large enough m

$$F_{m,0}(t_1, \Omega, t_{2...n}) = F_{n,0}(t_{1...n}) - p(t_1) \cdot \sum_{i=n}^{m-1} F_{i,0}(\Omega, t_{2...n})$$
$$= F_{n-1,0}(t_1+t_2, t_{3...n}) - p(t_1) \cdot F_{n-1,0}(t_{2...n}) \cdot \sum_{s=0}^{m-n} (1-L)^s.$$

Let $m \to \infty$ in this. We get

(44) $\quad \lim_{m \to \infty} F_{m,0}(t_1, \Omega, t_{2...n}) = F_{n-1,0}(t_1 + t_2, t_{3...n}) - q(t_1) \cdot F_{n-1,0}(t_{2...n})$

$\qquad\qquad\qquad\qquad\qquad = F_{n,1}(t_{1...n}),$

by equation (42); this holds for all $n \geqslant 1$.

For all positive integers r let P_r be the proposition

(45) $\qquad$ For all $n \geqslant r, \quad F_{n,r}(t_{1...n}) = \lim_{m \to \infty} F_{m,r-1}(t_{1...r}, \Omega, t_{r+1...n}).$

Then by equation (44) P_1 holds. But from equation (42)

$$\lim_{m \to \infty} F_{m,r-1}(t_{1...r}, \Omega, t_{r+1...n}) = \lim_{m \to \infty} F_{m-1,r-2}(t_1 + t_2, t_{3...r}, \Omega, t_{r+1...n})$$

$$- q(t_1) \cdot \lim_{m \to \infty} F_{m-1,r-2}(t_{2...r}, \Omega, t_{r+1...n}).$$

If P_{r-1} holds, the right-hand side of this equals

$$F_{n-1,r-1}(t_1 + t_2, t_{3...n}) - q(t_1) \cdot F_{n-1,r-1}(t_{2...n}) = F_{n,r}(t_{1...n}),$$

by equation (42). So by induction P_r holds for all $r \geqslant 1$. For all $r \geqslant 0$ let Q_r be the proposition that $F_{n,r} \geqslant 0$ for all $n \geqslant r$. Examination of equation (45) shows that for all $r \geqslant 1$, $P_r \cap Q_{r-1} \Rightarrow Q_r$. But Q_0 holds by hypothesis, so Q_r holds for all r. So

(46) $\qquad\qquad F_{n,n}(t_{1...n}) \geqslant 0 \quad$ for all $n \geqslant 1$ and $t_{1...n} > 0$.

We can go through a similar argument for the Gs. At the end of it we have the conclusion, analogous to formula (46), that

(47) $\qquad\qquad G_{n,n-1}(t_{1...n}) \geqslant 0 \quad$ for all $n \geqslant 1$ and $t_{1...n} > 0$.

This is not quite the desired statement, viz.,

(48) $\qquad\qquad G_{n,n}(t_{1...n}) \geqslant 0 \quad$ for all $n \geqslant 1$ and $t_{1...n} > 0$;

the reason for this is essentially the non-existence of a G_0. To get from formula (47) to formula (48) we note that from equations (37) and (38), $G_n(t_{1...n}; r)$ is a sum of signed products of the values of $r(\cdot)$ at some of the t-values given by the sum (40). Equations (34), (39), (38) and inspection of G_n then show that each of its summand products contains precisely one factor of the form

$$r(t_n + t_{n-1} + \ldots + t_s) \quad \text{for some } s \text{ such that} \quad 1 \leqslant s \leqslant n.$$

Hence, by the definition (41), $G_{n,n}(t_{1...n}) = G_{n,n-1}(t_{1...n})/L$, so that formula (48) follows from formula (47). Formula (48), (46) and the obvious continuity of q at $t = 0$ prove the proposition.

2.4.8 ARITHMETICAL PROPERTIES OF $\mathscr{P}$: CONSTRUCTION OF SIMPLES IN $\mathscr{P}$

Let $\mathscr{S}\mathscr{P}$ be the set of simples in $\mathscr{P}$, and let $\mathscr{Q}$ be the set of stable elements of $\mathscr{P}$—that is, those that have a finite derivative at $t = 0$. By consideration of the derivative at $t = 0$ we see that $\mathscr{Q}$ is an hereditary sub-semigroup (for the definition of this see Section 2.3.4 of [2]) of $\mathscr{P}$; considerations of continuity at $t = 0$ show that $\mathscr{P}$ is in its turn hereditary in $\mathscr{G}$.

We now recall some standard results (Kingman [5], Kendall [4]) on elements of $\mathscr{Q}$. For p_a in $\mathscr{Q}$ put $q_a = -p'_a(0)$. To any p_a in $\mathscr{Q}$ there corresponds a set W_a of independent non-negative random variables $X_{a,n}(1 \leqslant n < \infty)$ such that every $X_{a,2n-1}$ is exponentially distributed with mean $1/q_a$, and every $X_{a,2n}$ has a distribution F_a which has no atom at $t = 0$ and is determined by the relation

$$(49) \qquad r_a(s) = \int_0^\infty e^{-st} p_a(t)\, dt = (s + q_a - q_a \cdot \int_0^\infty e^{-sx}\, dF_a(x))^{-1}$$

for all $s > 0$. Define

$$T_{a,0} = 0;\, T_{a,n} = \sum_{r=1}^n X_{a,r}$$

for $n \geqslant 1$. Let

$$S_a = \bigcup_{r=0}^\infty [T_{a,2r}, T_{a,2r+1}[.$$

Then $p_a(t) = \mathrm{pr}\,(t \in S_a)$. Now let p_a and p_b be in $\mathscr{Q}$, and let W_a and W_b be taken as independent. If $p_c = p_a \cdot p_b$, then

$$p_c(t) = \mathrm{pr}\,(S_a \cap S_b \ni t),$$

implying that $p_c \in \mathscr{G}$. Since p_c is differentiable at $t = 0$ with finite derivative (in fact, $q_c = q_a + q_b$), p_c is in $\mathscr{Q}$. If we define the $T_{c,n}$ to be the endpoints, taken in order, of $S_c = S_a \cap S_b$, and then define for all $n \geqslant 1$

$$X_{c,n} = T_{c,n} - T_{c,n-1},$$

the $X_{c,n}$ correspond to p_c as did the $X_{a,n}$ to p_a and the $X_{b,n}$ to p_b. (This is by proposition 15 of Kingman [5].) The immediate application of this is that the distribution of $X_{c,2}$ is F_c as given by equation (49) (reading there c for a).

We now start breaking new ground with a definition: p_a in $\mathscr{Q}$ will be called R-bounded (by x) when there is a finite positive x such that $F_a(x) = 1$: equivalently, $X_{a,2n} \leqslant x$ a.s.

Proposition 12. *If p_c in $\mathcal{Q}$ is R-bounded, then p_c is in $\mathcal{SP}$.*

Proof. Suppose that $p_c = p_a \cdot p_b$, where neither of p_a, p_b is the identity e ($e(t) = 1$ for all $t \geqslant 0$). We have that p_a and p_b are in $\mathcal{Q}$; and by the non-triviality of the factorization, $q_a > 0$ and $q_b > 0$. Also $q_a + q_b = q_c$. By Kingman's [5] Theorem 6 we have

$$\lim_{t \to \infty} p_c(t) = (1+q_c) \cdot \int_0^\infty x \cdot dF_c(x))^{-1} \geqslant (1+q_c x_c)^{-1} > 0.$$

Hence

$$\lim_{t \to \infty} p_a(t) > 0 \quad \text{and} \quad \lim_{t \to \infty} p_b(t) > 0,$$

which facts imply that F_a and F_b must each put some mass on the finite part of the positive half-line. Accordingly let d_a, d_b and k be chosen such that $d_a > (13k/4)$, $d_b > (13k/4)$ and

$$f_a = F_a(d_a + \tfrac{1}{4}k) - F_a(d_a - \tfrac{1}{4}k) > 0, \quad f_b = F_b(d_b + \tfrac{1}{4}k) - F_b(d_b - \tfrac{1}{4}k) > 0.$$

Define $s_a = d_a - 5k/4, s_b = d_b - 5k/4$. Then

(50) $$s_a > 2k, \quad s_b > 2k$$

and

(51) $$\mathrm{pr}\,(s_z + k \leqslant X_{z,2r} < s_z + 3k/2) = f_z > 0 \quad \text{for } z = a \text{ and } z = b.$$

Let t^* be arbitrary and fixed such that $t^* > k$. Choose n so large that

$$n(s_a + s_b) - k > x_c.$$

Define W_a- and W_b-measurable events E_a and E_b:

$$E_a = \left\{ \begin{array}{l} \displaystyle\bigcap_{r=1}^{n+1} \{t^* + (r-1)(s_a + s_b) - k \leqslant T_{a,2r-1} < t^* + (r-1)(s_a + s_b) - \tfrac{1}{2}k\} \\[2ex] \displaystyle\cap \bigcap_{r=1}^{n} \{t^* + r \cdot s_a + (r-1)s_b \leqslant T_{a,2r} < t^* + r \cdot s_a + (r-1)s_b + k\} \end{array} \right\};$$

$$E_b = \left\{ \begin{array}{l} \displaystyle\bigcap_{r=1}^{n} \{t^* + r \cdot s_a + (r-1)s_b - k \leqslant T_{b,2r-1} < t^* + r \cdot s_a + (r-1)s_b - \tfrac{1}{2}k\} \\[2ex] \displaystyle\cap \bigcap_{r=1}^{n} \{t^* + r(s_a + s_b) \leqslant T_{b,2r} < t^* + r(s_a + s_b) + k\} \end{array} \right\}.$$

Then by equation (51) we have that

(52) $$\mathrm{pr}\,(E_a) \geqslant \tfrac{1}{2}kq_a \cdot \exp\,(-q_a t^*) \cdot f_a^n \cdot (\tfrac{1}{2}kq_a \cdot \exp\,(-q_a s_b))^n > 0,$$

and

(53) $$\operatorname{pr}(E_b) \geqslant \exp(-q_b t^*) . f_b^n . (\tfrac{1}{2} k q_b . \exp(-q_b s_a))^n > 0.$$

By the independence of W_a and W_b, then,

(54) $$\operatorname{pr}(E_a \cap E_b) = \operatorname{pr}(E_a) . \operatorname{pr}(E_b) > 0.$$

But now by the choice of n, if both E_a and E_b occur then $X_{c,2} > x_c$. (This can be checked, tediously, from the definitions of E_a and E_b.) So from equation (54) we have $\operatorname{pr}(X_{c,2} > x_c) > 0$, which contradicts the definition of x_c as the R-bound of p_c. So p_c must be in $\mathcal{SP}$.

2.4.9 THE CATEGORY OF $\mathcal{SP}$ IN $\mathcal{P}$

Proposition 13. *$\mathcal{SP}$ is dense in $\mathcal{P}$.*

Proof. Kingman shows (in the proof of his Theorem 4 in [5]) that $\mathcal{Q}$ is dense in $\mathcal{P}$, so all we have to do is show that $\mathcal{SP}$ is dense in $\mathcal{Q}$. Let then p be a member of $\mathcal{Q}$; if we define

$$r(s) = \int_0^\infty e^{-st} p(t) \, dt \quad \text{for all } s > 0$$

we have from equation (49) of the last section

(55) $$(r(s))^{-1} = s + q - q . \int_0^\infty e^{-sx} \, dF(x),$$

where $q = -p'(0)$ and F is the distribution function of a probability measure on $]0, \infty]$. Define p_n in $\mathcal{Q}$ for all positive integers n by

$$q_n = -p_n'(0) = q,$$

and $F_n(x) = F(x)$ if $x < n$; $F_n(x) = 1$ if $x \geqslant n$. Then by Proposition 12 p_n is in $\mathcal{SP}$ for each n, and by the Volterra equation of Kingman [5] and the definition of p_n we have $p_n(t) = p(t)$ for all $t \leqslant n$. Thus the sequence (p_n) T-converges to p, implying that $\mathcal{SP}$ is dense in $\mathcal{Q}$.

Proposition 14. *$\mathcal{SP}$ is residual in $\mathcal{P}$.*

Proof. Let $(\mathcal{G}, T)$ be the semigroup of all p-functions (standard or not) endowed with the topology of pointwise convergence. Let C be the metric of uniform convergence on compacta on $\mathcal{P}$; from Section 2.4.6, C and T are equivalent topologies on $\mathcal{P}$. Now $\mathcal{G}$ is compact under T, and $\mathcal{P}$ is an hereditary sub-semigroup of $\mathcal{G}$. So Proposition 3 of [2] tells us that $\mathcal{SP}$ is a G_δ in $\mathcal{P}$ under T; by the equivalence of T and C on $\mathcal{P}$, $\mathcal{SP}$ is a G_δ in $\mathcal{P}$ under C also. Now Proposition 13 shows that $\mathcal{SP}$ is T-dense (and so also C-dense) in $\mathscr{P}$. Under C, $\mathscr{P}$ is a complete metric space (Kendall [4]); so Baire's category theorem and the fact just proved, that $\mathscr{S}\mathscr{P}$ is a dense G_δ in $\mathscr{P}$, yield the proposition.

Proposition 15. $\mathscr{Q}$ *is meagre in* $\mathscr{P}$.

Proof. For each positive integer k define

$$\mathscr{Q}_k = (p_a\colon p_a \text{ is in } \mathscr{Q} \text{ and } q_a = -p_a'(0) \leqslant k),$$

so that

$$\mathscr{Q} = \bigcup_{k \geqslant 1} \mathscr{Q}_k.$$

Let (p_a) be a net of members of $\mathscr{Q}_k$. Since (p_a) is then a net of members of $\mathscr{G}$, and $\mathscr{G}$ is compact under T, there is a subnet (p_b) T-converging to a limit p^* (say) in $\mathscr{G}$. Since we have for all p_a in $\mathscr{Q}_k$

$$|p_a(s) - p_a(t)| \leqslant k|s - t| \quad \text{for all } s, t \geqslant 0,$$

the same holds for p^* which must thus lie in $\mathscr{Q}_k$. So $\mathscr{Q}_k$ is compact under T; T being Hausdorff, $\mathscr{Q}_k$ is thus T- (and so C-) closed.

On the other hand, let p be any member of $\mathscr{Q}_k$. Define, for each positive integer n, p_n by

$$p_n(t) = \exp(-\min(1/n, \sqrt{t})) \quad \text{for all } t \geqslant 0.$$

Then p_n is logarithmically convex, bounded between zero and unity, and continuous; and $p_n(0) = 1$. So by equation (4) p_n is (i.d.) in $\mathscr{P}$ for each n. Now it is clear that for all n, $(p \cdot p_n)'(0) = -\infty$, so $p \cdot p_n$ is in $\mathscr{P} - \mathscr{Q}$ for each n. Also, we have that $p_n(t) \to 1$ uniformly on the entire non-negative half-line, so the sequence $(p \cdot p_n)$ T-converges to p as $n \to \infty$. So $\mathscr{Q}_k$ has no interior; since $\mathscr{Q}_k$ is closed, this shows that $\mathscr{Q}_k$ is nowhere-dense in $\mathscr{P}$. Accordingly $\mathscr{Q}$, being the countable union of the nowhere-dense sets $\mathscr{Q}_k$, is meagre in $\mathscr{P}$.

Corollary. *There exist* p *in* $\mathscr{S}\mathscr{P}$ *with* $q = -p'(0) = \infty$.

It would be interesting to know whether $\mathscr{P}\mathscr{M}$, the set of diagonal Markov functions, contains any of the simples of $\mathscr{P}$. The answer would be Yes if we could show that $\mathscr{P}\mathscr{M}$ was a G_δ in $\mathscr{P}$; it would also be Yes if we could show that $\mathscr{P}\mathscr{M}$, which is a sub-semigroup of $\mathscr{P}$, was hereditary in $\mathscr{P}$.

2.4.10 THE I_0-PROBLEM FOR $\mathscr{P}$

Recall that I_0 is the set of i.d. elements (in this case of $\mathscr{P}$) all of whose factors are i.d. We call p in $\mathscr{P}$ exponential if there is a finite non-negative constant k such that

$$p(t) = e^{-kt} \quad \text{for all } t \geq 0.$$

Proposition 16. $I_0(\mathscr{P})$ *contains all the exponential p in $\mathscr{P}$.*

Proof. We note that

(a) p in $\mathscr{P}$ is exponential if and only if all its discrete skeleta are geometric;

(b) a geometric member of $\mathscr{R}^+$ (the set of positive renewal sequences) has only geometric factors (from Proposition 11 of [2]).

Suppose then that $p = q . r$, that p is exponential and that q and r are in $\mathscr{P}$. Were q not exponential, by (a) there would be a non-geometric skeleton of q; then (b) would be contradicted. So both q and r must be exponential; and, since exponential members of $\mathscr{P}$ are i.d., p has only i.d. factors and so lies in I_0. (Note. Kendall [4] has a proof of this proposition using the supermultiplicativity of the members of $\mathscr{P}$.)

Obviously, in the attack on the I_0-problem for any Delphic semigroup G one looks first at the extremals among the i.d. members of G (which are a convex cone under the semigroup multiplication). Knowledge of which extremals are in I_0 will not in general solve the problem (though it does in the case $G = \mathscr{R}^+$); for in general $I_0(G)$ is not a semigroup. However, it is clear that if u, i.d. in G, is not in I_0, and v, i.d. in G, has u as a factor, then v is not in I_0 either. So if we find many extremals that are not in I_0, we are making good progress with the description of it. Now (Kendall [4]) the extremal i.d. elements of $\mathscr{P}$ are the $p_{s,a}$ given, for $0 < s \leq \infty$ and $0 \leq a < \infty$ by

$$p_{s,a}(t) = \exp(-a . \min(s, t)) \quad \text{for all } t \geq 0.$$

From our experience (in Section 2.3.8 of [2]) with $\mathscr{R}^+$ we might hope that we could, for each (finite, positive) s and a, find some s' and a' such that $p_{s',a'}$ was a factor of $p_{s,a}$. But a result of Kingman [6], that

$$D^+ p(t) \geq D^- p(t) \quad \text{for all } t > 0$$

(where D^+, D^- are the right- and left-hand derived numbers) puts this possibility of division out of court. As a result I have as yet been able only partially to solve the I_0-problem for $\mathscr{P}$. We have two results: the first is an analogue of Cramér's [1] theorem in the case of the semigroup of probability laws on the line.

Proposition 17. *If the density of the absolutely continuous (henceforth written a.c.) part of the spectral measure of p, an i.d. member of $\mathscr{P}$, is uniformly exponentially large at infinity, then p is not in I_0.*

Before proceeding to proof, we elucidate the proposition a little. We take as canonical the x-function $x(t) = -\log(p(t))$. If p is in $\mathscr{P}$ we have

(i) $x(0) = 0$; $x(t) \geqslant 0$ for all $t \geqslant 0$; x is continuous.
Should p be i.d., we have in addition

(ii) x is concave.

As in Section 2.4.4, conditions (i) and (ii) are also sufficient for x to be the x-function of some i.d. p in $\mathscr{P}$.

Kendall [4] shows that there is a biunique correspondence between i.d. p in $\mathscr{P}$ and totally finite measures m on $]0, \infty]$ given by

$$(56) \qquad x(t) = -\log(p(t)) = \int_0^\infty \frac{\min(s,t)}{1-e^{-s}} \, . \, dm(s);$$

addition of the measures m corresponds to multiplication of the p-functions. Let the a.c. part of m be m^*, and let the corresponding density be $M^*(s)$. The condition that M^* be uniformly exponentially large at infinity is just that there exist constants $A^* > 0, B^* > 0$ and $S \geqslant 0$ such that

$$(57) \qquad \frac{M^*(s)}{1-e^{-s}} \geqslant A^* \, . \, e^{-B^*s} \quad \text{for almost all } s > S.$$

Proof of the proposition. We shall throw away the non-a.c. part of m, for we have $p = p_s \, . \, p^*$, where p^* corresponds to the a.c. part m^* of m and p_s to the non-a.c. part. If we show that p^* is not in I_0 it will follow that p is not in I_0 either.

We assume first that $S = 0$. Then by formulae (56), (57) and the absolute continuity of $m^*, x^{*\prime\prime}$ exists a.e., $x^{*\prime}$ is the integral of its derivative and

$$(58) \qquad -x^{*\prime\prime}(t) = \frac{M^*(t)}{1-e^{-t}} \geqslant A^* \, . \, e^{-B^*t} \quad \text{for almost all } t \geqslant 0.$$

Consider the two-state renewal process on the states X and X', whose times in state X have exponential distribution with mean a, and the distribution of whose times in state X' is the convolution of two exponential laws each of which has mean b. Put $c = b/a$.

The event 'the process is in state X' is an 'event of type B' in the sense of Kingman [5]. So the function $p_{b,c}$ given by

$$p_{b,c}(t) = \mathrm{pr} \left\{ \begin{array}{l} \text{the process is in state } X \text{ at time } t, \\ \text{given that it was in state } X \text{ at time } 0 \end{array} \right\}$$

is in $\mathscr{P}$. We put on the condition that

$$(59) \qquad b < 4a; \quad \text{so} \quad 0 < c < 4.$$

From the representation (55) we have

(60) $$r_{b,c}(s) = \int_0^\infty e^{-st} p_{b,c}(t)\, dt = b/\{bs + c(1 - (1+bs)^{-2})\};$$

if we put $x = bs$, this becomes

(61) $$(x/b) \cdot r_{b,c}(x/b) = (1+x)^2/((1+x)^2 + c(2+x)).$$

Let w, w' be the roots of the equation

(62) $$z^2 + (2+c)z + (1+2c) = 0.$$

By formula (59) w and w' are complex conjugates, $w' = \bar{w}$ say. Let u, v be the real and imaginary parts of w; we can suppose that $v \geqslant 0$. Define $u^* = 1 + u$. From equation (62) we have $u^* = -\tfrac{1}{2}c$ and $v = \sqrt{(c(1-\tfrac{1}{4}c))}$, so that

(63) $$u = u^* - 1 < -1.$$

Put $A = (1+w)^2/(w(w-\bar{w}))$. Then by equations (61) and (62) we have

(64) $$r_{b,c}(s) = A \cdot (s - w/b)^{-1} + \bar{A} \cdot (s - \bar{w}/b)^{-1} + (sw\bar{w})^{-1}.$$

On inversion,

(65) $$p_{b,c}(t) = A \cdot e^{wt/b} + \bar{A} \cdot e^{\bar{w}t/b} + (w\bar{w})^{-1}.$$

Let $x_{b,c}$ be the x-function of $p_{b,c}$. By definition we have

$$p_{b,c}^2(t) \cdot x''_{b,c}(t) = p'^2_{b,c}(t) - p''_{b,c}(t) \cdot p_{b,c}(t)$$

at each point t where $p_{b,c}$ is twice differentiable. From equation (65), then,

(66) $$p_{b,c}^2(t) \cdot x''_{b,c}(t) = -(u^2+v^2) \cdot b^{-2} \cdot \exp(ut/b)$$
$$\times (C(u,v) \cdot \cos(vt/b) + S(u,v) \cdot \sin(vt/b) + E(u,v) \cdot \exp(ut/b)),$$

where

$$C(u,v) = u^{*2} - v^2 + 2uu^*, \quad S(u,v) = (u/v)(u^{*2} - v^2) - 2vu^*,$$
$$E(u,v) = -(u^{*2} + v^2)^2 \quad \text{and as before} \quad u^* = 1 + u.$$

If we apply equation (63) to substitute for u, u^* and v in the definitions of $C(u,v)$ and $S(u,v)$, we see that C and S vanish simultaneously only when $c = 0$. Hence, since by equation (63) (u/b) is strictly negative, equation (66) shows that for any b and c satisfying condition (59),

(67) $$x''_{b,c}(t) \quad \text{oscillates in sign as} \quad t \to \infty.$$

Let us choose $b = 1/B^*$: with this choice of b call $p_{b,c}$, p_c. From equation (63) and their definitions $C(u,v)$, $S(u,v)$ and $E(u,v)$ all tend to zero with c; whereas $u^2 + v^2 \to 1$ as $c \downarrow 0$. So from equation (66) there is a $c_0 > 0$ such that for all $c \leqslant c_0$

(68) $|p_c^2(t) . x_c''(t)| \leqslant \frac{1}{4} A^* . e^{-B^* t}$ for all $t \geqslant 0$.

Now

$$\lim_{t \to \infty} p_c(t) = (w\bar{w})^{-1}   \text{from equation (64),}$$

$$= 1/(1+2c)   \text{from equation (62).}$$

So from the Kingman inequality

$$p_c(s+t) \leqslant 1 + p_c(s) . p_c(t) - p_c(s)$$

we have, letting $s \to \infty$,

$$p_c(t) \geqslant \left(2\left(\frac{1}{1+2c}\right) - 1 \right) \Big/ \frac{1}{1+2c} = 1 - 2c   \text{for all } t \geqslant 0.$$

So there is a $c_1 > 0$ such that for all $c \leqslant c_1$

(69) $\min_{t \geqslant 0} p_c(t) \geqslant \max \left(\frac{1}{2}, \lim_{t \to \infty} p^*(t)\right).$

Let $c^* = \min(c_0, c_1)$. Then from formulae (68) and (69) we have

$$|x_{c*}''(t)| \leqslant A^* . e^{-B^* t}   \text{for all } t \geqslant 0.$$

From formula (58), then,

(70) $x^{*\prime\prime}(t) - x_{c*}''(t) \leqslant 0$ for almost all $t \geqslant 0$.

Define $x^{**}(t) = x^*(t) - x_{c*}(t)$ and $p^{**}(t) = e^{-x^{**}(t)}$. Obviously x^{**} is continuous, and $x^{**}(0) = 0 - 0 = 0$. From inequality (69) we have

(71) $\lim_{t \to \infty} x^{**}(t) \geqslant 0.$

Now $x^{**\prime}(t)$ exists for all t and is absolutely continuous, and $x^{**\prime\prime}(t) \leqslant 0$ for almost all t by inequality (70). So $x^{**\prime}(t)$ decreases in t; inequality (71) then shows that $x^{**\prime}(t) \geqslant 0$ for all $t \geqslant 0$, so that x^{**} is an increasing concave function of t. From this and the fact that $x^{**}(0) = 0$ we see that x^{**} is non-negative; whence p^{**} is i.d. in $\mathscr{P}$. Further by definition we have $p^* = p^{**} . p_{c*}$, and p_{c*} is in $\mathscr{P}$ by its construction. But by condition (67) p_{c*} is not i.d., so has simple factors (or is itself simple); so p^* cannot be in $I_0(\mathscr{P})$.

We have now shown that if p^* is i.d. in $\mathscr{P}$ with a.c. spectral measure and $-x^{*\prime\prime}(t) \geqslant A^* . e^{-B^* t}$ for almost all $t \geqslant 0$ and some $A^* > 0$ and $B^* > 0$, then p^* is not in $I_0(\mathscr{P})$.

Now let p_1 be i.d. in $\mathscr{P}$, and let the a.c. part of its spectral measure be such that $x_1''(t) \leqslant -A . e^{-Bt}$ for some $A, B > 0$ and almost all $t > a > 0$. Then

p_1 has p_2 as a factor, where the spectral measure of p_2 is a.c., and the x-function x_2 of p_2 is given by

$$\lim_{t \to \infty} x_2'(t) = 0$$

and

$$x_2''(t) = 0 \quad \text{if } t \leqslant a; \quad x_2''(t) = -A . e^{-Bt} \quad \text{if } t > a.$$

(It is easily seen that $p_1 = p . p_2$, where p is i.d. in $\mathscr{P}$.) We have

$$q_2 = -p_2'(0) = x_2'(0) = x_2'(a) = (A/B) . e^{-Ba} < \infty,$$

so p_2 is in $\mathscr{Q}$. In fact we have

(72) $$p_2(t) = e^{-Ct} \quad \text{if } t < a; \quad p_2(t) = e^{-Ca} . p^*(t-a) \quad \text{if } t \geqslant a,$$

where $C = (A/B) . e^{-Ba}$ and p^* is given by its x-function x^*:

$$x^*{}''(t) = -A . e^{-Ba} . e^{-Bt} \quad \text{for all } t > 0; \quad \lim_{t \to \infty} x^*{}'(t) = 0.$$

We note that p^* is i.d. in $\mathscr{P}$, and is, by what we have already proved, not in I_0. Now Kendall [4] shows that if p_b is in $\mathscr{Q}$ with $p_b'(0) = -b$ say, then the function $T_a p_b$ given by

$$T_a p_b(t) = e^{-bt} \quad \text{when } t < a; \quad T_a p_b(t) = e^{-ba} . p_b(t-a) \quad \text{when } t \geqslant a$$

is in $\mathscr{Q}$ also, and is i.d. if and only if p_b is i.d. It is easy to see that if $p_b = p_{b'} . p_{b''}$, then $T_a p_b = T_a p_{b'} . T_a p_{b''}$; this implies that if p_b is not in I_0, no more is $T_a p_b$. But now equation (72) says that $p_2 = T_a p^*$, and p^* is not in I_0; so p_2 is not in I_0. Thus p_1, having p_2 as a factor, is not in I_0.

Example 5. *If $p(t) = 1/(1+t)$, then p is i.d. in $\mathscr{P}$, but is not in I_0.*

Example 6. *With the exception of the exponential members, the 'multiplicatively stable' i.d. p in $\mathscr{P}$ are not in I_0.* (Kendall [4] *calls p in $\mathscr{P}$ multiplicatively stable when $p(t) = \exp(-a . t^b)$ for some a and b such that $0 \leqslant a < \infty$ and $0 < b \leqslant 1$. Such p are i.d.*)

Example 7. $p(t) = \exp(-(1-e^{-t}))$ *is i.d. in $\mathscr{P}$, but is not in I_0.*

We now have a partial result about I_0 and the extremals

$$p_{s,a}(t) = \exp(-a . \min(s, t)) \quad (0 < a < \infty; \; 0 < s < \infty)$$

among the i.d. members of $\mathscr{P}$. Clearly by the change of time-scale $t' = t/s$ we have $p_{s,a}(t) = p_{1,sa}(t')$, so we need only consider the case $s = 1$.

Proposition 18. $p_{1,d}$ *is not in I_0 if $d > 1$.*

Proof. We have

$$p_{1,d}(t) = e^{-dt} \quad \text{if } t \leqslant 1; \quad p_{1,d}(t) = e^{-d} \quad \text{if } t > 1.$$

Let us put $a = \frac{1}{2}(1+d)$, so that $1 < a < d$. Define

(73) $p(t) = e^{-at}$ if $t \leqslant 1$; $\quad p(t) = e^{-a}(1 + b(1 - e^{-a(t-1)}))$ if $t > 1$,

where the *positive* constant b is to be determined later. Define also p^* by $p_{1,d}(t) = p(t) \cdot p^*(t)$ for all $t \geqslant 0$, so that

(74) $\qquad p^*(t) = e^{-(d-a)t}$ if $0 \leqslant t \leqslant 1$

$\qquad\qquad\qquad = e^{-(d-a)}/(1 + b(1 - e^{-a(t-1)}))$ if $t > 1$.

We shall show that it is possible to choose $b > 0$ such that both p and p^* are in $\mathscr{P}$. Let x^* be the x-function of p^*. Then $x^*(0) = 0$, and x^* is continuous. Also

$$x^{*\prime}(t) = d - a \quad \text{if } 0 \leqslant t \leqslant 1;$$

$$= ab \cdot e^{-a(t-1)}/(1 + b(1 - e^{-a(t-1)})) \quad \text{if } t > 1.$$

We see therefore that $x^{*\prime}$ is non-negative and decreasing if and only if

(75) $1 + b \leqslant d/a = (2a-1)/a;$

so p^* is i.d. in $\mathscr{P}$ if and only if equation (75) holds. We turn now to p (given by equation (73)). We call a function $L(s)$ of the non-negative real variable s an LT (LTF, LTP) if it is the Laplace transform of a non-negative (finite, probability) measure on $[0, \infty]$. If we put as usual

$$r(s) = \int_0^\infty e^{-st} p(t) \, dt$$

and define $L(s)$ by

(76) $(r(s))^{-1} = s + a - a \cdot L(s),$

then by Kingman's representation p (given by equation (73)) is in $\mathscr{P}$ if and only if $L(s)$ is an LTP and the probability concerned has no atom at the origin. Now from equation (73) we have for all $s > 0$

(77) $r(s) = (s + aA \cdot e^{-s})/(s(a+s)),$

where $A = e^{-a}(1+b)$; hence for all $s > 0$

(78) $L(s) = A \cdot e^{-s}(a+s)/(s + aA \cdot e^{-s}).$

Define $f(s) = e^s \cdot L(s)/A$. We shall show that $f(s)$ is an LT. By equation (78),

$$f(s) = (s+a)/(s+1-(1-aA \cdot e^{-s}));$$

if we make the proviso

(79) $1 + b \leqslant e,$

then there is a $k < 1$ such that for all $s \geqslant 0$

$$|g(s)| \leqslant k,$$

where $g(s) = (1 - aA \cdot e^{-s})/(1 + s)$. Hence

(80) $\qquad f(s) = (1 + (a - 1)/(1 + s)) \cdot \sum_{n=0}^{\infty} (g(s))^n \quad$ for all $s > 0$.

Now $(1 + (a - 1)/(1 + s))$ is clearly an LTF since $a > 1$. We can invert $g(s)$ to find that it is an LTF if and only if

(81) $\qquad\qquad\qquad 1 + b \leqslant e^{-1}/(a \cdot e^{-a})$.

The total mass of the measure corresponding to $g(s)$ is then $1 - aA < 1$; so if conditions (81) and (79) are satisfied, $f(s)$ is an LTF. Since

$$L(s) = A e^{-s} f(s),$$

under these conditions $L(s)$ is also an LTF. But equation (78) shows that $L(s) \to 1$ as $s \downarrow 0$, so if $L(s)$ is an LTF it is in fact an LTP. Thus conditions (81) and (79) hold, $L(s)$ is an LTP; and since $L(s) \to 0$ as $s \to \infty$, the probability concerned has no atom at the origin, so conditions (81) and (79) imply that p is in $\mathscr{P}$. Define now

$$b^* = \min (e^{-1}/(a \cdot e^{-a}) - 1, e - 1, (a - 1)/a).$$

Then formula (75) is satisfied as well as conditions (81) and (79) for this $b = b^*$, so both p and p^* are in $\mathscr{P}$; and $p_{1,d} = p \cdot p^*$ therefore has p as a factor in $\mathscr{P}$. But because $b^* > 0$, p is not monotone, much less i.d.; so $p_{1,d}$ cannot be in I_0.

The question is, What about the case $d \leqslant 1$? Firstly, it is possible by inversion of equation (78) into a differential-difference equation actually to prove that the choice of p in equation (73) does not yield an element of $\mathscr{P}$ when $a \leqslant 1$ (except, of course, in the unhelpful case $b = 0$). We do not have such an explicit non-theorem for other choices of p; but we have the following suggestive argument. Suppose we take

(82) $p(t) = e^{-at}$ if $0 \leqslant t \leqslant 1$; $\ p(t) = e^{-a}(1 + b \cdot f(t - 1))$ if $t > 1$, where $f(0) = 0$, f is twice differentiable, f'' is absolutely integrable over $[0, \infty]$ and $0 < f'(0) < \infty$.

If we put for all $s > 0$

$$g(s) = a \cdot e^{-a} + b \cdot e^{-a}(s + a) \cdot \int_0^{\infty} f'(t) e^{-st} \, dt,$$

then p is in $\mathscr{P}$ if and only if

$$L(s) = (s + a) g(s)/(s + e^{-s} g(s))$$

7

is an LT. We can apply the method of the last proposition to get

$$L(s) = \frac{s+a}{s+c} \cdot g(s) \cdot \sum_{n=0}^{\infty} \left(\frac{c}{s+c} \right)^n \cdot \left(1 - \frac{e^{-s} g(s)}{c} \right)^n$$

for all c in some interval $[C, C']$; but if $a \leqslant 1$ it is not possible (with $b > 0, f'(0) > 0$) to find a c such that

$$\frac{s+a}{s+c} \quad \text{and} \quad \frac{c}{s+c} \cdot \left(1 - \frac{e^{-s} g(s)}{c} \right)$$

are simultaneously LTs, so the method of the proposition appears to fail. Also, as we saw at the start of this apology, the method of the proposition, at any rate in the elementary case where $g(s)$ is a constant, does not waste anything. (It is easy to check that if equation (82) is replaced by equation (73) then $g(s)$ becomes a constant.) So it looks at present as if $p_{1,d}$ may be in I_0 if $d \leqslant 1$. If this is so it is very remarkable, for it will show that I_0 need not only not be a semigroup (it is not a semigroup in the case of the convolution semigroup of probability laws on the line), but need not even have the property that if p is in I_0 then p^2 is in I_0 also.

I am indebted to Professor D. G. Kendall for his constant help and encouragement in the course of the research leading to this paper and [2].

REFERENCES

1. H. Cramér, "On the factorization of probability distributions", *Ark. Mat.* **1** (1949), 61–65.
2. R. Davidson, "Arithmetic and other properties of certain Delphic semigroups, I", *Z. Wahrsch'theorie & verw. Geb.* **10** (1968), 120–145. (2.3 of this book.)
3. D. G. Kendall, "Renewal sequences and their arithmetic", *Symposium on Probability Methods in Analysis* (*Lecture Notes in Mathematics* 31), Springer, Berlin, Heidelberg and New York (1967). (2.1 of this book.)
4. ——, "Delphic semigroups, infinitely divisible regenerative phenomena, and the arithmetic of *p*-functions", *Z. Wahrsch'theorie & verw. Geb.* **9** (1968), 163–195. (2.2 of this book.)
5. J. F. C. Kingman, "The stochastic theory of regenerative events", *Z.Wahrsch'-theorie & verw. Geb.* **2** (1964), 180–224.
6. ——, "Some further analytical results in the theory of regenerative events", *J. math. Analysis Appl.* **11** (1965), 422–433.
7. I. P. Natanson, *Theory of Functions of a Real Variable*, vol. 1, Ungar, New York (1955).
8. R. R. Phelps, *Lectures on Choquet's Theorem*, Van Nostrand, Princeton, (1966).
9. A. P. Robertson and W. J. Robertson, *Topological Vector Spaces*, Cambridge University Press (1964).

2.5

More Delphic Theory and Practice

R. Davidson

[Reprinted from *Z. Wahrsch'theorie & verw. Geb.* **13** (1969), 191–203.]

2.5.1 SUMMARY

We find conditions on semigroups satisfying Kendall's [5] Delphic postulates A and B such that they then satisfy also postulate C (the central limit theorem). These conditions are of the type that the semigroups possess enough continuous homomorphisms (each into the additive reals or circle group) to separate its points. We show that the classical Delphic semigroups (of probability laws on the line, renewal sequences, and p-functions) satisfy our conditions, and thus get the classical results as consequences of the abstract theorems.

We also find some curious 'spiral' Delphic semigroups for counterexamples to do with the I_0-problem.

2.5.2 DEFINITIONS AND PRELIMINARIES

Let $\mathfrak{R}^+$ be the non-negative reals, $\mathfrak{Z}^+$ the positive integers, $\mathfrak{R}$ the reals, and $\mathfrak{T}$ the multiplicative group of complex numbers of unit modulus, under the Euclidean distance on the plane.

A semigroup G with elements $u, v, \ldots$ will be called an $L\frac{1}{2}$gp if G is commutative and has an identity e, and there is a set $M \subset G^{3^+}$ and a map L of M into G such that

(*i*) if $\{u_n\} \in M$ and $\{n'\} \subset \{n\}$, then $\{u_{n'}\} \in M$ and $L\{u_{n'}\} = L\{u_n\}$;

(*ii*) if $\{u_n\}$ is such that there exists $u \in G$ such that for all $\{n'\} \subset \{n\}$ there is $\{n''\} \subset \{n'\}$ with $\{u_{n''}\} \in M$ and $L\{u_{n''}\} = u$, then $\{u_n\} \in M$ and $L\{u_n\} = u$;

183

(iii) $\{u_n\} \in M$ and $\{v_n\} \in M$ imply $\{u_n . v_n\} \in M$ and

$$L\{u_n . v_n\} = L\{u_n\} . L\{v_n\};$$

(iv) if $u_n = u$ for each $n \in 3^+$, then $\{u_n\} \in M$ and $L\{u_n\} = u$.

For $u \in G$ let $F_G(u)$ be the set of factors of u in G. An element of G is called i.d. (infinitely divisible) if for each $k \in 3^+$ it has a kth root in G; it is called *simple* if it is not e and has no factors bar itself and e.

We give now a list of properties that an L$\frac{1}{2}$gp *may* have.

A. For all $u \in G$, $F_G(u)$ is sequentially compact.

A′. If $\{u_n\}$ is any element of M, and if, for each $n \in 3^+, v_n \in F_G(u_n)$, then there is $\{n'\} \subset \{n\}$ such that $\{v_{n'}\} \in M$ (and then by (i) and (iii) above, $L\{v_{n'}\} \in F_G(L\{u_n\}))$.

B. There is an s.c. (sequentially continuous) homomorphism

$$D: G \to (\Re^+, +),$$

such that if $D(u) = 0$ then $u = e$.

CLT. Let B hold. Let the triangular array $\{u_{i,j}: 1 \leqslant j \leqslant i < \infty\}$ be null: i.e., let

$$\lim_{i \to \infty} \max_{1 \leqslant j \leqslant i} D(u_{i,j}) = 0.$$

Suppose also that $\{u_i\} \in M$, where

$$u_i = \prod_{j=1}^{i} u_{i,j};$$

then $L\{u_i\}$ is i.d.

RCLT. Let B hold. Let the null triangular array $\{u_{i,j}: 1 \leqslant j \leqslant i < \infty\}$ satisfy

$$\prod_{j=1}^{i} u_{i,j} = u$$

for all $i \in 3^+$. Then u is i.d.

DCLT. Let B hold. Let u be such that for all $k \in 3^+$ there are $u_{k,j}$ $(1 \leqslant j \leqslant k)$ with

$$\prod_{j=1}^{k} u_{k,j} = u$$

and $D(u_{k,j}) = D(u)/k$ for $1 \leqslant j \leqslant k$. Then u is i.d.

T2. If $u \in G$ is not simple and no factor of u is simple, u is i.d.

H^*. *For each $i \in 3^+$ there exists an s.c. homomorphism $D_i: G \to (\Re^+, +)$, such that if $D_1(u) = 0$ then $u = e$, and if $D_i(u) = D_i(v)$ for every i, then $u = v$.*

H. *There is an s.c. homomorphism $D: G \to (\Re^+, +)$. Also, for each $i \in 3^+$ there is a group S_i (with identity e_i), which may be either $(\Re, +)$ or $\mathfrak{T}$, and an s.c. homomorphism $D_i: G \to S_i$ such that for all $\varepsilon > 0$ there exists $\delta_i > 0$ such that $|D_i(u) - e_i| \leqslant \varepsilon$ whenever $u \in G$ and $D(u) \leqslant \delta_i$. If $D(u) = D(v)$ and $D_i(u) = D_i(v)$ for all $i \in 3^+$, then $u = v$.*

If an $L\frac{1}{2}$gp G satisfies A and CLT, it is said to be sequentially Delphic. (Cf. [1] and Kendall's basic paper [6]. We recall that a Delphic semigroup is sequentially Delphic if its topology is first countable, and a sequentially Delphic semigroup is Delphic if its limiting operation L is induced by a metrizable topology.)

We now state and pass comments on our main theorems, valid for all $L\frac{1}{2}$gps G:

Theorem 1. *A and H^* imply B and $T2$;*

Theorem 2. *A and H imply B and $RCLT$;*

Theorem 3. *A' and H imply B and CLT, i.e. that G is sequentially Delphic.*

First we have trivially

Lemma 1. *A' implies A;*
each of H and H^* implies B;
CLT implies RCLT, which in its turn implies DCLT.

Further, in [1] we proved

Lemma 2. *If G satisfies A, B and $DCLT$, it satisfies $T2$ also.*

Thus it will be seen that the conclusions are strengthened as we pass from Theorem 1 up to Theorem 3, and the hypotheses as we pass from Theorems 2 to 3.

Theorem 3 is the most important. It gives us a quite general method of proving CLT for a semigroup whose Delphic nature we are trying to establish. We shall see immediately that for many purposes $RCLT$ or even $T2$ will do, but CLT is a beautiful theorem and, since we shall prove it with no knowledge of the i.d. elements of the semigroup, it gives us a way of constructing them.

Theorem 2 differs only slightly from Theorem 3, both in statement and proof. But there are semigroups which satisfy the hypotheses of Theorem 2 and for which the conclusion of Theorem 3 does not hold; the next section is devoted to an example of this.

Theorem 1 differs from Theorems 2 and 3 in that its main conclusion—that the property $T2$ holds—is purely algebraic. Now if we want to study the arithmetic or algebra of some G, an $L\frac{1}{2}$gp satisfying A and B, the only two known results which are to hand if we know that CLT holds for G are (Kendall [6]) $T2$ and what we may call

$T3$. Each $u \in G$ can be written $u = w \cdot \prod v_i$, where w is i.d. without simple factors (i.e. is in 'I_0'), and each v_i (of which there are at most countably many) is simple.

The point is that $T3$ follows from $T2$, A and B (Kendall [6]); and Lemmas 2 and 1 show that $T2$ follows as well from $RCLT$ as from CLT. So for arithmetic and algebraic purposes we are apparently just as well off with the conclusions of our Theorems 1 and 2 as we are with that of Theorem 3.

We state Theorem 1 as well as Theorem 2 because one can see, from the forms of H and H^*, that the latter will be easier to verify than the former, which has the rather awkward 'uniform (over G) smallness of D_i with respect to D' condition. On the other hand, I must admit that I know of no $L\frac{1}{2}$gp satisfying A and H^* but not H; nor do I know whether there is an $L\frac{1}{2}$gp satisfying A, B and $T2$ but not $RCLT$.

2.5.3 *A* AND *H* DO NOT ENTAIL *CLT*

We construct a sub-semigroup G of $(\Re^+, +)$ with the required properties. For $i \in 3^+$ put $x_i = (\log(p_i))/p_i$, where p_i is the ith prime positive integer ($p_1 = 2$). Define

$$G = \{0\} \cup \left\{ \sum_{i=1}^{n} r_i x_{k_i} : n, r_i \, (1 \leqslant i \leqslant n), k_i \, (1 \leqslant i \leqslant n) \in 3^+ \right\}.$$

The semigroup operation on G is addition, and G is topologized as a subspace of $\Re^+$ in the Euclidean topology. For $x \in G$, put $D(x) = x$; then B holds for G (and in fact H and H^* hold also).

Now it is easy to see that the set $\{x_i : i \in 3^+\}$ is linearly independent over the rationals. It follows from the definition of G that each $x \in G$ has precisely one representation

$$x = \sum_{i=1}^{\infty} r_i x_i$$

with the non-negative integers r_i vanishing for all but finitely many i. The set of factors of any $x \in G$ is therefore finite (so that A holds), and each x_i is simple. On the other hand, given any $x \in G$ we can easily construct a null triangular array whose row products converge to x; but no x (save $x = 0$)

is i.d., so *CLT* fails spectacularly. *T*2 holds trivially and *RCLT* vacuously, concluding the proof of the example.

It should be noted that (*i*) $(G, +)$ is algebraically isomorphic to $(3^+, \times)$, and (*ii*) the above example disproves the possible conjecture that in an $L\frac{1}{2}$gp G, A implies A'.

2.5.4 THE THEOREMS

First we need two lemmas on real numbers.

Lemma 3a. *Let there be given* rk $(r, k \in 3^+)$ *objects* $a_1, ..., a_{rk} \in \mathfrak{R}$, *such that* $|a_s| \leqslant c$ *for every* s; *define*

$$T = \sum_{s=1}^{rk} a_s.$$

Let $0 \leqslant l \leqslant k$. *Then we may divide the objects into two sets,* A_l *of* rl *objects and* A_{k-l} *of* $r(k-l)$ *objects, such that*

$$\left| \sum (a_s: a_s \in A_l) - Tl/k \right| \leqslant c$$

(and then of course also $\left| \sum (a_s: a_s \in A_{k-l}) - T(k-l)/k \right| \leqslant c$*).*

Proof. We may suppose that $l \leqslant k-l$, and that the as have been arranged in ascending order. Let S_1, S_2 be respectively the sums of the first and last rl as; then $S_1 \leqslant Tl/k \leqslant S_2$. If we exchange successively a_s with a_{kr-s+1} for $1 \leqslant s \leqslant rl$, ultimately $S_1 \geqslant Tl/k \geqslant S_2$. Since $|a_s| \leqslant c$ for all s, at some intermediate stage of the exchanging we must have $|S_1 - Tl/k| \leqslant c$, which is what we wanted to prove.

Lemma 3b. *Let there be given* rk $(r, k \in 3^+)$ *objects* $a_1, ..., a_{rk} \in \mathfrak{R}$, *such that* $|a_s| \leqslant c$ *for all* s; *define*

$$T = \sum_{s=1}^{rk} a_s.$$

Then we may divide the objects into k *sets* A_j *of* r *objects each, such that for all* j

$$\left| \sum (a_s: a_s \in A_j) - T/k \right| \leqslant 4c.$$

Proof. We may suppose $k \geqslant 2$. Let $l \geqslant 2$ be any integer $\leqslant k$, and suppose that we are dealing with a reduced system of rl as, totalling T_l say ($T_k = T$). If l is even, we can (by Lemma 3a) split this reduced system into two sets of $\frac{1}{2}rl$ objects each, such that their totals $T'_{\frac{1}{2}l}$ and $T''_{\frac{1}{2}l}$ differ from $\frac{1}{2}T_l$ by at most c. Similarly, if l is odd, we can split the system into two sets B (of r objects) and C (of $r(l-1)$ objects), such that the sum of the objects in B

differs from T_l/l by at most c and the sum of the objects in C differs from $(l-1)T_l/l$ by at most c.

So if k is even, we split the system into two sets of size $\frac{1}{2}kr$, and if k is odd, into two sets of sizes r and $(k-1)r$, with maximum inaccuracy in the totals as described above; and then iterate the process on the reduced systems. The number of iterations required to reduce the system to k sets each of r objects is exactly k; and each individual set has then been operated on a maximum of $2t$ times, where $t = \max(n \in 3^+: 2^n \leqslant k)$. It is then easy to see that the sum of the members of one of the final sets differs from T/k by at most $c(2+2.2^{-1}+2.2^{-2}+\ldots+2.2^{-(l-1)}) \leqslant 4c$.

(*Remark*. When we use this lemma it will be seen that the bound of $4c$ is unimportant. All we need is that the bound is independent of r and goes to zero with c. A similar statement to Lemma 3b holds when the objects lie in $\Re^N$ and have norm $\leqslant c$; the bound then $= c.f(N,k)$, still independent of r. If the objects lie in a general real inner-product space, we can only say that the bound $= O(r^{\frac{1}{2}})$ (e.g. in l^2 it $\sim r^{\frac{1}{2}}$).)

Theorem 1. *If A and H* hold for an $L\frac{1}{2}gp$ G, then B and T2 hold.*

Proof. By Lemma 1, B holds. For all $u \in G$ put $H_1(u) = D_1(u)$; for all $i \geqslant 1$, $H_{i+1}(u) = D_{i+1}(u) + H_i(u)$. Then u is determined by its H-values as well as by its D-values, and each H_i is an s.c. homomorphism into $(\Re^+, +)$ and has trivial kernel. Let then $u \in G$ satisfy the hypothesis of $T2$. By the argument leading to Theorem 2 of [1], we find that u satisfies the hypothesis of $DCLT$; in fact, for each separate $i \in 3^+$, the following is true:

(1)   For each $k \in 3^+$, there is a decomposition $u = \prod\limits_{j=1}^{k} v_{j,k}$

such that $H_i(v_{j,k}) = H_i(u)/k$ for $1 \leqslant j \leqslant k$.

For $1 \leqslant l \leqslant i \in 3^+$ let $P_{i,l}$ be the proposition that for each $k \in 3^+$ there is a decomposition

$$u = \prod_{j=1}^{k} v_{j,k}$$

such that $H_m(v_{j,k}) = H_m(u)/k$ for $l \leqslant m \leqslant i$ and $1 \leqslant j \leqslant k$. By statement (1) $P_{i,i}$ holds for all $i \in 3^+$; we want to prove $P_{i,1}$ for all $i \in 3^+$. Let then $i \geqslant 2$ be given and suppose that we have $P_{i,l}$ for some $l > 1$. Let k be arbitrary in 3^+. We know that for each $r \in 3^+$ there is a decomposition

$$u = \prod_{n=1}^{kr} v_{n,kr}$$

such that $H_m(v_{n,kr}) = H_m(u)/kr$ for $l \leqslant m \leqslant i$ and $1 \leqslant n \leqslant kr$. Now H_{l-1} is by its definition dominated by H_i; so we may apply Lemma 3b, with therein $a_s = H_{l-1}(v_{s,kr})$ and $c = H_i(u)/kr$, to conclude that the $v_{n,kr}$ may be divided into k products $w_{j,k}(r)$ of r vs each, such that

$$u = \prod_{j=1}^{k} w_{j,k}(r), \quad H_m(w_{j,k}(r)) = H_m(u)/k$$

for $1 \leqslant j \leqslant k$ and $l \leqslant m \leqslant i$, and $|H_{l-1}(w_{j,k}(r)) - H_{l-1}(u)/k| \leqslant 4H_i(u)/kr$ for $1 \leqslant j \leqslant k$.

Consider the sequence of such decompositions for $r \in 3^+$. By A there exists a convergent subsequence of decompositions, with limit

$$u = \prod_{j=1}^{k} w_{j,k}$$

say; and $H_m(w_{j,k}) = H_m(u)/k$ for $1 \leqslant j \leqslant k$ and $l-1 \leqslant m \leqslant i$. This holds for all k, so $P_{i,l}$ implies $P_{i,l-1}$ if $l > 1$; since $P_{i,i}$ holds for all $i \in 3^+$ so then does $P_{i,1}$.

Consider now for arbitrary fixed $k \in 3^+$ a sequence of decompositions

$$u = \prod_{j=1}^{k} w_{j,k}(i)$$

with $H_m(w_{j,k}(i)) = H_m(u)/k$ for $1 \leqslant j \leqslant k$ and $1 \leqslant m \leqslant i \in 3^+$; we have just shown that such decompositions exist. Again by A we can pick out a subsequence (of $\{i\}$) such that the corresponding decompositions converge, to

$$u = \prod_{j=1}^{k} w_{j,k}^*$$

say; and $H_m(w_{j,k}^*) = H_m(u)/k$ for $1 \leqslant j \leqslant k$ and all $m \in 3^+$. Thus, by the unicity part of H^*, u has a kth root for every $k \in 3^+$ and so is i.d. So $T2$ does indeed hold.

We now state Theorems 2 and 3. They follow at once from Propositions 1 and 2 below.

Theorem 2. *If A and H hold for an $L\frac{1}{2}gp$ G, then B and $RCLT$ also hold for G.*

Theorem 3. *If A' and H hold for an $L\frac{1}{2}gp$ G, then B and CLT also hold for G, so that G is sequentially Delphic.*

Proposition 1. *Let G be an $L\frac{1}{2}gp$ for which $DCLT$ holds. Then (i) $A \Rightarrow RCLT$; (ii) $A' \Rightarrow CLT$.*

Proof (of (ii); that of (i) is very similar and is omitted). Let D be the homomorphism whose existence is demanded, through B, by $DCLT$. We are given a null triangular array $\{u_{i,l}: 1 \leqslant l \leqslant i < \infty\}$ whose row products

$$u_i = \prod_{l=1}^{i} u_{i,l}$$

converge to u, and we must show that u is i.d.

Let $k \in \mathfrak{Z}^+$ be given. Because of the nullity of the array, given $n \in \mathfrak{Z}^+$ there exists $i(n)$ so large that we can group together the $u_{i(n),l} (1 \leqslant l \leqslant i(n))$ into products $v_j(n)$ say $(1 \leqslant j \leqslant k)$, such that

$$u_{i(n)} = \prod_{j=1}^{k} v_j(n)$$

and $D(v_j(n)) - D(u)/k \,|\leqslant 1/n$ for $1 \leqslant j \leqslant k$. Then by A' there is $\{n'\} \subset \{n\}$ such that the corresponding sequence of decompositions (of $u_{i(n')}$) converges to a limit decomposition of u,

$$u = \prod_{j=1}^{k} v_j, \text{ say,}$$

which has $D(v_j) = D(u)/k$ for $1 \leqslant j \leqslant k$. *DCLT* now yields *CLT* at once.

It should be noticed that the above proposition does not make use of either of the strengthened postulates H or H^*, and is thus of some independent interest.

Proposition 2. *Let G be an $L\frac{1}{2}gp$ for which H and A hold. Then B and DCLT also hold for G.*

Proof. By Lemma 1, B holds. Put $D_0 = D$. Let u satisfy the hypothesis of *DCLT*. For non-negative integers l let P_l be the proposition that for each $k \in \mathfrak{Z}^+$ there is a decomposition

$$u = \prod_{j=1}^{k} v_{j,k,l}$$

with $D_m(v_{j,k,l}) = D_m(v_{j',k,l})$ for $1 \leqslant j, j' \leqslant k$ and $0 \leqslant m \leqslant l$. We use induction on l, having P_0 by hypothesis.

Suppose then that P_l holds. Define for each $i \in \mathfrak{Z}^+$ and all $\delta > 0$ the (possibly infinite-valued) function

$$f_i(\delta) = \sup (|D_i(u) - e_i| : u \in G \text{ and } D(u) \leqslant \delta).$$

(Recall that e_i is the identity of $(\mathfrak{R}, +)$ or $\mathfrak{T}$ as appropriate.) We have from H that f_i is finite-valued for all small enough δ, and $f_i(\delta) \to 0$ as $\delta \to 0$. Now let k be arbitrarily fixed in $\mathfrak{Z}^+$. Put $g_{l+1}^{(k)}(r) = f_{l+1}(D(u)/kr)$; then $g_{l+1}^{(k)}(r)$ is defined for all $r \geqslant r_0(k)$, say, and $\to 0$ as $r \to \infty$. From now on we assume that the range of D_{l+1} is $\mathfrak{T}$; the case where the range is $\mathfrak{R}$ is easier and we omit it.

By hypothesis we have for each $r \geq r_0(k)$ a decomposition

$$u = \prod_{n=1}^{kr} v_{n,kr,l}$$

with $D_m(v_{n,kr,l}) = D_m(v_{n',kr,l})$ for $1 \leq n, n' \leq kr$ and $0 \leq m \leq l$, and

$$|D_{l+1}(v_{n,kr,l}) - 1| \leq g_{l+1}^{(k)}(r) \quad \text{for} \quad 1 \leq n \leq kr$$

($e_{l+1} = 1$ since the range of D_{l+1} is $\mathfrak{T}$). For $1 \leq n \leq kr$ define $D_n^* \in \mathfrak{R}$ by

$$iD_n^* = \log(D_{l+1}(v_{n,kr,l})),$$

the principal value of the logarithm being taken. For $1 \leq j \leq k$ let W_j be the set $\{v_{n,kr,l} : r(j-1)+1 \leq n \leq rj\}$, and let w_j be the product of the members of W_j. Define $D^*(W_j) = \sum (D_n^* : r(j-1) \leq n \leq rj)$, and

$$D^{**}(W_j) = \exp(iD^*(W_j)).$$

Then clearly $D^{**}(W_j) = D_{l+1}(w_j)$.

Lemma 3b now applies, with objects $D_n^* (1 \leq n \leq rk)$ and c equal to $(\pi/2)g_{l+1}^{(k)}(r)$. We conclude that we can group the vs into k sets W_j of r vs each, with

$$|D^*(W_j) - D^*(W_{j'})| \leq 2.4(\pi/2)g_{l+1}^{(k)}(r) \quad \text{for} \quad 1 \leq j, j' \leq k.$$

Immediately,

$$|D_{l+1}(w_j) - D_{l+1}(w_{j'})| \leq 4\pi g_{l+1}^{(k)}(r) \quad \text{for} \quad 1 \leq j, j' \leq k.$$

Hence for all $r \geq r_0(k)$ we have a decomposition

$$u = \prod_{j=1}^{k} w_{j,k,l}(r)$$

such that

$$D_m(w_{j,k,l}(r)) = D_m(w_{j',k,l}(r)) \quad \text{for } 0 \leq m \leq l \text{ and } 1 \leq j, j' \leq k,$$

and

$$|D_{l+1}(w_{j,k,l}(r)) - D_{l+1}(w_{j',k,l}(r))| \leq 4\pi g_{l+1}^{(k)}(r) \quad \text{for } 1 \leq j, j' \leq k.$$

Now by A there is a subsequence $\{r'\}$ of $\{r\}$ such that the corresponding sequence of decompositions converges, and the limit decomposition

$$u = \prod_{j=1}^{k} v_{j,k,l+1}$$

has $D_m(v_{j,k,l+1}) = D_m(v_{j',k,l+1})$ for $1 \leq j, j' \leq k$ and $0 \leq m \leq l+1$. Since k was arbitrary we have established P_{l+1} and with it the induction.

Let now $k \in \mathfrak{Z}^+$ be given. For each $l \in \mathfrak{Z}^+$ we can find a decomposition

$$u = \prod_{j=1}^{k} v_j(l)$$

with $D_m(v_j(l)) = D_m(v_{j'}(l))$ for $1 \leqslant j, j' \leqslant k$ and $0 \leqslant m \leqslant l$. Again by A there is a subsequence $\{l'\}$ of $\{l\}$ such that the corresponding sequence of decompositions converges to a limit,

$$u = \prod_{j=1}^{k} v_j, \text{ say,}$$

with $D_m(v_j) = D_m(v_{j'})$ for $1 \leqslant j, j' \leqslant k$ and all $m \in \mathfrak{Z}^+$. The unicity condition in H now tells us that u has a kth root; but k was arbitrary, so u is i.d.

2.5.5 SEMIGROUPS TO WHICH THE FOREGOING THEORY APPLIES

A sub-semigroup G of a commutative semigroup H is said to be *hereditary* in H (see [1]) if for all $u \in G$, $u = v \cdot w$ with $v, w \in H$ implies $v, w \in G$.

Lemma 4. *If an $L\frac{1}{2}gp\ G$ is hereditary in a sequentially compact $L\frac{1}{2}gp\ H$ and inherits the limit operation of H, then A' holds for G.*

Proof. Trivialization of proof of next lemma.

Lemma 5. *Let the $L\frac{1}{2}gp\ G$ be hereditary in a compact Hausdorff semigroup H, and let the limit operation of G be induced by the topology of G as a subspace of H; suppose that this topology (on G) is first countable. Then A' holds for G.*

Proof. Suppose we are given a sequence $\{u_n\}$ of elements of G converging to some $u \in G$, and that $v_n \in F_G(u_n)$ for each n; we have to show that there is a subsequence $\{r\}$ of $\{n\}$ such that $\{v_r\} \to v \in F_G(u)$.

For each n choose $w_n \in G$ such that $u_n = v_n \cdot w_n$. Since $H \times H$ is compact,

(2)   there is a subnet α of $\{n\}$ such that $v_\alpha \to v \in H$ and $w_\alpha \to w \in H$.

But also $u_\alpha \to u \in H$, so $v \cdot w = u$ since H is a Hausdorff topological semigroup. Since actually $u \in G$, and G is hereditary in H, $v \cdot w = u$ implies

(3) $v \in F_G(u) \subset G$.

Now the topology of G is first countable; this and conditions (2) and (3) imply that we can pick a subsequence $\{r\}$ of α such that $\{v_r\} \to v \in F_G(u)$. But $\{r\}$ is a subsequence of $\{n\}$.

Example 1. *Let $\mathscr{R}^+$ be the semigroup of all positive renewal sequences under term-by-term multiplication and convergence. Let a typical element of $\mathscr{R}^+$ be $u = \{u_n: 0 \leqslant n < \infty\}$. Define the homomorphism D_n by $D_n(u) = -\log(u_n)$. It is easy to see that $\mathscr{R}^+$ satisfies Lemma 4 and all the conditions of Section 2.5.2 (see, e.g., [1]).*

Example 2. *Let $\mathscr{P}$ be the semigroup of all standard p-functions of Kingman [7] under pointwise multiplication and convergence. Let a typical element of $\mathscr{P}$ be $p = \{p(t): t \in \mathfrak{R}^+\}$. Define the homomorphism D_n by*

$$D_n(p) = -\log(p(t_n)),$$

where $\{t_n: n \in \mathfrak{Z}^+\}$ is an enumeration of the binary rationals; put

$$D(p) = -\log(p(1)).$$

Now (Kingman [7]) the semigroup of all (standard or not) p-functions is compact under pointwise convergence, and (see [2]) $\mathscr{P}$ is hereditary in this larger semigroup and the subspace topology is metrizable. So Lemma 5 applies, yielding A' for $\mathscr{P}$. Also (see Proposition 6 of [2]) a base of neighbourhoods of the identity e ($e(t) \equiv 1$) is given by $\{p: D(p) < 1/n\}$ $(n \in \mathfrak{Z}^+)$, so H holds for $\mathscr{P}$; H^* also holds, of course.

Example 3. *Let $\mathscr{L}$ be the semigroup of probability laws a (on the circle), under term-by-term multiplication and convergence of the Fourier sequences $\{a_n: 0 \leqslant n < \infty\}$. Let $\mathscr{L}'$ be the sub-semigroup of laws whose first Fourier coefficient a_1 is real and non-negative; then $\mathscr{L}'$, with the limit operation of $\mathscr{L}$, is a sequentially compact $L\frac{1}{2}$gp. Let $\mathscr{L}^+$ be the semigroup of those $a \in \mathscr{L}'$ such that $a_1 > 0$, and let $\mathscr{L}^{++}$ be the semigroup of those $a \in \mathscr{L}^+$ none of whose Fourier coefficients vanish; let $\mathscr{L}^+$ and $\mathscr{L}^{++}$ have the limit operation of $\mathscr{L}$ also. Since $\mathscr{L}^+$ and $\mathscr{L}^{++}$ are hereditary in $\mathscr{L}'$, Lemma 4 tells us that A' holds for both of them.*

Define $D_n: \mathscr{L}^{++} \to (\mathfrak{R}^+, +)$ and $D'_n: L^{++} \to \mathfrak{X}$ by $D_n(a) = -\log(|a_n|)$ and $D'_n(a) = a_n/|a_n|$. Then D_1 will do for D in H, and (by the increments inequality of Loève [10]) H holds for $\mathscr{L}^{++}$, which is accordingly Delphic.

Define $D: \mathscr{L}^+ \to (\mathfrak{R}^+, +)$ by $D(a) = -\log(|a_1|)$. Then by the increments inequality, B holds for $\mathscr{L}^+$; we already have A', so to prove $\mathscr{L}^+$ Delphic we only have to demonstrate CLT. Let then $a \in \mathscr{L}^+$ be the limit of row products of a null triangular array. Then we can use the increments inequality to show that a lies in $\mathscr{L}^{++}$ and satisfies the hypothesis of DCLT, with triangular array Δ say. Because $\mathscr{L}^{++}$ is hereditary in $\mathscr{L}^+$, all the elements of Δ lie in $\mathscr{L}^{++}$, so that a is i.d. in $\mathscr{L}^{++}$ (and so in $\mathscr{L}^+$ also). $\mathscr{L}^+$ is thus Delphic. (The argument of this last paragraph is given at greater length in Proposition 6 in the next section, where we deduce the Delphic nature of $\mathscr{W}^1/Q$ from that of $\mathscr{W}/Q$.)

There are other examples of semigroups where the methods of Section 2.5.4 work (e.g., Lamperti's [8] semigroups R_m), but we omit them all except the convolution semigroup $\mathscr{W}^*$ of probability laws on the line, dealt with now.

2.5.6 APPLICATION OF THE THEORY TO $\mathscr{W}^*$

The elements of $\mathscr{W}^*$ are to be regarded as characteristic functions, and are written $f, g, \dots$ Let $\mathscr{W} \subset \mathscr{W}^*$ be the set of those f which vanish nowhere on $\mathfrak{R}$. Then $\mathscr{W}$ is an hereditary sub-semigroup of $\mathscr{W}^*$ (under the semigroup operation of pointwise multiplication). The units of $\mathscr{W}^*$ are e_λ: $e_\lambda(t) = e^{i\lambda t}, \lambda \in \mathfrak{R}$. Define the equivalence relation Q on $\mathscr{W}^*$ by fQg if there is a unit e such that $g = e.f$; then it is clear that Q is an equivalence relation on $\mathscr{W}$ also, and, indeed, an equivalence relation on every $\mathscr{W}^a$ (the set of those $f \in \mathscr{W}^*$ which do not vanish in $[-a, a]$). We shall write the equivalence class of $f \in \mathscr{W}^*$ as Qf, and the elements of $\mathscr{W}^*/Q$ in general as u, v, w etc.

For $f \in \mathscr{W}$, define
$$D_i(f) = -\log(|f(r_i)|) \quad (i \in \mathfrak{Z}^+),$$
where $\{r_i : i \in \mathfrak{Z}^+\}$ is some enumeration of the positive binary rationals; and
$$D_{m,n}(f) = \frac{f(m.2^{-n})}{|f(m.2^{-n})|} \left\{\frac{f(2^{-n})}{|f(2^{-n})|}\right\}^{-m} \quad (m, n \in \mathfrak{Z}^+).$$

Then it is clear that each $D_{m,n}$ and D_i is constant on the Q-equivalence classes in $\mathscr{W}$, so that $D_{m,n}$ and D_i are homomorphisms from $\mathscr{W}/Q$ to $\mathfrak{X}$ and $(\mathfrak{R}^+, +)$ respectively. Our first task is to show that collectively they separate the points of $\mathscr{W}/Q$.

Let β be the topology on $\mathscr{W}$ of convergence on the binary rationals.

Proposition 3. *Let $f, g \in \mathscr{W}$ be such that $D_i(f) = D_i(g)$ for every i, and $D_{m,n}(f) = D_{m,n}(g)$ for every m and n. Then fQg.*

Proof. Since $D_i(f) = D_i(g)$ for every $i, |f(t)| = |g(t)|$ for all $t \in \mathfrak{R}$, by continuity. Also for each $n \in \mathfrak{Z}^+$ there is a unit $e_{\lambda_n} \in \mathscr{W}$ such that
$$(e_{\lambda_n}.f)(2^{-n}) = g(2^{-n});$$
because $D_{m,n}(f) = D_{m,n}(g)$ this implies that $(e_{\lambda_n}.f)(m.2^{-n}) = g(m.2^{-n})$ for all $m \in \mathfrak{Z}^+$. Thus g lies in the β-closure of Qf. Lemma 6 below yields the proposition at once.

Lemma 6. *For every $f \in \mathscr{W}$, Qf is β-closed.*

Proof. We have to show that the quotient topology on $\mathscr{W}/Q$ is T_1; we in fact show that it is Hausdorff. For this it suffices that Q be closed in $\mathscr{W} \times \mathscr{W}$ and that the projection q of $\mathscr{W}$ onto $\mathscr{W}/Q$ be open (see e.g. Kelley [4], p. 98).

We see easily, by considering basic neighbourhoods, that $\mathscr{W}$ is a topological semigroup under β. So if e_λ is any unit in $\mathscr{W}$, the map E_λ of $\mathscr{W}$ onto itself given by $E_\lambda(f) = e_\lambda . f$ is continuous, and has the continuous inverse $E_{-\lambda}$, so is a homeomorphism. Now q is open if and only if for every open set $U \subset \mathscr{W}$, the set $\{g:$ there exists $f \in U$ such that $gQf\}$ is open. This set is, of course, just $\bigcup\{e_\lambda U: \lambda \in \mathfrak{R}\}$. But each E_λ is a homeomorphism, so each $e_\lambda U$ is open and the same goes for their union. q is thus indeed an open map.

Let $((x_a, y_a): a \in A)$ be a net in $\mathscr{W} \times \mathscr{W}$ converging to $(x, y) \in \mathscr{W} \times \mathscr{W}$, such that $x_a Q y_a$ for each a. The characteristic functions x and y are continuous and non-zero, so their quotient is also continuous. For each binary rational t_n we have $x_a(t_n)/y_a(t_n) \to x(t_n)/y(t_n)$, from which it follows that since each x_a/y_a (*qua* function on $\mathfrak{R}$) is non-negative-definite, x/y is non-negative-definite on the binary rationals, that is,

$$\sum_{i,j=1}^{n} \xi_i \bar{\xi}_j (x/y)(t_i - t_j) \geqslant 0$$

for all complex ξ_i $(1 \leqslant i \leqslant n)$, binary rational t_i $(1 \leqslant i \leqslant n)$, and $n \in \mathfrak{Z}^+$. Continuity then implies non-negative-definiteness on $\mathfrak{R}$. It is trivial that $|(x/y)(t)| = 1$ for all $t \in \mathfrak{R}$, so (see Loève [10], pp. 207 and 202) x/y is the characteristic function of a degenerate distribution, i.e. xQy. So Q is closed in $\mathscr{W} \times \mathscr{W}$. Thus the quotient topology on $\mathscr{W}/Q$ is indeed Hausdorff.

Define a further homomorphism $D: \mathscr{W}/Q \to (\mathfrak{R}^+, +)$:

$$D(u) = -\int_0^1 \log(|f(s)|)\,ds,$$

where $f \in \mathscr{W}$ is such that $u = Qf$. We observe that D is finite-valued and does not depend on the particular f.

For $u, v \in \mathscr{W}^*/Q$ define a distance $d^*(u, v)$ by

$$d^*(u, v) = \inf(d(f, g): Qf = u, Qg = v),$$

where $d(f, g)$ is the Lévy metric. It is easy to check that d^* is a metric, and that $\{u_n\} \to u \in \mathscr{W}^*/Q$ under d^* if and only if there is $\{f_n\} \in \mathscr{W}^{*3^+}$ with $u_n = Qf_n$ for each n, and $f \in \mathscr{W}^*$ with $u = Qf$, such that $\{f_n\} \to f$ under d.

Lemma 7. $\mathscr{W}^*/Q$ (*and similarly* $\mathscr{W}/Q$) *is an* $L\frac{1}{2}gp$ *under pointwise multiplication of representatives and the limiting operation on sequences induced by* d^*.

Proof trivial.

Proposition 4. H *(of Section* 2.5.2) *holds for* $\mathcal{W}/Q$.

Proof. We proved in Proposition 3 the unicity part of H. It is obvious (since the Lévy metric induces uniform convergence on compacta) that D is continuous with respect to d^*, and the same goes for every D_i and $D_{m,n}$. Thus all we have to show is that each D_i and $D_{m,n}$ is uniformly (over $\mathcal{W}/Q$) small with respect to small D. Let then $u \in \mathcal{W}/Q$, and choose $f \in \mathcal{W}$ such that $u = Qf$. Using inequalities given by Loève ([10], pp. 303 and 306), we find that for all $s \in \Re^+$,

$$1 - |f(s)|^2 \leqslant sc + 2\int_{|x| \geqslant c} d\bar{F} \leqslant sc + 2(1 + c^{-2})\int_{|x| \geqslant c} x^2/(1 + x^2)\,d\bar{F}$$

$$\leqslant sc + 2(1 + c^{-2})k\int_0^1 (1 - |f(t)|^2)\,dt,$$

where $\bar{F}$ is the symmetrized d.f. of f, c is any positive number, and k is a positive absolute constant. We put $c = (D(u))^{\frac{1}{3}}$, and find at once that

(4) $$1 - |f(s)|^2 \leqslant s(D(u))^{\frac{1}{3}} + 4k(D(u) + (D(u))^{\frac{1}{3}}),$$

from which it follows that every D_i is uniformly small w.r.t. small D.

Now characteristic functions are non-negative-definite; so all for $f \in \mathcal{W}$, positive binary-rational t, complex ξ_1 and ξ_2, and $m \in \mathfrak{Z}^+$,

$$\sum_{i,j=1,2} \xi_i \bar{\xi}_j f(t_i - t_j) \geqslant 0,$$

where $t_1 = t$ and $t_2 = mt$. From this we deduce that for such t and m,

$$\cos\{\arg(f(mt)) - [\arg(f((m-1)t)) + \arg(f(t))]\}$$

$$\geqslant \frac{|f(mt)|^2 + |f((m-1)t)|^2 + |f(t)|^2 - 1}{2|f(mt)||f((m-1)t)||f(t)|}.$$

Uniform smallness of each $D_{m,n}$ w.r.t. small D follows from this and inequality (4).

Mr. J. G. Basterfield, who was the first to show (quoting the central limit theorem, which we deduce at the end of this section) that $\mathcal{W}^*/Q$ is covered by (hereditary) Delphic semigroups, has proved that A (of Section 2.5.2) holds for $\mathcal{W}^*/Q$. The next result extends this slightly.†

† *Note added in proof.* Professor D. G. Kendall has drawn my attention to a note by R. Cuppens (*Comptes Rendus Acad. Sci. Paris*, **256** (A) (1963)), 3560—3561 who proves (in our notation) that

$$f_n \to f \in \mathcal{W}^*, \quad g_n \in F_{\mathcal{W}^*}(f_n)\,\forall n \Rightarrow \lim_{n \to \infty} \inf\{d(g_n, h) : h \in F_{\mathcal{W}^*}(f)\} = 0.$$

To establish this, he effectively proves our Proposition 5.

Proposition 5. *A' holds for $\mathscr{W}^*/Q$.*

Proof. Let $\{u_n\} \to u \in \mathscr{W}^*/Q$ under d^*, and let v_n be a factor of u_n for each n. Then for each n there are distribution functions F_n and G_n, with characteristic functions f_n and g_n, such that $u_n = Qf_n, v_n = Qg_n, G_n$ is a $\mathscr{W}^*$-factor of F_n and is centred at a median, and $\{F_n\} \to F$ under d, where F has characteristic function f and $u = Qf$. For each n, let F_n^s and G_n^s be the symmetrized d.f.s of F_n and G_n respectively.

By the symmetrization inequalities of Loève ([10], p. 245) we have, for all $c > 0$,

$$(5) \qquad G_n(\mathfrak{R}\backslash] - c, c[\,) \leqslant 2G_n^s(\mathfrak{R}\backslash] - c, c[\,) \leqslant 4F_n^s(\mathfrak{R}\backslash] - c, c[\,)$$
$$\leqslant 8F_n((\mathfrak{R}\backslash] - \tfrac{1}{2}c, \tfrac{1}{2}c[\,).$$

Since $\{F_n\} \to F$ under d, the set $\{F_n : n \in 3^+\}$ of distribution functions is tight. Together with formula (5) this implies that the set $\{G_n : n \in 3^+\}$ is also tight, and so there is $\{n'\} \subset \{n\}$ such that $\{G_{n'}\} \to G$, say, under d, where G is an honest distribution function. Thus $\{v_{n'}\} \to v = Qg$, where g is the ch.f. of G. Iterating the argument, we see that there is $\{n''\} \subset \{n'\}$ such that $\{w_{n''}\} \to w$, say, where w_n is some cofactor of v_n in u_n. By continuity of multiplication we have $v.w = u$, so v is a factor of u.

Corollary 1. *A' holds for all hereditary sub-semigroups of $\mathscr{W}^*/Q$ (and so in particular for $\mathscr{W}/Q$).*

Corollary 2. *$\mathscr{W}/Q$ is a sequentially Delphic (and, because the limiting operation is induced by a metric, Delphic) semigroup.*
This follows from the above corollary, Proposition 4, Lemma 7 and the general theory.

Recall that $\mathscr{W}^a$ was to be the sub-semigroup of $\mathscr{W}^*$ consisting of those f such that $f(t) \neq 0$ for $|t| \leqslant a$ $(a > 0)$; we remarked earlier that Q was an equivalence relation on $\mathscr{W}^a$. It is clear that $\mathscr{W}/Q$ is an hereditary sub-semigroup of each $\mathscr{W}^a/Q$, and that each $\mathscr{W}^a/Q$ is an hereditary sub-semigroup of $\mathscr{W}^*/Q$. For $u \in \mathscr{W}^a/Q$ we define

$$D^a(u) = -\int_0^a \log(|f(t)|)\,dt,$$

where f is such that $u = Qf$.

Lemma 8. *For each a, $\mathscr{W}^a/Q$ is (under the limit operation induced by d^*) an $L\frac{1}{2}gp$ for which A' and B (of Section 2.5.2) hold (with $D = D^a$).*

Proof. See Corollary 1 to Proposition 5 for A'; B is trivial.

Proposition 6. *For each a, $(\mathscr{W}^a/Q, d^*, D^a)$ is Delphic.*

Proof. The correspondence $f(t) \in \mathscr{W}^1 \sim f(at) \in \mathscr{W}^a$ induces a topological isomorphism $\mathscr{W}^1/Q \cong \mathscr{W}^a/Q$, so it suffices to consider the case $a = 1$. In view of Lemma 8 and the fact that the limit operation in $\mathscr{W}^1/Q$ is induced by the metric d^*, it suffices to prove *CLT* for $\mathscr{W}^1/Q$.

Let then $\{u_{i,j}: 1 \leqslant j \leqslant i < \infty\}$ be a triangular array in $\mathscr{W}^1/Q$ such that

$$\lim_{\substack{i \to \infty}} \max_{1 \leqslant j \leqslant i} D^1(u_{i,j}) = 0 \quad \text{and} \quad \{u_i\} \to u \in \mathscr{W}^1/Q,$$

where

$$u_i = \prod_{j=1}^{i} u_{i,j}.$$

Then (as in Proposition 1) given any $k \in 3^+$, for all $n \in 3^+$ there is an $i(n) \in 3^+$ so large that we can make a decomposition

$$u_{i(n)} = \prod_{j=1}^{k} v_{j,k,n}$$

such that

$$\left| D^1(v_{j,k,n}) - D^1(u)/k \right| < 1/n \quad \text{for } 1 \leqslant j \leqslant k.$$

Using A' we see that there is a decomposition

$$u = \prod_{j=1}^{k} v_{j,k}$$

with

(6) $$D^1(v_{j,k}) = D^1(u)/k \quad \text{for } 1 \leqslant j \leqslant k.$$

Now D^1 is the D of Proposition 4, and so, because of inequality (4), for every $s \in \mathfrak{R}$ there is a $k \in 3^+$ so large that $|g_{j,k}(s)| > 0$ for $1 \leqslant j \leqslant k$, where $v_{j,k} = Qg_{j,k}$. Thus $f(s) \neq 0$ for all $s \in \mathfrak{R}$, where $u = Qf$, so that $u \in \mathscr{W}/Q$. Now $\mathscr{W}/Q$ is hereditary in $\mathscr{W}^1/Q$, so $u \in \mathscr{W}/Q$ implies that the components of the decomposition in (6) all lie in $\mathscr{W}/Q$, whatever be $k \in 3^+$. By Proposition 2, then, u is i.d. in $\mathscr{W}/Q$, and so also in $\mathscr{W}^1/Q$.

$\mathscr{W}^*/Q$ is, of course, covered by the semigroups $\mathscr{W}^a/Q$ as a runs through the positive reals. Since each $\mathscr{W}^a/Q$ is hereditary in $\mathscr{W}^*/Q$, is open in $\mathscr{W}^*/Q$ (by the definition of the compact-open topology, generated on $\mathscr{W}^*$ by d), and is now seen to be Delphic, we can apply Propositions 5 and 6 of [1] to deduce the Khintchine theorems and the central limit theorem for $\mathscr{W}^*/Q$.

2.5.7 THE I_0-PROBLEM AND SPIRAL SEMIGROUPS

When G is a Delphic, or sequentially Delphic, semigroup, let $I(G)$ be the sub-semigroup of G consisting of its i.d. members, and let $I_0(G)$ be the set of those elements of $I(G)$ all of whose factors in G are i.d. It is usually quite easy, given G, to conjecture what $I(G)$ may be (though it may not be easy to prove the conjecture); the problem of finding $I_0(G)$ is in general much harder, for I_0 is not bound to be a sub-semigroup of G. In fact, the relations of G and its I and I_0 can take practically any form, subject, of course, to I being a sub-semigroup of G and I_0 being hereditary in G.

For example:

In $\mathscr{R}^+$, I_0 is a non-trivial $(I_0 \neq \{e\}$ or $I)$ sub-semigroup of I. (See [1]. In that paper we showed also that $\mathscr{R}^+$ could be represented as the direct product of two Delphic semigroups, to wit $I_0(\mathscr{R}^+)$ and another called $\mathscr{R}^+/D$; and $I_0(I_0(\mathscr{R}^+)) = I_0(\mathscr{R}^+)$, while $I_0(\mathscr{R}^+/D) = \{e\}$, and $I(\mathscr{R}^+/D)$ is a non-trivial sub-semigroup of $\mathscr{R}^+/D$.)

In $\mathscr{W}/Q$, I_0 is not a semigroup (Lévy [9]). But (Ostrowski [11]) every element of $I(\mathscr{W}/Q)$ can be represented as an at most countable product of members of $I_0(\mathscr{W}/Q)$.

In $\mathscr{L}^+$, I_0 is not a semigroup, nor does it generate I (Davidson [3]).

In $(\mathfrak{Z}^+, \times)$, $I_0 = I = \{1\}$.

These examples still leave open the conjectures which follow.

Conjecture 1. *If G is Delphic, then $u \in I_0$ implies $u^2 \in I_0$ (i.e., I_0 is star-shaped*);

Conjecture 2. *If G is Delphic and $I_0 = I$, then $I = \{e\}$ or $I = G$.*

We debunk both these conjectures with the spiral semigroups below.

Example 4. *Let S consist of two equiangular spirals S_0 and S_1: in the complex plane, $S_0 = \{2^{-a} \cdot e^{a\pi i} : a \in \mathscr{R}^+\}$, and $S_1 = \{2^{-a-1} \cdot e^{a\pi i} : a \in \mathscr{R}^+\}$. The semigroup operation is complex-number multiplication, and S is to have the metric topology induced on it as a subset of the complex plane. Now S is hereditary in its closure in the plane, which is sequentially compact, so that A' of Section 2.5.2 holds; S is trivially an $L\frac{1}{2}gp$. For $u \in S$ put $D(u) = -\log(|u|)$ and $D_1(u) = u/|u|$; it is then easy to check that H is satisfied (the restriction that there be an infinite sequence of homomorphisms is irrelevant, and anyway we can satisfy it by putting $D_n(u) = D(u)$ for all $n \geqslant 2$). Theorem 3, and the fact that the limit operation on S is induced by a metric, now tell us that S is Delphic.*

Clearly $I(S) \supset S_0$. *On the other hand, let* $\{u_{i,j}: 1 \leqslant j \leqslant i < \infty\}$ *be a convergent null triangular array. By the nullity, for all large enough* i *all the* $u_{i,j}$ *for* $1 \leqslant j \leqslant i$ *must be in* S_0, *so the limit of the array is in* S_0. *So* $I(S) = S_0$. *It is trivial that* S *has precisely one simple element, the point of* S_1 *which has* $a = 0$. *Since* $2^{-2} \cdot e^{2\pi i} = (2^{-0-1} \cdot e^{0\pi i})^2$, $I_0(S)$ *must lie in the part of* S_0 *which has* $a < 2$. *But it is easy to see that the unique simple is not a factor of any element of* S_0 *which has* $a < 2$, *so* $I_0(S)$ *is just the part of* S_0 *with* $a < 2$. *Conjecture* 1 *is thus seen to be false.*

Example 5. *Let* S^* *be the sub-semigroup of the complex plane under multiplication generated by* S_0 *of the last example and the complex number* e^{-1}, *so that geometrically* S^* *is an infinite set of similar equiangular spirals, with starting points* e^{-n} *for all non-negative integers* n. *In the same way as before we see that* S^* *is a Delphic semigroup and that* $I(S^*) = S_0$. *As in the last example,* S^* *has exactly one simple, in this case the point* e^{-1}; *since no integral power of this simple lies on* S_0 *we find that all of* S_0 *lies in* $I_0(S^*)$, *so that* $I_0(S^*) = I(S^*)$, *and yet* $I(S^*)$ *is a non-trivial sub-semigroup of* S^*. *Conjecture* 2 *is thus seen to be false.*

I am very grateful to Professor J. F. C. Kingman for proposing, at my Ph.D. interview, the problems examined here.

REFERENCES

1. R. Davidson, "Arithmetic and other properties of certain Delphic semigroups, I", *Z. Wahrsch'theorie & verw. Geb.* **10** (1968), 120–145. (2.3 of this book.)
2. ——, "Arithmetic and other properties of certain Delphic semigroups, II", *Z. Wahrsch'theorie & verw. Geb.* **10** (1968), 146–172. (2.4 of this book.)
3. ——, Ph. D. thesis, Cambridge University (1968).
4. J. L. Kelley, *General Topology*, Van Nostrand, Princeton (1955),
5. D. G. Kendall, "Delphic semigroups", *Bull. Am. math. Soc.* **73**(1) (1967), 120–121.
6. ——, "Delphic semigroups, infinitely divisible regenerative phenomena, and the arithmetic of *p*-functions", *Z. Wahrsch'theorie & verw. Geb.* **9** (1968), 163–195. (2.2 of this book.)
7. J. F. C. Kingman, "The stochastic theory of regenerative events", *Z. Wahrsch'theorie & verw. Geb.* **2** (1964), 180–224.
8. J. Lamperti, "The arithmetic of certain semigroups of positive operators", *Proc. Cam. phil. Soc.* **64** (1968), 161–166.
9. P. Lévy, "Sur les exponentielles de polynômes ...", *Ann. sci. Ecole norm. sup.* III Sér. **54** (1937), 231–292.
10. M. Loève, *Probability theory*, 3rd ed., Van Nostrand, Princeton (1963).
11. I. V. Ostrowski, "On factorizations of infinitely divisible laws without Gaussian components" (in Russian), *Doklady Akad. Nauk USSR* **161** (1965), 48–52.

2.6

Sorting Vectors

R. DAVIDSON

[Reprinted from *Proc. Cambridge Philos. Soc.* **68** (1970), 153–157.]

2.6.1 INTRODUCTION

In what follows, X will be a real inner-product linear space (inner product (x,y), norm $|x| = (x,x)^{\frac{1}{2}}$); E^n will be n-dimensional real Euclidean space. We shall be dealing with sets $\{a_s: 1 \leqslant s \leqslant rk; r \geqslant 1, k \geqslant 2\}$ of elements of X; we write $\mathbf{a}$ for the vector $(a_1, ..., a_{rk})$, and put

$$T = \sum_{s=1}^{rk} a_s.$$

We suppose that $|a_s| \leqslant \varepsilon (>0)$ for every s.

We want to sort the rk elements of $\mathbf{a}$ into k sets A_j $(1 \leqslant j \leqslant k)$ of r each, so that the (vector) sums of the elements of each A_j are, as nearly as possible, equal (to T/k). Let then $B(k, r, \mathbf{a}, X)$ be the minimum, over all possible choices of $\{A_j: 1 \leqslant j \leqslant k\}$, of

$$\max_{1 \leqslant j \leqslant k} \{|\sum (a_s \in A_j) - T/k|\}.$$

We want to say that $B(k, r, \mathbf{a}, X)$ is small over all $\mathbf{a}$ with $|a_s| \leqslant \varepsilon (1 \leqslant s \leqslant rk)$; that is, that

$$B(k, r, \varepsilon, X) = \sup\{B(k, r, \mathbf{a}, X): \mathbf{a} \in X^{rk}, |a_s| \leqslant \varepsilon \text{ for each } s\}$$

is small. Here, k and X are to be thought of as fixed, and r and ε as variable, ε small, r large.

In a former paper [2] we considered this problem for $X = E^1$, and obtained

Theorem 0. $B(k, r, \varepsilon, E^1) \leqslant 4\varepsilon.$

We give a résumé of the simplest case of the main theorem of [2], to show where Theorem 0 is needed.

201

Let S be a first countable Hausdorff commutative topological semigroup with identity, possessing a countable separating increasing family of continuous homomorphisms $D_i (i \geqslant 1)$ into the non-negative reals under addition. Suppose that for all $u \in S$, the set of factors of u is compact. Given $u \in S$, we define, for positive integers k, j and $m \leqslant j$, the proposition $P(k, j, m)$ thus:

'There exist $v_1, \ldots, v_k \in S$ such that

$$u = \prod_{i=1}^{k} v_i$$

and $D_l(v_i) = D_l(u)/k$ for $1 \leqslant i \leqslant k$ and $m \leqslant l \leqslant j$.'

Suppose we know that, for our u, $P(k, j, j)$ holds for all k and j. It is desired to show that u has a kth root for each positive integer k (a sort of central limit theorem); for this it suffices to prove $P(k, j, 1)$ for all k and j. To do this, we fix j and prove $P(k, j, m)$ for all k and decreasing values of m, by induction on increasing $j - m$. Thus, given the induction up to $j - m$, and a fixed k, we split u into kr factors (r a large positive integer), each with the same value of D_l for $m \leqslant l \leqslant j$. It is then necessary to make a statement about D_{m-1}. This is done by grouping the kr factors into k sets of r each. Since there are r members of each set, the D_l-value of the product of each set is the same ($m \leqslant l \leqslant j$), and because of Theorem 0 we can make the D_{m-1}-values of the products more or less equal. Let $r \to \infty$; then ε (of Theorem 0) $\to 0$; and thus, *because there is no dependence of the bound on r* in Theorem 0, we get the induction (by the assumption of compactness of sets of factors).

It is thus seen that what we want out of the bound $B(k, r, \varepsilon, X)$ is that it go to zero with ε (in fact any reasonable bound will be linear in ε) and should not depend on r.

The proof of Theorem 0 depends essentially on the ordering of E^1, so it is interesting to see what can be done for general X and, in particular, for E^n. It will incidentally emerge from Theorems 1 and 3 below that the induction outlined above could be circumvented (we could obtain $P(k, j, 1)$ directly, using Theorems 1 and 3 on E^j).

2.6.2 RESULTS

Theorem 1. *Define* $B(r, \varepsilon, X) = B(2, r, \varepsilon, X)$. *For each k there is a finite positive $K(k)$ such that for all r, ε, X,*

$$B(k, r, \varepsilon, X) \leqslant K(k) B(r, \varepsilon, X).$$

Theorem 2. *For all r, ε, X,*

$$B(r, \varepsilon, X) \leqslant \varepsilon . r^{\frac{1}{2}};$$

but there exists X such that for all r and ε,

$$B(r, \varepsilon, X) \geqslant \varepsilon(\tfrac{1}{2}r)^{\frac{1}{2}}.$$

Theorem 3. *For each n there exists $R_n > 0$ such that for all r and ε,*

$$B(r, \varepsilon, E^n) \leqslant \varepsilon \sqrt{R_n}.$$

Further, we have the following estimates:

$$R_1 = 1, R_2 \leqslant 2, R_3 \leqslant 6;$$

and for all $n \geqslant 4$,

$$R_n \leqslant [\tfrac{1}{2} N_n^*],$$

where

$$N_n^* = \sqrt{\pi}\Gamma(\tfrac{1}{2}(n-1))\bigg/\bigg\{2\sqrt{2}\Gamma(\tfrac{1}{2}n)\int_0^{\frac{1}{4}\pi}(\cos\theta - \cos\tfrac{1}{4}\pi)\sin^{n-2}\theta\,d\theta\bigg\}.$$

Proof of Theorem 1. Given r, ε, X, put $B = B(r, \varepsilon, X)$. In the sequel we shall always write $T\ldots$ for $\sum(a_s \in A\ldots)$, where $\ldots$ is any set of affixes and $A\ldots$ is any set of r elements of **a**.

Given k, divide, quite arbitrarily, the rk elements of **a** into k sets $A_j^{(0)}$ $(1 \leqslant j \leqslant k)$ of r each. We define now a canonical process for obtaining a new partition $A_j^{(t+1)}$ $(1 \leqslant j \leqslant k)$ from the previous one $A_j^{(t)}$ $(1 \leqslant j \leqslant k)$ (where t is any non-negative integer):

Sort the members of $A_1^{(t)} \cup A_2^{(t)}$ into new sets A_1', A_2', such that

$$\big|T_i' - \tfrac{1}{2}(T_1^{(t)} + T_2^{(t)})\big| \leqslant B \quad (i = 1, 2);$$

this can be done, by the definition of B. For $2 \leqslant j \leqslant k - 1$, sort the members of

$$A_j' \cup A_{j+1}^{(t)} \quad \text{into new sets} \quad A_j^{(t+1)}, A_{j+1}',$$

such that $\big|T_j^{(t+1)} - \tfrac{1}{2}(T_j' + T_{j+1}^{(t)})\big| \leqslant B$ and $\big|T_{j+1}' - \tfrac{1}{2}(T_j' + T_{j+1}^{(t)})\big| \leqslant B$. Lastly, sort the members of $A_k' \cup A_1'$ into new sets $A_k^{(t+1)}, A_1^{(t+1)}$, such that

$$\big|T_i^{(t+1)} - \tfrac{1}{2}(T_k' + T_1')\big| \leqslant B \quad (i = k, 1).$$

At the end of this process we shall have

(1) $$\mathbf{T}^{(t+1)} = \mathbf{PT}^{(t)} + \mathbf{e}^{(t+1)},$$

where $\mathbf{T}^{(t+1)}$ and $\mathbf{T}^{(t)}$ are the column $(k \times 1)$ vectors of totals, $\mathbf{P}$ is a certain fixed $(k \times k)$ matrix (whose elements are given in equation (3) below), and $\mathbf{e}^{(t+1)}$ is a $(k \times 1)$ vector of errors $e_j^{(t+1)}$ $(1 \leqslant j \leqslant k)$, which depend on **a** but must in any case satisfy

(2) $$\big|e_j^{(t+1)}\big| \leqslant 5B/2 \quad (1 \leqslant j \leqslant k) \quad \text{and} \quad \sum_{j=1}^{k} e_j^{(t+1)} = 0.$$

Let the elements of $\mathbf{P}$ be $P_{i,j}$ $(1 \leqslant i, j \leqslant k)$. Then we have, from the definition of the process above, and elementary calculations, that

$$
\begin{aligned}
(3) \qquad \text{for} \quad j = 1 \text{ or } k, \quad P_{i,j} &= 2^{-2} + 2^{-k} \quad (i = 1, 2) \\
&= 2^{-2+i-k} \quad (i \geqslant 3); \\
\text{for} \quad 1 < j < k, \quad P_{i,j} &= 2^{-j} \quad (i = 1) \\
&= 2^{-j+i-1} \quad (2 \leqslant i \leqslant j+1) \\
&= 0 \quad (i > j+1).
\end{aligned}
$$

Thus (in the language of the theory of finite Markov chains; see e.g. Cox and Miller [1]), the transpose $\mathbf{P}'$ of $\mathbf{P}$ is the one-step transition matrix of an *irreducible aperiodic* (or *primitive*) Markov chain; also it is easy to check that $\mathbf{P}'$ is in fact *doubly stochastic*. From this it follows ([1], pp. 118ff.) that there are constants $K > 0$ and $\rho \in \,]0, 1[$, depending only on the matrix $\mathbf{P}$ (which is independent of r, $\mathbf{a}$ and X), such that for all positive integers t,

$$
(4) \qquad |P_{i,j}^{(t)} - 1/k| \leqslant K\rho^t, \quad \text{uniformly in } i \text{ and } j,
$$

where $P_{i,j}^{(t)}$ is the (i, j)th entry in the matrix $\mathbf{P}^t$.

Now from equation (1) we have, for all positive integers t,

$$
(5) \qquad \mathbf{T}^{(t)} = \mathbf{P}^t \mathbf{T}^{(0)} + \sum_{\tau=1}^{t} \mathbf{P}^{t-\tau} \mathbf{e}^{(\tau)}.
$$

Let $\mathbf{Z}$ be the $(k \times k)$ matrix with all entries equal to $1/k$, and put $\mathbf{Q}^{(t)} = \mathbf{P}^t - \mathbf{Z}$. Then we may rewrite equation (5) thus

$$
\mathbf{T}^{(t)} = \mathbf{Z}\mathbf{T}^{(0)} + \mathbf{Q}^{(t)} \mathbf{T}^{(0)} + \sum_{\tau=1}^{t} (\mathbf{Z} + \mathbf{Q}^{(t-\tau)}) \mathbf{e}^{(\tau)}
$$

$$
= \mathbf{T}/k + \mathbf{Q}^{(t)} \mathbf{T}^{(0)} + \sum_{\tau=1}^{t} \mathbf{Q}^{(t-\tau)} \mathbf{e}^{(\tau)},
$$

(where $\mathbf{T}$ is the $(k \times 1)$ vector all of whose elements equal T), by equation (2) and the definition of $\mathbf{Z}$. Estimating via equations (2) and (4), we obtain

$$
|T_i^{(t)} - T/k| \leqslant K \cdot \rho^t \sum_{j=1}^{k} |T_j^{(0)}| + \sum_{\tau=1}^{t} K \cdot \rho^{t-\tau} \sum_{j=1}^{k} |e_j^{(\tau)}|, \quad \text{for all } i;
$$

$$
\leqslant K\rho^t kr\varepsilon + \sum_{\tau=1}^{t} Kk\rho^{t-\tau}(5B/2).
$$

Since $0 < \rho < 1$, we see from this that for all positive integers m there is a t_m so large that

$$
(6) \qquad |T_i^{(t_m)} - T/k| \leqslant 1/m + 5BKk/2(1-\rho), \quad \text{for all } i.
$$

Now consider the sequence of partitions $\{A_j^{(f_m)}\,(1\leqslant j\leqslant k)\}$, as m runs through the positive integers. Because there is only a finite number of partitions possible, some partition must occur infinitely often in this sequence; let this partition be

$$\{A_j\,(1\leqslant j\leqslant k)\}.$$

Then in virtue of inequality (6) we have

$$|T_j - T/k| \leqslant 5BKk/2(1-\rho), \quad \text{for all } j.$$

But this is the statement of the theorem, with $K(k) = 5Kk/2(1-\rho) < \infty$.

Proof of Theorem 2. For any partition of the elements of $\mathbf{a}$ into two sets A, A' each of r members, we define the *deviation* $\delta_r(A, \mathbf{a}) = \sum (a_s \in A) - \tfrac{1}{2}T$. There will exist a partition, $(A_\mathbf{a}, A'_\mathbf{a})$ say, such that this deviation is of minimum norm; we put then $\mu_r(\mathbf{a}) = \delta_r(A_\mathbf{a}, \mathbf{a})$. We have then, in the notation of Section 2.6.1,

$$\sup\{|\mu_r(\mathbf{a})|: a_s \in X \text{ and } |a_s| \leqslant \varepsilon \text{ for } 1 \leqslant s \leqslant 2r\} = B(r, \varepsilon, X)$$

$$= B_r, \text{ say.}$$

The theorem is obviously true for $r = 1$. Suppose then that $r \geqslant 2$, and that we have been given $\mathbf{a}$. Let s, t be any positive integers such that $r = s + t$. We split the elements of $\mathbf{a}$ into two sets of size $2s$ and $2t$, quite arbitrarily. Call the elements of the $2s$-set $b_1, \ldots, b_{2s}$, and those of the $2t$-set $c_1, \ldots, c_{2t}$. Let us sort these optimally, into (B, B') and (C, C'), say, obtaining deviations $\delta_s(B, \mathbf{b}) = \mu_s(\mathbf{b}) = \mu_s$, say, and $\delta_t(C, \mathbf{c}) = \mu_t(\mathbf{c}) = \mu_t$, say. Then either $(\mu_s, \mu_t) \leqslant 0$ or $(\mu_s, \mu_t) > 0$; if the latter obtains, we take the complementary position of the cs, which will have as its deviation $\delta_t(C, \mathbf{c}) = -\mu_t(\mathbf{c}) = -\mu_t$. Thus we can always arrange simultaneously that the partitions of the bs and cs are optimal and that $(\mu_s, \mu_t) \leqslant 0$. Then we combine these partitions into a partition of the as, say

$$(A, A') \quad (A = B \cup C, \ A' = B' \cup C'),$$

which has

$$|\delta_r(A, \mathbf{a})|^2 = |\mu_s|^2 + |\mu_t|^2 + 2(\mu_s, \mu_t) \leqslant |\mu_s|^2 + |\mu_t|^2 \leqslant B_s^2 + B_t^2.$$

Thus also $|\mu_r(\mathbf{a})|^2 \leqslant B_s^2 + B_t^2$; since this holds for all $\mathbf{a} \in X^{2r}$ with $|a_i| \leqslant \varepsilon$ for all i, we deduce that $B_r^2 \leqslant B_s^2 + B_t^2$. Hence $B_r^2 \leqslant rB_1^2$, i.e. $B_r \leqslant \varepsilon\sqrt{r}$, concluding the first part of the proof.

To prove the second part, consider l^2 (real square-summable sequences). Let $a_s\,(1 \leqslant s \leqslant 2r)$ be the vector with ε in the sth place and zeros elsewhere. Then the norm of the deviation is the same for all partitions, equal to $\varepsilon\sqrt{(\tfrac{1}{2}r)}$.

Before proceeding to the proof of Theorem 3, we give the following

Lemma. *For all positive integers n there is an integer R_n such that for all $r > R_n$ and all sets $\{x_i : 1 \leqslant i \leqslant r\}$ of vectors in E^n, there exist distinct suffixes i, j $(1 \leqslant j \leqslant r)$ such that either*

$$|x_i + x_j|^2 \leqslant \max(|x_i|^2, |x_j|^2) \quad \text{or} \quad |x_i + (-x_j)|^2 \leqslant \max(|x_i|^2, |-x_j|^2).$$

Proof. Let $S_x = \{x_1, ..., x_r, -x_1, ..., -x_r\}$. If S_x contains a zero vector, the lemma clearly holds. If S_x does not contain a zero vector, we remark that, since an open cap of semi-vertical angle $\frac{1}{6}\pi$ of an $(n-1)$-sphere in E_n has positive $(n-1)$-dimensional measure, only a finite number of such caps may be placed on the sphere without overlapping. Thus, provided r is large enough ($> R_n$ say), some two elements of S_x bearing distinct suffixes meet at an angle $\leqslant \frac{1}{3}\pi$. Let these elements be x_i and $\pm x_j$ $(i \neq j)$. We have then that x_i and $\mp x_j$ meet at an angle $\geqslant 2\pi/3$, and so

$$|x_i \mp x_j|^2 = |x_i|^2 + |\mp x_j|^2 + 2(x_i, \mp x_j) \leqslant \max(|x_i|^2, |\mp x_j|^2).$$

Proof of Theorem 3. Let us divide the $2r$ vectors $a_1, ..., a_{2r}$ into r sets of two each, pairing, say, a_s with a_{r+s} $(1 \leqslant s \leqslant r)$. Then, in the notation of the proof of Theorem 2, we have the deviations $\delta_1(\{a_s\}; a_s, a_{r+s}) = \delta(a_s, a_{r+s})$ say. Then of course

$$(7) \qquad\qquad \delta(a_s, a_{r+s})| \leqslant \varepsilon \quad \text{for each } s,$$

and

$$(8) \qquad\qquad \delta(a_s, a_{r+s}) = -\delta(a_{r+s}, a_s) \quad \text{for each } s.$$

Let A be the set $\{a_s : 1 \leqslant s \leqslant r\}$; then

$$\delta_r(A; \mathbf{a}) = \sum_{s=1}^{r} \delta(a_s, a_{r+s}).$$

By the lemma, if $r > R_n$ there are two δs, one if necessary reversed (as in equation (8)), which meet at an angle $\geqslant 2\pi/3$. We combine them into a joint deviation, which, by the lemma and inequality (7), still has norm $\leqslant \varepsilon$, and thus obtain a new total deviation $\delta_r(A_1, \mathbf{a})$ say. If $r - 1 > R_n$, we can repeat the process; reversal of one of the intermediate deviations being effected by simultaneous reversal of its component δs (as in equation (8)). We then get a new deviation $\delta_j(A_2; \mathbf{a})$ say, and proceed iteratively; at each stage, all the component deviations have norm $\leqslant \varepsilon$, by the lemma. After at most $r - R_n$ iterations we have a total deviation $\delta_r(A_{r-R_n}; \mathbf{a}) = \delta_r(\mathbf{a})$ say, which is the vector sum of precisely R_n vectors each of norm $\leqslant \varepsilon$.

Because of the possibility of reversing any of these, we have

$$|\delta_r(a)|^2 \leqslant R_n \, \varepsilon^2,$$

which is the assertion of the theorem.

With regard to numerical estimation of R_n: that $R_1 = 1$ and $R_2 \leqslant 2$ is obvious. We have $R_3 \leqslant 6$ since one can place twelve caps of semi-vertical angle $\frac{1}{6}\pi$ symmetrically on a sphere so that no two overlap, but a thirteenth cap can in no circumstances be accommodated. For $n \geqslant 4$, we use the estimate given by Rankin [3] for the maximum number of spherical caps of semi-vertical angle $\frac{1}{6}\pi$ that can be placed on a sphere of $(n-1)$ dimensions so that no two overlap; one easily checks that his result yields the estimate given. (Using it, we have $R_4 \leqslant 16$, $R_5 \leqslant 30$, $R_6 \leqslant 53$, $R_7 \leqslant 92$ and $R_8 \leqslant 154$.)

REFERENCES

1. D. R. Cox and H. D. Miller, *The Theory of Stochastic Processes*, Methuen, London (1965).
2. R. Davidson, "More Delphic theory and practice", *Z. Wahrsch'theorie & verw. Geb.* **13** (1969), 191–203. (2.5 of this book.)
3. R. A. Rankin, "The closest packing of spherical caps in *n* dimensions", *Proc. Glasgow Math. Assoc.* **2** (1955), 139–144.

2.7

Sur un Problème Relatif aux Suites Récurrentes Monotones

R. DAVIDSON

[Reprinted from *C. R. Acad. Sc. Paris* **268** (1969), 549–551.]

Nous étudions les problèmes suivants, posés par M. A. Tortrat.

(*a*) Quelles sont exactement les suites récurrentes monotones (décroissantes, bien entendu)? et, surtout,

(*b*) sont-elles toutes indéfiniment divisibles (voir Kendall [3])?

Dans la suite, $u = \{u_n\}$ désignera une suite récurrente, à *f*-suite (répartition des probabilités du temps du premier retour) $f = \{f_r\}$. Nous écrirons $u \in \mathcal{R}^+$ si $u_1 > 0$ (et, par suite, $u_n > 0$ pour tout n). $\mathcal{R}^+$ est un semigroupe commutatif métrique pour la multiplication et la convergence des suites élément par élément; et, avec l'homomorphisme $D(u) = -\log(u_1)$, il est Delphique (voir [1], Section 2.3.5). Évidemment, si la suite récurrente u est monotone, $u \in \mathcal{R}^+$ ou bien $u_n = 0$ pour tout $n \geqslant 1$. Soit donc $\mathcal{M}$ la classe des éléments monotones, et $\mathcal{I}$ la classe des éléments indéfiniment divisibles (voir [3]) de $\mathcal{R}^+$. Alors $\mathcal{M}$ est un sous-semigroupe fermé de $\mathcal{R}^+$, et $\mathcal{M} \supset \mathcal{I}$, car (voir [3]), pour que $u(\in \mathcal{R}^+) \in \mathcal{I}$, il faut et il suffit que u_{n+1}/u_n croisse avec n, ainsi ce rapport ne peut pas dépasser 1. Parce que $\mathcal{R}^+$ est Delphique, ces faits montrent que $\mathcal{M}$ est aussi un semigroupe Delphique (on vérifie sans peine les axiomes donnés dans [4]).

Théorème 1. *Il existe un $u \in \mathcal{R}^+$ tel que: (i) u est convexe (donc $u \in \mathcal{M}$); (ii) $u \notin \mathcal{I}$, et (iii) u est positivement régulier.*

Preuve. Dans [2] nous avons (Example 1) démontré que $u \in \mathcal{R}^+$, où

$$u_0 = 1, \quad u_1 = a + 3c, \quad u_2 = a + 2c, \quad u_r = a + c \quad \text{pour tout } r \geqslant 3,$$

et a, c sont tels que $0 < a = 1 - 18c < 1$. Nous voyons que (i)–(iii) sont satisfaits par u.

Une méthode naïve pour construire des éléments de $\mathcal{M}$ consiste à prendre n'importe quel élément de $\mathcal{R}^+$ et à le multiplier par un élément fortement décroissant de $\mathcal{R}^+$; si nous avons de la chance, le produit sera élément de $\mathcal{M}$. Soit $\mathcal{B} \subset \mathcal{R}^+$ la classe des u pour lesquels il n'existe aucun $v \in \mathcal{R}^+$ tel que $u.v \in \mathcal{M}$.

Théorème 2. *Pour que $u \in \mathcal{B}$, il faut et il suffit que*

$$S(u) = \sup\left(\frac{u_{n+1}}{u_n}\right) = \infty.$$

Preuve. Soit d'abord $S(u) < \infty$. Posons $a = \min(1, 1/S(u))$. Alors $v(\infty, a) \in \mathcal{R}^+$ (où $v_n(\infty, a) = a^n$ pour tout n); et évidemment $u.v(\infty, a) \in \mathcal{M}$, donc $u \notin \mathcal{B}$.

Soit, au contraire, $S(u) = \infty$. Alors pour tout $v \in \mathcal{R}^+$ et $n \geqslant 0$, $v_{n+1} \geqslant v_1 v_n$. Puisque $v_1 > 0$, cela démontre que les rapports v_{n+1}/v_n sont bornés loin de zéro. Donc pour tout $v \in \mathcal{R}^+$ nous avons $S(u.v) = \infty$, ainsi $u \in \mathcal{B}$.

Corollaire 1. *$u \notin \mathcal{B}$ si u est positivement régulier.*

Corollaire 2. *$\mathcal{B}$ est un idéal premier dans $\mathcal{R}^+$.*

Théorème 3. *$\mathcal{B} \neq \varnothing$.*

Preuve. Pour tout $u \in \mathcal{R}^+$ nous avons

(1) $\sum\limits_{r=1}^{\infty} f_1 < 1$ implique $u_n \to 0$ quand $n \to \infty$;

(2)   pour tout n, u_n est fonction seulement de (en effet un polynôme fixe dans) $f_1, \ldots, f_n$.

Construisons un $u \in \mathcal{B}$ ainsi qu'il suit: soit $\{r_i; \; i \geqslant 1\}$ une suite fortement croissante d'entiers positifs. Posons $f_{r_i} = 2^{-i}$, et $f_r = 0$ si r n'égale aucun r_i. Choisissons $r_1 = 1$, et $r_i \, (i \geqslant 2)$ par induction, tels que $u_{r_i-1} \leqslant 2^{1-2i}$; conditions (1) et (2) nous assurent que ceci est possible. Donc pour tout i nous avons

$$\frac{u_{r_i}}{u_{r_i-1}} \geqslant \frac{2^{-i}}{2^{1-2i}} = 2^{i-1},$$

ainsi $S(u) = \infty$ et u est vraiment élément de $\mathcal{B}$.

Remarques. Pour $u \in \mathcal{R}^+$ posons

$$q_n = \log u_{n+1} - 2 \log u_n + \log u_{n-1} \quad (n \geqslant 1).$$

Alors pour que $u \in \mathscr{I}$ il faut et il suffit que $q_n \geqslant 0$ pour tout n. Dans [1], nous avons démontré que $u \in \mathscr{R}^+$ est quotient de deux éléments positivement réguliers de $\mathscr{I}$ si

(3)
$$\sum_{n=1}^{\infty} n \cdot |q_n| < \infty;$$

il n'est pas difficile de démontrer que cette condition est aussi nécessaire. De plus, pour que $u \in \mathscr{R}^+$ soit quotient de deux éléments de $\mathscr{I}$, il faut et il suffit que

(4)
$$\sum_{n=1}^{\infty} |q_n| < \infty.$$

Ces conditions résultent toutes les deux de l'identité

$$\frac{u_1 u_n}{u_{n+1}} = \exp\left(-\sum_{r=1}^{n} q_r\right).$$

Évidemment, pour que $u \notin \mathscr{B}$, il faut et il suffit que les sommes partielles de $\sum q_n$ soient bornées. Une modification triviale de l'exemple du Théorème 3 montre qu'il existe un $u \notin \mathscr{B}$ ne satisfaisant pas condition (4). C'est-à-dire qu'il existe un $u \in \mathscr{R}^+$ tel que:

(5)   il existe un $v \in \mathscr{I}$, $w \in \mathscr{M}$, avec $u \cdot v = w$ (voir Théorème 2), et

(6)   il n'existe aucun v, $w \in \mathscr{I}$, avec $u \cdot v = w$.

Mais je ne sais pas s'il existe $u \in \mathscr{R}^+$ tel que condition (5) ait lieu, mais que

(7)   il n'existe aucun $v \in \mathscr{R}^+$ (N.B.), $w \in \mathscr{I}$, avec $u \cdot v = w$;

parce que je ne connais aucune condition analytique, analogue à conditions (3) ou (4), pour que condition (7) ait lieu.

Enfin, un problème. Est-il vrai que, pour toutes les suites récurrentes u et tous les entiers $n, m \geqslant 0$,

$$\frac{1}{2}(1 + u_n) \geqslant \sum_{r=0}^{n} {}^nC_r \left(\frac{1}{2}\right)^n u_{m+r}?$$

(C'est vrai si $n = 1$ ou 2; et, pour u et n (ou m) donnés, pourvu que m (resp. n) soit assez grand.)

REFERENCES

1. R. Davidson, "Arithmetic and other properties of certain Delphic semigroups: I", *Z. Wahrsch'theorie & verw. Geb.* **10** (1968), 120–145. (2.3 of this book.)

2. R. Davidson, "Arithmetic and other properties of certain Delphic semigroups: II", *Z. Wahrsch'theorie & verw. Geb.* **10** (1968), 146–172. (2.4 of this book.)

3. D. G. Kendall, "Renewal sequences and their arithmetic", *Symposium on Probability Methods in Analysis* (*Lecture Notes in Mathematics* **31**) Springer, Berlin, Heidelberg and New York (1967). (2.1 of this book.)

4. ——, *Bull. Amer. Math. Soc.* **73** (1) (1967), 120–121.

2.8

Amplification of Some Remarks of Lévy concerning the Wishart Distribution

R. DAVIDSON

The Wishart distribution is the distribution, in three dimensions, of the triad of random variables $(A^2, 2AB, B^2)$, where A and B are independent normal $(0, 1)$ random variables. Let the triad be written alternatively (U, V, W); then the important properties of the triad are

(*i*) that U and W are independent;

(*ii*) that $V^2 = 4UW$;

(*iii*) that, conditional on U and W, V takes the values $\pm 2(UW)^{\frac{1}{2}}$ with probabilities $\frac{1}{2}$ each.

As well as these structural properties, the following also obtain:

(*iv*) for all $\alpha > 0$, $\mathrm{pr}(U < \alpha) > 0$ and $\mathrm{pr}(W < \alpha) > 0$;

(*v*) for all $\beta < \infty$, $\mathrm{pr}(U > \beta) > 0$.

Lévy states that the Wishart distribution is *simple* (i.e. indecomposable, see Lévy [1]) and goes some way to proving it. Here we show that subsets of the above properties suffice to establish the result.

Theorem. *Let (U, V, W) be a triad of random variables such that U and W are both non-negative. If (ii), (iii) and (iv) hold or if (i), (ii), (iv) and (v) hold then (U, V, W) is simple.*

Proof. We suppose that $(U, V, W) = (X, Y, Z) + (X', Y', Z')$, where the two triads on the right-hand side are independent; and we regard (X, Z) and (X', Z') as coordinate variables, each in R^2. The ranges of the coordinate variables are to be regarded as the supports of the random

212

variables. Since U and W are non-negative the same is true of X, X', Z, Z', as follows from (*iv*). Thus the range of (X,Z) lies in the first quadrant and so does that of (X',Z'). (Lévy appears to assume that these ranges are the whole of the first quadrant, or at least contain open sets.)

Now we have that $(Y+Y')^2 = 4(X+X')(Z+Z')$ by (*ii*), and so, regarding Y and Y' as random functions of (X,Z) and (X',Z') respectively, we deduce that

(1) $\qquad f(x,z)+f'(x',z')+e(x,z)+e'(x',z') = \pm 2\{(x+x')(z+z')\}^{\frac{1}{2}},$

provided that (x,z) and (x',z') lie in the relevant ranges. Here f and f' are deterministic functions of their arguments, and e and e' are the superimposed random errors. Now e and e' are independent, and obviously neither of them can take more than two values. But suppose that, for some fixed (x,z), e does take two values. Then, because there is but one $\pm$ available, e' must take only one value for each (x',z'), which we may assume to be zero (absorbing the constant into f'). Similarly by suitably redefining f we may assume that the values taken by e are 0 and some other value, possibly, which if present we may take to be positive.

Suppose that this other value does exist for some (x,z). Then we have equation (1) for all (x',z'), and so, subtracting,

(2) $\qquad\qquad 4\{(x+x')(z+z')\}^{\frac{1}{2}} = e(x,z) > 0.$

Remember that in this x and z are constants. Since equation (2) holds for all (x',z') in the relevant range, this range must be contained in the locus of (x',z') defined by equation (2), which is a rectangular hyperbola with centre $(-x,-z)$ and asymptotes parallel to the axes. Now the point is that the range of (x',z') must contain $(0,0)$, because (U,W) can by (*iv*) come arbitrarily close to $(0,0)$ and the ranges of (x,z) and (x',z') are contained in the first quadrant. Thus the rectangular hyperbola under consideration must pass through $(0,0)$. But if it does then by inspection it has no other points in the first quadrant, so that $X' = Z' = 0$, which does not yield a proper factorization. For then Y' is also a constant, because $e'(0,0)$ is a constant.

Thus we have shown, using (*ii*) and (*iv*) alone, that the assumption, that e takes two values for some (x,z), is disastrous to the factorization. On the other hand, (*iii*) demands that e takes two values, so the first implication of the theorem is proved.

For the second implication of the theorem, then, suppose (for a non-trivial factorization) that for no (x,z) does e take two values and that the range of (x,z) contains two distinct points, (x_1,z_1) and (x_2,z_2) say. Then, using equation (1) for these two values of (x,z) and any one value of

8

(x', z'), we find

(3) $f(x_1, z_1) - f(x_2, z_2) = \pm\{(x_1 + x')(z_1 + z')\}^{\frac{1}{2}} \pm \{(x_2 + x')(z_2 + z')\}^{\frac{1}{2}}$,

where we have omitted, without loss, the factors of 2 on the right-hand side. From equation (3) and property (*iv*) it follows fairly readily that $x_1 \neq x_2$ and $z_1 \neq z_2$ (so that (X, Z) is not degenerate only if both X and Z are not).

Now equation (3) is to be regarded as the equation of a locus in (x', z'), and, as such, appears to describe a curve of the fourth degree. But, if we put $b^2 = (f(x_1, z_1) - f(x_2, z_2))^2$, we find, on squaring, that the cubic and quartic terms disappear, and that the equation of the locus is

$$
\begin{aligned}
(4) \quad & x'^2(z_1 - z_2)^2 + 2x'z'((x_1 - x_2)(z_1 - z_2) - 2b^2) + z'^2(x_1 - x_2)^2 \\
& + 2x'((z_1 - z_2)(x_1 z_1 - x_2 z_2) - b^2(z_1 + z_2)) \\
& + 2z'((x_1 - x_2)(x_1 z_1 - x_2 z_2) - b^2(x_1 + x_2)) \\
& + b^4 - 2b^2(x_1 z_1 + x_2 z_2) + (x_1 z_1 - x_2 z_2)^2 = 0.
\end{aligned}
$$

Again, this locus must pass through the origin, so that the constant term vanishes, and we evaluate $b^2 = x_1 z_1 + x_2 z_2 \pm 2(x_1 z_1 x_2 z_2)^{\frac{1}{2}}$. With this value of b^2 it is easily checked that the locus described by equation (4) is an hyperbola with centre $(-\frac{1}{2}(x_1 + x_2), -\frac{1}{2}(z_1 + z_2))$. So we conclude firstly that the range of (x', z') lies on the parts of a certain hyperbola which lie in the first quadrant, and secondly, from inspection of the form of any such hyperbola, that for all x_0' there exists a $g'(x_0')$ such that, when $x' > x_0'$, $z' > g'(x_0')$, and g' increases to ∞ with x_0'.

Now we may apply the same argument in reverse, for there must be at least two points (x_1', z_1') and (x_2', z_2') in the range of (x', z'); so for all x_0 there is a $g(x_0)$ such that, for all $x > x_0$, $z > g(x_0)$, and g increases to ∞ with x_0.

But now U can be large, by (v), in which case one at least of X and X' must be large; then, in view of what we have just done, one at least of Z and Z' must be large and W also must be large. But U and W are, according to (i), independent; so we have reached a contradiction.

REFERENCE

1. P. Lévy, "The arithmetical character of the Wishart distribution", *Proc. Cam. phil. Soc.* **44** (1948), 295–297.

[*Editorial note* (by D. G. K.). The above is a lightly edited version of a manuscript found among Davidson's papers. It is clearly not in a finished state, and the following remarks indicate how he might have reformulated the argument before final publication. Let us recall the definition of the support supp(μ) of a *probability* measure μ in k dimensions; it consists of those points each of whose neighbourhoods has positive mass. Thus supp(μ) is closed and non-vacuous; its complement is the largest open set of zero mass. If π denotes projection onto a coordinate axis (or plane, etc.), then the relation

(5) $$\text{supp}\,(\mu\pi^{-1}) = \text{Cl}\,(\pi\,\text{supp}\,(\mu))$$

(where Cl(.) denotes topological closure) is the most that can be said concerning the connection between the supports of a measure and of one of the associated 'marginal' measures $\mu\pi^{-1}$; it is easily proved. The position as regards convolutions is slightly more delicate. If $\lambda*\mu$ denotes the convolution of the probability measures λ and μ, then it is not hard to show that supp$(\lambda)+$supp(μ) is a subset of supp$(\lambda*\mu)$, whence so is its closure (here $+$ is vector addition). In fact we have

(6) $$\text{supp}\,(\lambda*\mu) = \text{Cl}\,(\text{supp}(\lambda)+\text{supp}\,(\mu)),$$

as the following argument shows. Let z belong to supp$(\lambda*\mu)$, i.e. let z have a shrinking sequence $(G_n: n = 1, 2, ...)$ of open neighbourhoods with the singleton $\{z\}$ as intersection, where each G_n has positive mass with respect to $\lambda*\mu$. Let us think of λ and μ as the distribution-measures of random variables X and Y, and $\lambda*\mu$ as the distribution-measure of $Z = X+Y$ (X and Y being independent). Then, in an obvious notation, we must have

$$\int \mu(G_n-x)\,\lambda(dx)>0 \quad \text{for each } n.$$

Now a function which vanishes over the support of a measure will have zero integral, so for each n there exists x_n in supp(λ) such that

$$\mu(G_n-x_n)>0.$$

But G_n-x_n is open, so it contains a point y_n in supp(μ); that is, each G_n contains a point x_n+y_n in the vector sum, supp$(\lambda)+$supp(μ), as required (for this shows that z must lie in the closure of the said vector sum).

We now turn to the Lévy–Davidson problem, using Davidson's notation. It is my impression that (i) must be added to the hypotheses of the first part of Davidson's theorem, or alternatively that (iv) should be strengthened to

$(iv)^*$ for all $\alpha>0$, pr(both of U and $W<\alpha)>0$.

As (i) and (iv) imply $(iv)^*$, we shall just use $(iv)^*$ in what follows, and ignore (i) for the time being. Identifying the supports of random variables with those of their distribution-measures, we know from the 'preamble' to the theorem that $U = X + X'$ and $W = Z + Z'$ have supports in $[0, \infty)$. It is clear that X and X' cannot *both* have negative numbers in their supports, for then there would be a negative number in $\text{supp}(U)$, so let $\text{supp}(X) \subset [0, \infty)$. It then becomes clear that $\text{supp}(X')$ must be bounded on the left, so let α be the minimal left-hand point of $\text{supp}(X)$, and let α' be the minimal left-hand point of $\text{supp}(X')$; then evidently $\alpha + \alpha'$ will be the minimal left-hand point of $\text{supp}(U)$. But from $(iv)^*$ (or indeed from (iv)), this must be zero, so $\alpha' = -\alpha$. If we now replace X by $X - \alpha$, and X' by $X' - \alpha'$, then the problem is unchanged, save that now both X and X' have zero as minimal left-hand support-points. We similarly redefine Z and Z' to achieve a comparable normalization, and now note that $(iv)^*$ (and here the * *is* needed) tells us that $(0, 0)$ belongs to $\text{supp}((U, W)) = \text{supp}((X, Z) + (X', Z'))$, so belongs to the closure of the vector sum, $\text{supp}((X, Z)) + \text{supp}((X', Z'))$. But from equation (5), and what we have already proved, we know that each of $\text{supp}((X, Z))$ and $\text{supp}((X', Z'))$ lies in the closed non-negative quarter-plane, whence so does their vector sum. A little argument with equation (6) and the partial ordering of the plane then shows that $(0, 0)$ must belong to *each* of $\text{supp}((X, Z))$, and $\text{supp}((X', Z'))$, and this fact is vital for the ensuing argument.

Adjoining part of what we have learnt to (ii), we see that $\text{supp}((U, V, W))$ must be a closed subset of that sheet of the cone $v^2 = 4uw$ which lies in the quarter-space $u \geqslant 0$, $w \geqslant 0$, and from conic considerations and what we already know we can then infer that this closed subset must contain the point $(0, 0, 0)$.

Now let (x, y, z), resp. (x', y', z') be points in $\text{supp}((X, Y, Z))$, resp. $\text{supp}((X', Y', Z'))$. Then from equation (6), $(x + x', y + y', z + z')$ must lie on the cone, and so for *given* (x, z) and (x', z') there are at most two values possible for $y + y'$, whence at most two for each of y and y', and certainly not two for *both*. If possible, let y have two distinct possible values, and let $e(x, z)$ (x and z fixed) be their positive difference. Then as Davidson observes, equation (2) has to hold *for all* (x', z') in $\pi \text{supp}((X', Y', Z'))$, where here π is projection onto the (x, z)-plane, and so

$$16(x' + x)(z' + z) = (e(x, z))^2 > 0$$

also hold (by continuity) for all (x', z') in the closure of this πsupp, i.e. throughout $\text{supp}((X', Z'))$. Davidson's argument about rectangular hyperbolas now applies, and tells us that $\text{supp}((X', Z')) = \{(0, 0)\}$, and so that (X', Y', Z') lies almost surely on the line $x' = z' = 0$. But y' can only have one value, as y has been allowed two, and so the support of

(X', Y', Z') must be a singleton and the decomposition is improper. That is, (*ii*) and (*iv*)* tell us that: *either* (U, V, W) is indecomposable; *or*, in every proper factorization, the supports of the two factors have the form

$$\{(x, a(x, z), z): (x, z) \in A\}$$

and

$$\{(x', b(x', z'), z'): (x', z') \in B\},$$

where a and b are *single-valued* mappings. To obtain a Lévy–Davidson theorem, then, we must add to (*ii*) and (*iv*)* sufficiently many additional hypotheses to contradict the alternative possibility associated with a, b, A and B. (Note that we cannot take it for granted that the mappings a and b are, say, continuous.)

Davidson's intention was that his axiom (*iii*) should play this last role, but I do not find the argument entirely convincing. One can vary (*iii*) to bring it into closer contact with the 'support' approach; for example one might consider, instead,

(*iii*)*   there exists a pair $(u, \pm v, w)$ of distinct points in supp $((U, V, W))$.

This would imply, of course, that $u > 0$, $w > 0$ and $v = \pm 2\sqrt{(uw)} \neq 0$. These two points would then have to lie in the *closure* of the vector sum of the supports of (X, Y, Z) and (X', Y', Z'). From the non-negativity conditions we have already obtained from (*ii*) and (*iv*)*, and a compactness argument, this would imply the existence in A and B of pairs (x_n, z_n) and (x'_n, z'_n) for $n = 1, 2, \ldots$, such that

$$x_n \to x, \quad z_n \to z, \quad x'_n \to x', \quad z'_n \to z',$$

$$x + x' = u, \quad z + z' = w \quad \text{and} \quad a(x_n, z_n) + b(x'_n, z'_n) \to v,$$

and also the existence of another pair of sequences having all the above properties save for the switch from v to $-v$. Without some extra information about the mappings a and b it seems difficult to proceed further. The reader may like to speculate on what alternative additional axiom might clinch the argument (or, indeed, he may be able to see that Davidson's (*iii*) suffices).

We now turn to the second part of Davidson's theorem, where he assumes (in our present version) both (*ii*) and (*iv*)*, so that a proper decomposition must be of the form described above, and he further assumes that supp $((X, Z))$ contains at least two points (x_1, z_1) and (x_2, z_2), say. This last is in fact no restriction at all, for if the support of (X, Z) were a singleton then, from equation (5), all the members of supp $((X, Y, Z))$ would be of the form $(x, ., z)$ with x and z fixed, and so (in the special situation in which we imagine ourselves to be here) would reduce to the singleton $(x, a(x, z), z)$, so that the decomposition could not have been a proper one, after all. On writing β for the modulus of the difference between $a(x_1, z_1)$ and $a(x_2, z_2)$ (and throwing away 2s as he did) we obtain Davidson's identity (4) (with β in the place of his b), which has to hold whenever $(x', b(x', z'), z')$ lies in $\mathrm{supp}((X', Y', Z'))$. The b-function does not occur in this identity, which from equation (5) and continuity must hold whenever (x', z') lies in $\mathrm{supp}((X', Z'))$, this being a closed subset of the non-negative quarter-plane which contains at least the point $(0, 0)$. This last fact, as Davidson remarks, ensures that the 'constant' term in the identity (4) must vanish.

Now Davidson adduces axioms (i) and (v) to get a contradiction; his argument can be reformulated as follows. Because of (v), $\mathrm{supp}(U)$ is unbounded on the right, and so one at least of $\mathrm{supp}(X)$ and $\mathrm{supp}(X')$ must be unbounded on the right (because of equation (6)). Say it is $\mathrm{supp}(X')$; then from equation (5), $\mathrm{supp}((X', Z'))$ must contain a sequence of points (x'_n, z'_n) with $x'_n \to \infty$, and these must all satisfy the identity (4). Whence (4), which at least contains one of the terms in x'^2 and z'^2, and so is not vacuous, cannot merely represent a compact locus (an ellipse or a single point), and must therefore be a line-pair, a doubled line, a parabola or a hyperbola. If it is a parabola we can conclude at once that z'_n also $\to \infty$, so suppose that we are in one of the other cases. Then we can evade $z'_n \to \infty$ only if (a) there is a linear factor of the form $(z' - \gamma)$, or (b) we have a hyperbola with an asymptote of the form $z' = \gamma$, where in each case $\gamma \geqslant 0$.

Davidson asserts that we can choose both $x_1 \neq x_2$ *and* $z_1 \neq z_2$, but I have been unable to follow the implied argument here, so we shall not use this assertion. It seems to me essential to arrange that $z_1 \neq z_2$; let us us see what happens if this cannot be done. That would mean that all zs in $\mathrm{supp}((X, Y, Z))$ were the same, in which case (using equation (5)) we should find that $\mathrm{supp}(Z)$ was a singleton, necessarily $\{0\}$, so $z_1 = z_2 = 0$. From our earlier argument it will then follow that (as the zs cannot now be made distinct) we can choose $x_1 \neq x_2$, and then equation (3) with $z' \to 0$ will show that $\beta = 0$, so that equation (4) becomes

$$z'^2(x_1 - x_2)^2 = 0;$$

that is, we would be able to deduce that not only Z but also Z' was zero a.s. To avoid this, I think that we must assume

(vi) W is not a.s. zero.

If this is done, then we can certainly take $z_1 \neq z_2$. It then becomes profitable to follow through Davidson's analysis of the geometric meaning of equation (4) which, because of the alternative sign in β^2, represents not *one* conic, but rather *two* concentric ones, with the common centre noted by him. As this centre lies in the $(-, -)$ quarter-plane, we are only interested (in view of (v)) in infinite branches of one or other conic which move out to infinity in the closed $(+, +)$ quarter-plane, and which permit x' to increase towards $+\infty$. We *know* (because of (v)) that there is at least *one* such. All we have to assure ourselves of is that it is not parallel ('ultimately') to $z' = 0$.

Now the behaviour of infinite branches of these conics is wholly controlled by the purely quadratic parts of equation (4), and there will be a branch ultimately parallel to $z' = 0$ if and only if the term in x'^2 is absent. This, however, is impossible, because $(z_1 - z_2)^2 > 0$. Thus (a) as there *is* a branch with ultimate direction in the closed first quadrant, and (b) as it cannot have the direction of the x'-axis, it *must* be such that

$$z' \to \infty \quad \text{as} \quad x' \to \infty.$$

Davidson's own arguments now take over. As he points out, the same argument can be used when the dashed and undashed factors are exchanged. Thus, as U can have indefinitely large values, we are entitled to ask what we can then deduce about W, when U is known to be large. From U large we deduce X' large (or X large), and from $X'(X)$ large we deduce that $Z'(Z)$ is large also, whence (since both of Z and Z' are non-negative), we can infer that $W = Z + Z'$ is large when U is large. As W *can* also be near zero, this contradicts (i) (the independence of U and W), and the augmented Davidson proof winds to a close. Thus (i), (ii), (iv), (v) *and* (vi) *imply indecomposability*.

It seems clear that in such a complex situation many other sets of axioms (other than (i), (ii), (iv), (v) and (vi)) will suffice to prove the Wishart measure indecomposable, by merely noting that it has such simple qualitative properties. It seems highly probable that the present set may not be the most elegant or instructive one, and virtually certain, in fact, that Davidson would have completed his solution in a far more satisfactory way than that laboriously effected here.

The reader may find this a challenge to his ingenuity and insight. In view of the immense *statistical* importance of the Wishart measure, such a challenge deserves a response.

We close with the comment that a weaker result than that given above may be found more attractive; *it is now clear that the (U, V, W)-measure is indecomposable if it satisfies axioms $(i), (ii), (iv)$ and $(v)^*$, where*

$(v)^*$ for all $\beta > 0$, both $\mathrm{pr}\,(U > \beta)$ and $\mathrm{pr}\,(W > \beta)$ are > 0.

This formulation has the merit of being symmetrical in the demands made upon U and W.]

3

REGENERATIVE PHENOMENA AND THEIR p-FUNCTIONS

3.1

Another Approach to one of Bloomfield's Inequalities

R. DAVIDSON

1. For *stable p* we can write (* denoting convolution)

$$(1) \qquad p(t) = [e(\alpha) * dF * p](t) + e^{-\alpha t}.$$

Here $\alpha = q = \mu(0, \infty]$, $e(\alpha)(t) = \alpha e^{-\alpha t}$ and $\alpha F(t) = \mu(0, t]$ in the accepted notation of [3]. Suppose that we could find a function $\tilde{p}$ such that

$$0 \leqslant \tilde{p}(t) \leqslant 1$$

and

$$(2) \qquad \tilde{p}(t) \leqslant [e(\alpha) * dF * \tilde{p}](t) + e^{-\alpha t}$$

for all $t \geqslant 0$; then by iteration we would obtain a sequence of upper bounds for $\tilde{p}$ which converge to the iteration expansion for p as limit, and so we could conclude that

$$(3) \qquad p(t) \geqslant \tilde{p}(t).$$

We shall obtain a well known inequality of Bloomfield [1] by using an argument of this form.

2. We choose $\tilde{p}(t) = e^{-A(t)}$, where (in Kingman's notation)

$$A(t) = \int_0^t m(s) \, ds = \alpha t - \alpha B(t)$$

and

$$B(t) = \int_0^t F(s) \, ds.$$

Obviously $\tilde{p}$ is bounded by zero and unity, so that we have only to prove

223

inequality (2); that is, we want

$$e^{-A(t)} \leqslant \int_0^t \alpha e^{-\alpha(t-s)} ds \int_0^s e^{-A(s-u)} dF(u) + e^{-\alpha t},$$

and on transforming from As to Bs this is equivalent to

(4) $$e^{\alpha B(t)} \leqslant \int_0^t \alpha\, ds \int_0^s e^{\alpha B(s-u)+\alpha u} dF(u) + 1.$$

It will thus suffice to show that

(5) $$\alpha F(t) e^{\alpha B(t)} \leqslant \int_0^t \alpha e^{\alpha B(t-u)+\alpha u} dF(u),$$

because we can then integrate this to obtain inequality (4); note that inequality (4) is trivially true (with equality) at $t = 0$, and that the left-hand side of inequality (4) is absolutely continuous and has a derivative equal almost everywhere to the left-hand side of inequality (5).

3. In order to prove inequality (5) we rewrite it as

$$\alpha F(t) e^{-\alpha t+\alpha B(t)} \leqslant \int_0^t \alpha e^{-\alpha(t-u)+\alpha B(t-u)} dF(u),$$

or more concisely as

$$F(t) e^{-A(t)} \leqslant \int_0^t e^{-A(t-u)} dF(u).$$

But that this is true follows immediately from the fact that $A(t)$ increases, and so $e^{-A(t)}$ decreases, as t increases. Thus Bloomfield's inequality,

(6) $$p(t) \geqslant e^{-A(t)},$$

is established for stable p-functions. To prove it in general we have only to observe (cf. [3] and [2]) that with any standard p-function we can associate a sequence of stable p-functions (p_n) in such a way that $p_n(t) \to p(t)$ and $\mu_n \to \mu$, so that

$$\begin{aligned}
A_n(t) &= \int_0^t m_n(s)\, ds = \int_0^t ds \int_{(s,\infty]} \mu(dx) \\
&= \int_{(0,\infty]} \min(1,x)\, \mu_n(dx) \\
&= \int_{[0,\infty]} \frac{\min(1,x)}{1-e^{-x}} \lambda_n(dx) \\
&\to \int_{[0,\infty]} \frac{\min(1,x)}{1-e^{-x}} \lambda(dx) = A(t).
\end{aligned}$$

Here $\lambda(dx) = (1-e^{-x})\mu(dx)$.

[*Editorial note* (by D. G. K). This neat argument was found among Davidson's papers, in contracted form. Observe that the remainder terms in the iteration series tend to zero because both $p(t)$ and $\tilde{p}(t) \leqslant 1$.]

REFERENCES

1. P. Bloomfield, "Lower bounds for renewal sequences and p-functions", *Ztschr. Wahrsch'theorie & verw. Geb.* **19** (1971), 271–273. (See also his contribution 3.2 to this book.)
2. D. G. Kendall, "Delphic semigroups, infinitely divisible regenerative phenomena, and the arithmetic of p-functions", *Ztschr. Wahrsch'theorie & verw. Geb.* **9** (1968), 163–195. (2.2 of this book.)
3. J. F. C. Kingman, "The stochastic theory of regenerative events", *Ztschr. Wahrsch'theorie & verw. Geb.* **2** (1964), 180–224.

3.2

Stochastic Inequalities for Regenerative Phenomena

PETER BLOOMFIELD

3.2.1 SUMMARY

Two stochastic inequalities for regenerative phenomena are deduced from constructions in which corresponding inequalities hold almost surely. Some consequences of these inequalities are developed, and a new proof of the characterization of infinitely divisible p-functions is given.

3.2.2 INTRODUCTION

The study of recurrent events in discrete time is a useful way to approach the behaviour of the transition matrix of a Markov chain. In continuous time, the corresponding theory is that of regenerative phenomena, which was studied in some detail by Kingman [5, 6].

Suppose that $(\Omega, \mathscr{A}, \mathrm{pr})$ is a probability space, and that $\mathscr{E} = \{E(t); t > 0\}$ is a family of elements of $\mathscr{A}$. If $0 < t_1 < t_2 < \ldots < t_k$ implies that

$$\mathrm{pr}\{E(t_1) \cap E(t_2) \cap \ldots \cap E(t_k)\} = \mathrm{pr}\{E(t_1)\}\,\mathrm{pr}\{E(t_2 - t_1) \cap \ldots \cap E(t_k - t_1)\},$$

then $\mathscr{E}$ is a regenerative phenomenon.

For any regenerative phenomenon, we may define a stochastic process $\{X(t): t > 0\}$ by

$$X(t, \omega) = \begin{cases} 1, & \omega \in E(t), \\ 0, & \omega \notin E(t), \end{cases}$$

$t > 0, \omega \in \Omega$. The regenerative phenomenon is itself determined by this process, for clearly

$$E(t) = \{\omega: X(t, \omega) = 1\}.$$

We shall not distinguish between $\mathscr{E}$ and $\{X(t)\}$.

Clearly many of the probabilistic properties of $\mathscr{E}$ are determined by its p-function,

$$p(t) = \text{pr}\{E(t)\} = \text{pr}\{X(t) = 1\} \quad (t > 0).$$

A p-function is said to be standard if $p(t) \to 1$ as $t \downarrow 0$, and the class of these p-functions, denoted by $\mathscr{P}$, is easily seen to be a semigroup. A canonical representation for $\mathscr{P}$ was found by Kingman [5]. If $p \in \mathscr{P}$, and r is the Laplace transform of p, then there exists a measure μ with

$$\int_{(0,\infty]} (1 - e^{-x}) \mu(dx) < \infty,$$

such that

$$\frac{1}{r(s)} = s + \int_{(0,\infty]} (1 - e^{-sx}) \mu(dx) \quad (s > 0).$$

The measure μ is determined by either of the functions

$$m(t) = \mu(t, \infty] \quad (t > 0),$$

or

(1) $$A(t) = \int_0^t m(u)\, du \quad (t > 0),$$

which are used frequently below.

When $m(0+) = \mu(0, \infty]$ is finite, p is said to be stable, and we shall write

(2) $$m(t) = \alpha\{1 - F(t)\},$$

where $\alpha = m(0+)$. Thus F is non-decreasing, $F(0+) = 0$, and $F(\infty) \leqslant 1$.

Suppose that $\{U_n : n = 1, 2, ...\}$ are independent exponential variables with mean α^{-1}, and that $\{V_n : n = 1, 2, ...\}$ are independent, each with the distribution F. Note that this distribution is defective if $F(\infty) < 1$, when $\text{pr}(V_n = +\infty) = 1 - F(\infty)$. Let $T_0 = 0, T_{2n+1} = T_{2n} + U_n, T_{2n+2} = T_{2n+1} + V_n$, $n \geqslant 0$. Define a stochastic process $\{X(t) : t > 0\}$ by

$$X(t) = \begin{cases} 1, & \text{if } T_{2n} < t < T_{2n+1}, \\ 0, & \text{if } T_{2n+1} \leqslant t \leqslant T_{2n+2}, \end{cases}$$

for $n \geqslant 0$. That is, we have an alternating system of intervals, and $X(t)$ takes the value 1 on those with the exponential distribution. If $F(\infty) < 1$, then one of $\{V_n\}$ is sure to be infinite, and if the first such is V_N, then $X(t) = 0$ for $t \geqslant T_{2N+1}$, as $T_n = +\infty$ for $n > 2N+1$. It is easily verified that $\{X(t)\}$ is a regenerative phenomenon, and that its canonical measure μ is determined by $\mu(t, \infty] = m(t) = \alpha\{1 - F(t)\}$. This type of phenomenon was discussed by Bartlett [1, p. 57], and is referred to as a regenerative phenomenon of type B.

3.2.3 A SIMPLE INEQUALITY

Consider an infinite server queue, with the following features. Arrivals are Poisson with rate α, service times have the distribution function F, $F(0+) = 0$, and the queue is initially empty. There is a principal server who serves the first arrival and subsequently serves the first customer to arrive after he finishes each service. Thus the principal server is idle for exponentially distributed intervals, with mean α^{-1}.

Let

$$X(t) = \begin{cases} 1, & \text{if the principal server is idle,} \\ 0, & \text{otherwise,} \end{cases}$$

for $t > 0$. Clearly $\{X(t)\}$ is identical to the stable regenerative phenomenon constructed in the previous section. Now the principal server is idle whenever the queue is empty, and thus

$$p(t) = \text{pr}\{X(t) = 1\} \geqslant \text{pr}\{\text{queue is empty at time } t\}.$$

But this last probability was shown by Takács [7] to be $e^{-A(t)}$, where

$$A(t) = \alpha \int_0^t \{1 - F(u)\}\,du = \int_0^t m(u)\,du,$$

as in equation (1). Thus we have for stable p-functions the inequality,

$$(3) \qquad\qquad\qquad p(t) \geqslant e^{-A(t)}.$$

This was first shown by Bloomfield [2] in the more general context of standard p-functions, using a skeletal argument. Rollo Davidson, in a personal communication (see 3.1 of this book), showed that it could be derived for stable p-functions from the integral equation

$$p(t) = e^{-\alpha t} + \int_0^t p(t-u) \int_0^u \alpha e^{-\alpha(u-v)}\,dF(v)\,du.$$

Davidson also stated that the result, once proved for stable p-functions, could be extended to $\mathscr{P}$ by approximation. For suppose that p is not stable, so that $\mu(0, \infty] = +\infty$. Let

$$m_n(t) = \begin{cases} m(t), & t \geqslant \dfrac{1}{n}, \\[2ex] m\left(\dfrac{1}{n}\right), & 0 < t < \dfrac{1}{n}. \end{cases}$$

Then m_n is the tail function of a measure μ_n, which coincides with μ except that it places no mass in the interval $(0, 1/n]$. Since m has a finite integral, $m_n(0+) = m(1/n) < \infty$ for each n. If we define a measure λ by

$$\lambda(dt) = (1 - e^{-t}) \mu(dt)$$

and similarly $\{\lambda_n : n = 1, 2, \ldots\}$, then it is easily seen that $\lambda_n \to \lambda$ weakly. By Theorem 9 of Kendall [4], this implies that $p_n(t) \to p(t)$ as $n \to \infty$, $t > 0$. From the construction of m_n, we see that

$$A(t) = \int_0^t m(u) \, du \geqslant \int_0^t m_n(u) \, du = A_n(t)$$

$$\geqslant \int_{1/n}^t m_n(u) \, du = \int_{1/n}^t m(u) \, du = A(t) - A\left(\frac{1}{n}\right),$$

so that $A_n(t) \to A(t)$ as $n \to \infty$, $t > 0$. But since each p_n is stable, we have that

$$p_n(t) \geqslant e^{-A_n(t)},$$

and, letting $n \to \infty$, we obtain the general result.

The inequality (3) has two simple consequences. It is known (Kingman [5]) that

$$p(\infty) = \lim_{t \to \infty} p(t)$$

always exists, and that

$$p(\infty) = \frac{1}{1 + A(\infty)}.$$

Since A is non-decreasing, it follows that

(4) $\qquad p(t) \geqslant e^{-A(t)} \geqslant e^{-A(\infty)} = \exp[-\{p(\infty)^{-1} - 1\}] \quad (0 < t < \infty).$

This gives a positive uniform lower bound for p whenever $p(\infty) > 0$. Clearly if $p(\infty) = 0$, the only uniform lower bound is zero.

Suppose that the service times in the queueing model were almost surely equal to a constant, say τ. Then for $0 < t \leqslant \tau$, the principal server is idle if and only if the queue is empty. Thus in this case, we have equality in the inequality (3) for $0 < t < \tau$, and in particular $p(\tau) = e^{-A(\tau)}$. This may also be seen from the integral equation

(5) $\qquad\qquad 1 - p(t) = \int_0^t p(t - u) m(u) \, du$

(Kingman [5]). For here we have

$$m(t) = \begin{cases} \alpha, & 0 < t < \tau, \\ 0, & \tau \leqslant t, \end{cases}$$

whence $p(t) = e^{-\alpha t}$ and $A(t) = \alpha t$ for $0 < t \leqslant \tau$, and $A(t) = A(\tau)$ for $t > \tau$. It follows that $p(\tau) = e^{-A(\tau)} = e^{-A(\infty)}$, and thus the lower bound in inequality (4) is attained. Hence the inequality (4) is the best possible uniform lower bound for p in terms of $p(\infty)$.

The second application is to the (M, m) diagram, introduced by Davidson [3]. Let

$$M = p(1)$$

and

$$m = \inf_{0 < t < 1} p(t).$$

Davidson shows that if $M > \frac{3}{4}$, then

$$m \geqslant \tfrac{1}{2} + \sqrt{(M - \tfrac{3}{4})}.$$

Now for $0 < t \leqslant 1$,

$$p(t) \geqslant e^{-A(t)} \geqslant e^{-A(1)}$$

and thus

$$m \geqslant e^{-A(1)}$$

or

$$A(1) \geqslant -\log m.$$

Also, from the integral equation (5), we have that

$$1 - M = 1 - p(1) = \int_0^1 p(1 - u)\, m(u)\, du$$

$$\geqslant m \int_0^1 m(u)\, du = mA(1) \geqslant -m \log m.$$

This gives rise to a further restriction on the attainable pairs (M, m).

3.2.4 PRODUCTS OF p-FUNCTIONS

Suppose that a p-function $p(t)$ can be written as the product $p_1(t) p_2(t)$ of two factors, and that $p(\infty) > 0$. Then from inequality (4) we have that

$$p(t) \geqslant \exp\left[-\{p(\infty)^{-1} - 1\}\right] \quad (0 < t < \infty).$$

Applying inequality (4) separately to p_1 and p_2, and multiplying the results, however, we find that

$$p(t) \geqslant \exp\left[-\{p_1(\infty)^{-1} + p_2(\infty)^{-1} - 2\}\right] \quad (0 < t < \infty),$$

which is easily shown to be a better bound, except in the trivial case where one of the factors is identically equal to one.

Since inequality (4) was derived from inequality (3), this suggests that

$$p(t) = p_1(t) p_2(t) \geqslant \exp\left[-\{A_1(t) + A_2(t)\}\right]$$

may be an improvement on inequality (3), that is that $A_1(t) + A_2(t) \leqslant A(t)$. We now show that this indeed is true. In fact, we show the stronger result that

(6) $$m_1(t) + m_2(t) \leqslant m(t).$$

We no longer assume that $p(\infty) > 0$.

Suppose first that p_1 and p_2 are stable, and that $\{X_1(t)\}$ and $\{X_2(t)\}$ are independent regenerative phenomena of type B, having the p-functions p_1 and p_2 respectively. If $X(t) = X_1(t) X_2(t)$, $t > 0$, then $\{X(t)\}$ is also of type B, and its p-function is $p(t) = p_1(t) p_2(t)$. Since this is a superposition, we have that $\alpha = \alpha_1 + \alpha_2$.

Now consider the first interval, of length V, during which $X(t) = 0$, and the corresponding intervals, of the lengths V_1 and V_2, for $\{X_1(t)\}$ and $\{X_2(t)\}$. Clearly V is at least as large as whichever of V_1 and V_2 corresponds to the interval which starts the earliest, say W. But the distribution of W is a mixture of the distributions of V_1 and V_2, in the ratio $\alpha_1 : \alpha_2$. Thus

$$1 - F(t) = \mathrm{pr}\{V > t\} \geqslant \mathrm{pr}\{W > t\} = [\alpha_1\{1 - F_1(t)\} + \alpha_2\{1 - F_2(t)\}]/(\alpha_1 + \alpha_2).$$

Rearranging and using equation (2), inequality (6) follows. As before, the result may be extended to $\mathscr{P}$ by approximation.

3.2.5 INFINITELY DIVISIBLE p-FUNCTIONS

An infinitely divisible p-function is one for which $p(t)^{1/n}$ is a p-function for each integer $n \geqslant 1$. Arguing as before, we deduce from inequality (4) that if p is infinitely divisible, then

$$p(t) \geqslant \exp\left[-n\{p(\infty)^{-1/n} - 1\}\right] \quad (0 < t < \infty),$$

which converges to $p(\infty)$ as $n \to \infty$. This suggests that the corresponding sequence of minorants may also perform increasingly well. We show now that in fact they converge to p itself.

Let $p_n(t) = p(t)^{1/n}$, so that from inequality (3),

$$p(t) = p_n(t)^n \geqslant e^{-n A_n(t)}$$

or

(7) $$n A_n(t) \geqslant -\log p(t).$$

Now the inequality (6) may be extended to products of more than two factors, and in particular

$$n m_n(t) \leqslant m(t),$$

whence
$$nA_n(t) \leqslant A(t),$$

which implies that, for fixed t, $A_n(t) = O(n^{-1})$. Furthermore, the inequality (3) and the integral equation (5) may be combined to give
$$p_n(t) \leqslant 1 - A_n(t) e^{-A_n(l)},$$

whence
$$-\log p(t) = -n \log p_n(t) \geqslant -n \log\{1 - A_n(t) e^{-A_n(l)}\}$$
$$= nA_n(t) + O(n^{-1}) \quad \text{as } n \to \infty.$$

This, together with inequality (7), shows that
$$nA_n(t) \to -\log p(t),$$

or
$$e^{-nA_n(l)} \to p(t),$$

as $n \to \infty$, as required.

Note that
$$nA_n(t) = n \int_0^t \mu_n(u, \infty]\, du$$

$$= \int_{(0,\infty]} \min(t, u)\, n\mu_n(du),$$

whence
$$-\log p(t) = \lim_{n \to \infty} \int_{(0,\infty]} \min(t, u)\, n\mu_n(du) \quad (0 < t < \infty),$$

and that
$$\int_{(0,\infty]} (1 - e^{-u})\, n\mu_n(du) \leqslant \int_{(0,\infty]} \min(1, u)\, n\mu_n(du)$$

$$= nA_n(1) \leqslant A(1).$$

These imply the existence of a limit measure ν with
$$\int_{(0,\infty]} (1 - e^{-u})\, \nu(du) < \infty,$$

such that
$$-\log p(t) = \int_{(0,\infty]} \min(t, u)\, \nu(du).$$

This result was first shown by Kendall [4], using a different argument.

REFERENCES

1. M. S. Bartlett, *An Introduction to Stochastic Processes*, Cambridge University Press (1955).
2. P. Bloomfield, "Lower bounds for renewal sequences and p-functions", *Ztschr. Wahrsch'theorie & verw. Geb.* **19** (1971), 271–273.
3. R. Davidson, "Arithmetic and other properties of certain Delphic semigroups, II", *Ztschr. Wahrsch'theorie & verw. Geb.* **10** (1968), 146–172. (2.4 of this book.)
4. D. G. Kendall, "Delphic semigroups, infinitely divisible regenerative phenomena, and the arithmetic of p-functions", *Ztschr. Wahrsch'theorie & verw. Geb.* **9** (1968), 163–195. (2.2 of this book.)
5. J. F. C. Kingman, "The stochastic theory of regenerative events", *Ztschr. Wahrsch'theorie & verw. Geb.* **2** (1964), 180–224.
6. ——, "Some further analytical results in the theory of regenerative events", *J. Math. Anal. Appl.* **11** (1965), 422–433.
7. L. Takács, "On a probability problem arising in the theory of counters", *Proc. Cam. Phil. Soc.* **52** (1956), 488–498.

3.3

Smith's Phenomenon, and 'Jump' p-functions

R. DAVIDSON

(*from an incomplete manuscript, edited by* J. F. C. KINGMAN, G. E. H. REUTER *and* D. S. GRIFFEATH)

3.3.1 PROPERTIES OF p-FUNCTIONS

G. Smith [6] produced an example of a Markov chain with a state 0 such that the diagonal Markov function $p_{00}(t)$ had the properties

(1) $$p_{00}'(0) = -\infty; \quad \limsup_{t \to 0} p_{00}'(t) = +\infty.$$

Another construction was given by Freedman [3]. Both constructions are long and difficult. Here we shall use the theory of p-functions given by Kingman [5] to produce a diagonal Markov function with the property (1); the chain itself will not be explicitly constructed, but is similar to the Blackwell chain [1]. (Our construction owes a lot to Freedman's: the complication of his is due to his not using general p-functions.)

Let $\mathscr{P}$ be the class of standard (i.e. continuous) p-functions, and $\mathscr{PM}$ the subclass of diagonal Markov functions. We have

Lemma 1.
 (*i*) *Each* $p \in \mathscr{P}$ *has a limit, Lp say, as* $t \to \infty$ ([5], p. 198).
 (*ii*) $\mathscr{P}$ *and* $\mathscr{PM}$ *are semigroups under pointwise multiplication* (Kendall [4]).
 (*iii*) *For all* $p \in \mathscr{P}$ *and* $t \geq 0$, $p(t) \geq 2Lp - 1$ ([5], p. 199).

Lemma 2. *Let* $p \in \mathscr{P}$, $\varepsilon > 0$ *and* $s > 0$ *be given. Then there exists* $q \in \mathscr{PM}$ *such that* $|q(t) - p(t)| < \varepsilon$ *for all* $t \leq s$, $Lq = Lp$ *and* $-q'(0) \leq -p'(0)$.

234

Proof. We have ([5], p. 222) that for all $k > 0$,

$$p_k(t) = e^{-kt} \sum_{n=0}^{\infty} p(n/k)(kt)^n/n! \in \mathscr{PM},$$

and $p_k(t) \to p(t)$ for each fixed t as $k \to \infty$. We verify easily that $Lp_k = Lp$, and, by differentiation term by term (justified by uniform convergence of the derivative), that

$$-p_k'(0) = k(1 - p(1/k)) \leqslant k(1 - \exp(-\alpha/k)) \quad (\alpha = -p'(0))$$
$$\leqslant \alpha.$$

Now Kendall ([4], Theorem 10) tells us that pointwise convergence implies uniform convergence on compacta, whence the result.

3.3.2 CONSTRUCTION OF A SMITH FUNCTION

First we outline the steps of the construction. (*i*) For each positive integer n we find a $p_n \in \mathscr{P}$ which has a very large ($> n$, say) derivative near b_n ($b_n \to 0$). (*ii*) We approximate p_n well in $\mathscr{PM}$, by q_n, say. (*iii*) We let

$$q = \prod_{n=1}^{\infty} q_n,$$

and show first that $q \in \mathscr{P}$, second that q satisfies property (1) (i.e. is a Smith function), and third that q corresponds to a Markov chain, so is in $\mathscr{PM}$.

First then (*i*). We use the 'event of type B' described in [5], p. 193. A particle, initially in state 0, holds there for an exponential time with parameter a_n, after which it moves to state 1. It stays there for a fixed time b_n and then moves back to state 0 and begins again. The event of being in state 0 is regenerative, with p-function $p_n(t)$ satisfying

Lemma 3.

(*i*) $p_n(t) = \exp(-a_n t) \quad (t \leqslant b_n)$,

$\qquad = \exp(-a_n t) + a_n(t - b_n)\exp(-a_n(t - b_n)) \quad (b_n \leqslant t \leqslant 2b_n)$.

(*ii*) $p_n'(0) = -a_n$,

$\qquad p_n'(b_n+) = a_n(1 - \exp(-a_n b_n)) \sim a_n^2 b_n$.

(*iii*) $Lp_n = (1 + a_n b_n)^{-1}$.

Proofs. (*i*) and (*ii*): direct computation; (*iii*): from [5], p. 199.

It is then clear that (a) in order not to allow the infinite product of p-functions to collapse to zero, we must have $\sum c_n < \infty$ $(c_n = a_n b_n)$;

(b) given c_n, if we let $a_n \to \infty$ sufficiently fast, we can have both

$$b_n = c_n/a_n \to 0 \quad \text{and} \quad p'_n(b_n+) \sim a_n^2 b_n \to \infty.$$

If then we approximate p_n by q_n as in Lemma 2, it will be possible to make q'_n very large near b_n; and also we shall have, since

$$Lq_n = Lp_n = (1+c_n)^{-1},$$

that for all t, $q_n(t) \geqslant 2(1+c_n)^{-1}-1 \geqslant 1-2c_n$ (by Lemmas 3(*iii*) and 1(*iii*)). Thus if we put

$$q(t) = \prod_{n=1}^{\infty} q_n(t),$$

the assumptions $\sum c_n < \infty$, $c_n < \frac{1}{2}$ for all n, tell us that the product converges uniformly on the whole half-line, so that $q(t) > 0$ for all t and $q(t) \to 1$ as $t \to 0$. This, since the class of (standard or not) p-functions is pointwise closed, tells us that $q \in \mathscr{P}$.

We now turn to the precise choice of a_n, b_n and q_n, i.e. part (*ii*) of the construction. For definiteness choose $c_n = 3^{-n}$ $(n \geqslant 1)$.

Suppose that we have chosen q_r already for all $r < n$ $(n \geqslant 1)$, such that we have x_r, y_r $(r = 1, \ldots, n-1)$, $0 < x_{n-1} < y_{n-1} < x_{n-2} < \ldots < y_2 < x_1 < y_1$, with

(2) $$Q_r^*(y_r) - Q_r^*(x_r) > r(y_r - x_r) \quad (1 \leqslant r < n),$$

(3) $$|q_r^*(y_s) - q_r^*(x_s)| < 2^{-r}(y_s - x_s) \quad (1 \leqslant s < r < n),$$

where

$$Q_r(t) = \prod_{s=1}^{r} q_s(t),$$

and an asterisk denotes that the logarithm is taken. (The idea of condition (3) is that, having arranged a large derivative in condition (2) for some s, we must not subsequently disturb it.)

With the given c_n we take now $p_{n,d}$ with $a_n = 1/d, b_n = c_n d$, where $d > 0$. For $t < x_{n-1}$ we know, from the general theory of p-functions and Lemma 2, that

$$Q_{n-1}^{*\prime}(t) \geqslant - \sum_{r=1}^{n-1} a_r.$$

So, using Lemma 3(*ii*), for all sufficiently small d we may find x'_n and y'_n $(0 < x'_n < y'_n < x_{n-1})$ such that $R_n^*(y'_n) - R_n^*(x'_n) > n(y'_n - x'_n)$, where $R_n(t) = Q_{n-1}(t)p_{n,d}(t)$. Also, by Lemma 1(*i*), we can, for all sufficiently small d, satisfy condition (3) with $r = n$ and $p_{n,d}$ for q_n. Using Lemma 2 we may transfer these statements from $p_{n,d} \in \mathscr{P}$ to $q_n \in \mathscr{PM}$, and we deduce that there exist x_n and y_n $(0 < x_n < y_n < x_{n-1})$, and a q_n of the type described, satisfying conditions (2) and (3) with n for $n-1$. Choosing q_1 presents no problems—we only have to satisfy condition (2) for it—and so the induction is established: we can find an infinite sequence of q_n satisfying conditions (2) and (3). Let q be their infinite product. From the work at the end of section (*i*) of the construction we see that $q \in \mathscr{P}$; and from conditions (2) and (3), for all $r \geqslant 1$, $q^*(y_r) - q^*(x_r) \geqslant (r-1)(y_r - x_r)$, so that, getting rid of the logarithms,

$$q(y_r) - q(x_r) \geqslant (r-1)(y_r - x_r) \prod_{s=1}^{\infty} (1 - 2/3^s).$$

From the definition of the sequence x_n, y_n it is seen that q is a Smith function.

It remains to show (*iii*), that the state-space of the process we associate with q is countable. *A priori*, we have countable state-spaces S_n associated with the q_n; S_n contains a state 0 and the regenerative phenomenon of q_n is that of being in the state 0. The state-space of the Markov process we associate with q is just

$$\prod_{n=1}^{\infty} S_n,$$

the motions of different coordinates of the process being independent. The state-space just defined is uncountable, but we have

Lemma 4. *For any time t, with probability* 1 *the variable of the q-process lies in*

$$X = \bigcup_{C} \left\{ \prod_{n \in C} S_n \times \prod_{n \notin C} \{0\} \right\},$$

where C runs over all finite subsets of positive integers. X is thus countable and the essential state-space of the process, which is accordingly a Markov chain. So $q \in \mathscr{PM}$.

Proof. The last two sentences in the statement are clear. For the first, let the state at time t be $\mathbf{x}(t) = \{x_i(t)\}$, where $x_i(t)$ is the state of the chain corresponding to q_i. The probability that x_i is not in its state 0 is $1 - q_i(t)$. But the infinite product of the q_i converges, so $\sum (1 - q_i(t)) < \infty$. Then by the first Borel–Cantelli lemma all but a finite number of the $x_i(t)$ must be in their states 0.

3.3.3 OSCILLATIONS OF p-FUNCTIONS AND THE MARKOV GROUP PROBLEM

Let $\{P_t : t \geqslant 0\}$ be a strongly continuous Markov semigroup on l_1. We study the question: suppose $\|P_1 - I\| < 1$; is it true that $\|P_t - I\| \to 0$ as $t \downarrow 0$? (For previous work, see D. Williams [7].)

If we put $g(t) = \inf_i p_{ii}(t)$, then $\|P_t - I\| = 2(1 - g(t))$, so the above question becomes: if $g(1) > \frac{1}{2}$, does $g(t) \to 1$ as $t \downarrow 0$? Just now I think the question will be resolved by the use of the Banach algebra numerical range, but here we are concerned with a p-function approach.

We have the theorem ([7], p. 284) that if $g(1) > \frac{1}{2}$, then $\liminf_{t \downarrow 0} g(t) = 0$ or 1. Suppose that we can prove that for all $\varepsilon > 0$ there are $\delta_1, \delta_2 > 0$ such that for all $p \in \mathcal{P}$,

(4) $\qquad p(1) \geqslant \frac{1}{2} + \varepsilon \quad$ implies $\quad p(t) \geqslant \delta_1 \quad$ for all $t \leqslant \delta_2$.

Then the result just quoted will give the answer Yes to our question; for it would bound $g(t)$ out of a small box near the origin. We have some partial results.

Theorem 1. *The implication* (4) *holds if* $\frac{1}{2} + \varepsilon > \frac{2}{3}$, *with* $\delta_1 = \frac{1}{2}$.

Proof. We have the Kingman inequalities

(5) $\quad \begin{cases} 1 - p(t) - p(1) + p(t)p(1-t) \geqslant 0, \\ 1 - p(t) - p(1-t) - p(1) + p(t)p(1-2t) + p(t)p(1-t) \\ \quad + p(1-t)p(t) - p(t)p(1-2t)p(t) \geqslant 0. \end{cases}$

If we put $x = p(t)$, $y = p(1)$, $z = p(1-2t)$, we find, if $x \leqslant \frac{1}{2}$, by elimination of $p(1-t)$ from these inequalities, that

(6) $\qquad\qquad\qquad y \leqslant 1 - x + zx(1-x)$.

Now if $1 - 2t \geqslant t$, we may use the inequality of [2] (Proposition 6), to say that $z \leqslant \frac{3}{4}$, since $x \leqslant \frac{1}{2}$. Also, since p is continuous, there must be a point between 0 and t at which p is equal to $\frac{1}{2}$; applying inequality (6) at this point, we find that $y \leqslant \frac{11}{16}$.

Then we may use inequality (6) again, with a bound of $\frac{11}{16}$ on z, instead of $\frac{3}{4}$, provided that now $(1-2t) - 2t \geqslant t$, to conclude that $y \leqslant \frac{43}{64}$. An induction is, indeed, easily established to show that for all positive integers n, if $p(t) \leqslant \frac{1}{2}$ for some $t \leqslant 1/(2n+1)$, then $y = p(1) \leqslant \frac{2}{3} + \frac{1}{3}.4^{-n-1}$; the theorem is thus established.

Our other result—which we shall here merely state, because of its limited applicability and tedious proof [in Appendix A]—was derived from the following line of attack. The result of the last theorem bears a striking similarity to that easily derived from inequality (5), that if we let 1 and t go to infinity in such a way that $1 - t$ remains fixed, equal to s say, then (in the notation of the first section) $p(s) \geqslant (2Lp - 1)/Lp$. If $Lp \geqslant \frac{2}{3}$, then, $p(s) \geqslant \frac{1}{2}$. What we need is to convert the right-hand side of this inequality into an expression involving $p(1)$ (where $s \leqslant 1$), instead of the limit Lp. The idea that suggests itself is of averaging: one conjectures that for each positive integer n,

$$(7) \qquad p(nt) \geqslant 2^{-n+1} \sum_{r=0}^{n} {}^{n}C_{r} \, p(1 - rt) - 1.$$

If this were true, then on letting $n \to \infty$ with $t = s/n$ we would get $p(s) \geqslant 2p(1 - \frac{1}{2}s) - 1$, from the continuity of p, which would be enough.

In fact we have

Theorem 2.

(i) *Inequality* (7) *holds* (*trivially from inequality* (5)) *when* $n = 1$, *and* (*from computations on the third p-function inequality*) *when* $n = 2$.

(ii) *Inequality* (7) *holds for all* n, *provided that* $p(u) = \mathrm{e}^{-\alpha u}$ *for some* α *and all* $u \leqslant nt$.

Proof. From the p-function inequalities. [See Appendix A.]

It is conceivable that inequality (7) will be proved generally by arguments on renewal sequences, but the obvious induction fails.

REFERENCES

1. D. Blackwell, "Another countable Markov process with only instantaneous states", *Ann. Math. Statist.* **29** (1958), 313–316.
2. R. Davidson, "Arithmetic and other properties of certain Delphic semigroups, II", *Ztschr. Wahrsch'theorie & verw. Geb.* **10** (1968), 146–172. (2.4 of this book.)
3. D. A. Freedman, *Approximating Countable Markov Chains*, Holden Day, San Francisco (1971).
4. D. G. Kendall, "Delphic semigroups, infinitely divisible regenerative phenomena, and the arithmetic of p-functions", *Ztschr. Wahrsch'theorie & verw. Geb.* **9** (1968), 163–195. (2.2 of this book.)
5. J. F. C. Kingman, "The stochastic theory of regenerative events", *Ztschr. Wahrsch'theorie & verw. Geb.* **2** (1964), 180–224.
6. G. Smith, "Instantaneous states of Markov processes", *Trans. Am. Math. Soc.* **110** (1964), 185–195.
7. D. Williams, "On operator semigroups and Markov groups", *Ztschr. Wahrsch'theorie & verw. Geb.* **13** (1969), 280–285.

APPENDIX A: EDITED VERSION OF DAVIDSON'S OUTLINE OF A PROOF OF THEOREM 2

1. The case $n = 1$ follows trivially from the second-order inequalities. To deal with $n = 2$, consider the following special cases of the third-order inequalities:

(1) $\qquad 1 - p(1-2t) - p(1-t) - p(1) + p(1-2t)p(t) + p(1-2t)p(2t)$

$$+ p(1-t)p(t) - p(1-2t)p(t)p(t) \geqslant 0,$$

(2) $\qquad p(1) - p(t)p(1-t) - p(1-t)p(t) + p(t)p(1-2t)p(t) \geqslant 0.$

Adding twice inequality (1) to inequality (2),

$$2 - 2p(1-2t) - 2p(1-t) - p(1) + 2p(1-2t)p(t) + 2p(1-2t)p(2t)$$

$$- p(1-2t)p(t)^2 \geqslant 0,$$

whence

$$\sum_{r=0}^{2} {}^{2}C_r\, p(1-rt) - 2 \leqslant 2p(1-2t)\,p(2t) - [1 - p(t)]^2\, p(1-2t)$$

$$\leqslant 2p(2t),$$

as required.

2. Turning now to the case $n \geqslant 3$, assume that

(3) $\qquad\qquad\qquad p(u) = \mathrm{e}^{-\alpha u} \quad (u \leqslant nt \leqslant 1).$

For $r + s \leqslant nt$, the inequality

$$p(r+\tau+s) - p(r)p(\tau+s) - p(r+\tau)p(s) + p(r)p(\tau)p(s) \geqslant 0$$

becomes

(4) $\qquad p(r+\tau+s) - \mathrm{e}^{-\alpha r}p(\tau+s) - \mathrm{e}^{-\alpha s}p(r+\tau) + \mathrm{e}^{-\alpha(r+s)}p(\tau) \geqslant 0.$

Hence, if $m_j\ (j = 0, 1, 2, ..., n)$ is defined by

(5) $\qquad\qquad\qquad m_j = p(1-jt)\,\lambda^j, \quad \lambda = p(t) = \mathrm{e}^{-\alpha t},$

then setting $r = s = t,\ \tau = 1 - jt$ in inequality (4) gives

$$m_j - 2m_{j-1} + m_{j-2} \geqslant 0 \quad (2 \leqslant j \leqslant n),$$

so that (m_j) is a convex sequence. Similarly, the inequality

$$p(t+\tau) - p(t)p(\tau) \geqslant 0$$

shows that (m_j) is non-increasing.

Now consider the probability

$$\Phi = \Phi(1 - nt, 1 - (n-1)t, ..., 1 - t, 1; p)$$

that the regenerative phenomenon with p-function p does not occur at any of the points $1 - jt$ $(0 \leqslant j \leqslant n)$. Using equation (3) it is easy to compute that

$$(6) \qquad \Phi = 1 - (1 - \lambda) \sum_{r=1}^{n} p(1 - rt) - p(1),$$

and since $\Phi \geqslant 0$,

$$p(nt) = \lambda^n \geqslant \lambda^n p(1 - nt)$$

$$\geqslant -\Phi + \lambda^n p(1 - nt)$$

$$= p(1) + \sum_{r=1}^{n-1} (1 - \lambda) p(1 - rt) + (1 - \lambda + \lambda^n) p(1 - nt) - 1.$$

Hence we have only to prove that

$$p(1) + \sum_{r=1}^{n-1} (1 - \lambda) p(1 - rt) + (1 - \lambda + \lambda^n) p(1 - nt) \geqslant 2^{-n+1} \sum_{r=0}^{n} {}^{n}C_r p(1 - rt),$$

and this inequality is of the form

$$(7) \qquad \sum_{r=0}^{n} k_r m_r \geqslant 0,$$

where m_r is defined by equation (5), and

$$k_0 = 2^{n-1} - 1,$$

$$k_r = [2^{n-1}(1 - x^{-1}) - {}^{n}C_r] x^r \quad (1 \leqslant r \leqslant n-1),$$

$$k_n = [2^{n-1}(1 - x^{-1} + x^{-n}) - 1] x^n,$$

$$x = \lambda^{-1}.$$

Because (m_r) is non-negative, non-increasing and convex, it is a non-negative linear combination of the sequences

$$(1, 1, 1, \ldots, 1),$$

$$(1, 0, 0, \ldots, 0),$$

$$(2, 1, 0, \ldots, 0),$$

$$\vdots$$

$$(n, n-1, n-2, \ldots, 0).$$

Hence inequality (7) will hold if it holds with (m_r) replaced with each of these sequences, i.e. if

$$(8) \qquad \sum_{r=0}^{n} k_r \geqslant 0$$

and

$$(9) \qquad \sum_{r=0}^{j-1}(j-r)k_r \geqslant 0 \quad (j=1,2,...,n).$$

The first of these conditions is easily verified, since

$$\sum_{r=0}^{n}k_r = \sum_{r=0}^{n}[2^{n-1}(1-x^{-1})-{}^nC_r]x^r+2^{n-1}x^{-1}+2^{n-1}$$

$$= 2^{n-1}(1+x^n)-(1+x)^n \geqslant 0$$

by Jensen's inequality. To prove the second, note that

$$\sum_{r=0}^{j}(j-r)k_r = 2^{n-1}j+2^{n-1}(1-x^{-1})\sum_{r=1}^{j-1}(j-r)x^r-\sum_{r=0}^{j-1}(j-r)\,{}^nC_r x^r$$

$$= 2^{n-1}\frac{x^j-1}{x-1}-P_{nj}(x),$$

where

$$(10) \qquad P_{nj}(x) = \sum_{r=0}^{j-1}{}^nC_r(j-r)x^r.$$

Hence the proof will be complete if we can show that

$$(11) \qquad P_{nj}(x) \leqslant 2^{n-1}\frac{x^j-1}{x-1}$$

for $1 \leqslant j \leqslant n$ and $x \geqslant 1$.

First note that inequality (11) holds trivially when $j=1$ and also when $j=n$ since

$$P_{nn}(x) = n(x+1)^{n-1}$$

and differentiation shows that

$$\frac{(x+1)^{n-1}(x-1)}{x^n-1}$$

is non-increasing. We fill in the remaining values of j by induction on n, using the relation

$$P_{nj}(x) = P_{n-1,j}(x)+xP_{n-1,j-1}(x).$$

Thus if inequality (9) holds (for all j) for any value of n, it holds also for

$n+1$ since (except in the cases $j = 1$ and $j = n+1$ already dealt with)

$$P_{n+1,j}(x) \leqslant 2^{n-1}\frac{x^j-1}{x-1} + x2^{n-1}\frac{x^{j-1}-1}{x-1}$$

$$= 2^n\frac{x^j-1}{x-1} - 2^{n-1} < 2^{(n+1)-1}\frac{x^j-1}{x-1}.$$

Hence the proof is complete.

[J. F. C. KINGMAN]

APPENDIX B: NOTES ON SECTIONS 3.3.1–3.3.3 OF DAVIDSON'S MANUSCRIPT

1. [p. 236, ll. 3, 4]. Lemma 2 does not say that q'_n approximates p'_n, but does guarantee that we can make q_n have a large difference quotient near b_n.

2. [p. 236, 25 ff.]. Note that $p_{n,d}(t) = p_{n,1}(t/d)$.

3. [p. 237, ll. 12, 13]. The construction of x_n, y_n shows that

$$\limsup q'(t) = +\infty;$$

$q'(0+)$ must then be $-\infty$, otherwise continuity of $q'(.)$ at $t = 0+$ would imply $\lim q'(t) = q'(0) < +\infty$. Alternatively, direct calculation gives $q'(0+) = \sum q'_n(0+) = -\infty$.

4. [p. 238, l. 17]. Theorem 1 might be restated as

$$p(1) \geqslant \tfrac{2}{3} + \varepsilon \quad implies \quad p(t) \geqslant \tfrac{1}{2} \quad for \ t \leqslant \delta_2(\varepsilon).$$

In the proof, the inequality (6) should read

(6') $$y \leqslant 1 - x + x^2 z.$$

The slip is harmless because inequalities (6) and (6') coincide when $x = \tfrac{1}{2}$. The statement that, for all integers $n \geqslant 0$,

(6'') if $p(t) \leqslant \tfrac{1}{2}$ for some $t \leqslant 1/(2n+1)$, then

$$y = p(1) \leqslant \tfrac{2}{3} + \tfrac{1}{3}.4^{-n-1},$$

can be proved by induction. To see this, note that for $n = 0$ the statement becomes the Davidson, Blackwell and Freedman theorem that

$$p(1) > \tfrac{3}{4} \ (= \tfrac{2}{3} + \tfrac{1}{3}.4^{-1})$$

implies $p(t) > \tfrac{1}{2}$ for all $t \leqslant 1$. Next, suppose statement (6'') established up to n, and suppose that $p(t) \leqslant \tfrac{1}{2}$ for some $t \leqslant 1/(2n+3)$. Choose $\tau \leqslant t$ with

$p(\tau) = \frac{1}{2}$ and note that $\tau \leqslant [1/(2m+1)] \, (1-2\tau)$. Now apply statement (6″) to the interval $[0, 1-2\tau]$ rather than $[0, 1]$, to get $p(1-2\tau) \leqslant \frac{2}{3} + \frac{1}{3} . 4^{-n-1}$; then inequality (6′), with $x = p(\tau)$, $y = p(1)$, $z = p(1-2\tau)$, gives

$$p(1) \leqslant \tfrac{1}{2} + \tfrac{1}{4}(\tfrac{2}{3} + \tfrac{1}{3} . 4^{-n-1}) = \tfrac{2}{3} + \tfrac{1}{3} . 4^{-n-2}$$

which is statement (6″) for $n+1$.

One might add a *Corollary* to Theorem 1:

$$p(1) \geqslant \tfrac{2}{3} + \varepsilon \quad \textit{implies that} \quad p(t) \geqslant c(\varepsilon) > 0 \quad \textit{for all } t \leqslant 1.$$

Proof. Take the smallest n with $\frac{1}{3} . 4^{-n-1} < \varepsilon$. Then statement (6″) implies that $p(t) > \frac{1}{2}$ for $t \leqslant 1/(2n+1)$; $p(u+v) \geqslant p(u)p(v)$ then implies that $p(t) > (\frac{1}{2})^{2n+1} = \frac{1}{2} . 4^{-n} \geqslant \frac{3}{2}\varepsilon$ for all $t \leqslant 1$.

[G. E. H. REUTER]

APPENDIX C: THE DOTTED CURVE $M = 1/(2-m)$ OF DAVIDSON'S DIAGRAM

1. We show that the locus $M = 1/(2-m)$ faintly sketched on the (M, m) diagram (Figure 1, p. 34) appears in two seemingly unrelated contexts connected with Davidson's last work on maximal oscillations of standard p-functions.

2. Davidson's Theorem 1 as given by Reuter in Appendix B may be refined to yield

Theorem 1. *Given $\varepsilon > 0$ and $M(p) = M > \frac{2}{3}$, then there exists a $\delta(M, \varepsilon) > 0$ such that for all such $p \in \mathscr{P}$,*

$$(1) \qquad\qquad p(t) > \frac{2M-1}{M} - \varepsilon \quad \textit{for } 0 \leqslant t \leqslant \delta.$$

In particular, if $M(p) > \frac{2}{3}$, then $(2M-1)/M > \frac{1}{2}$ and so $p(t) > \frac{1}{2}$ on some $[0, \delta]$, which is Davidson's result. But inequality (1) also identifies the curve $M = 1/(2-m)$, which Davidson naturally tried to extend from $m = \frac{1}{2}$ down to $m = 0$, as this would be just sufficient to answer the question he poses in connection with the problem of Markov groups. Before showing in our section 3 how this curve might be extrapolated, we give a proof of the theorem.

Proof. Write the second Kingman inequality (of Davidson's inequalities (5)) in the equivalent form

$$(2) \qquad\qquad 1 - M \geqslant p(t) \, [1 - p(t)] + \theta,$$

where

$$\theta = [1-p(t)]^2 [p(1-t)-p(t)p(1-2t)]$$
$$+ [p(t)]^2 [1-p(1-t)-p(1-2t)+p(t)p(1-2t)].$$

Note that all bracketed terms of θ are non-negative, from lower-order inequalities. Now for $p(t) \leqslant \frac{1}{2}$,

$$\theta = [1-2p(t)][p(1-t)-p(t)p(1-2t)]+[p(t)]^2 [1-p(1-2t)]$$
$$\geqslant [p(t)]^2 [1-p(1-2t)],$$

while for $p(t) \geqslant \frac{1}{2}$,

$$\theta \geqslant [1-p(t)]^2 [p(1-t)-p(t)p(1-2t)]$$
$$+[1-p(t)]^2 [1-p(1-t)-p(1-2t)+p(t)p(1-2t)]$$
$$= [1-p(t)]^2 [1-p(1-2t)].$$

Thus

(3)
$$1-M \geqslant p(t)[1-p(t)]+\min\{[p(t)]^2 [1-p(1-2t)], [1-p(t)]^2 [1-p(1-2t)]\}.$$

Here and below we use the 'min' notation for brevity; in each case the first term will be the minimum for $p(t) \leqslant \frac{1}{2}$ and the second term for $p(t) \geqslant \frac{1}{2}$.

If $t \leqslant \frac{1}{3}$, we have by the Davidson, Blackwell and Freedman inequality that $[1-p(1-2t)] \geqslant p(t)[1-p(t)]$, and hence

(4)
$$1-M \geqslant p(t)[1-p(t)]+\min\{[p(t)]^3 [1-p(t)], p(t)[1-p(t)]^3\}.$$

For $t \leqslant \frac{1}{5}$ we may then apply inequality (4) to inequality (3) as giving a better bound on $[1-p(1-2t)]$, by making the appropriate scale change. Indeed, an induction argument analogous to those of Davidson and Reuter establishes that for $t \leqslant 1/(2n+1)$,

(5)
$$1-M \geqslant \min\left\{[1-p(t)]\sum_{i=0}^{n}[p(t)]^{2i+1}, p(t)\sum_{i=0}^{n}[1-p(t)]^{2i+1}\right\}$$
$$= \min\left\{\frac{p(t)}{1+p(t)}-\frac{p(t)^{2n+3}}{1+p(t)}, \frac{1-p(t)}{2-p(t)}-\frac{[1-p(t)]^{2n+3}}{2-p(t)}\right\}.$$

If $p(t) \leqslant \bar{\alpha}$ for some $t \leqslant 1/(2n+1)$, then by continuity and since $p(0)=1$, $p(t_0)=\alpha=\max\{\frac{1}{2},\bar{\alpha}\}$ for some t_0 in this interval, and applying inequality (5) at t_0,

(6)
$$M \leqslant \frac{1}{2-\alpha}+\frac{(1-\alpha)^{2n+3}}{2-\alpha}.$$

This bound is sharpest at $\alpha = \frac{1}{2}$, where inequality (6) reduces to Reuter's inequality (6''). Unfortunately, for $\bar{\alpha} < \frac{1}{2}$ no better result is available.

9

Equivalent to inequality (6) we have

(7)   if for $\frac{1}{2} \leqslant \alpha \leqslant 1, M > \dfrac{1}{2-\alpha} + \dfrac{(1-\alpha)^{2n+3}}{2-\alpha}$, then for all such $p \in \mathscr{P}$,

$$p(t) > \alpha \text{ on } [0, 1/(2n+1)],$$

from which the theorem readily follows. For suppose that $M > \frac{2}{3}$ and that $0 < \varepsilon \leqslant (3M-2)/(2M)$, so that $\frac{1}{2} \leqslant \alpha \leqslant 1$ if $\alpha = [(2M-1)/M] - \varepsilon$. Then the left-hand side of inequality (6) will be less than M for $n \geqslant n_0(M, \varepsilon)$, and applying inequality (7), we find that

$$p(t) > \frac{2M-1}{M} - \varepsilon \quad \text{on } [0, \delta]$$

where $\delta(M, \varepsilon) = 1/(2n_0+1)$, as was to be shown.

The method which Davidson used here is a sharpening of the Davidson–Blackwell–Freedman result, derived from a third-order Kingman inequality. By considering higher-order inequalities it may be possible to improve further the bound (1), but due to symmetry this type of argument breaks down for $m < \frac{1}{2}$. A technique applicable to all $m \in [0, 1]$ was the objective of Davidson's second line of attack.

3. Davidson's combinatorial conjecture, a stronger version of which is

(8) $$1 + p(nt)p(1-nt) \geqslant 2 \sum_{r=0}^{n} b(n,r)p(1-rt),$$

where $b(n,r) = {}^n C_r / 2^n$, represented an effort to extend the inaccessibility boundary curve $M = 1/(2-m)$ down to $m = 0$. Before showing how inequality (8) would accomplish this, we remark that inequality (8) is in fact the natural result which Davidson wanted to prove, though the weaker inequality ((7) on p. 239) would be sufficient for the Markov group application. Note also that inequality (8) may be formulated entirely within the framework of Feller's discrete renewal theory: can we show for any renewal sequence $\{u_n\}$ and non-negative m that for all non-negative n,

(9) $$1 + u_n u_m \geqslant 2 \sum_{r=0}^{n} b(n,r) u_{m+r} \ ?$$

If either of the inequalities (8) or (9) could be verified, then taking $s = nt$ fixed and letting $n \to \infty$ as Davidson suggested, we would obtain (using the continuity of p)

$$1 + p(s)p(1-s) \geqslant 2p\left(1 - \frac{s}{2}\right) \quad (0 \leqslant s \leqslant 1),$$

and after a scale change

$$(10) \qquad 1 + p(t)p\left(1 - \frac{t}{2}\right) \geqslant 2p(1) \quad (0 \leqslant t \leqslant 2).$$

Immediately we would have

$$(11) \qquad p(1) \leqslant \frac{1 + p(t)}{2} \quad (0 \leqslant t \leqslant 2),$$

a result saying something about how far p can descend from M in time 1 as well as how far it can ascend. The relation (10) is sufficient to solve the problem relating to Markov groups and shows why Davidson's inequality (7) 'would be enough', but a better bound is easily found.

For $t \leqslant 1$ we deduce from inequality (11) that

$$p\left(1 - \frac{t}{2}\right) \leqslant \frac{1 + p(t)}{2},$$

so that using inequality (10),

$$(12) \qquad p(1) \leqslant \frac{2 + p(t) + p^2(t)}{4} \quad (0 \leqslant t \leqslant 1).$$

After a scale change inequality (12) may then be used to obtain a better bound for $p(1 - \frac{1}{2}t)$ in inequality (10), and proceeding in this manner we establish inductively (along the same lines as the inductive argument of our section 2) that for $t \leqslant 2/n$,

$$(13) \qquad p(1) \leqslant \left[\sum_{r=0}^{n} \frac{p^r(t)}{2^{r+1}}\right] + \frac{p^n(t)}{2^{n+1}} = \frac{1}{2 - p(t)} + \left[\frac{p(t)}{2}\right]^n \left[\frac{1 - p(t)}{2 - p(t)}\right].$$

From inequalities (8) or (9) we could therefore deduce that given $\varepsilon > 0$ and $M(p) > \frac{1}{2}$ there exists a $\delta(M, \varepsilon) > 0$ such that for all $p \in \mathscr{P}$,

$$p(t) \geqslant [(2M - 1)/M] - \varepsilon$$

on $[0, \delta]$, the desired extrapolation of Theorem 1. We note in closing that both Theorem 1 and the above argument yield the boundary locus $M = 1/(2 - m)$ only in the local sense (i.e. on some $[0, \delta]$). The best result which the above calculation would legitimately add to the (M, m) diagram is inequality (12), the second estimate in the iterative derivation of inequality (13), from which we would obtain

$$(14) \qquad m(p) \geqslant \frac{1}{2}[\sqrt{(16M(p) - 7)} - 1].$$

[D. S. GRIFFEATH]

3.4

On Partly Exponential p-functions, and 'Identifying' Skeletons

R. DAVIDSON and D. G. KENDALL

1. [Davidson and I intended to write a joint note combining my results on 'exponential starts' with his on 'identifying skeletons'. In putting these together now it is worth while to point out the link with his paper [1], where an 'exponential start' appears as an assumption in one of the theorems. In paragraphs 2–9 we show that 'exponential starts' are a consequence of what at first sight would seem to be a very much weaker assumption, although they are, of course, very far from being the general rule. The discovery that *some* skeletons identify a *p*-function which carries them, is surprising, especially in view of Jane Speakman's earlier discovery that two distinct standard Markov *p*-functions can have a common skeleton, and indeed that two standard Markov *processes* can have a common skeleton [4]. Finally the closing remarks of paragraph 10, taken essentially verbatim from Davidson's thesis, suggest a line of thought leading to his remarkable paper [1]. D. G. K.]

2. We have in mind *standard p*-functions throughout, but this is more than is necessary for the first arguments, which will work provided that our *p*-function is *everywhere positive*.

3. We know that $p(s+t) \geqslant p(s)p(t)$. An early member of Kingman's series of *p*-function-characterizing inequalities further tells us that

(1) $$p(s+t) - p(s)p(t) \geqslant p(x)\{p(s-x+t) - p(s-x)p(t)\} \geqslant 0,$$

if $0 \leqslant x \leqslant s$, and this last property of *p*-functions can of course also be proved very simply directly from the definition. We shall say that (U, V) is a *Cauchy pair* for a given *p*-function if it happens to be true that

$$p(U+V) = p(U)p(V).$$

It is of course trivial that $(U, 0)$ and $(0, V)$ are always Cauchy pairs, and our purpose here is to see what more we can learn if we start with the non-trivial hypothesis that (U, V) is a Cauchy pair for some $U > 0$ and $V > 0$.

4. If we put $s = U$ and $t = V$ in inequality (1), we see immediately (because $p(x) > 0$) that $(U - x, V)$ is a Cauchy pair if $0 \leqslant x \leqslant U$, and then by symmetry that $(U, V - y)$ is a Cauchy pair if $0 \leqslant y \leqslant V$, and on combining these two results we find that (u, v) must be a Cauchy pair if

$$0 \leqslant u \leqslant U \quad \text{and} \quad 0 \leqslant v \leqslant V.$$

5. We can do better than this, however. Let $0 \leqslant w \leqslant \min(U, V)$. Then on applying the partial results obtained in the last paragraph, we see that

$$p((U - w) + (V + w)) = p(U + V) = p(U)p(V) = p(U - w)p(w) . p(V)$$
$$= p(U - w) . p(w)p(V) = p(U - w)p(w + V),$$

so that $(U - w, V + w)$ is a Cauchy pair if

$$-\min(U, V) \leqslant w \leqslant \min(U, V).$$

6. If $U = V$, then a single application of the last result shows that (u, v) is a Cauchy pair whenever $u + v = U + V$. If $U \neq V$ (say $U < V$), then a single application of the last result gives us the Cauchy pairs (u, v), where $u + v = U + V$ and $0 \leqslant u \leqslant 2U$. On iterating this argument a finite number of times we discover that

$$\left(\frac{U + V}{2}, \frac{U + V}{2} \right)$$

is a Cauchy pair, and then the first argument of this paragraph can be applied to show that (u, v) is a Cauchy pair whenever $u + v = U + V$.

7. We now know (combining the results of paragraphs (4) and (6)) that if (U, V) is a Cauchy pair, then $p(.)$ satisfies Cauchy's functional equation on the segment $[0, U + V]$. All p-functions are bounded, so that we can reject the non-measurable solutions, and there follows

Theorem 1. *If (U, V) is a Cauchy pair for the positive p-function $p(.)$, then $p(.)$ is exponential on the closed segment $[0, U + V]$.*

Corollary. *If the positive p-function $p(.)$ has a geometric skeleton, then it is exponential.*

Proof of the corollary. We can find a sequence of Cauchy pairs (U_n, U_n), where $2U_n \to \infty$. Now apply the theorem.

8. Theorem 1 plainly cannot be improved, because

$$p(t) = e^{-\min(t,1)} \quad (0 \leqslant t < \infty)$$

is a (standard) p-function. If, however, $p(.)$ is a standard *Markov* p-function, then the existence of a non-trivial Cauchy pair implies a 'Cauchy start', and then we can appeal to a result of Kingman [2] which tells us that for standard Markov p-functions an exponential start implies that the function is exponential, *tout court*. Later Kingman showed that an exponential 'start' is in fact the *only* 'start' which uniquely determines the further course of the standard *Markov p*-function [3].

9. A related question which we might ask is whether there is any skeleton, other than geometric, which compels the p-function (*not necessarily Markov, now*) to have a specific form as a function of t in $[0, \infty)$. In the light of Kingman's work one might expect the answer to be negative. [In his thesis Davidson proved the surprising result that the answer is, in fact, affirmative. I do not think he ever published this, so we close with a section in quotation marks, entirely due to him. D. G. K.]

10. 'Let $a \neq 0$ satisfy $a/2 = 1 - e^{-a}$, so that a is given uniquely and $2 - a = 2e^{-a} > 0$. Define

$$(2) \qquad u_0 = 1, \quad u_1 = e^{-a/2}, \quad u_2 = e^{-a}, \quad u_3 = e^{-a/2}.$$

We shall show that there is a unique p in $\mathscr{P}$ such that $p(\tfrac{1}{2}k) = u_k$ for $0 \leqslant k \leqslant 3$.

First, there is such a p. Take, for example,

$$(3) \qquad p_a(t) = \sum_{k=0}^{[t/b]} e^{-a(t-kb)} a^k (t - kb)^k / k!,$$

where as usual $[t]$ means the largest integer not exceeding t. In it we put $b = 1$, and a is to be the a of equations (2) above. It is easily seen that $p_a(\tfrac{1}{2}k) = u_k$ for $0 \leqslant k \leqslant 3$.

Any p which has $p(\tfrac{1}{2}k) = u_k$ for $0 \leqslant k \leqslant 3$ must (by the theorem on Cauchy pairs) have

$$(4) \qquad p(t) = e^{-at} \quad \text{for } 0 \leqslant t \leqslant 1, \text{ at least.}$$

That is, p corresponds to 'an event of type B'; and the distribution F of the return time to the neutral state puts no mass on the interval $[0, 1)$. So for $1 \leqslant t \leqslant 2$ we may write

$$p(t) = e^{-at} + \int_{1-}^{t} e^{-a(t-s)} a(t-s) \, dF(s),$$

so that

$$(5) \qquad p(\tfrac{3}{2}) = e^{-3a/2} - \int_0^{\tfrac{1}{2}+} e^{-as} as \, dF(\tfrac{3}{2} - s).$$

In the range of integration of the integral in equation (5), *as* e^{-as} is largest when $s = \tfrac{1}{2}$, for $a/2 < 1$, and *as* $e^{-as} < (a/2) \, e^{-(a/2)}$ if $s < \tfrac{1}{2}$. Thus

$$p(\tfrac{3}{2}) \leqslant e^{-3a/2} + (a/2) \, e^{-a/2} \quad (= e^{-a/2}),$$

with equality if and only if F puts all its mass at the point 1. So we can complete the sequence (2) to a u in the class of positive renewal sequences *which uniquely determines its parent p in $\mathscr{P}$.*

The example is unimportant, if curious. Its main interest lies in the argument around formula (5): if we could press that argument further, it might reduce the area of nescience in the (M, m) diagram; this in its turn might be useful in the theory of Markov groups.'

REFERENCES

1. R. Davidson, "Smith's phenomenon, and 'jump' *p*-functions", 3.3 of this book.
2. J. F. C. Kingman, "An approach to the study of Markov processes", *J. R. statist. Soc. B* **28** (1966) 417–447. (The reference is to Kingman's reply to the discussion.)
3. ——, "Markov transition probabilities, V", *Ztschr. Wahrsch'theorie & verw. Geb.* **17** (1971), 89–103.
4. Jane M. O. Speakman, "Two Markov chains with a common skeleton", *Ztschr. Wahrsch'theorie & verw. Geb.* **7** (1967), 224.

3.5

Multiplicative Semigroups and p-functions

A. M. SYKES

3.5.1 INTRODUCTION

The theory of discrete- and continuous-time p-functions is now well established. Its contribution to the theory of continuous-time Markov processes has been displayed by Kingman [12], whilst the arithmetic of p-functions has been thoroughly explored by Kendall [7], [8] and Davidson [4], [5].

Relatively little progress has been made in attempts to describe and display the range of possible oscillatory behaviour of p-functions (in particular standard p-functions). This problem, worthy of attention in its own right, relates to such problems as the embedding problem [10] and the Markov group problem [9]. The purpose of this paper is to state some miscellaneous results on this problem.

The paper is divided into five sections.

3.5.2. Quasi-Poisson renewal sequences.

3.5.3. A class of oscillatory renewal sequences.

3.5.4. Oscillations of standard p-functions; the T_λ diagram.

3.5.5. Applications to the boundary set $\mathscr{P}_0$.

3.5.6. Oscillations of standard p-functions with prescribed behaviour in $(0, 1)$.

Throughout, we use the terminology adopted in [10], so that, for example, $\mathscr{P}$ denotes the class of standard p-functions, if $p \in \mathscr{P}$ then μ is its canonical measure on $(0, \infty]$ and $m(s) = \mu(s, \infty]$. If $\{u_n\}$ is a renewal sequence, $\{f_n\}$

252

will normally denote its canonical probability measure, and $\{q_n\}$ is the 'tail sequence' of this measure, i.e.

$$q_n = f_{n+1} + f_{n+2} + \dots.$$

3.5.2 QUASI-POISSON RENEWAL SEQUENCES

Definition 1. A renewal sequence $\{u_n\}$ is *quasi-Poisson* if there exists a positive integer m such that

$$u_n = a \quad (0 < a < 1)$$

for all $n \geqslant m$; the smallest such positive integer is called the *index*. (The 'Poisson' renewal sequence is of course $\{1, a, a, a, a, \dots\}$.)

Some examples of such sequences are well known. For $n = 1, 2, \dots$, the sequences

$$v(n, p) = \{1, p, p^2, \dots, p^{n-1}, p^n, p^n, p^n, \dots\}$$

occur as 'extremal elements' from which all aperiodic infinitely divisible renewal sequences can be constructed [8]. Davidson constructs other examples in his discussion of the I_0-problem of $\mathscr{R}^+$ in [4].

Quasi-Poisson renewal sequences are characterized in the following proposition which is little more than a suitably revised form of the standard representation of renewal sequences due to Feller [6].

Proposition 1. *A sequence* $\{1, u_1, u_2, \dots, u_{m-1}, u_m, u_m, \dots\}$ *is a quasi-Poisson renewal sequence if and only if the rational function*

$$Q(z) = \left[\sum_{i=0}^{m-1} (u_i - u_{i+1}) z^i \right]^{-1}$$

has a Taylor expansion

$$\sum_{n=0}^{\infty} q_n z^n$$

about the origin such that

$$1 = q_0 \geqslant q_1 \geqslant q_2 \geqslant \dots \geqslant q_n \geqslant q_{n+1} \dots > 0.$$

The point of studying such sequences is that by imposing the condition that a renewal sequence should be quasi-Poisson, the behaviour of the sequence is limited to such an extent that sensible questions about the range of behaviour can be asked and, in some cases, answered.

Proposition 1 is useful, not so much for indicating when a sequence is a renewal sequence, but in conjunction with Pringsheim's theorem ([3], [15]) in telling us when a sequence fails to qualify. (If the denominator of the rational (non-integral) function $Q(z)$ (regular at $z = 0$, and real on the real axis) has no positive real zero, then at least one of the coefficients in its Taylor expansion must be negative.)

Definition 2. For $n = 2, 3, 4, \dots$ let $w(n, a, b)$ be the sequence

$$w_0(n, a, b) = 1, \quad w_r(n, a, b) = a \quad (1 \leqslant r \leqslant n-1)$$

and

$$w_r(n, a, b) = b \quad (r \geqslant n).$$

We ask the question: for what values of a, b is $w(n, a, b)$ a renewal sequence?

Definition 3. Let $S_n = \{(a, b): w(n, a, b) \text{ is a renewal sequence}\}$.

Proposition 2. S_n *is a multiplicative semigroup of* $[0, 1] \times [0, 1]$ *consisting of all points* (a, b) *in the unit square such that*

$$a - a(1-a)^{n-1} \leqslant b \leqslant a + (1-a)^n \frac{(n-1)^{n-1}}{n^n}.$$

Proof. Let (a, b) be 'attainable' if it belongs to S_n. Because the term-by-term product sequence $\{u_n u'_n\}$ of two renewal sequences $\{u_n\}, \{u'_n\}$ is again a renewal sequence, it follows that if (a, b) and (a', b') are attainable, then so is (aa', bb'), hence S_n is a multiplicative semigroup under the binary operation

$$(a, b)(a', b') = (aa', bb').$$

The proof of Proposition 2 requires two constructions which we state as lemmas.

Lemma 1. *If* $\{u_n\}$ *is a renewal sequence with canonical probability distribution* $\{f_n\}$, *then for each* $m = 2, 3, \dots$ *the sequence* p_{mm} *defined by*

$$p_{mm}(0) = 1, \quad p_{mm}(j) = 0 \quad (1 \leqslant j < m)$$

and

$$p_{mm}(m+n) = \sum_{j=0}^{n} u_j f_{m+n-j} \quad (n \geqslant 0)$$

is a renewal sequence.

Sketch proof. Define the infinite matrix $P = (p_{ij})$ $(i, j = 1, 2, \dots)$ by

$$p_{1j} = f_j \quad (j = 1, 2, \dots),$$
$$p_{j,j-1} = 1 \quad (j = 2, 3, \dots),$$
$$p_{ij} = 0 \quad \text{(for all other } i, j).$$

Let $p_{ij}(n)$ be the ijth element of P^n. It is clear that $p_{11}(n) = u_n$, and after a little calculation, the transition function $p_{mm}(n)$ corresponding to a jump from state m to state m in n steps, is as defined in the lemma.

It is easy to show that if the original sequence $\{u_n\}$ is quasi-Poisson, then all the diagonal transition functions are quasi-Poisson. For our purpose, however, we need only one special manifestation of this: let $\{u_n\}$ be the Poisson renewal sequence $\{1, \alpha, \alpha, ...\}$ for which $f_n = \alpha(1-\alpha)^{n-1}$, and hence p_{mm} is the renewal sequence $w(m, 0, \alpha(1-\alpha)^{m-1})$. Since

$$\max_{0\leqslant\alpha\leqslant1} \alpha(1-\alpha)^{m-1} = \frac{(m-1)^{m-1}}{m^m}$$

we have the important

Corollary 1. *The sequence $w(n, 0, b)$ is a renewal sequence if*

$$0 \leqslant b \leqslant \frac{(n-1)^{n-1}}{n^n}.$$

Lemma 2. *If $\{p_n\}$ is a renewal sequence and $0 < \alpha < 1$, then $\{p_\alpha(n)\}$ is a renewal sequence, where*

$$p_\alpha(n) = 1 - \alpha(1-p_1) - \alpha^2(p_1-p_2) - ... - \alpha^n(p_{n-1}-p_n).$$

Sketch proof. Let $\{p_n\}$ have the tail sequence $\{q_n\}$. Define a new tail sequence $\{q_n \alpha^n\}$; the new renewal sequence is $\{p_\alpha(n)\}$. (I am indebted to Professor D. Williams for this suggestion.)

Combining Lemma 2 with Corollary 1 we have

Corollary 2. *With $0 \leqslant a, b \leqslant 1$, $w(n, a, b)$ is a renewal sequence if*

$$a < b \leqslant a + (1-a)^n \frac{(n-1)^{n-1}}{n^n}.$$

The major part of Proposition 2 is now proved. Two tidying-up operations are required. First, that Corollary 2 lists all attainable (a, b) for which $a < b$, may be seen by using Pringsheim's theorem in conjunction with Proposition 1.

Second, if $b \leqslant a$ the canonical sequence $\{f_j\}$ of $w(n, a, b)$ satisfies

$$f_j = a(1-a)^{j-1} \quad (1 \leqslant j < n),$$
$$f_n = b - a[1 - (1-a)^{n-1}]$$

and

$$f_{m+n-1} = f_{m+n}(1-a) + f_{m+1}(a-b) \quad (m \geqslant 0).$$

From this, it is not difficult to show that a necessary and sufficient condition for $w(n, a, b)$ to be a renewal sequence if $b \leq a$ is that $f_n \geq 0$, i.e.

$$b \geq a - a(1-a)^{n-1}$$

and so Proposition 2 is now proved.

The semigroup S_n is sketched in Figure 1, where it is assumed that $n > 2$. Define α_n to be the maximum vertical distance from the line $a = b$ to the

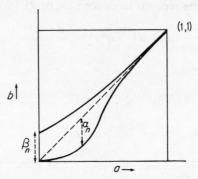

Figure 1. The semigroup S_n $(n > 2)$

lower boundary of S_n, and β_n to be the intercept of the upper boundary of S_n on the line $a = 0$. We note that S_n has the curious property

$$\alpha_n = \beta_n = \frac{(n-1)^{(n-1)}}{n^n} \quad (n = 2, 3, ...).$$

3.5.3 A CLASS OF OSCILLATORY RENEWAL SEQUENCES

It has been remarked that there is a need to exhibit classes of renewal sequences, particularly ones behaving in an unpleasant manner [11]. The following class appears to have escaped notice for some time.

Proposition 3. *Let $\{a_n\}$ be a sequence of non-negative numbers such that*

$$\sum_{i=1}^{\infty} a_i \leq 1.$$

Let $c_n = \sum a_{i_1} a_{i_2} \dots a_{i_n}$, where the summation is over all combinations of $i_1, i_2, ..., i_n$ such that $i_1 < i_2 < ... < i_n$. Then

$$u_n = 1 - c_1 + c_2 - ... + (-1)^n c_n$$

is the nth term of a renewal sequence.

Proof. Let $\sum^{(n)}$ denote summation over all n-tuples $(i_1, i_2, ..., i_n)$ such that $1 \leqslant i_1 \leqslant i_2 ... \leqslant i_n$.

Define

$$q_n = \sum^{(n)} a_{i_1} a_{i_2} ... a_{i_n}.$$

It is not difficult to show that $\{q_n\}$ is the tail sequence of a probability measure and the corresponding renewal sequence is $\{u_n\}$.

For example, let $a_n = 2^{-n}$: then $\{u_n\}$ is

$$\left\{ 1, 0, \frac{1}{3}, \frac{1}{3} - \frac{1}{3.7}, \frac{1}{3} - \frac{1}{3.7} + \frac{1}{3.7.15}, \frac{1}{3} - \frac{1}{3.7} + \frac{1}{3.7.15} - \frac{1}{3.7.15.31}, ... \right\}.$$

It is of course clear that if for $n > m$, $a_n = 0$, whilst for $n \leqslant m$, $a_n > 0$, then the resulting renewal sequence is quasi-Poisson of index m. For instance let $a_1 = a_2 = ... = a_m = 1/m$ and $a_j = 0, j > m$. The corresponding renewal sequence $\{u_m(n)\}$ is given by

$$u_m(n) = \sum_{i=0}^{n} \binom{m}{i} \frac{(-1)^i}{m^i} \quad (n \leqslant m)$$

$$= u_m(m) \quad (n > m).$$

Moreover,

$$\lim_{m \to \infty} u_m(n) = u(n),$$

where

$$\{u(n)\} = \left\{ 1, 0, \frac{1}{2!} - \frac{1}{3!}, \frac{1}{2!} - \frac{1}{3!} + \frac{1}{4!}, ... \right\},$$

showing that the sequence of partial sums of the Taylor expansion of e^{-1} is in fact a renewal sequence. (We use here the fact that the pointwise limit of a sequence of p-functions is again a p-function, see [10].) Finally, by using Lemma 2, it follows easily that the sequence of partial sums of the Taylor series of e^{-a} is a renewal sequence if and only if $0 \leqslant a \leqslant 1$.

3.5.4 OSCILLATIONS OF STANDARD p-FUNCTIONS: THE T_λ DIAGRAM

There are a number of well known results indicating restrictions on how badly standard p-functions can behave. The (M, m) diagram due to Davidson [5] and the inequalities displayed in [1] are both cases in point. In this section the problem is placed in a new context by a formulation due to Kingman.

Definition 4. For $1 < \lambda \leqslant \infty$ put

$$T_\lambda = \{(x, y): \text{there exists a } p \in \mathscr{P} \text{ such that } p(1) = x \text{ and } p(\lambda) = y\}.$$

The set T_λ is the collection of all points $(p(t), p(\lambda t))$ for all $t > 0$ and p in $\mathscr{P}$ and our ultimate aim is to display T_λ for all λ.

Proposition 4. T_λ *is a multiplicative semigroup of* $[0, 1] \times [0, 1]$ *with the properties:*

(i) *if* $(x, y) \in T_\lambda$ *and* $0 < a \leqslant 1$, *then* $(ax, ay) \in T_\lambda$;

(ii) $T_n \cap \{(x, y): x \geqslant y\} = \{(x, y): 0 < x^n \leqslant y \leqslant x \leqslant 1\}$.

Proof. $\mathscr{P}$ is a multiplicative semigroup, and hence if (x, y), (x', y') are both in T_λ, then (xx', yy') is also in T_λ, thus T_λ is a multiplicative semigroup under the binary operation

$$(x, y)(x', y') = (xx', yy').$$

Property (i) is a consequence of the semigroup property of T_λ, for the function

$$p(t) = e^{t \log a} \quad (t \leqslant 1)$$

$$= a \quad (t > a)$$

is a member of $\mathscr{P}$ if $0 < a \leqslant 1$; hence for such a, the point (a, a) is 'attainable'.

Property (ii) states that for $\lambda = 2, 3, \ldots$, the region of T_λ below the line $x = y$ is known precisely to be $\{(x, y): 0 < x^n \leqslant y \leqslant x \leqslant 1\}$. Clearly, because for all p in $\mathscr{P}$

$$p(1)^n \leqslant p(n),$$

the specified region must include all attainable (x, y) with $x \geqslant y$. That all such points are indeed attainable is proved by constructing a suitable logarithmically convex standard p-function.

Rephrasing, property (ii) states that the lower boundary of T_n is the curve $y = x^n$ $(0 < x \leqslant 1)$. It is curious that, for λ non-integral, the lower boundary of T_λ is no longer the curve $y = x^\lambda$ $(0 < x \leqslant 1)$ [14].

Unfortunately, even for λ integral, the upper boundary of T_λ remains elusive. In the following proposition we extract information in [5] relevant to the semigroup T_2. (Similar statements are easily made for T_n, $n = 3, 4, \ldots$.)

Proposition 5. *If for* $0 < x \leqslant 1$, $t_2(x) = \sup\{y: (x, y) \in T_2\}$, *then*

$$t_2(ab) \geqslant t_2(a)\, t_2(b)$$

and

$$L(x) \leqslant t_2(x) \leqslant U(x),$$

where

$$U(x) = \tfrac{3}{4} \quad (0 < x \leqslant \tfrac{1}{2})$$

$$= 1 - x + x^2 \quad (\tfrac{1}{2} \leqslant x \leqslant 1)$$

and

$$L(x) = e^{-1} + x^2 \quad (0 < x \leqslant e^{-1})$$

$$= x^2 - x \log x \quad (e^{-1} < x \leqslant 1).$$

Proof. The fact that

$$t_2(ab) \geqslant t_2(a) t_2(b)$$

is an easy consequence of T_2 being a semigroup.

The upper bound function $U(x)$ is simply a restatement of Theorem 3 of [1] or Proposition 7 of [5], in which the second Kingman inequality

$$p(s+t) \leqslant 1 + p(s)p(t) - p(t)$$

is synthesized with

$$\lim_{t \to 0+} p(t) = 1.$$

The lower bound function $L(x)$ is constructed by using the well known example of a member in $\mathscr{P}$ whose canonical measure consists of one atom. Specifically, for all $q > 0$, there exist standard p-functions $p_q(t)$ such that

$$p_q(t) = e^{-qt}, \quad t \leqslant \max(1, 2 - q^{-1}) = \delta \quad (\text{say}),$$

$$= e^{-qt} + q(t - \delta) e^{-q(t-\delta)}, \quad \delta \leqslant t \leqslant 2\delta.$$

It is trivial to verify that the points $(p_q(1), p_q(2))$ are related by the function L as q varies from 0 to ∞.

As a corollary, it is clear that

$$\limsup_{x \to 0+} t_2(x) \geqslant e^{-1}$$

and, although it is generally conjectured that equality actually holds, this still remains an unsolved problem.

We end this section by stating briefly what is known about T_∞.

Proposition 6. $T_\infty \cap \{(x,y): x \geqslant y\} = \{(x,y): 0 < y \leqslant x < 1\} \cup \{(1,1)\}$, *and if*

$$(x, y) \in T_\infty$$

then

$$y \leqslant [1 - \log x]^{-1}.$$

Proof. We have already remarked that for $0 < a \leqslant 1$, (a, a) is attainable in T_λ for every λ. If $p \in \mathscr{P}$ is such that $p(t) \to y$ as $t \to \infty$, then clearly all points (x, y) with $1 > x > y$ are in T_∞.

A result due to Bloomfield [2] states that

$$p(t) \geqslant e^{-M(t)},$$

where

$$M(t) = \int_0^t \mu(s, \infty] \, ds.$$

Hence,

$$p(t) \geqslant e^{-M(\infty)}$$
$$= e^{1 - p(\infty)^{-1}},$$

that is,

$$x \geqslant e^{1 - y^{-1}},$$

from which

$$y \leqslant [1 - \log x]^{-1}.$$

Whether or not this is an efficient upper bound is not known.

We display the pertinent features of T_2 and T_∞ in Figures 2 and 3 respectively.

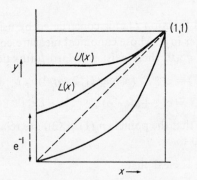

Figure 2. Concerning the semigroup T

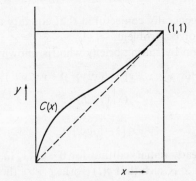

Figure 3. Concerning the semigroup T_∞

3.5.5 APPLICATIONS TO THE BOUNDARY SET $\mathscr{P}_0$

The concern for information about $t_2(x)$ for small x is not without purpose. It is known [13] that the limit of any sequence of standard p-functions is either almost-everywhere zero or a constant multiple of a member of $\mathscr{P}$. Hence, in the topology of pointwise convergence, the closure of $\mathscr{P}$ (written $\overline{\mathscr{P}}$) is of the form

$$\overline{\mathscr{P}} = \mathscr{P}_a \cup \mathscr{P}_0,$$

where $\mathscr{P}_a$ consists of all members of $\mathscr{P}$ and their multiples by a constant in $[0, 1)$, and $\mathscr{P}_0$ consists of a select set of almost-everywhere zero p-functions.

For example, it is well known [5] that the non-standard p-function

$$p(t) = e^{-n} \frac{n^n}{n!} \quad (t = n),$$

$$= 0 \quad \text{(otherwise)},$$

is a member of $\mathscr{P}_0$, whereas if $X_1, X_2, X_3, \ldots$ is a sequence of independent identically distributed random variables on $(0, \infty]$, then the function

$$p(t) = \mathrm{pr}\,(X_1 + X_2 + \ldots + X_n = t \text{ for some } n),$$

although a p-function, need not necessarily be in $\mathscr{P}_0$. For example, suppose $\mathrm{pr}\,(X_i = 2) = \delta > \frac{3}{4}$, and $\mathrm{pr}\,(X_i = 3) = 1 - \delta$, then the function $p(t)$ is such that

$$p(1) = 0, \quad p(2) = \delta \ (> \tfrac{3}{4}),$$

and from Proposition 5, such a function cannot be in $\mathscr{P}_0$. The following proposition gives new information about $\mathscr{P}_0$.

Proposition 7. *Let F be any probability measure on $(0, \infty]$ and let $X_1, X_2, \ldots$ be a sequence of independent identically distributed (according to F) random variables. Then the function*

$$p(t) = \sum_{n=1}^{\infty} e^{-n} \frac{n^n}{n!} \mathrm{pr}\,(X_1 + X_2 + \ldots + X_n = t)$$

is a member of $\mathscr{P}_0$.

We omit the proof of this result (see [14]), but remark that the construction of the appropriate sequence of p-functions mimics the well-known special case of this result (when $F\{1\} = 1$) by sliding suitable canonical measures to the right.

3.5.6 OSCILLATIONS OF p-FUNCTIONS WITH PRESCRIBED BEHAVIOUR IN $(0, 1)$

The definition of T_2 asks the question: given a member of $\mathscr{P}$ with specified value at $t = 1$, what values are possible at $t = 2$? Such a general question is unlikely to have an easy solution and so we pose a more specific problem and demonstrate how this may be solved.

For illustration, suppose that

$$p(t) = e^{-qt} \quad (0 < t \leqslant 1).$$

Then, using the fact that

$$p(t) = e^{-qt} + \sum_{n=1}^{\infty} \int_0^t \frac{q^n (t-s)^n}{n!} e^{-q(t-s)} dF^{*(n)}(s)$$

(where qF is the canonical measure of p and $F^{*(n)}$ denotes the nth convolution of F with itself), it follows that

$$p(2) = e^{-2q} + \int_0^2 q(2-s) e^{-q(2-s)} dF(s)$$

(because F attributes no mass to $(0, 1)$) and hence

$$e^{-2q} \leqslant p(2) \leqslant e^{-2q} + \delta e^{-\delta},$$

where

$$\delta = \min(q, 1).$$

(This simple and elegant proof is due to Rollo Davidson—private communication.)

It should be pointed out that the bounds for $p(2)$ are the best possible in the sense that every point in the interval $[e^{-2q}, e^{-2q} + \delta e^{-\delta}]$ may occur as $p(2)$ for some p in $\mathscr{P}$ with canonical measure of total mass q concentrated on $[1, \infty]$.

The following proposition shows how best possible bounds for $p(2)$ may be constructed given the behaviour of p in $(0, 1)$.

Proposition 8. *Let p be a fixed member of $\mathscr{P}$ with canonical measure μ. Define $\mathscr{F}_p$ to be those members of $\mathscr{P}$ which coincide with p in $[0, 1]$, and let*

$$\mathscr{I}_p = \inf_{q \in \mathscr{F}_p} q(2), \quad \mathscr{S}_p = \sup_{q \in \mathscr{F}_p} q(2).$$

Then

$$\mathscr{I}_p = p(1)^2 + \int_{(0,1)} d\mu(w) \int_{1-w}^1 p(u) p(2-w-u) du$$

and

$$\mathscr{S}_p = \mathscr{I}_p + \mu[1,\infty] \max_{0<w\leqslant1} \int_0^w p(u)p(w-u)\,du.$$

Proof. Lemma 2 of paper 3 in the series [12] (with $s = t = 1$) gives the representation,

$$p(2)-p(1)^2 = \int_0^1 \int_0^1 p(1-u)p(1-v)\,dv(u,v),$$

where v is a measure on $(0,\infty) \times (0,\infty)$ defined by

$$v(B) = (\mu \times m)\{(w+v,v): (w,v) \in B\}$$

for all Borel sets of $(0,\infty) \times (0,\infty)$, m denoting one-dimensional Lebesgue measure.

The double integral may be changed into a repeated integral, by use of Fubini's theorem, with the result that

$$p(2)-p(1)^2 = \int_{(0,1)} d\mu(w) \int_{1-w}^1 p(u)p(2-w-u)\,du$$

$$+ \int_{(0,1]} d\mu(2-w) \int_0^w p(u)p(w-u)\,du.$$

Whereas the first term involves values of p and μ in $(0,1)$, the second term involves p in $(0,1)$ but μ in $[1,2)$. Hence given $p \in \mathscr{P}$, the first term is fixed, whilst the second term may be altered by changing values of μ in $[1,2)$, without moving outside the class $\mathscr{F}_p$.

Thus, to minimize $p(2)-p(1)^2$, we move all the mass of μ in $[1,2)$ to the right of the point $x = 2$, making the second term zero and thus getting the stated value of $\mathscr{I}_p$.

To maximize $p(2)-p(1)^2$, we need to concentrate the weight of measure at our disposal, i.e. $\mu[1,\infty]$, at the point $2-x$, where x is the point in $(0,1]$ at which

$$\int_0^w p(u)p(w-u)\,du$$

attains its maximum; hence

$$\mathscr{S}_p = \mathscr{I}_p + \mu[1,\infty] \max_{0<w\leqslant1} \int_0^w p(u)p(w-u)\,du.$$

For example, let $p(t) = \frac{1}{2} + \frac{1}{2}e^{-2t}$. Using Proposition 8, we may show easily that if $q \in \mathcal{F}_p$ then $q(2)$ must lie in the interval

$$[p(2) - \tfrac{1}{4}e^{-1}(1 - e^{-2}), p(2) + \tfrac{1}{4}e^{-1}(1 + e^{-2})]$$

and moreover every point in this interval occurs as $q(2)$ for some q in $\mathcal{F}_p$.

As it stands, Proposition 8 unfortunately adds no new information to the semigroups T_2. But it may be possible to 'integrate out' the dependence of this result on the behaviour of p in $(0, 1)$ to yield a more general result which would be applicable to T_2.

ACKNOWLEDGEMENTS

I am indebted to Professor J. F. C. Kingman for his constant support throughout a course of research sponsored by the Science Research Council, and to Professor D. Williams for many helpful comments in the preparation of this paper.

REFERENCES

1. D. Blackwell and D. Freedman, "On the local behaviour of Markov transition probabilities", *Ann. Math. Stat.* **39** (1968), 6, 2123–2127.
2. P. Bloomfield, "Lower bounds for renewal sequences and *p*-functions", *Ztschr. Wahrsch'theorie & verw. Geb.* **19** (1971), 271–273.
3. E. T. Copson, *Theory of Functions of a Complex Variable*, Clarendon Press, Oxford (1935).
4. R. Davidson, "Arithmetic and other properties of certain Delphic semigroups, I", *Ztschr. Wahrsch'theorie & verw. Geb.* **10** (1968), 120–145. (2.3 of this book.)
5. ——, "Arithmetic and other properties of certain Delphic semigroups, II", *Ztschr. Wahrsch'theorie & verw. Geb.* **10** (1968), 146–172. (2.4 of this book.)
6. W. Feller, *An Introduction to Probability Theory and its Applications*, vol. 1, Wiley, New York (3rd edn., 1968).
7. D. G. Kendall, "Delphic semigroups, infinitely divisible regenerative phenomena and the arithmetic of *p*-functions", *Ztschr. Wahrsch'theorie & verw. Geb.* **9** (1968), 163–195. (2.2 of this book.)
8. ——, "Renewal sequences and their arithmetic", *Symposium on Probability Methods in Analysis* (*Lecture Notes in Mathematics* **31**) (1967), 147–175, Springer. (2.1 of this book.)
9. ——, "On Markov groups", *Proceedings of the Fifth Berkeley Symposium on Mathematical Statistics and Probability*, vol. II, part II (1966), pp. 165–173.
10. J. F. C. Kingman, "The stochastic theory of regenerative events", *Ztschr. Wahrsch'theorie & verw. Geb.* **2** (1964), 180–224.
11. ——, "An approach to the study of Markov processes", *J. Roy. Statist. Soc., Ser. B*, **28** (1966), 417–447.

12. ——, "Markov transition probabilities 1 to 5", *Ztschr. Wahrsch'theorie & verw. Geb.* **7, 9, 10, 11** and **17** (1967–71).
13. ——, "On measurable *p*-functions", *Ztschr. Wahrsch'theorie & verw. Geb.* **11** (1968), 1–8.
14. A. M. Sykes, *p-Functions*, D.Phil. Thesis, University of Sussex (1970).
15. E. C. Titchmarsh, *Theory of Functions*, Clarendon Press, Oxford (2nd edn., 1939), pp. 214–215.

3.6

Limit Theorems for a Class of Markov Processes: Some Thoughts on a Postcard from Kingman

N. H. BINGHAM

3.6.1 SUMMARY

We obtain analogues of certain limit theorems in renewal theory due to Dynkin and Lamperti for Kingman's regenerative phenomena. The results are related to the author's regenerative analogues of the Darling–Kac theorems.†

3.6.2 INTRODUCTION

Our object here is to explore the connections between two classical limit theorems in probability theory, namely

(*i*) the Darling–Kac theorems [7] on the limit behaviour (for large time) of occupation-times for Markov processes, and

(*ii*) the Dynkin–Lamperti theorems ([8], [20]) in renewal theory, on the limit behaviour of the lapses of time between the generic time-point t and the first renewal-point after t, and between t and the last renewal-point before t.

In [2] we generalized the Darling–Kac theorems from convergence of one-dimensional distributions to weak convergence. In [3], the methods so developed were applied in the contexts of regenerative phenomena [16], recurrent events and renewal theory. Here we obtain analogues of the

† For the sake of compactness in the writing of these formulae, we adopt throughout this paper the convention that abc/xyz means $abc/(xyz)$, etc.

266

Dynkin–Lamperti theorems for regenerative phenomena. The results obtained are closely connected with those of [3], which they complement and help to explain. Indeed, all the limit processes obtained here and in [3] (or in [7], [8] and [20]) are obtainable from the stable subordinator of index $\alpha \in (0, 1)$ in ways which arise simply and naturally in the regenerative theory.

In Section 3.6.3 we introduce the various processes we shall need to consider, in the renewal-theoretic case. We formulate a somewhat extended and simplified version of the Dynkin–Lamperti theorem as Theorem 1. In Section 3.6.4 we formulate the analogous problem in the regenerative theory, and state our main result, Theorem 2. The connections between this and the results of [3] are discussed, and the limit processes obtained are identified (Proposition 1). We also formulate a 'solidarity theorem', Proposition 2, which shows that the limit properties of Section 3.6.4 and [3] are in fact class properties in the Markov chain case (and indeed more generally [17]). Theorem 2 involves a certain tetrad of Markov processes associated naturally with a regenerative phenomenon Φ. Properties of this Markov tetrad are considered in Section 3.6.5. The limit processes are discussed in Section 3.6.6, where the connections with the regenerative phenomenon Φ_α of [3] are considered. The proofs of Theorems 1 and 2 follow in Section 3.6.7. In Section 3.6.8 we prove the solidarity result, Proposition 2. We close with a discussion of various questions related to this work in Section 3.6.9.

3.6.3 RENEWAL THEORY

Let $F(dx)$ be a probability distribution on $(0, \infty)$, and let $\{X_n(\omega)\}_{n=1}^\infty$ be a sequence of independent random variables with distribution $F(dx)$; write

$$S_0(\omega) = 0, \quad S_n(\omega) = \sum_{r=1}^n X_r(\omega) \quad (n = 1, 2, \ldots).$$

Write

(1) $$F(t, \omega) = \inf\{S_n(\omega) - t \colon S_n(\omega) \geqslant t\},$$

(2) $$B(t, \omega) = \inf\{t - S_n(\omega) \colon t \geqslant S_n(\omega)\}.$$

The processes $F = \{F(t, \omega) \colon t \geqslant 0\}$, $B = \{B(t, \omega) \colon t \geqslant 0\}$ are Markovian, and indeed the bivariate process $(F, B) = \{(F(t, \omega), B(t, \omega)) \colon t \geqslant 0\}$ is Markovian. The transition probabilities are stationary and are specified in terms of the one-dimensional distributions below.

Conditioning on the value of the largest partial sum in $[0, t-x)$, we obtain

(3) $$\mathrm{pr}\{B(t, \omega) > x \,|\, B(0, \omega) = 0\} = \int_{[0, t-x)} [1 - F(t-y)] \, N(dy),$$

where as usual $F(x)$ denotes $F((0, x])$ and

$$N(x) = \sum_{n=0}^{\infty} F^{*(n)}(x)$$

is the renewal function of $F(dx)$. Then conditioning on the presence and whereabouts of the first partial sum in $[s, s+t]$, we have

(4) $$\mathrm{pr}\{B(s+t, \omega) \leqslant x \,|\, B(s, \omega) = y\} = \frac{[1 - F(y+t)]}{[1 - F(y)]} E_0(x-y-t)$$

$$+ \int_{[0,t]} \mathrm{pr}\{B(t-u, \omega) \leqslant x \,|\, B(0, \omega) = 0\} \, F(y+du)/[1 - F(y)]$$

(where E_x here denotes not an expectation but the distribution function with unit mass at x). We pose the ergodic limit problem ([2], Section 1) for the B-process. Under what conditions do we obtain a non-degenerate limit process B_∞ for suitably normed processes $B_\lambda = \{B(\lambda t, \omega)/b(\lambda): t \geqslant 0\}$, and what are the possible limit processes and norming functions?

Lamperti [20] showed that the ergodic limit problem for the B-process has a solution if and only if $F(dx)$ satisfies

(5a) $$[1 - F(x)] \sim L(x)/x^\alpha \quad (x \to \infty) \quad (0 < \alpha < 1)$$

for some function L varying slowly at infinity, or equivalently [10]

(5b) $$N(x) \sim \sin \pi \alpha \; x^\alpha / \pi \alpha L(x) \quad (x \to \infty).$$

The limit process $Y_\alpha = \{Y_\alpha(t, \omega): t \geqslant 0\}$ is the stationary Markov process with one-dimensional distributions given by

(6) $$\mathrm{pr}\{Y_\alpha(t, \omega) \leqslant x\} = G_\alpha(x/t) = \pi^{-1} \sin \pi \alpha \int_0^{\min(1, x/t)} \frac{du}{u^\alpha (1-u)^{1-\alpha}},$$

and transition probabilities given by

(7) $$\mathrm{pr}\{Y_\alpha(t+s, \omega) \leqslant x \,|\, Y_\alpha(s, \omega) = y > 0\}$$

$$= [y/(t+y)]^\alpha E_0(x-y-t) + y^{-1} \alpha t \int_0^1 G_\alpha(x/t(1-u)) \, [y/(y+ut)]^{1+\alpha} \, du$$

$$= P_\alpha(x, t \,|\, y).$$

This convolution expression for P_α may be simplified; $P_\alpha(., t \,|\, y)$ is the distribution with an atom of mass $[y/(t+y)]^\alpha$ at $t+y$, and an absolutely continuous component on $[0, t]$ with density

(8) $\pi^{-1} \sin \pi\alpha \ (t-x)^\alpha/x^\alpha(y+t-x) \quad (x \in [0, t])$.

Thus as $y \to 0+$, $P_\alpha(., t|y)$ clearly converges to the arc-sine law (6). One may take $b(t) \equiv t$ for the norming function. One then has an invariance principle: $\{B(\lambda t, \omega)/\lambda: t \geq 0\}$ converges to the Lamperti process Y_α as $\lambda \to \infty$, weakly under the Stone–Skorohod J_1-topology ([2], Sections 1, 2).

   Dynkin's related work [8] involves the ergodic limit problem for the F-process. The one-dimensional distributions of the F-process are given by

(9) $\mathrm{pr}\{F(t, \omega) > x \,|\, F(0, \omega) = 0\} = \int_{(0,t)} [1 - F(t+x-y)] \, N(dy)$,

and the transition probabilities by

$$\mathrm{pr}\{F(s+t, \omega) \leqslant x \,|\, F(s, \omega) = y > 0\} \quad (x \geqslant 0),$$

which is equal to

$$1, \quad \text{if } t \leqslant y \leqslant t+x,$$

$$0, \quad \text{if } t+x < y,$$

(10) $\mathrm{pr}\{F(t-y, \omega) \leqslant x \,|\, F(0, \omega) = 0\}, \quad \text{if } y < t.$

The ergodic limit problem for the F-process has a solution if and only if the relation (5) holds. The limit process $X_\alpha = \{X_\alpha(t, \omega): t \geq 0\}$ is a Markov process with one-dimensional distributions

(11) $\mathrm{pr}\{X_\alpha(t, \omega) \leqslant x \,|\, X_\alpha(0, \omega) = 0\} = \pi^{-1} \sin \pi\alpha \int_0^{x/t} du/u^\alpha(1+u)$

$$= H_\alpha(x/t), \quad \text{say.}$$

The transition probabilities are then given by

(12) $\mathrm{pr}\{X_\alpha(s+t, \omega) \leqslant x \,|\, X_\alpha(s, \omega) = y > 0\} = Q_\alpha(x, t|y)$,

where $Q_\alpha(x, t|y)$ is equal to

$$1, \quad \text{if } t \leqslant y \leqslant t+x,$$

$$0, \quad \text{if } t+x < y,$$

and

$$H_\alpha(x/(t-y)), \quad \text{if } y < t.$$

One may take the norming function as $f(t) \equiv t$. Again the invariance principle applies; we have weak convergence under the J_1-topology.

In fact one may obtain equally satisfactory results for the bivariate Markov process (F, B). The sample functions of the (F, B)-process lie in the function-space $D[0, \infty) \times D[0, \infty)$, where $D[0, \infty)$ denotes the space of right-continuous functions on $[0, \infty)$ with left-hand limits on $(0, \infty)$. We impose the Stone–Skorohod J_1-topology on $D[0, \infty)$ (as in [2]), and the corresponding product topology on $D[0, \infty) \times D[0, \infty)$.

Theorem 1. *Condition* (5) *is necessary and sufficient for convergence of normed one-dimensional distributions, and for convergence of normed transition probabilities, of each of the Markov processes F, B and* (F, B), *to non-degenerate limits.*

When condition (5) *holds, the process* $\{(\lambda^{-1} F(\lambda t, \omega), \lambda^{-1} B(\lambda t, \omega)) : t \geqslant 0\}$ *converges as* $\lambda \to \infty$ *to a bivariate Markov process* $\{Z_\alpha(t, \omega) : t \geqslant 0\}$, *weakly under the Stone–Skorohod* J_1-*topology on* $D[0, \infty) \times D[0, \infty)$. *The corresponding marginal processes of* Z_α *are the univariate Markov processes* X_α, Y_α.

In view of the theorem, we shall write the bivariate limit process Z_α as (X_α, Y_α). We shall not write down the transition probabilities of (X_α, Y_α) explicitly, thereby subsuming relations (7) and (12), since in Section 5 we shall obtain more convenient ways of specifying the process. However, the one-dimensional distributions of (X_α, Y_α) are simply related to those of X_α, Y_α by

$$(13) \qquad \mathrm{pr}\{X_\alpha(t, \omega) > x, Y_\alpha(t, \omega) > y\} = 1 - G_\alpha((x+y)/(t-y))$$
$$= 1 - H_\alpha((x+y)/(t+x)).$$

The Markov processes $X_\alpha, Y_\alpha, Z_\alpha$ arise here as ergodic limits of standard processes in renewal theory. They do not themselves, however, appear within the context of renewal theory. Lamperti [20] showed how his process Y_α was connected with the stable subordinator with index α (and, when $\alpha \in (0, \frac{1}{2})$, with the symmetric stable process of index $\beta = 1/(1-\alpha)$). In view of the occurrence of subordinators in the theory of regenerative phenomena, the limit processes are thus regenerative in character. We proceed to formulate and solve ergodic limit problems in the regenerative theory analogous to those considered here in renewal theory. The limit processes $X_\alpha, Y_\alpha, (X_\alpha, Y_\alpha)$ arise naturally in this context; the results obtained complement those of [3].

3.6.4 REGENERATIVE PHENOMENA

We take a standard separable regenerative phenomenon Φ with indicator or Z-process $\{Z(t, \omega) : t \geqslant 0\}$ and canonical measure μ. We define the *forward Markov process* or *F*-process of Φ by

(14) $\qquad F(t,\omega) = \inf\{s \geqslant 0 : Z(s+t,\omega) = 1\} \quad (t \geqslant 0)$

and the *backward Markov process* or *B*-process by

(15) $\qquad B(t,\omega) = \inf\{s \geqslant 0 : Z(t-s,\omega) = 1\} \quad (t \geqslant 0)$

(of course by the standardness assumption, $Z(0,\omega) = 1$, and thus $B(t,\omega) \in [0,t]$). We call the bivariate process $\{(F(t,\omega), B(t,\omega)): t \geqslant 0\}$ the (F,B)-process of Φ. Our main concern is to solve the ergodic limit problem for the (F,B)-process, and hence for the *F*- and *B*-processes also.

Theorem 2a. *The bivariate (F,B)-process of Φ has a non-degenerate ergodic limit if and only if the canonical measure μ of Φ satisfies*

(∗) $\qquad m(x) \equiv \mu(x, \infty] \sim 1/L(x)\, x^{\alpha}\, \Gamma(1-\alpha) \quad (0 < \alpha < 1) \quad (x \to \infty),$

where L varies slowly at infinity. The ergodic limit process is then the bivariate Markov process (X_{α}, Y_{α}) of Theorem 1.

The univariate F- and B-processes each have a non-degenerate ergodic limit if and only if (∗) holds. The corresponding limit processes are then the Dynkin process X_{α} and the Lamperti process Y_{α}. The norming functions may be taken as $f(\lambda) = b(\lambda) \equiv \lambda \ (\lambda > 0)$.

The process (X_{α}, Y_{α}) is not itself the (F,B)-process of a regenerative phenomenon. However, it is a bivariate Markov process associated with a regenerative phenomenon Φ_{α} (see [3], Section 7) in a way we shall now explain.

In [3] we defined the following tetrad of processes associated with Φ:

(16) $\qquad \begin{cases} H(t,\omega) = \displaystyle\int_0^t Z(u,\omega)\,du, \\[2ex] T(v,\omega) = \sup\{t: H(t,\omega) \leqslant v\}, \\[2ex] U(v,\omega) = T(v,\omega) - v, \\[2ex] V(t,\omega) = \sup\{v: U(v,\omega) \leqslant t\}. \end{cases}$

It was shown that the *V*-process satisfies

$$V(t,\omega) = \int_0^{\infty} Z(u,\omega)\, I\{u: u - H(u,\omega) \leqslant t\}\, du,$$

and that

$$\sup\{u: u - H(u,\omega) \leqslant t\} = t + V(t,\omega).$$

Thus

(17) $$V(t, \omega) = H(t + V(t, \omega), \omega) \quad (t \geqslant 0).$$

Relation (17) suggests that to pass from the H-process to the V-process (or what equations (16) show to be equivalent, to remove the drift), one should pass from time-point t to $t + V(t, \omega)$. We consider the bivariate process

(18) $$\{(F(t + V(t, \omega), \omega), B(t + V(t, \omega), \omega)) : t \geqslant 0\}$$

From continuity of H, equations (16) give right-continuity of T, U, and left-continuity of V. The V-process will have a jump discontinuity of size $a > 0$ at time t whenever the U-process takes the value t throughout an interval of length a. Almost surely, this does not occur when Φ is instantaneous, and so the V-process of an instantaneous phenomenon is almost surely continuous. Since the V-process is monotone, one obtains a right-continuous process $\{\tilde{V}(t, \omega) : t \geqslant 0\}$ by putting

$$\tilde{V}(t, \omega) = V(t + 0, \omega) = \lim_{s \to 0+} V(t + s, \omega).$$

(The V- and $\tilde{V}$-processes of any instantaneous phenomenon coincide almost surely.) These observations motivate the following definition. We write

(19) $$\begin{cases} F^*(t, \omega) = F(t + \tilde{V}(t, \omega), \omega) \\ B^*(t, \omega) = B(t + \tilde{V}(t, \omega), \omega) \end{cases} \quad (t \geqslant 0),$$

and form the bivariate process (F^*, B^*). The ergodic limit problem for these processes has the following solution:

Theorem 2b. *The assertions of Theorem 2a apply when F, B are replaced by F^*, B^*.*

In view of Theorem 2, we have a complete solution to the ergodic limit problem for the two bivariate processes $(F, B), (F^*, B^*)$, and for the corresponding four univariate processes. We shall refer to the tetrad $\{F, B; F^*, B^*\}$ as the *Markov tetrad* of Φ (we shall see that (F, B) and (F^*, B^*) are in fact Markovian). Theorem 2 thus complements the results of [3] on the ergodic limit problem for the tetrad $\{H, T, U, V\}$.

Proposition 1. *The process (X_α, Y_α) is the (F^*, B^*)-process of the phenomenon Φ_α with canonical measure*

$$\mu_\alpha(dx) = \alpha \, dx / x^{1+\alpha} \Gamma(1 - \alpha) \quad (\alpha \in (0, 1), x > 0).$$

The processes X_α, Y_α are the F^*- and B^*-processes of Φ_α.

We shall show in Section 3.6.5 that, after the exclusion of a null-set from our probability space,

(20) $\{\omega: F(t, \omega) = 0\} = \{\omega: Z(t, \omega) = 1\}$ for each $t \geqslant 0$.

The H-process is thus given by equations (16) and (20) as the occupation-time process of the Markov process $\{F(t, \omega): t \geqslant 0\}$ in the set $\{0\}$. Such processes are of Darling–Kac type and were considered in [2]. If $p(x \mid dy, s)$ denotes

$$\int_0^\infty e^{-st} \operatorname{pr}\{F(t, \omega) \in dy\} dt,$$

Fubini's theorem shows that

(21) $\displaystyle\int_{-\infty}^\infty I_{\{0\}}(y) p(x \mid dy, s) = \int_0^\infty e^{-st} I_{\{0\}}(F(t, \omega)) dt,$

$$= \int_0^\infty e^{-st} Z(t, \omega) dt \quad \text{(by equation (20)),}$$

$$= \int_0^\infty e^{-st} p(t) dt = r(s)$$

in the notation of [16], [3]. When the phenomenon Φ is not transient (i.e. when μ has no atom at ∞),

$$\int_0^\infty p(t) dt = \lim_{s \to 0} r(s) = \infty$$

(observe that when condition (∗) of Theorem 2 holds, Φ is not transient). So in the non-transient case, the Darling–Kac Condition (A) is satisfied [7]:

(22) $\displaystyle [r(s)]^{-1} \int_{-\infty}^\infty e^{-st} p(x \mid dy, s) \to 1 \quad (r(s) \to \infty) \quad (s \to 0),$

and the results of [2] are applicable. In Theorem 1 of [3], we obtained a regenerative analogue of Theorem 1 of [2], giving a simple and self-contained proof and pointing out that the methods used in [2] gave this result also. Our remarks above show that Theorem 1 of [3] is in fact a direct consequence of the *statement*, and not just of the *proof*, of Theorem 1 of [2]. However, it is not the case that Theorem 1 of [3] can itself be obtained from our main result here by the continuous mapping theorem of weak convergence theory [1], as will be seen later.

It is of interest to consider our results in the context of linked systems of regenerative phenomena [17]. Let $(p_{ij}(t))$ $(t > 0, i, j = 1, 2, \ldots)$ be the matrix of p-functions of such a system, and let $(r_{ij}(s))$ be the corresponding matrix of Laplace transforms. Let μ_k denote the canonical measure of the regenerative phenomenon Φ_k given by occupation of the kth state. There arises the question of linking the ergodic behaviour of two phenomena Φ_i, Φ_j for which the corresponding states i, j intercommunicate. It turns out that the ergodic limit problems for Φ_i, Φ_j have identical solutions (up to a constant factor c_{ij} in the norming functions). This is because one such state satisfies condition (∗) if and only if the other does, as follows from Proposition 2 below. From this (together with Theorem 2 and Theorem 1 of [3]), it follows that the ergodic properties of each of the eight processes of the two tetrads $\{H_i, T_i, U_i, V_i\}, \{F_i, B_i; F_i^*, B_i^*\}$ are in fact 'class properties'. One thus obtains solidarity theorems concerning ergodic behaviour.

Let f_{ij}, g_{ij} be the first-entrance and last-exit functions appearing in the decompositions ([17])

$$(23) \qquad p_{ij}(t) = \int_0^t f_{ij}(u) p_{jj}(t-u)\, du,$$

$$(24) \qquad p_{ij}(t) = \int_0^t p_{ii}(u) g_{ij}(t-u)\, du.$$

Then we have our solidarity result:

Proposition 2. *Let i, j be two intercommunicating states in a linked system of regenerative phenomena. Then if*

$$c_{ij} = \int_0^\infty g_{ij}(u)\, du \Big/ \int_0^\infty f_{ij}(u)\, du,$$

we must have $c_{ij} \in (0, \infty)$. If L is a function varying slowly at infinity, the following conditions are equivalent:

$$(25a) \qquad r_{ii}(s) \sim L(1/s)/s^\alpha \quad (s \to 0+) \quad (0 < \alpha < 1),$$

$$(25b) \qquad m_i(x) = \mu_i(x, \infty] \sim 1/L(x)\, x^\alpha \Gamma(1-\alpha) \quad (x \to \infty),$$

$$(26a) \qquad r_{jj}(s) \sim c_{ij} L(1/s)/s^\infty \quad (s \to 0+),$$

$$(26b) \qquad m_j(x) = \mu_j(x, \infty] \sim c_{ij}/L(x)\, x^\alpha \Gamma(1-\alpha) \quad (x \to \infty).$$

One obtains the corresponding result for Markov chains by specialization. In the case where the states i, j are stable, this result can be obtained by considering the first-passage distributions and the corresponding taboo distributions (Chung [6], Part II, Sections 11–13), using essentially renewal-theoretic methods. The analogous result for regular Markov renewal processes in the Pyke theory was obtained by Teugels [27].

3.6.5 THE MARKOV TETRAD

We now discuss certain properties of the Markov tetrad of Φ which we shall need in the proof of our limit theorem. We first recall some notation from [3]. We write $S(\omega) = \{t: Z(t, \omega) = 1\}$ for the random set on which Φ occurs. If $r(s)$ is the Laplace transform of the p-function $p(t)$, we write $[r(s)]^{-1} = s + \Lambda(s)$, where

$$\Lambda(s) = \int_{(0,\infty]} (1 - e^{-xs}) \mu(dx),$$

and $W(dx)$ for the measure with Laplace–Stieltjes transform (LST) $1/\Lambda(s)$. If

$$m(x) = \mu(x, \infty]$$

is the tail-function of μ and

$$M(x) = \int_0^x m(y)\,dy,$$

let $\tilde{M}(s)$ be the LST of $M(dx)$. Then $\Lambda(s) = s\tilde{M}(s)$ ([3], (4.12)), and so

(27) $$[r(s)]^{-1} = s[1 + \tilde{M}(s)].$$

Laplace inversion of equation (27) yields

(28) $$\int_{[0,t]} m(t-u)\,W(du) = 1$$

for almost all $t \geqslant 0$. Kesten [15] has shown that in fact equation (28) holds for *all* $t \geqslant 0$.

Kendall [13] obtains the probability of non-occurrence of Φ throughout an interval as follows:

(29) $$\mathrm{pr}\{S(\omega) \cap [t, t+x] = \varnothing\} = \int_0^t m(t+x-v)p(v)\,dv.$$

A double Laplace transform shows that equation (29) is equivalent to

(30) $$\int_0^\infty \int_0^\infty e^{-\alpha t} e^{-\beta x}\,dt\,dx\,\mathrm{pr}\{S(\omega) \cap [t, t+x] = \varnothing\}$$

$$= (\alpha - \beta)^{-1} r(\alpha)\{\beta^{-1}\Lambda(\beta) - \alpha^{-1}\Lambda(\alpha)\}$$

$$= [\tilde{M}(\beta) - \tilde{M}(\alpha)]/\alpha(\alpha - \beta)[1 + \tilde{M}(\alpha)] \quad (\alpha \neq \beta).$$

Lemma 1. *If* $t \geqslant 0$,

(i) $$\{F(t, \omega) > 0\} = \{Z(t, \omega) = 0\} \quad a.s.,$$

(ii) $$\{B(t, \omega) > 0\} = \{Z(t, \omega) = 0\} \quad a.s.$$

Proof. We first prove (*i*). If Φ occurs at t, $Z(t, \omega) = 1$ and equation (14) shows that $F(t, \omega) = 0$. Also, equation (29) gives

(31) $\mathrm{pr}\{S(\omega) \cap [t, t+x) = \varnothing \,|\, Z(t, \omega) = 0\} = \mathrm{pr}\{F(t, \omega) \geqslant x \,|\, Z(t, \omega) = 0\}$

$$= \int_0^t p(v)\, m(t + x - v)\, dv / [1 - p(t)].$$

By Kingman's Volterra equation ([16]),

(32) $$\int_0^t p(v)\, m(t - v)\, dv = 1 - p(t).$$

By equations (31) and (32) and monotone convergence,

(33) $$\mathrm{pr}\{F(t, \omega) \geqslant x \,|\, Z(t, \omega) = 0\} \to 1 \quad (x \to 0+).$$

By formula (33) and the continuity theorem,

(34) $$\mathrm{pr}\{F(t, \omega) > 0 \,|\, Z(t, \omega) = 0\} = 1,$$

which proves (*i*). The proof of (*ii*) is similar.

The exceptional null-sets in Lemma 1 may depend on $t \geqslant 0$. However, the assumption that Φ is separable enables us to choose a fixed null-set outside which the assertions (*i*) and (*ii*) hold simultaneously for all $t \geqslant 0$. We shall reduce the proof of Theorem 2 to proving convergence of normed transition probabilities of the Markov processes in question, and these are not affected by the behaviour on the exceptional null-set. We thus lose no generality by excluding this exceptional null-set from our probability space. Accordingly, we shall suppose this done.

We may now write

(35) $\quad S(\omega) = \{t \colon Z(t, \omega) = 1\} = \{t \colon F(t, \omega) = 0\} = \{t \colon B(t, \omega) = 0\}$,

and thus

(36) $$H(t, \omega) = \int_0^t Z(u, \omega)\, du = \int_0^t I_{\{0\}}(F(u, \omega))\, du$$

$$= \int_0^t I_{\{0\}}(B(u, \omega))\, du.$$

Lemma 2.

(*i*) $\qquad F(t, \omega) = T(H(t, \omega), \omega) - t = \inf\{S(\omega) \cap [t, \infty)\}$,

(*ii*) $\qquad B(t, \omega) = t - T(H(t, \omega) - 0, \omega) = \sup\{S(\omega) \cap [0, t]\}$,

(*iii*) $\qquad F^*(t, \omega) = U(\tilde{V}(t, \omega), \omega) - t$,

(*iv*) $\qquad B^*(t, \omega) = t - U(\tilde{V}(t, \omega) - 0, \omega)$.

Proof. From the definition of the T-process as the right-continuous inverse of the H-process,

(37) $T(H(t, \omega) - 0, \omega) \leqslant t \leqslant T(H(t, \omega), \omega),$

with equality in the right-hand inequality if and only if t is a point of right increase of the H-process, i.e. if and only if t lies in the closure of $S(\omega) \cap (t, \infty)$. Also, the T-process has a jump of size $\varepsilon > 0$ at the point v if and only if the H-process has the value v throughout some maximal open interval $(t_1, t_1 + \varepsilon)$. When this is the case, and $t \in (t_1, t_1 + \varepsilon)$, one has

$$t + F(t, \omega) = t_1 + \varepsilon,$$

$$t - B(t, \omega) = t_1.$$

Assertions (*i*) and (*ii*) follow readily from this.

For (*iii*), note that

$$
\begin{aligned}
F^*(t, \omega) &= F(t + \tilde{V}(t, \omega), \omega) \quad \text{(by relation (18)),} \\
&= T(H(t + \tilde{V}(t, \omega), \omega), \omega) - t - \tilde{V}(t, \omega) \quad \text{(by (i)),} \\
&= T(\tilde{V}(t, \omega), \omega) - t - \tilde{V}(t, \omega) \quad \text{(by equation (17)),} \\
&= U(\tilde{V}(t, \omega), \omega) - t \quad \text{(by equations (16)),}
\end{aligned}
$$

proving (*iii*). The proof of (*iv*) is similar.

The Markov property for each of the bivariate processes (F, B), (F^*, B^*), and the corresponding univariate processes, now follows from Lemma 2 and the independent increments property of the T- and U-processes of Φ.

By continuity of the H-process, $H(t, \omega) = \inf\{v: T(v, \omega) > t\}$. So $H(t, \omega)$ is the hitting time of the interval $(t, \infty]$ for the T-process, and

$$t + F(t, \omega) = T(H(t, \omega), \omega)$$

is the position of the T-process at the time of first hitting $(t, \infty]$. Similarly, $\tilde{V}(t, \omega)$ is the hitting time of $(t, \infty]$ for the U-process, and

$$t + F^*(t, \omega) = U(\tilde{V}(t, \omega), \omega)$$

the corresponding position.

One sees from Lemma 2 that for each ω and each $t \geqslant 0$, $B(t, \omega) \in [0, t]$ and $B^*(t, \omega) \in [0, t]$.

Lemma 3a. *If $t > 0$,*

(*i*) $\text{pr}\{F(t, \omega) > x\} = \displaystyle\int_0^t m(t + x - v)\, p(v)\, dv \quad (x \geqslant 0),$

$\text{pr}\{F(t, \omega) = 0\} = p(t);$

10

(ii) $\mathrm{pr}\{B(t,\omega) > x\} = \displaystyle\int_0^{t-x} m(t-v)\,p(v)\,dv \quad (x \in [0,t]),$

$\mathrm{pr}\{B(t,\omega) = 0\} = p(t);$

(iii) $\mathrm{pr}\{F^*(t,\omega) > x\} = \displaystyle\int_{[0,t]} m(t+x-v)\,W(dv) \quad (x \geqslant 0);$

(iv) $\mathrm{pr}\{B^*(t,\omega) > x\} = \displaystyle\int_{[0,t-x]} m(t-v)\,W(dv) \quad (x \in [0,t]).$

Lemma 3b. *If* $t > 0$, $x_1 > 0$, $x_2 \in [0,t]$,

(i) $\mathrm{pr}\{F(t,\omega) > x_1, B(t,\omega) > x_2\} = \displaystyle\int_0^{t-x_2} m(t+x_1-v)\,p(v)\,dv;$

(ii) $\mathrm{pr}\{F^*(t,\omega) > x_1, B^*(t,\omega) > x_2\} = \displaystyle\int_{[0,t-x_2]} m(t+x_1-v)\,W(dv).$

Proof. By equation (35) and Lemma 2, if $x \geqslant 0$,

(38) $\{\omega: F(t,\omega) > x\} = \{\omega:$ range of T-process avoids $[t,t+x]\}$

$= \{\omega: S(\omega) \cap [t,t+x] = \varnothing\}.$

By right-continuity of $m(u)$ and Kendall's result (29), equation (38) gives assertion (i). Alternatively, (i) follows from results of Kesten (Lemma 6.1 and Proposition 6 of [15]); (ii) is proved similarly. Assertion (iii) on the F^*-process is Kesten's result (Lemma 6.1 of [15]) applied to a driftless subordinator; (iv) follows similarly. Lemma 3b follows from Lemmas 2 and 3a.

Corollary.

(i) $\displaystyle\int_0^\infty \int_0^\infty e^{-\alpha t} e^{-\beta x}\,dt\,dx\,\mathrm{pr}\{F(t,\omega) > x\}$

$= [\tilde{M}(\beta) - \tilde{M}(\alpha)]/\alpha(\alpha - \beta)\,[1 + \tilde{M}(\alpha)] \quad (\alpha \neq \beta).$

(ii) $\displaystyle\int_0^\infty \int_0^\infty e^{-\alpha t} e^{-\beta x}\,dt\,dx\,\mathrm{pr}\{F^*(t,\omega) > x\}$

$= [\tilde{M}(\beta) - \tilde{M}(\alpha)]/\alpha(\alpha - \beta)\,\tilde{M}(\alpha) \quad (\alpha \neq \beta).$

Proof. Statement (i) is due to Kendall [13], and follows from equations (30) and (38). The proof of (ii) is similar.

In view of the analogy between the assertions of the lemma and its corollary for the (F,B)- and (F^*,B^*)-processes, we may say that the assertions on (F^*, B^*) are obtained from those on (F, B) by removing the drift.

We shall need a further result on subordinators.

Lemma 4. *Let* $L(\omega)$ *be the length of an interval* $(a(\omega), b(\omega))$, *which avoids the range of a subordinator but is such that both its end-points are limit-points of the range. Then if* μ *denotes the canonical measure of the subordinator,*

$$\mathrm{pr}\{L(\omega) > y \geqslant x > 0 \mid L(\omega) > x\} = \mu(y, \infty]/\mu(x, \infty].$$

Proof. Since the drift plays no role in determining the distribution of the length of such an interval, it suffices to consider only a subordinator of unit drift, which may be regarded as the T-process of a regenerative phenomenon Φ. Then for $y \geqslant x > 0$

$$\mathrm{pr}\{L \geqslant y \mid L \geqslant x\}$$

$$= \lim_{\varepsilon \to 0} \frac{\mathrm{pr}\{\text{Range of } T \text{ meets } (a - \varepsilon, a) \text{ and avoids } [a, a + y)\}}{\mathrm{pr}\{\text{Range of } T \text{ meets } (a - \varepsilon, a) \text{ and avoids } [a, a + x)\}}$$

$$= \lim_{\varepsilon \to 0} \int_{a-\varepsilon}^{a} m(a + y - v) p(v) \, dv \bigg/ \int_{a-\varepsilon}^{a} m(a + x - v) p(v) \, dv$$

(by Lemmas 2 and 3),

$$= m(y - 0)/m(x - 0)$$

since $p(t) \to 1$ as $t \to 0+$, Φ being standard. Hence

$$\mathrm{pr}\{L > y \mid L > x\} = m(y)/m(x) \quad (y \geqslant x > 0),$$

as required.

Using Lemmas 3 and 4, we can now write down the transition probabilities of the univariate and bivariate Markov processes in question. These are sufficiently exemplified by

Lemma 5.

$$\mathrm{pr}\{B(s+t, \omega) \leqslant x \mid B(s, \omega) = y > 0\} = [m(y+t)/m(y)] E_0(x - y - t)$$

$$+ \int_{[0, t)} \mathrm{pr}\{B(t - u, \omega) \leqslant x\} \mu(d(y + u))/m(y).$$

Proof. The atom of mass $m(t + y)/m(y)$ at $x = y + t$ corresponds to the possibility that the range of the T-process avoids $[s, s + t)$, by Lemma 4. The convolution integral is obtained by conditioning on

$$u = \inf\{(\text{range of } T) \cap [s, s + t)\} - s$$

in the remaining case.

Finally, let us consider the bivariate process (18). We write $\sum(\mu)$ for the additive semigroup generated by the atoms of a Lévy measure μ; then $\sum(\mu)$ is (at most) countable. The one-dimensional distributions of the process (18) coincide with those of the (F^*, B^*)-process given in Lemma 3, except when Φ is stable and $t \in \sum(\mu)$; this may be shown using results of Kesten [15]. In his main result (Theorem 1 of [15]) Kesten considers the hitting probabilities

$$h(t) = \mathrm{pr}\{X(v, \omega) = t \text{ or } X(v-, \omega) = t \text{ for some } v \in (0, \infty)\};$$

for, in particular, a subordinator $\{X(v, \omega): v \geqslant 0\}$. He shows that $h(t)$ vanishes identically, except for the case of a driftless subordinator with bounded Lévy measure μ, when $h(t) > 0$ if and only if $t = 0$ or $t \in \sum(\mu)$. In our terminology, $h(t) \equiv 0$ except for the U-process of a stable phenomenon, when $\{t: h(t) > 0\} = \{0\} \cup \sum(\mu)$. These exceptional cases may be dealt with using renewal-theoretic methods. One can show that whenever condition (*) holds, the ergodic limit of the processes (18) and (F^*, B^*) coincide. As a simple (and typical) example of a phenomenon exhibiting such exceptional cases, we mention the stable phenomenon with canonical measure $\mu = E_1$, (where E_x denotes the probability measure with unit mass at x), for which $\sum(\mu) = \{1, 2, ...\}$. The W-measure is counting measure, while the U-process is the Poisson process with unit rate. Trivially, the hitting probability $h(t)$ satisfies

$$h(t) = 1 \quad (t \in \{0, 1, 2, ...\} = \{0\} \cup \sum(\mu)),$$
$$= 0 \quad (\text{otherwise}).$$

3.6.6 THE LIMIT PROCESSES

Using results of Blumenthal and Getoor [4], Lamperti ([20], Theorem 4.2) obtains a connection between his process Y_α and the driftless stable subordinator of index $\alpha \in (0, 1)$. In our notation, this latter process is the U-process U_α of the regenerative phenomenon Φ_α with canonical measure $\mu_\alpha(dx) = \alpha \, dx/x^{1+\alpha} \Gamma(1-\alpha)$ ([3], Section 7). From this and Lemma 2, one can identify the Lamperti process Y_α with our process B_α^*. Similarly, the Dynkin process X_α can be identified with our process F_α, and the bivariate process (X_α, Y_α) with our (F_α, B_α), thereby establishing Proposition 1.

We thus see that our limit process (X_α, Y_α)—which we write henceforth as (F_α^*, B_α^*)—is indeed regenerative in character, as we claimed in Section 3.6.3. As in [3], we summarize Theorem 2 by saying that Φ_α is the ergodic limit phenomenon of Φ when (*) holds, and that the class of Φ satisfying (*) for a given $\alpha \in (0, 1)$ is the domain of ergodic attraction of Φ_α.

Lamperti also showed that if $\{\xi_\gamma(t, \omega): t \geqslant 0\}$ is a (right-continuous, separable) symmetric stable process with index $\gamma \in (1, 2]$ and $\xi_\gamma(0, \omega) = 0$, then $b_\gamma(t, \omega) = t - \sup\{u \leqslant t: \xi_\gamma(u, \omega) = 0\}$ defines the same process as B_α^* with $\alpha = 1 - \gamma^{-1} \in (0, \frac{1}{2}]$. Similarly,

$$f_\gamma(t, \omega) = \inf\{u \geqslant t: \xi_\gamma(u, \omega) = 0\} - t$$

defines the same process as F_α^*, and (f_γ, b_γ) the same process as (F_α^*, B_α^*). These results display a close connection between the symmetric stable process ξ_γ and the limit phenomena Φ_α, for $\gamma \in (1, 2]$, $\alpha \in (0, \frac{1}{2}]$, $\alpha = 1 - \gamma^{-1}$. (In particular, for $\gamma = 2$, $\alpha = \frac{1}{2}$, one has a connection between Brownian motion and the 'Brownian regenerative phenomenon' $\Phi_{\frac{1}{2}}$ with

$$\mu_{\frac{1}{2}}(dx) = dx/2\pi^{\frac{1}{2}} x^{\frac{3}{2}}$$

([3], Section 7)). This connection extends to the processes $\{H_\alpha, T_\alpha, U_\alpha, V_\alpha\}$ of [3]. Stone [23] showed that the process V_α can be identified with the local time at zero of ξ_γ; that is, there exists a two-parameter stochastic process $\{L(t, x, \omega): t \geqslant 0, x \text{ real}\}$ with L continuous in (t, x) a.s., and

$$(39) \qquad \int_0^t I_B(\xi_\gamma(u, \omega)) \, du = \int_B L(t, x, \omega) \, dx \quad \text{a.s.}$$

for all Borel sets B. Then one can realize V_α as

$$(40) \qquad\qquad L(t, 0, \omega) = V_\alpha(t, \omega) \quad (t \geqslant 0).$$

We regard equations (39) and (40) as setting up a representation of Φ_α ($\alpha \in (0, \frac{1}{2}]$) in terms of the symmetric stable processes.

One can also [23] obtain all the processes V_α ($\alpha \in (0, 1)$) as local times at zero of a certain two-parameter family of semi-stable processes (in the sense of [21]), obtainable from the symmetric stable processes by a construction of the type used in diffusion theory.

By Lemma 2 and Proposition 1, we have

$$\{F_\alpha^*(t, \omega) > x_1, B_\alpha^*(t, \omega) > x_2\} = \{F_\alpha^*(t - x_2, \omega) > x_1 + x_2\}$$
$$= \{B_\alpha^*(t + x_1, \omega) > x_1 + x_2\}$$
$$= \{\text{closure of range of } U_\alpha \text{ avoids } [t - x_2, t + x_1]\}.$$

Expressed in terms of the distributions G_α, H_α of Section 3.6.3, this reduces to

$$\text{pr}\{F_\alpha^*(a, \omega) \geqslant b - a\} = \text{pr}\{B_\alpha^*(b, \omega) \geqslant b - a\},$$

or to

$$\int_{(b-a)/a}^\infty dx/x^\alpha(1 + x) = \int_{(b-a)/b}^1 dy/y^\alpha(1 - y)^{1-\alpha}$$

(as follows from the bilinear transformation $x = y/(1 - y)$).

If we write

$$\Gamma(\alpha, x) = \int_x^\infty e^{-u} u^{\alpha-1} du$$

for the incomplete gamma-integral, the LST of the Dynkin distribution G_α is given by ([12], p. 941)

$$\pi^{-1} \sin \pi\alpha \int_0^\infty e^{-sx} dx/x^\alpha(1+x) = e^s \Gamma(\alpha, s)/\Gamma(\alpha),$$

or

$$\int_0^\infty e^{-sx} d_x \operatorname{pr}\{F_\alpha^*(t, \omega) \leqslant x\} = [\Gamma(\alpha)]^{-1} e^{ts} \int_{ts}^\infty e^{-u} u^{\alpha-1} du.$$

A further Laplace transform subsumes the one-dimensional distributions of the process F_α^* into the formula

$$\int_0^\infty e^{-\sigma t} dt \int_0^\infty e^{-sx} d_x \operatorname{pr}\{F_\alpha^*(t, \omega) \leqslant x\} = (\sigma^\alpha - s^\alpha)/\sigma^\alpha(\sigma - s) \quad (\sigma \neq s).$$

Note that when $\Phi = \Phi_\alpha$, $W = W_\alpha$ is absolutely continuous with density $w_\alpha(t) = 1/t^{1-\alpha} \Gamma(\alpha)$ ([3], Section 7). Also, $m(x) = m_\alpha(x) = 1/x^\alpha \Gamma(1-\alpha)$. Lemmas 2 and 3 show that

$$\operatorname{pr}\{\text{range of } U_\alpha \text{ avoids } [t, t+x]\} = \int_0^t m_\alpha(t+x-v) w_\alpha(v) dv$$

$$= [\Gamma(\alpha) \Gamma(1-\alpha)]^{-1} \int_0^t dv/v^{1-\alpha}(t+x-v)^\alpha$$

$$= \pi^{-1} \sin \pi\alpha \int_0^{t/(t+x)} du/u^{1-\alpha}(1-u)^\alpha$$

$$= \pi^{-1} \sin \pi\alpha \int_{x/t}^\infty dy/y^\alpha(1+y),$$

using the bilinear transformation $u = 1/(1+y)$. This shows how the Dynkin distributions G_α arise in connection with Φ_α and provides another proof of a result of Blumenthal and Getoor ([4], Lemma 3.1).

Consider now the transition probabilities of the Lamperti process B_α^*. These are given by equation (7), and have an atomic component and an absolutely continuous component given by a convolution integral. Write $F(x, t | y)$ for this integral. We claim that $F(., t | y)$ has a density given by equation (8), namely

$$\partial F(x, t | y)/\partial x = \pi^{-1} \sin \pi\alpha (t-x)/x^\alpha(y+t-x) \quad (x \in [0, t]).$$

To prove this, we differentiate equation (7) with respect to x under the integral sign. In view of equation (6) we obtain, after some manipulation,

$$\frac{\pi}{\sin \pi\alpha} \frac{x^\alpha(y+t-x)}{\alpha y^\alpha} \frac{\partial F(x,t\,|\,y)}{\partial x} = \int_0^{(t-x)/(y+t-x)} \frac{dv}{v^{1-\alpha}(1-v)^{1+\alpha}}.$$

The x-derivative of the right-hand side is $-1/y^\alpha(t-x)^{1-\alpha}$. Integrating the resulting differential equation for $\partial F/\partial x$, we obtain equation (8), as required.

That the total mass of the absolutely continuous component is

$$1 - [y/(y+t)]^\alpha$$

follows from equations (7) and (8). Analytically, this is easily seen to be equivalent to

$$\pi^{-1} \sin \pi\alpha \int_0^1 du/u^\alpha(1-u)^{1-\alpha}(1+a-u) = 1/a^{1-\alpha}(1+a)^\alpha \quad (0<\alpha<1, a>0).$$

This may be proved by contour integration.

The surprisingly simple functional form of the density in equation (8) does not seem to have occurred in the literature, either in relation to the Dynkin–Lamperti theorem or in other contexts. However, the author gathers from Professor J. G. Wendel that these densities were known to him through his work on [28].

3.6.7 THE LIMIT THEOREM

We proceed now to the proof of Theorem 2. Suppose that $(f(t), b(t))$ is a possible pair of norming functions for the (F, B)-process.

Lemma 6. (i) *We may take* $f(t) \equiv b(t) \equiv t$ *as a possible pair of norming functions for the (F, B)-process, whenever a non-degenerate ergodic limit exists.*

(ii) *Statement* (i) *remains valid with (F, B) replaced by (F^*, B^*).*

Proof. If the (F, B)-process has a non-degenerate ergodic limit, then in particular the one-dimensional distributions of the F-process satisfy

$$(41) \qquad \mathrm{pr}\,\{F(t, \omega)/f(t) \leqslant x\} \to G(x) \quad (t \to \infty)$$

for some norming function f and some non-degenerate distribution G. By Lemma 3, formula (41) gives

$$(42) \qquad t \int_0^1 m(t(1+xf(t)t^{-1}-u))p(tu)\,du \to 1-G(x) \quad (t \to \infty).$$

If we assume

$$\liminf_{t\to\infty} f(t)/t = 0 \quad \text{or} \quad \limsup_{t\to\infty} f(t)/t = \infty,$$

then letting $t \to \infty$ through suitable subsequences in formula (42), we see that $1 - G(x)$ is independent of $x > 0$, contrary to the hypothesis of non-degeneracy. Write $L(t)$ for $f(t)/t$. There remains the case

$$(43) \qquad 0 < c = \liminf_{t\to\infty} L(t) \leqslant \limsup_{t\to\infty} L(t) = C < \infty.$$

Choose $\varepsilon > 0$. By monotonicity of the tail-function m, for all sufficiently large t we have

$$m(t(1+x(c+\varepsilon)-u)) \leqslant m(t(1+xL(t)-u)) \leqslant m(t(1+xc-u)).$$

Multiply by $tp(tu)$, integrate over $u \in [0,1]$ and let $t \to \infty$ along a subsequence in which the limit inferior in formula (43) is approached. We obtain

$$(44) \qquad \limsup_{t\to\infty} \int_0^1 tp(tu)\, m(t(1+x(c+\varepsilon)-u))\, du$$

$$\leqslant 1 - G(x) \leqslant \liminf_{t\to\infty} \int_0^1 tp(tu)\, m(t(1+xc-u))\, du.$$

One obtains similar inequalities by passing to the limit along a sequence in which the limit superior in formula (43) is approached. These inequalities combine with formula (43) to give

$$1 - G(x/C) \leqslant 1 - G(x/(c+\varepsilon)).$$

Since this holds for each $x \geqslant 0$, and G is assumed non-degenerate, we easily obtain $C \leqslant c + \varepsilon$. Since this holds for each $\varepsilon > 0$, formula (43) shows that $c = C$ and $f(t)/t \to c$ as $t \to \infty$. Thus $f(t) \sim ct$ and we lose nothing by taking $f(t) \equiv t$. Similar arguments show that we may take $b(t) \equiv t$.

This completes the proof of (i). For (ii), we replace $p(t)\,dt = P(dt)$ in formula (42) by $W(dt)$ and argue similarly.

The next result establishes an invariance principle by reducing the problem from proving weak convergence to proving convergence of normed transition probabilities.

Lemma 7. (i) *Suppose that the transition probabilities of the process $\{(\lambda^{-1} F(\lambda t, \omega), \lambda^{-1} B(\lambda t, \omega)): t \geqslant 0\}$ converge as $\lambda \to \infty$ to non-degenerate limits. Then the process converges weakly under the Stone–Skorohod J_1-topology to a non-degenerate limit process whose transition probabilities are given by these limits.*

(ii) Statement (i) remains valid if (F, B) is replaced by (F^*, B^*).

Proof. The finite-dimensional distributions of a Markov process are obtained from its transition probabilities by iteration. Thus the assumption is equivalent to existence of non-degenerate limits for the normed finite-dimensional distributions; these limits automatically satisfy the Daniell–Kolmogorov consistency conditions.

There thus exists a non-degenerate process whose finite-dimensional distributions are given by these limits. Since the limiting finite-dimensional distributions are obtained from the limits of the transition probabilities of the normed (F, B)-process by iteration, one sees that the limit process is in fact a Markov process with the stated transition probabilities.

It remains only to deduce weak convergence from convergence of finite-dimensional distributions. This follows just as in [20] from the Stone–Skorohod theorem (see also [2]). We need only show that the Skorohod Δ-condition is satisfied. But this is immediate, since the sample-functions of the F- and B-processes have gradient 1 at points of continuity, and the norming functions are $f(t) \equiv b(t) \equiv t$; and similarly for the F^*- and B^*-processes.

Lemma 8. *If*

$$(*) \qquad m(x) = \mu(x, \infty] \sim 1/L(x)\, x^\alpha\, \Gamma(1 - \alpha) \quad (x \to \infty) \quad (0 < \alpha < 1),$$

then the normed transition probabilities of the (F, B)- and (F^*, B^*)-processes converge to those of the process $(X_\alpha, Y_\alpha) = (F_\alpha^*, B_\alpha^*)$.

Proof. We shall consider only the (F, B)-process; the argument for the (F^*, B^*)-process is similar. We consider the normed transition probabilities in the form

(45)
$$\mathrm{pr}\{F(\lambda(s+t), \omega) > \lambda x_1, B(\lambda(s+t), \omega) > \lambda x_2 \,|\, F(\lambda s, \omega) = \lambda y_1, B(\lambda s, \omega) = \lambda y_2\}.$$

This is,

(46) pr {closure of range of T-process avoids $[\lambda(s + t - x_2), \lambda(s + t + x_1)] \,|$
  the maximal open interval containing λs and free of the range of
  the T-process is $(\lambda(s - y_2), \lambda(s + y_1))\}$.

This expression is independent of $s > 0$, by stationarity. For fixed $t > 0$, however, the functional form of the expression changes as $(x_1, x_2; y_1, y_2)$ change, giving rise to various cases. By considering the dependence on y_1, it suffices to consider terms exemplified by

(47) $\mathrm{pr}\{B(\lambda(s + t + x_1), \omega) > \lambda(x_1 + x_2) \,|\, B(\lambda s, \omega) = \lambda y_2\},$

which reduces the problem to consideration of the normed transition probabilities of the B-process.

When (∗) holds, then

(48) $\quad \begin{cases} m(\lambda x)/m(\lambda y) \to (y/x)^\alpha \quad (\lambda \to \infty), \\ \mu(\lambda\,du)/m(\lambda y) \to \alpha y^\alpha\,du/u^{1+\alpha} \quad (\lambda \to \infty). \end{cases}$

We next prove that when (∗) holds, the one-dimensional distributions of the B-process converge to the generalized arc-sine law,

(49)

$$\mathrm{pr}\,\{B(\lambda t, \omega) \leqslant \lambda x\} \to \pi^{-1}\sin\pi\alpha \int_0^{x/t} dy/y^\alpha(1-y)^{1-\alpha} = H_\alpha(x/t) \quad (\lambda \to \infty).$$

To prove the relation (49), observe that by Lemma 3,

(50) $\qquad \mathrm{pr}\,\{B(\lambda t, \omega) > \lambda x\} = \int_0^{t-x} m(\lambda(t-u))\,P(\lambda\,du).$

Under (∗), using formula (48) and the Tauberian result of [3] (Proposition 2)

(51) $\quad \begin{cases} m(\lambda x) \sim x^\alpha/L(\lambda)\,\lambda^\alpha\,\Gamma(1-\alpha) \\ P(\lambda t) \sim \lambda^\alpha\,t^\alpha L(\lambda)/\Gamma(1+\alpha) \end{cases} \quad (\lambda \to \infty),$

uniformly for t, x in compact subsets of $(0, \infty)$. Choose $\delta > 0$; consider a partition of $[\delta, t-x]$ into subintervals by points

$$\delta = u_0 < u_1 < \ldots < u_n = t-x,$$

and the corresponding Riemann sums. We have

(52) $\qquad \dfrac{P(\lambda u_r) - P(\lambda u_{r-1})}{\lambda^\alpha/\Gamma(1+\alpha)} \to u_r^\alpha - u_{r-1}^\alpha \quad (\lambda \to \infty)$

uniformly over all choices of partition points. One may verify that given $\varepsilon > 0$, one may choose δ so small that the contributions of the integrals over $[0, \delta]$ are uniformly less than ε for sufficiently large λ. Also, the Riemann sums for the integrals over $[\delta, t-x]$ tend to

$$\sum_k (u_k^\alpha - u_{k-1}^\alpha)/\Gamma(1+\alpha)\,\Gamma(1-\alpha)\,(t-u_k)^\alpha,$$

which is a Riemann sum for

$$\int_\delta^{t-x} du/\Gamma(\alpha)\,\Gamma(1-\alpha)\,u^{1-\alpha}(t-u)^\alpha.$$

It readily follows that

$$\operatorname{pr}\{B(\lambda t,\omega)>\lambda x\} \to \pi^{-1}\sin\pi\alpha\int_0^{t-x} du/u^{1-\alpha}(t-u)^\alpha$$

$$= \pi^{-1}\sin\pi\alpha\int_{x/t}^1 dy/y^\alpha(1-y)^{1-\alpha}$$

$$= 1 - H_\alpha(x/t) \quad (x\in[0,t]) \quad (\lambda\to\infty),$$

which proves the relation (49).

If we use formulae (48) and (49) in the expression of Lemma 5 for the transition probabilities of the B-process, we see that convergence as $\lambda\to\infty$ to Lamperti's convolution expression (7) is at least formally indicated. Convergence of such Stieltjes convolution integrals where both the integrand and the integrator are regularly varying has been considered by Lamperti [20] (proof of Theorem 2.1) and Teugels [25]. The passage to the limit can be justified as in these cases. By the remarks above, this completes the proof of the lemma.

Lemma 9. (*i*) *If* $F(t,\omega)/t$ *has a non-degenerate limit distribution* G *as* $t\to\infty$, *then* (∗) *holds and* $G = G_\alpha$, *where*

$$G_\alpha(x) = \pi^{-1}\sin\pi\alpha\int_0^x du/u^\alpha(1+u) \quad (x\geqslant 0).$$

(*ii*) *If* $B(t,\omega)/t$ *has a non-degenerate limit distribution* H *as* $t\to\infty$, *then* (∗) *holds and* $H = H_\alpha$, *where*

$$H_\alpha(x) = \pi^{-1}\sin\pi\alpha\int_0^x dy/y^\alpha(1-y)^{1-\alpha} \quad (x\in[0,1]).$$

Similar statements hold with F, B *replaced by* F^*, B^*.

Proof. We prove only the assertion on the F-process; the remaining parts are proved similarly. Our assumption is

$$\operatorname{pr}\{F(\lambda t,\omega)\leqslant\lambda x\}\to 1 - G(x/t) \quad (\lambda\to\infty) \quad (x,t\geqslant 0).$$

Taking the double Laplace transform and using dominated convergence,

$$(53) \quad \int_0^\infty e^{-\sigma t}\,dt\int_0^\infty e^{-sx}\,dx\,\operatorname{pr}\{F(\lambda t,\omega)>\lambda x\}$$

$$\to \int_0^\infty e^{-\sigma t}\,dt\int_0^\infty e^{-sx}\,dx[1 - G(x/t)] = g(s,\sigma), \quad \text{say } (\lambda\to\infty).$$

Taking the double Laplace transform of (*i*) of Lemma 3a, the left-hand side of formula (53) is

(54) $$[\tilde{M}(s/\lambda) - \tilde{M}(\sigma/\lambda)]/\sigma(\sigma - s)[1 + \tilde{M}(\sigma/\lambda)].$$

As $\lambda \to \infty$, $\tilde{M}(\sigma/\lambda)/[1 + \tilde{M}(\sigma/\lambda)]$ tends to 1 if M is unbounded and $M(\infty)/[1 + M(\infty)]$ if M is bounded. Thus formulae (53) and (54) give

(55) $$\tilde{M}(s/\lambda)/\tilde{M}(\sigma/\lambda) \to h(s, \sigma) \quad (\lambda \to \infty),$$

where $h(s, \sigma)$ can be written down from formula (53). But it follows from formula (55) ([10], Section VIII. 8) that $\tilde{M}(s)$ varies regularly. Hence by Karamata's Tauberian theorem for the LST, so does $M(x)$, and by the monotone density theorem ([10], Section XIII. 5) so does its density $m(x) = \mu(x, \infty]$. If the exponent of regular variation of m is $-\alpha$, formula (55) shows that $h(s, \sigma) = \sigma^{1-\alpha}/s^{1-\alpha}$. Also, since m is decreasing we have $\alpha \geqslant 0$, and since M is increasing and varies regularly with exponent $1 - \alpha$, we have $\alpha \leqslant 1$. If $\alpha = 1$, $h(s, \sigma) \equiv 1$, giving a deterministic limit distribution, which is excluded by the assumption of non-degeneracy. So $\alpha \in [0, 1)$, and M is unbounded. Hence

(56) $$g(s, \sigma) = (s^\alpha \sigma - \sigma^\alpha s)/\sigma^{\alpha+1} s(\sigma - s) \quad (\sigma \neq s).$$

When $\alpha = 0$, $g(s, \sigma) = 1/s\sigma$, which again gives a degenerate limit distribution. Thus $m(x)$ varies regularly with exponent $-\alpha$, where $\alpha \in (0, 1)$, and so (absorbing a scale-factor $\Gamma(1 - \alpha)$ into the corresponding slowly varying function), condition (∗) holds. Lemma 8 then shows that the limit laws are indeed the Dynkin distributions G_α, as required (this can also be deduced from equation (56) and the results of Section 3.6.6).

Since convergence of normed transition probabilities implies convergence of normed one-dimensional distributions, Lemmas 8 and 9 show that condition (∗) is necessary and sufficient for convergence of normed transition probabilities to non-degenerate limits. By Lemma 7, this is equivalent to weak convergence. The possible norming functions are specified by Lemma 6. So condition (∗) is necessary and sufficient for the processes $(F, B), (F^*, B^*)$ to have ergodic limits, the limit processes being given by Lemma 7 as (F_α^*, B_α^*). This completes the proof of Theorem 2. Those assertions of Theorem 1 not included in the work of Dynkin [8] and Lamperti [20] are established in like manner.

3.6.8 THE SOLIDARITY THEOREM

We now come to the proof of our solidarity result, Proposition 2. Let $\tilde{f}_{ij}(s), \tilde{g}_{ij}(s)$ denote the Laplace transforms of $f_{ij}(t), g_{ij}(t)$. Then by equations (23) and (24),

(57) $$r_{ij}(s) = r_{ii}(s)\tilde{g}_{ij}(s)/\tilde{f}_{ij}(s).$$

As $s \to 0+$,

$$\tilde{f}_{ij}(s) \to \int_0^\infty f_{ij}(t)\, dt \in [0, 1].$$

Since state j is accessible from state i, one has

$$\tilde{f}_{ij}(0+) = \int_0^\infty f_{ij}(t)\, dt \in (0, 1].$$

So to prove that

$$c_{ij} = \lim_{s \to 0+} \{\tilde{g}_{ij}(s)/\tilde{f}_{ij}(s)\} \in (0, \infty),$$

it suffices to show that

(58) $$\tilde{g}_{ij}(0+) \in (0, \infty).$$

We shall use the notation of Kingman's representation theorem for linked systems ([17, Theorem 7]), where matrices (q_{ij}), (a_{ij}) and matrix measures (μ_{ij}) are introduced. By relabelling the states if necessary, we may suppose that $i = 2, j = 1$. Kingman ([17], proof of Theorem 9) shows that

(59) $$g_{21}(t) = q_{21}\tilde{p}(t) + \int_{[0,t]} \tilde{p}(t-x)\,\mu_{21}(dx),$$

where $q_{21} \in [0, \infty)$, μ_{21} is a bounded measure on $(0, \infty)$ and $\tilde{p}(t)$ is the p-function of a regenerative phenomenon. Taking Laplace transforms of equation (59) and writing $\tilde{r}(s)$ for

$$\int_0^\infty e^{-st}\tilde{p}(t)\, dt,$$

(60) $$\tilde{g}_{21}(s) \sim \tilde{r}(s)\{q_{21} + \int_{(0,\infty)} \mu_{21}(dx)\} \quad (s \to 0+).$$

so the relation (58) is equivalent to

(61) $$\tilde{r}(0+) \in (0, \infty).$$

Also ([17], cf. (63))

(62) $$[\tilde{r}(s)]^{-1} = s + a_{11} + \int_{(0,\infty)} (1 - e^{-xs})\,\mu_{11}(dx)$$

for some $a_{11} \geqslant 0$ and some canonical measure μ_{11}. So the relation (58) is equivalent to $a_{11} > 0$.

We have ([17], proof of Theorem 8)

(63) $\displaystyle\int_0^\infty e^{-st} f_{12}(t)\, dt$

$$= -\left\{ a_{12} + \int (1 - e^{-sx}) \mu_{12}(dx) \right\} \Big/ \left\{ s + a_{11} + \int (1 - e^{-sx}) \mu_{11}(dx) \right\}.$$

To establish the relation (58) by contradiction, we assume $a_{11} = 0$. Then letting $s \to 0+$ in equation (63) and using

$$\int_0^\infty f_{12}(t)\, dt \in [0, 1],$$

we obtain $a_{12} = 0$. But it follows from (38) of [17] that if $a_{11} = 0$, then $\mu_{1k} \equiv 0$ for all $k \neq 1$. Thus both a_{12} and μ_{12} in equation (63) vanish, and so $f_{12}(t)$ vanishes almost everywhere on $(0, \infty)$. So the first passage distribution F_{12} whose absolutely continuous component has density f_{12} on $(0, \infty)$ is concentrated at infinity. But this contradicts the assumption that state $i = 2$ is accessible from state $j = 1$; thus the relation (58) is established.

Using the relation (58), we now have $c_{ij} \in (0, \infty)$ and

(64) $r_{jj}(s) \sim c_{ij} r_{ii}(s) \quad (s \to 0+).$

Assertions (25) and (26) follow from the relation (64) and the Tauberian result ([3], Proposition 2) for regenerative phenomena.

3.6.9   COMPLEMENTS

Our main result, Theorem 2 and Theorem 1a of [3] are complementary. Together they show the fundamental role played by condition (∗) and solve completely the ergodic limit problem for the two regenerative tetrads $\{H, T, U, V\}, \{F, B; F^*, B^*\}$. It is natural to ask whether Theorem 1a of [3] can in fact be deduced from Theorem 2 and the continuous mapping theorem [1] via equation (36). This, however, is not the case, since the map

$$f(.) \to \int_0^\cdot I_B(f(u))\, du$$

is not J_1-continuous. Indeed, were this the case, we could take $B = \{0\}$ and use Lemma 1(*iii*) to deduce that under (∗), the ergodic limit of the H-process vanished identically, contradicting Theorem 1a of [3].

In equation (36), our treatment of the H-process is brought within the scope of occupation-times of Markov processes satisfying the Darling–Kac condition (A). However, similar situations arise naturally in which condition (A) must be replaced by a weaker condition (A*). For instance, let Φ be a transient regenerative phenomenon which is *stationary* [18]. We can begin to observe Φ at an arbitrary origin in time, which may or may not be a regeneration point. Let

$$x(\omega) = F(0, \omega) = \inf\{s \geqslant 0: Z(s, \omega) = 1\}.$$

Conditional on $x(\omega) = x > 0$, we can observe the resulting phenomenon (which we shall call Φ_x) at subsequent times. For Φ_x we have, in an obvious adaptation of the notation of equation (21),

$$\int_{-\infty}^{\infty} I_{\{0\}}(y)p(x|dy,s) = \mathbf{E}_x \int_0^{\infty} e^{-st} Z(t, \omega)\,dt$$

$$= \int_x^{\infty} e^{-st} p(t)\,dt$$

$$= r(s) - \int_0^x e^{-st} p(t)\,dt.$$

In the non-transient case, $r(s) \to \infty$ $(s \to 0+)$, and so

$$[r(s)]^{-1} \int_{-\infty}^{\infty} I_{\{0\}}(y)p(x|dy,s) = 1 - [r(s)]^{-1} \int_0^x e^{-st} p(t)\,dt$$

$$\to 1 \quad (s \to 0+).$$

Although this resembles the Darling–Kac condition (A), it is in fact weaker than (A), because we do not have uniformity in x but only uniformity for x in compact sets.

Let us call the condition obtained from (A) by replacing the requirement of uniformity by that of uniformity on compacta condition (A*). It is natural to wonder to what extent the results of [7], [2] remain valid when (A) is replaced by (A*). For example, Darling and Kac used (A) to study limit theorems for occupation times of bounded measurable sets in planar Brownian motion. To what extent are the results of [7], [2] applicable to *unbounded* sets of finite planar Lebesgue measure?

Dwass and Karlin [9] consider the discrete-time versions of the Darling–Kac and Dynkin–Lamperti theorems (among others), and obtain conditioned limit theorems in these contexts. One may ask whether these results may be generalized from convergence of one-dimensional distributions to convergence of finite-dimensional distributions and weak convergence. One may also ask whether such extensions have regenerative analogues constituting conditioned versions of the results of this paper and [2]. We propose to consider such questions elsewhere.

Theorem 2 suggests that analogous results may hold in the context of Darling and Kac. Let $\{X(t, \omega): t \geqslant 0\}$ be a strong Markov process and B a Borel set such that

$$H(t, \omega) = \int_0^t I_B(X(u, \omega))\, du$$

satisfies the Darling–Kac condition (A) and has ergodic limit V_α. If we write

$$F(t, \omega) = \inf\{s \geqslant t: X(s, \omega) \in B\} - t,$$

$$B(t, \omega) = t - \sup\{s \leqslant t: X(s, \omega) \in B\},$$

one may ask whether, under suitable regularity conditions, the (F, B)-process has ergodic limit (F_α^*, B_α^*), and whether any other ergodic limit processes are possible here.

The distributions G_α, H_α of Section 3.6.3 have appeared in various related contexts. For instance, Taylor and Wendel [24] obtain results on the Hausdorff measure of the zero-set of a stable process with order in $(1, 2]$, using the Stone representation of Section 3.6.6. Limit theorems concerning zero-free intervals of stable and semi-stable processes have been obtained by Getoor [11] and Wendel [28] (see also Kesten [14]). Wendel also obtains conditioned limit theorems for semi-stable processes analogous to those of Dwass and Karlin [9].

3.6.10 ACKNOWLEDGEMENTS

The definitions of the regenerative F- and B-processes are due to Professor J. F. C. Kingman. The author is grateful to Professor Kingman for pointing out to him the connection between the results of [2], [3] mentioned in Section 3.6.6, and to Professor D. G. Kendall [13] for access to his previously unpublished work, now made available in this book.

REFERENCES

1. P. Billingsley, *Convergence of Probability Measures*, Wiley, New York (1968).
2. N. H. Bingham, "Limit theorems for occupation times of Markov processes", *Ztschr. Wahrsch'theorie & verw. Geb.* **17** (1971), 1–22.
3. ———, "Limit theorems for regenerative phenomena, recurrent events and renewal theory", *Ztschr. Wahrsch'theorie & verw. Geb.* **21** (1972), 20–44.
4. R. M. Blumenthal and R. K. Getoor, "The dimension and the set of zeros and the graph of a symmetric stable process", *Ill. J. Math.* **6** (1962), 308–316.
5. K. -L. Chung, "Sur une équation de convolution", *C. R. Acad. Sci. Paris* **260** (1965), 4665–4667 and 6794–6796.

6. ——, *Markov Chains with Stationary Transition Probabilities*, 2nd ed., Springer-Verlag, Berlin (1967).
7. D. A. Darling and M. Kac, "On occupation times of Markov processes", *Trans. Am. math. Soc.* **84** (1957), 444–458.
8. E. B. Dynkin, "Limit theorems for sums of independent random variables with infinite expectation", *Selected Translations (I.M.S.–A.M.S.)*, **1** (1961), 171–189.
9. M. Dwass and S. Karlin, "Conditioned limit theorems", *Ann. Math. Stat.* **34** (1963), 1147–1167.
10. W. Feller, *An Introduction to Probability Theory and its Application*, vol. II, Wiley, New York (1966).
11. R. K. Getoor, "The asymptotic distribution of the number of zero-free intervals of a stable process", *Trans. Am. math. Soc.* **106** (1963), 127–138.
12. I. S. Gradshteyn and I. M. Ryzhik, *Tables of Integrals, Series and Products*, 4th ed., Academic Press, New York (1965).
13. D. G. Kendall, "On the non-occurrence of a regenerative phenomenon in a given interval", published in extended form as 3.7 of this book.
14. H. Kesten, "Positivity intervals of stable processes", *J. Math. Mech.* **12** (1963), 391–410.
15. ——, "Hitting probabilities of single points for processes with stationary independent increments", *Mem. Am. math. Soc.* **93** (1969), 1–129.
16. J. F. C. Kingman, "The stochastic theory of regenerative events", *Ztschr. Wahrsch'theorie & verw. Geb.* **2** (1964), 180–224.
17. ——, "Linked systems of regenerative events", *Proc. Lond. math. Soc.* **15** (1965), 125–150.
18. ——, "Stationary regenerative phenomena", *Ztschr. Wahrsch'theorie & verw. Geb.* **15** (1970), 1–18.
19. J. Lamperti, "Limit theorems for stochastic processes", *J. Math. Mech.* **7** (1958), 433–450.
20. ——, "An invariance principle in renewal theory", *Ann. Math. Stat.* **33** (1962), 685–696.
21. ——, "Semi-stable stochastic processes", *Trans. Am. math. Soc.* **104** (1962), 62–78.
22. H. P. McKean, "An integral equation arising in connection with Markov chains", *Ztschr. Wahrsch'theorie & verw. Geb.* **8** (1967), 298–300 and **11** (1968), 82.
23. C. Stone, "The set of zeros of a semi-stable process", *Ill. J. Math.* **4** (1962), 114–126.
24. S. J. Taylor and J. G. Wendel, "The exact Hausdorff measure of the zero-set of a stable process", *Ztschr. Wahrsch'theorie & verw. Geb.* **6** (1966), 170–180.
25. J. L. Teugels, "Renewal theorems when the first or second moment is infinite", *Ann. Math. Stat.* **39** (1968), 1210–1219.
26. ——, "Exponential ergodicity in Markov renewal processes", *J. Appl. Prob.* **5** (1968), 387–400.
27. ——, "Regular variation of Markov renewal functions", *J. Lond. math. Soc.* **2** (1970), 179–190.
28. J. G. Wendel, "Zero-free intervals of semi-stable Markov processes", *Math. Scand.* **14** (1964), 21–34.

3.7

On the Non-occurrence of a Regenerative Phenomenon in Given Intervals

D. G. KENDALL

1. We shall be concerned with standard regenerative phenomena on the half-line, that is with probability triples $(\Omega, \mathscr{F}, \mathrm{pr})$ for which $\mathscr{F}$ contains a one-parameter family of events $\{E_t: 0 \leqslant t < \infty\}$ such that $E_0 = \Omega$,

$$\mathrm{pr}\left(\bigcap_{j=1}^{k} E_{t_j}\right) = \mathrm{pr}(E_{t_1})\,\mathrm{pr}\left(\bigcap_{j=2}^{k} E_{t_j - t_1}\right)$$

for $0 \leqslant t_1 \leqslant t_2 \leqslant \ldots \leqslant t_k$ (for all $k > 1$), and $\lim_{t \to 0} \mathrm{pr}(E_t) = 1$. Such objects were introduced by Kingman [8], who showed that their properties could be described in terms of a 'p-function' defined by $p(t) = \mathrm{pr}(E_t)$, and who found (among much else) a canonical representation for all such p-functions, namely

$$(1) \qquad r(s) = \int_0^\infty \mathrm{e}^{-st} p(t)\, dt = \left\{ s + \int_{(0,\infty]} \frac{1 - \mathrm{e}^{-sx}}{1 - \mathrm{e}^{-x}} \lambda(dx) \right\}^{-1},$$

where λ is an arbitrary totally finite measure on the Borel sets of $(0, \infty]$, the representation at (1) being unique. (Here $0 < s < \infty$.) Kingman also made it plain that there is a probabilistic representation for such a system in terms of 'local time' and a process with non-negative stationary independent increments and a positive drift. Here we set up this representation in a direction different from that taken by Kingman, and sharpen one or two details for the sake of the later calculations.

2. Let $\{U(u): 0 \leqslant u < \infty\}$ be a stochastic process with stationary non-negative independent increments and such that $U(0) = 0$ surely. Then we know (see, for example, Kendall [6]) that

$$(2) \qquad \mathbf{E}(\mathrm{e}^{-sU(u)}) = \exp\{-u\,\Lambda(s)\} \quad (0 < s < \infty),$$

294

where

$$\Lambda(s) = \alpha s + \int_{(0,\infty]} \frac{1-e^{-sx}}{1-e^{-x}} \lambda(dx); \tag{3}$$

here λ (uniquely determined) can be any totally finite measure on the Borel subsets of $(0,\infty]$, and $\alpha \geqslant 0$. When $\alpha = 0$ we may call the process *driftless*. Let us agree to set up such a driftless process, starting with the finite-dimensional distributions as determined by equations (2) and (3), on the *Borel* subsets of the compact Hausdorff Cartesian-product space $(R_+ \cup \{\infty\})^{R_+}$, by the Kakutani procedure (see Nelson [10], or Kendall [4]). The Kakutani probability measure is regular, and it is easy to show that (by rejecting a measurable set of probability zero) we can ensure that $U(0) = U(0+) = 0$, that $U(t)$ is non-negative for all t and that $U(.)$ is non-decreasing. These properties are also enjoyed by the stochastic process $\{U(u+0): 0 \leqslant u < \infty\}$, which is moreover a 'standard modification' of the original one and which has the new feature of being right-continuous, and so process-measurable. We exchange the old process for this new one, and *fix it* in the following discussion. We now put back a drift by defining $T(u) = u + U(u)$ (the drift constant α turns out to be irrelevant, which is why we omit it here), so that $T(0) = 0$, $T(t)$ is non-negative for each t, $T(.)$ is right-continuous and

$$T(u+v) \geqslant T(u)+v \quad \text{for } 0 \leqslant u, v < \infty. \tag{4}$$

There is no difficulty in showing that $\{T(u): 0 \leqslant u < \infty\}$ has an appropriate form of the strong Markov property, and we shall make essential use of this.

3. Let us say that Φ occurs at time $t < \infty$ *when t is in the range of the function $T(.)$* (with domain $[0,\infty)$). We write E_t for this event, which we can also define (for $t > 0$) ($E_0 = \Omega$) by

$$E_t = \bigcap_{k \geqslant 1} \bigcup_{m \geqslant 1} \bigcup_{n \geqslant 0} \{\omega: T(n2^{-m}) < t \leqslant T((n+1)2^{-m}) < t + 2^{-k}\}. \tag{5}$$

A rather cumbersome calculation including a slightly tricky estimate enables one to evaluate the Laplace transform $r(s)$ of $\mathrm{pr}(E_t)$; it turns out to be given by equation (1). It is easily verified that $sr(s) \to 1$ as $s \to \infty$, because $\Lambda(s)/s \to \alpha = 1$ as $s \to \infty$.

We define, for each t in $[0,\infty)$,

$$H(t) = \inf_{\rho \geqslant 0} \{\rho: T(\rho) \geqslant t\} \quad (\rho \text{ rational}), \tag{6}$$

and observe that

(i) *for each t, $H(t)$ is an 'optional' random variable;*

(ii) Φ *happens at t if and only if* $T(H(t)) = t$.

(Note that when $T(u) < \infty$, then $H(T(u)) = u$.) These facts, together with the strong Markov theorem, make it easy to show that $\{E_t: 0 \leqslant t < \infty\}$ describes a regenerative phenomenon Φ.

We now prove Φ to be 'standard'. Whether it is standard or no, the function $p(t) = \mathrm{pr}\,(E_t)$ must satisfy Kingman's p-function inequalities (for which see [8]), and on taking the Laplace transform of one of these (which we may do because from equation (5) $p(.)$ is Lebesgue measurable) we find that

(7) $$sr(s)\,[1 - p(t)] \leqslant e^{st}\,[1 - sr(s)]$$

for all t in $[0, \infty)$ and all s in $(0, \infty)$. Now we have noted that $sr(s) \to 1$ as $s \to \infty$, so that $p(t) \to 1$ as $t \to 0$, as required. That is, *we obtain a model for the most general standard regenerative phenomenon by constructing a non-negative, increasing, right-continuous process with stationary independent increments and positive (say unit) drift, starting at 0, and by defining Φ to happen at a finite time t when t is in the range of the process.*

In this model, the random set

(8) $$S = \{t : \Phi \text{ happens at time } t\}$$

is *'right-perfect'*; that is, it coincides with the set consisting of its own right cluster-points. Also the random variables $H(t)$ satisfy

$$0 \leqslant H(t_2) - H(t_1) \leqslant t_2 - t_1 \quad (0 \leqslant t_1 \leqslant t_2).$$

4. From the important set-theoretic identity

(9) $$\{\omega : H(t) \leqslant u\} = \{\omega : T(u) \geqslant t\}$$

and a double Laplace transform operation we can now identify the distribution function

(10) $$F_t(c) = \mathrm{pr}\,\{H(t) \leqslant c\}$$

by

(11) $$\int_0^\infty e^{-st}\,dt \int_0^\infty e^{-\sigma c}\,d_c\,F_t(c) = s^{-1}[s + \Lambda(s)]/[\sigma + s + \Lambda(s)].$$

From this we can, for example, compute the moments of the random variable

(12) $$L(t) = \text{Lebesgue measure } (S \cap [0, t]);$$

thus we find that

$$\mathbf{E}(L(t)) = \int_0^t p(v)\,dv = \mathbf{E}(H(t)),$$

and because $H(.)$ and $L(.)$ are both continuous and $L(t) \geqslant H(t)$ we deduce that

$$(13) \qquad\qquad H(t) = L(t),$$

so that $H(t)$ *is in fact the 'local time' for the phenomenon* Φ.

From equation (11) we also obtain

$$(14) \qquad\qquad \operatorname{var}(L(t)) = 2(1 * p * p) - (1 * p)^2,$$

where $*$ denotes a Laplace convolution.

5. Now consider the event

$$(15) \qquad\qquad \{\omega \colon \Phi \text{ does } not \text{ occur in } [t, s)\},$$

where $0 < t < s < \infty$. This is equal to the event

$$(16) \qquad \bigcap_{m \geqslant 1} \bigcup_{n \geqslant 0} \{\omega \colon T(n2^{-m}) < t < s \leqslant T((n+1)2^{-m})\},$$

and so it is measurable with respect to the uncompleted product of $\mathscr{F}$ with the uncompleted Cartesian square of the σ-algebra of linear Borel sets. This fact, a double Laplace transformation and the Fubini theorem enable us to conclude that the probability of the event (15) is given by

Theorem 1. *For any standard regenerative phenomenon* Φ *with p-function p,*

$$(17) \qquad \operatorname{pr}\{\omega \colon \text{no } \Phi \text{ in } [t, s)\} = \int_0^t p(v) \, m^-(s - v) \, dv,$$

where $0 \leqslant t < s < \infty$, *and*

$$(18) \qquad\qquad m^-(v) = \int_{[v, \infty]} \frac{\lambda(dx)}{1 - \mathrm{e}^{-x}},$$

λ *being the measure in the representation* (1).

6. The original draft of the above argument was written some years ago (1968), when it was left incomplete. It is included in this book because of the cross-references to the papers [1] (by Bingham) and [5], and in elucidation of the latter connection we now add the following remarks.

First it is necessary to draw attention to the fact that the function $m^-(.)$ introduced at equation (18) is closely related to but is not the same as the function $m(.)$ of Kingman [8]; in fact Kingman's m is given by a formula differing from equation (18) in that the domain of integration is changed to $(v, \infty]$, but it is $m^-(.)$ which presents itself naturally in the proof of equation (17); the two-dimensional set of Lebesgue measure zero, which arises from the double use of Lerch's theorem in the inversion of the double Laplace transform, has to be declared vacuous, and this is achieved by observing that each side of equation (17) is

for fixed t, a left-continuous decreasing function of s > t,

and

for fixed s > 0, a right-continuous increasing function of t < s.

However, formula (17) is only claimed to hold for $t < s$, and so long as this inequality is satisfied we can modify it by replacing $m^-(s-v)$ by $m(s-v)$; these two functions differ at most on a countable set and so the integral in equation (17) is unaffected.

If we want to let $s \downarrow t$, we can do so after having changed from $m^-(.)$ to $m(.)$, because $m(.)$ is *right*-continuous. On doing this we find that

$$\int_0^t p(v)\, m(t-v)\, dv = \mathrm{pr}\{\omega: \text{no } \Phi \text{ at } t\} = 1 - p(t),$$

and so we recover Kingman's Volterra integral equation [8].

On returning to the $m^-(.)$-form of the result, we can easily derive

Theorem 2. *For any standard regenerative phenomenon* Φ *with p-function p, and* m^- *defined as at equation (18), we have*

$$(19) \qquad m^-(s) = \lim_{\varepsilon \downarrow 0} \frac{\mathrm{pr}\{\omega: \text{no } \Phi \text{ in } [\varepsilon, s)\}}{\varepsilon} \quad \text{for all } s > 0.$$

This provides us with an interesting probabilistic interpretation for the m-functions.

If we turn now to the ideas introduced in [5], we note that the set S defined at equation (8) is, for the present model, *a random right-perfect set*, and it is natural to ask whether this can be brought within the general framework of [5]. The right perfect character of S suggests the use of the upper limit topology for the real line, but this is very badly behaved from the point of view adopted in [5], because it is not second-countable. If we use finite non-vacuous open intervals as traps, then the associated linear topology will be the usual one for the real line, and we can only identify a set as far as its closure. This, however, is not so serious a drawback as might at first be supposed, for the following reasons.

(i) *If* S_1 *and* S_2 *are each right-perfect, then they coincide if and only if their closures do so.*

(ii) *If S is right-perfect, and we know its closure* $\bar{S}$, *then S can be recovered, because it consists exactly of those elements of* $\bar{S}$ *which are right or two-sided cluster-points of* $\bar{S}$.

(iii) *If F is a closed set, then it is the closure of a right-perfect set S if and only if the right and two-sided cluster-points of F are dense in F.*

Thus we are led to the idea that standard regenerative phenomena can be studied *by considering $\bar{S}$ as a random closed set in the ordinary topology for the line*. It is now helpful to notice that the following random events are identical:

$$\{\omega: \Phi \text{ does not occur in } [t,s)\};$$

$$\{\omega: \Phi \text{ does not occur in } (t,s)\};$$

$$\{\omega: \bar{S} \text{ avoids } (t,s)\}.$$

Thus

$$(20) \qquad A_1(t,s) = \int_0^t p(v)\, m(s-v)\, dv,$$

which is the probability of any one of these three random events, and in particular of the last one, is a specification of the value of the *first-order avoidance-function* for $\bar{S}$ on the trap (t,s). (This can now be seen to be continuous in each of s and t.)

We are now left with the question, can we extend this last result so as to obtain the avoidance function for the random closed set $\bar{S}$ *to all orders*? Let us look, for example, at the second-order avoidance function $A_2(t_1, s_1; t_2, s_2)$, where $0 \leqslant t_1 < s_1 \leqslant t_2 < s_2$. This is the probability that the random closed set will avoid *both* of the open intervals (t_1, s_1) and (t_2, s_2). Notice that in the present problem we can split up the event whose probability is the value of A_2 into two portions; the first is just $A_1(t_1, s_2)$, i.e. the probability that the random closed set avoids the whole of (t_1, s_2), while the second portion is the probability of the disjoint event, that $\bar{S}$ avoids each of (t_1, s_1) and (t_2, s_2), *but hits* $[s_1, t_2]$.

Notice further that in arguing about the value of this probability we can replace $\bar{S}$ by S, because $\bar{S}$ hits an open interval if and only if S does so. But then we see that we can sharpen $0 \leqslant t_1$ to $0 < t_1$, and further that we can sharpen $s_1 \leqslant t_2$ to $s_1 < t_2$, without altering the probability, because no point of S can be right-isolated. For the same reason we can take as the probability to be calculated, that of the event: Φ *does not* happen in $[t_1, s_1)$, *does* happen in $[s_1, t_2)$ and *does not* happen in $[t_2, s_2)$. This, however, is the event

$$(21) \qquad \bigcup_{M \geqslant 1} \bigcap_{m \geqslant M} \bigcup_{n \geqslant 0} \bigcup_{r \geqslant 2} \{\omega: T(n2^{-m}) < t_1 < s_1 \leqslant T((n+1)2^{-m}) < T((n+r)2^{-m})$$

$$< t_2 < s_2 \leqslant T((n+r+1)2^{-m})\}.$$

We can evaluate its probability by following exactly the course pursued in passing from the event (16) to equation (17) or, rather, what we 'easily find' is the quadruple Laplace transform of the probability with respect to the four variables $t_1, s_1 - t_1, t_2 - s_1, s_2 - t_2$.

The inversion of this would have presented a substantial challenge, but fortunately we can guess the answer, and then afterwards check that it does have the correct quadruple Laplace transform. We then make a four-fold application of Lerch's theorem. The four-dimensional set of measure zero which might otherwise spoil the result can be shown to be vacuous because, on putting back the term $A_1(t_1, s_2)$, each side of the resulting identity is decreasing and left-continuous in each of s_1 and s_2, and is increasing and right-continuous in each of t_1 and t_2. In this way we find that

$$\text{pr}\{\omega: \Phi \text{ occurs in neither of } [t_1, s_1) \text{ and } [t_2, s_2)\}$$

is equal to

$$(22) \qquad A_1(t_1, s_2) - \int_{s_1}^{t_2} A_1(t_2 - x, s_2 - x)\, A_1(t_1, dx).$$

Notice that the second term in the expression (22) is a positive addition to the first term, because the integral is negative; this is because $A_1(t_1, x)$ decreases as x increases.

How does one guess that the second term of the expression (22) is the reverse Laplace transform of the ungainly expression which is obtained by Laplace transformation of the probability of the event (21)? The answer to this is, that one writes down what the probability of the event (21) 'ought' to be, if an appropriate analogue of the Strong Markov Property were available; that is, one observes (from the right-perfectness of S, and the consequent right isolation of the points in its complement) that there must be a *first* occurence of Φ in $[s_1, t_2)$ unless S avoids this interval altogether. One takes this as a Markov time and 'point of regeneration for the process', from which point of view the second term in the expression (22) is quite natural.

Doubtless this process could with sufficient labour be made rigorous by itself, but if one wishes to evade the subtleties of the Strong Markov Property (or here one should rather say, the Strong *Regenerative* Property), then the procedure I have outlined gives a logically complete and technically less sophisticated proof.†

---

† The Strong Regenerative Property for S lies far deeper than the Strong Markov Property for $\{T(u): u \geqslant 0\}$. This is why we outflank the former, although we have already made use of the latter.

It is clear that an iteration of this procedure will permit the computation of the avoidance function for $\bar{S}$ to all orders in terms of $A_1(.,.)$; i.e. in terms of the function evaluated in Theorem 1.

The way is therefore now open for an analysis of the behaviour of the random right-perfect set associated with a standard regenerative phenomenon by the methods of [5].

Reference must of course be made to the powerful analysis of regenerative sets by Hoffmann-Jørgensen [3]; no attempt has yet been made to link up the present analysis with his work. Mention must also be made of the fact that a special case of Theorem 1 has been proved by Chung [2]; he considers only *Markovian* standard *p*-functions (as is quite natural in the context within which he is working), and for these he proves a result equivalent in that special case to ours, in his equation at the foot of p. 207 of his book. From the standpoint of Kingman's concept of 'depth', therefore (for which see [4]), we can say that our extension of Chung's result from $\mathscr{PM}$ to the whole of $\mathscr{P}$ shows that it is, so to speak, 'maximally shallow'. It is unfortunately true that shallow results in this field may still be painful to acquire.

7. My reason for previously suppressing paragraphs 1–5 of this paper was a temporary uncertainty about the canonical status of the main results. If we take the canonical model

$$(23) \qquad (2^{R_+}, \mathscr{B}_0, \mathrm{pr};\, 2, R_+;\, Z_t\, (t \in R_+))$$

for a regenerative phenomenon Φ, where 2 denotes $\{0, 1\}$, R_+ denotes the non-negative reals, $\mathscr{B}_0$ denotes the σ-algebra of Baire sets in 2^{R_+}, and Z_t is the tth coordinate random variable (interpreted so that Φ happens at 'time' t if and only if Z_t has the value 1), and if pr is the Kingman measure set up in the early paragraphs of his classic paper [8], then because pr is wholly determined by the associated *p*-function $p(.)$, we can say that $p(.)$ itself is the 'name' of the name-class of stochastic processes of which the model (23) is the canonical representative. This name-class will be continuous in probability because of the continuity of $p(.)$ (assumed throughout this paper to be a *standard p*-function).

The model we have been using in the preceding sections provides us with a version within this name-class, if we fix our attention on the random variables carried by it which correspond to the above random variables Z_t, and forget for the moment about the other stochastic structures which it carries in addition to this. Now the set whose probability is evaluated in Theorem 1 is a Borel set (indeed a compact set), and it is also what I have called a Doob set (see [7] for this). Thus we know (from [7]) that the probability of this set can be assigned a unique value determined by the model (23), and so by the specification of $p(.)$ alone, either by extending the model (23) to the Kakutani model, or alternatively by taking *any* dense denumerable set D in R and working with *any* version of the process which is complete and D-separable in the sense of Doob; all these different procedures are guaranteed to give the same answer. This does not, of itself, guarantee that the value assigned to the probability in Theorem 1 is this unique 'canonical' value; we know only too well that strange versions can assign strange probabilities to sets which are not Baire sets and which are not in the pr-completion of the Baire σ-algebra. It is true that the T-process with which we started is a separable one (for all paths were arranged to be right-continuous), but we cannot carelessly suppose that these good properties of the T-process are automatically inherited by the Z-process. We therefore proceed to investigate the separability of the latter, and we shall find that this investigation will lay our fears at rest.

We therefore return to the original model of paragraphs 1–5, which in the notation of [4] would be described by the symbol

$$(24) \quad (\Omega, \mathscr{B}(\Omega), P; (\bar{R}_+, R_+; T(u) \ (u \in R_+)), (S), (2, R_+; Z_t(t \in R_+))).$$

Here R_+ denotes the non-negative half-line, $\bar{R}_+$ its one-point compactification, Ω the complement of the $\mathscr{K}_{\sigma\delta}$-set of P-measure zero rejected at the beginning of paragraph 2, $\mathscr{B}(\Omega)$ the trace of the Borel class on Ω and P the trace of the Kakutani measure (previously written pr, but now renamed to prevent confusion with the model (23)). The remaining entries in the main parenthesis identify the various stochastic structures with which Ω has been endowed; in particular the entry (S) recalls that we have constructed S as a random right-perfect set. Our main concern, however, is with the third component of the stochastic structure, $(Z_t(t \in R_+))$, and we have to decide whether this is separable in the sense of Doob in relation to some dense denumerable subset D of R_+ which is happily at our choice. We agree to take D to consist of all rationals of the form $n/2^k$.

In testing for separability in the sense of Doob it will suffice to show that the sets

$$(25) \qquad\qquad \{\omega : Z_U(\omega) \subset K\}$$

and

$$(26) \qquad\qquad \{\omega : Z_{U \cap D}(\omega) \subset K\},$$

relative to the completion of $\mathscr{B}(\Omega)$ under P, are measurable and have the same P-measure, where U denotes any open interval (t, s) whose end-points are in D, *and where K is either the singleton $\{0\}$, or the singleton $\{1\}$.* (Here $Z_U = \{Z_t : t \in U\}$, etc.)

We start by looking at the state of affairs when $K = \{1\}$. Then if ω is in the set (26), Φ will occur everywhere on a dense subset of (t, s), and so *no* point of (t, s) can be in S^c, because S^c is a union of half-open intervals. Thus the set (26), always a cover of the set (25), in this case coincides with it.

We need only look, therefore, at the case $K = \{0\}$. We observe that we have already established the measurability and evaluated the P-measure of the set (25) in Theorem 1 and the following discussion, and that as the set (26) is a Baire set, we can evaluate its measure by working with the canonical model (23). The construction (24) will yield us a D-separable version of the Z-process if and only if these two probabilities turn out to be the same.

We thus find ourselves in the undignified position of having to rederive what is essentially the result of Theorem 1 by 'elementary' arguments, in order to prove the separability of our version and so the absolute character of results like equation (17).

If equation (17) were our only result of any interest, this would amount to saying that we can only be sure of the absolute status of equation (17) by proving it afresh, and the whole investigation would have been reduced to an absurdity. But that is not a fair summary; a direct proof of equation (17) will guarantee the legitimacy of employing the model (24) in calculating the probabilities of *any* events concerning the Z-process which are what we have called Doob-measurable. This means that the model (24) (since it also gives us a version of the Z-process which is process-measurable) is just as powerful and just as reliable as any other separable-measurable version of the Kingman Z-process, and this is highly satis-factory because in working with the model (24) *we can exploit to the full all aspects of the tripartite $T(u)$–S–Z_t structure.*

8. We now give the 'elementary' proof of equation (17) when each of t and s lie in $D = \{n/2^k : n = 0, 1, 2, ..., k = 1, 2, ...\}$, with $[t, s)$ replaced by (t, s), and m^- replaced by m. (We have already seen that this will suffice for our purposes.)

For $k \geqslant 1$, let $t = T/2^k$ and let $s = S/2^k$. (This use of T and S has no connection with the previous employment of these symbols. Note that t and s are fixed, and that k will eventually increase indefinitely; thus T and S depend on k although for ease of printing we do not show this explicitly. Finally note that k must exceed some fixed lower bound for t and s to be so expressible, but that it is not otherwise limited.)

We are going to use a 'skeleton' argument, and we shall try to set this out in such a way that the technique employed can readily be adapted to other similar situations.

We begin by recalling a portion of Kingman's proof of his integral representation in [8]. Let p denote a standard p-function, stable or not, and let $h = 2^{-k}$. Then

$$p(0), p(h), p(2h), \ldots$$

is a renewal sequence, and so is uniquely associated with a 'first-passage' distribution

$$f_1(h), f_2(h), f_3(h), \ldots, f_\infty(h),$$

where we have deliberately introduced the 'defect' $f_\infty(h)$ in order to make the f-distribution 'honest'. Let us now set up a totally finite measure ρ_k on the Borel subsets of $[0, \infty]$ as follows:

a mass $(1 - e^{-rh}) f_r(h)/h$ at $x = rh$ $(r = 1, 2, \ldots)$;

a mass $f_\infty(h)/h$ at $x = \infty$.

Then from the argument on page 189 of Kingman [8] it can be deduced that as $k \to \infty$,

$$(27) \qquad \int_{[0,\infty]} \frac{1 - e^{-sx}}{1 - e^{-x}} \rho_k(dx) \to \frac{1}{r(s)} = \int_{[0,\infty]} \frac{1 - e^{-sx}}{1 - e^{-x}} (\delta + \lambda)(dx),$$

where $0 < s < \infty$. Here δ denotes unit mass at $x = 0$, and λ is the measure which occurs in the Kingman integral formula (formula (1) of the *present* paper); it is *not* identical with Kingman's λ.

It is to be noticed that formula (27) can only be asserted after one has worked right through the argument given in [8] (which depends on the selection of a subsequence of $k = 1, 2, \ldots$); once this argument has been established, one can go back and assert formula (27) as it stands, without any reference to a subsequence.

On putting $s = 1$ in formula (27) we see that mass $(\rho_k) \to \text{mass}(\delta + \lambda)$, while on subtracting formula (27) from the same statement with s replaced by $s + 1$, we see that

$$(28) \qquad \int_{[0,\infty]} e^{-sx} \rho_k(dx) \to \int_{[0,\infty]} e^{-sx}(\delta + \lambda)(dx),$$

first for $s > 0$ and then (because of the convergence of total mass) *for $s = 0$ also*. Now the exponential polynomials of the form

$$\alpha + \sum_{j=1}^{r} \alpha_j \exp(-s_j x) \quad (0 \leqslant x \leqslant \infty) \quad (\text{all } s_j \geqslant 0)$$

are dense in the function-space $C[0, \infty]$ with the uniform metric, and so we have proved the

Lemma. *If p is any standard p-function, and if ρ_k and λ are as defined above, then*

$$\text{(29)} \qquad \lim_{k \to \infty} \rho_k = \delta + \lambda$$

in the sense of weak convergence for measures relative to bounded continuous functions on $[0, \infty]$.

This will be the basis of our skeleton argument.

We now work with the canonical Daniell–Kolmogorov model (23) and evaluate the probability of the Baire set of elements z in 2^{R_+} for which

$$z_\tau = 0 \quad \text{for } \tau = L/2^k \quad (L = T+1, T+2, \dots, S-1).$$

The probability of this Baire set is equal to

$$\sum_{m=0}^{T} h \, p(mh) \sum_{n=S-m}^{n=\infty} f_n(h)/h$$

$$= \sum_{n=S-T}^{S-1} f_n(h) \, h^{-1} \sum_{m=S-n}^{T} h \, p(mh) + \sum_{n \geqslant S}^{n=\infty} f_n(h) \, h^{-1} \sum_{m=0}^{T} h \, p(mh)$$

$$= \int_{[s-t, s-h]} \frac{1}{1 - e^{-x}} \int_{s-x}^{t} p(u) \, du \, \rho_k(dx)$$

$$+ \int_{[s, \infty]} \frac{1}{1 - e^{-x}} \int_{0}^{t} p(u) \, du \, \rho_k(dx)$$

$$+ \int_{[s-t, s-h]} \frac{1}{1 - e^{-x}} \left\{ \int_{s-x}^{t} (p_k(u) - p(u)) \, du + h \, p(t) \right\} \rho_k(dx)$$

$$+ \int_{[s, \infty]} \frac{1}{1 - e^{-x}} \left\{ \int_{0}^{t} (p_k(u) - p(u)) \, du + h \, p(t) \right\} \rho_k(dx),$$

where $p_k(u)$ denotes $p([2^k u]/2^k)$.

We must find the limit of this last expression when $k \to \infty$ (so that $h \to 0$). Now we already know that $\rho_k([s-t, s])$ and $\rho_k([s, \infty])$ are bounded (because mass $(\rho_k) \to 1 + \text{mass}(\lambda)$), and from the uniform continuity of $p(.)$ over $[0, \infty)$ it then follows that each of the last two integrals tends to zero.

We can rewrite and combine the first two double integrals so that together they are replaced by

$$(30) \qquad \int_{[0,\infty]} q(x)\,\rho_k(dx),$$

where

$$q(x) = 0 \quad \text{for } 0 \leqslant x \leqslant s - t,$$

$$q(x) = \int_{s-x}^{t} p(u)\,du/(1 - e^{-x}), \quad \text{for } s - t \leqslant x \leqslant s,$$

and

$$q(x) = \int_{0}^{t} p(u)\,du/(1 - e^{-x}), \quad \text{for } s \leqslant x \leqslant \infty.$$

This function $q(.)$ belongs to $C[0,\infty]$, and so the integral at (30) has the limit

$$\int_{[0,\infty]} q(x)\,(\delta + \lambda)\,(dx),$$

in virtue of the lemma. We have therefore shown, in relation to the canonical model (23), that

$$(31) \qquad \mathrm{pr}\,\{z: Z_{(t,s) \cap D} \subset \{0\}\} = \int_{[s-t,s)} \int_{s-x}^{t} p(u)\,du\,\frac{\lambda(dx)}{1 - e^{-x}}$$

$$+ \int_{[s,\infty]} \int_{0}^{t} p(u)\,du\,\frac{\lambda(dx)}{1 - e^{-x}}$$

$$= \int_{0}^{t} p(u) \int_{(s-u,\infty]} \frac{\lambda(dx)}{1 - e^{-x}}\,du,$$

$$= \int_{0}^{t} p(u)\,m(s-u)\,du.$$

From the discussion following Theorem 1, however, we already know that *with our original model*

$$\mathrm{pr}\,\{\omega: Z_{(t,s)} \subset \{0\}\} = \int_{0}^{t} p(u)\,m(s-u)\,du,$$

and so we have completed the proof that for our original model (completed or not) the implied version of the Z-process is separable; i.e. outside a set of P-measure zero, each Z-path is topologically D-separable, where D is the set of binary rationals.

It will perhaps be useful to summarize the properties we have established for the model (24) in

Theorem 3. (*i*) *The T-process is separable-and-measurable, with paths which are increasing, right-continuous, and satisfy* $T(0) = 0$ *and*

$$T(v) - T(u) \geqslant v - u$$

for all $u < v$, *and it has independent increments distributed as indicated at equations* (2) *and* (3) (*where* $U(u) = T(u) - u$).

(*ii*) *The random set S, being the set of all finite t in the range of the random function* $T(.)$, *is right-perfect, and it can be taken to be the set of times-of-occurrence of a standard regenerative phenomenon* Φ *associated with an arbitrary standard p-function identified via* λ *by equations* (1) *and* (3). *The set*

$$\{(\omega, t): \Phi \text{ occurs at } t\}$$

is product-measurable.

(*iii*) *If* $\{Z_t: t \geqslant 0\}$ *is the indicator process for the random set S, then it is separable-and-measurable.*

(*iv*) *The T-process can be reconstructed from the Z-process by means of the equations,*

$$L(t) = \text{Lebesgue measure } (S \cap [0, t]),$$

$$T(L(t)) = t \text{ for each } t \text{ in } S,$$

$$T(u) = \infty \text{ for } u \geqslant L(\infty).$$

9. Much of the detail has been omitted from this sketch, and it is hoped that it will be possible to present a complete account elsewhere. From the present point of view, my object has been to illustrate what happens when stochastic analysis, stochastic geometry, and 'the theory of versions' are jointly applied to the same problem.

[*Note.* Professor G. E. H. Reuter has pointed out to me that there is some overlap between this article and a preprint by C. C. Heyde ("On the distribution of the last occurrence time in an interval for a regenerative phenomenon").]

REFERENCES

1. N. H. Bingham, "Limit theorems for a class of Markov processes", 3.6 of this book.
2. K. L. Chung, *Markov Chains*, 2nd ed., Springer, Berlin (1967).
3. J. Hoffmann-Jørgensen, "Markov sets", *Math. Scand.* **24** (1969), 145–166.

4. D. G. Kendall, "An introduction to stochastic analysis", 1.1 of this book.
5. ——, "Foundations of a theory of random sets", 6.2 of *Stochastic Geometry*, Wiley, London (1973).
6. ——, "Extreme-point methods in stochastic analysis", *Ztschr. Wahrsch'-theorie & verw. Geb.* **1** (1963), 293–300.
7. ——, "Separability and measurability for stochastic processes: a survey", 5.4 of this book.
8. J. F. C. Kingman, "The stochastic theory of regenerative events", *Ztschr. Wahrsch'theorie & verw. Geb.* **2** (1964), 180–224.
9. ——, "Some further analytical results in the theory of regenerative events", *J. Math. Anal. Appl.* **11** (1965), 422–433.
10. E. Nelson, "Regular probability measures on function space", *Ann. Math.* **69** (1959), 630–643.

4

MARKOVIAN AND OTHER STOCHASTIC PROCESSES

4.1

An Expression for Canonical Entrance Laws

K. L. CHUNG

The purpose of this note is to give an expression for the canonical entrance law (relative to the minimal semigroup) attached to each passable boundary atom, in the formulation of [1] and [2] for a finite boundary set. Part of this result is stated without proof in [3]. We shall use the terminology and notation in these sources without explanation, but refer only to [3] for cited propositions except the one near the end.

Theorem 1. *If a is not a recurrent trap, then we have for every $j \in I$ and $t > 0$:*

(1)
$$\lim_{s \downarrow 0} \frac{\langle \xi^a(s), \Phi_{\cdot j}(t-s) \rangle}{1 - \langle \xi^a(s), L^a(\infty) \rangle} = \eta_j^a(t).$$

For any trap a (recurrent or not) we have

(2)
$$\lim_{s \downarrow 0} \frac{\langle \xi^a(s), \Phi_{\cdot j}(t-s) \rangle}{1 - \langle \xi^a(s), L^a(t-s) \rangle} = \frac{\eta_j^a(t)}{\langle \eta^a(t), 1 \rangle}.$$

If a is non-sticky, then either limit above may be evaluated by letting $s \downarrow 0$ separately in the numerator and denominator.

Proof. We have by (2) on p. 54 of [3]:

(3)
$$\langle \xi^a(s), \Phi_{\cdot j}(t-s) \rangle = \langle \rho^a(s), \Phi_{\cdot j}(t-s) \rangle$$
$$+ \sum_{b \in B} \int_0^s F^{ab}(dr) \langle \xi^b(s-r), \Phi_{\cdot j}(t-s) \rangle.$$

By Theorem 1 on p. 62 of [3], the first term on the right-hand side is equal

311

to

$$(4) \qquad \int_0^s E^a(dr)\langle \eta^a(s-r), \Phi_{.j}(t-s)\rangle = \int_0^s E^a(dr)\eta_j^a(t-r).$$

If a is a trap, then the second term on the right-hand side of equation (3) is absent, hence we have from equations (3) and (4):

$$(5) \qquad \lim_{s\downarrow 0} \frac{\langle \xi^a(s), \Phi_{.j}(t-s)\rangle}{E^a(s)} = \eta_j^a(t)$$

by the continuity of η_j^a. On the other hand, it follows from (1) on p. 83 of [3] that

$$(6) \qquad \lim_{s\downarrow 0} \frac{1-\langle \xi^a(s), L^a(t-s)\rangle}{E^a(s)} = \delta^a + \sigma^{aa}(t),$$

the change from $L^a(t)$ to $L^a(t-s)$ being trivial.

For the sake of explicitness, we shall state two lemmas.

Lemma 1. *We have for every $t\geqslant 0$,*

$$(7) \qquad \langle \eta^a(t), 1\rangle = \sum_{b\in \mathbf{B}} \sigma^{ab}(t) + \tilde{c}^a,$$

where

$$(8) \qquad \tilde{c}^a = \langle \eta^a(t), 1-L(\infty)\rangle = \lim_{t\to\infty}\langle \eta^a(t), 1\rangle.$$

As a consequence, if a is a trap then

$$(9) \qquad \langle \eta^a(t), 1\rangle = \delta^a + \sigma^{aa}(t).$$

Proof. Formula (7) is given between p. 77 and p. 78 of [3]. We have also $\tilde{c}^a = \delta^a c^a$ (p. 77) and $c^a = 1$ if a is a trap (p. 76), from which equation (9) follows.

We are now ready to prove equation (2). If a is a trap then the limit in equation (6) is also equal to $\langle \eta^a(t), 1\rangle$ by equation (9). Comparing equations (5) and (6) we obtain equation (2).

Lemma 2. *Let $\mathbf{B}_s$ be the set of sticky boundary atoms. If $b\in \mathbf{B}_s$ then we have for every j and $t>0$:*

$$(10) \qquad \lim_{\substack{s\to 0\\ s'\to 0}} \langle \xi^b(s'), \Phi_{.j}(t-s)\rangle = 0.$$

Proof. We have for $s'<t'<t$, as soon as $s<t-t'$:

$$(11) \qquad \langle \xi^b(s'), \Phi_{.j}(t-s)\rangle \leqslant 1-\langle \xi^b(s'), L^b(t-s)\rangle$$
$$\leqslant 1-\langle \xi^b(s'), L^b(t-s'-s)\rangle \leqslant 1-\langle \xi^b(s'), L^b(t'-s')\rangle.$$

Hence we may replace s' by s in equation (10) and then the lemma follows from the first inequality in (11) (with s' replaced by s) and Proposition 8 on p. 81 of [3].

We can now prove equation (1). Recall that by Proposition 7 on p. 78 of [3] and the remark after the corollary on p. 76 of [3], we have

$$(12) \qquad \lim_{s \downarrow 0} \frac{F^{ab}(s)}{E^a(s)} = d^{ab} \leqslant 1;$$

$$(13) \qquad d^{ab} = 0 \quad \text{if } b \notin \mathbf{B}_s;$$

(cf. p. 145 of [2]). Returning to equation (3), we have by Lemma 2 and formulae (12) and (13):

$$(14) \qquad \lim_{s \downarrow 0} \frac{1}{E^a(s)} \sum_{b \in \mathbf{B}} \int_0^s F^{ab}(dr) \langle \xi^a(s-r), \Phi_{\cdot j}(t-s) \rangle$$

$$\leqslant \sum_{b \in \mathbf{B}_s - \{a\}} d^{ab} \limsup_{0 < r < s \downarrow 0} \langle \xi^a(s-r), \Phi_{\cdot j}(t-s) \rangle = 0.$$

Combining formulae (3), (4) and (14) we obtain again equation (5). Now if $\delta^a = 1$, then by formulae (1) with $t = \infty$ on p. 83 of [3]:

$$(15) \qquad \lim_{s \downarrow 0} \frac{1 - \langle \xi^a(s), L^a(\infty) \rangle}{E^a(s)} = 1.$$

Hence equation (1) follows from equations (5) and (15).

It remains to prove the last assertion of the theorem. Since

$$\langle \xi^a(s), \Phi_{\cdot j}(t-s) \rangle$$

decreases as s decreases, the limit certainly exists but may be zero (cf. p. 40 of [1]). On the other hand, for $0 < t < \infty$:

$$(16) \qquad \lim_{s \downarrow 0} \langle \xi^a(s), L^a(t-s) \rangle = L^{aa}(t)$$

exists by Proposition 8 on p. 81 of [3]. Recall also from formula (1) on p. 83 that

$$(17) \qquad 1 - L^{aa}(t) = E^a(0) [\delta^a + \sigma^{aa}(t)].$$

Hence if $a \in \mathbf{B}_s$ and $\delta^a = 1$, then

$$(18) \qquad 1 - L^{aa}(\infty) = E^a(0) > 0$$

by Lemma 6 on p. 63 of [3]. This shows that the limit of the denominator in equation (1) is not zero. If $a \notin \mathbf{B}_s$ and $\delta^a = 0$, then the right-hand side of equation (17) reduces to $E^a(0) \sigma^{aa}(t)$, where by definition

$$(19) \qquad \sigma^{aa}(t) = \langle e^a, l^a(t) \rangle.$$

Now if $\delta^a = 0$, then we have for every $i \in I^a$ (see p. 67 of [3]):

$$e_i^a > 0, \quad L_i^a(\infty) = 1.$$

Hence $l_i^a(t) > 0$ for all $t > 0$ by the corollary on p. 76 of [1]. Thus

$$\sigma^{aa}(t) > e_i^a \, l_i^a(t) > 0$$

for all $t > 0$ and so the limit of the denominator in equation (2) is not zero. The proof of the theorem is therefore completed.

Remark. In general, the argument above shows that for any boundary atom a, either $\sigma^{aa}(.)$ vanishes identically or it never vanishes. The first case occurs when a is ephemeral as defined on p. 53 of [3], but it can occur otherwise such as when a can be hit at most once or when it can be hit only via another atom b. It is easy to see that $\sigma^{aa}(.) \equiv 0$ implies that $a \notin \mathbf{B}_g$, $\delta^a = 1$, and $E^a(.)$ is the unit mass at 0, so that $\rho^a \equiv \eta^a$.

REFERENCES

1. Kai Lai Chung, "On the boundary theory for Markov chains", *Acta Math.* **110** (1963), 19–77.
2. ——, "On the boundary theory for Markov chains, II", *Acta Math.* **115** (1966), 111–163.
3. ——, *Lectures on Boundary Theory for Markov Chains*, Princeton University Press (1970).

4.2

Some Algebraic Results and Problems in the Theory of Stochastic Processes with a Discrete Time Parameter

J. F. C. KINGMAN

4.2.1 INTRODUCTION

In connection with the problem of deciding whether a stochastic process (in discrete time, with a finite number of states) can be derived by lumping together the states in some finite Markov chain, Heller [2] has introduced an algebraic structure which he calls the 'stochastic module' associated with the process. His definition can be recast so as to apply to processes with arbitrary state spaces, and a number of the properties of such processes have simple expressions in the algebraic formulation. The theory is particularly simple when a certain vector space has finite dimension, and it is suggested that the class of processes with this property, which includes certain Markov processes studied by Runnenburg and Steutel, might repay further study.

4.2.2 HELLER'S STOCHASTIC MODULES

Heller's construction proceeds as follows. Let

$$(1) \qquad X = (X_1, X_2, X_3, \ldots)$$

be a sequence of random variables taking values in the finite state-space S. Write $A = A(S)$ for the free real associative algebra, with identity 1, generated by S. By definition, A is the (real) vector space whose basis consists of all formal products

$$(2) \qquad x_1 x_2 \ldots x_n$$

with $x_r \in S$, $n \geqslant 0$, where the product (2) is interpreted as 1 when $n = 0$. Multiplication is defined in A by the distributive laws and the rule

$$(3) \qquad (x_1 x_2 \ldots x_m)(x_{m+1} x_{m+2} \ldots x_{m+n}) = x_1 x_2 \ldots x_{m+n}.$$

A linear functional on A is determined by its values on the basis elements (2); Heller defines such a functional E (he calls it p) by

$$(4) \qquad \begin{cases} E(1) = 1, \\ E(x_1 x_2 \ldots x_n) = \mathrm{pr}\{X_1 = x_1, X_2 = x_2, \ldots, X_n = x_n\}, \end{cases}$$

where pr denotes probability. Clearly the pair (A, E) contains all the information about the finite-dimensional distributions of X.

An element ξ of A is positive ($\xi \geqslant 0$) if all its coordinates, with respect to the basis (2), are non-negative. Then E is a positive linear functional:

$$(5) \qquad E(\xi) \geqslant 0 \quad \text{if } \xi \geqslant 0.$$

An important role is played by the special element σ of A defined by

$$(6) \qquad \sigma = \sum_{x \in S} x.$$

If ξ is of the form (2), then

$$\begin{aligned} E(\xi\sigma) &= \sum_{x \in S} E(x_1 x_2 \ldots x_n x) \\ &= \sum_{x \in S} \mathrm{pr}\{X_1 = x_1, \ldots, X_n = x_n, X_{n+1} = x\} \\ &= \mathrm{pr}\{X_1 = x_1, \ldots, X_n = x_n\} \\ &= E(\xi). \end{aligned}$$

Hence by linearity,

$$(7) \qquad E(\xi\sigma) = E(\xi)$$

for all $\xi \in A$. Clearly formulae (5) and (7) summarize the Kolmogorov consistency conditions, and we have the following rather obvious theorem.

Theorem 1. *If E is a positive linear functional on $A = A(S)$ with $E(1) = 1$ and $E(\xi\sigma) = E(\xi)$ for all $\xi \in A$, then there exists a random sequence X satisfying equations (4), and its finite-dimensional distributions are determined by E.*

Equation (7) asserts that E annihilates the left ideal of A defined by

$$(8) \qquad A(\sigma - 1) = \{\xi(\sigma - 1); \; \xi \in A\}.$$

It follows that, if we form the quotient vector space

$$(9) \qquad A/A(\sigma - 1),$$

then E induces a linear functional on this vector space. However, the vector space (9) is rather large, and has infinite dimension when $|S| \geqslant 2$ (since the products x_1^n are linearly independent modulo $A(\sigma - 1)$) and this observation is of little value.

Heller noted, however, that for many processes there are larger ideals annihilated by E. The largest such is the left ideal

$$(10) \qquad J = \{\eta \in A; \ E(\xi\eta) = 0 \text{ for all } \xi \in A\},$$

and he considers the quotient vector space

$$(11) \qquad\qquad L = A/J.$$

If $\lambda: A \to L$ denotes the natural map, then $\lambda(\eta) = 0$ implies $E(\eta) = 0$, so that E factors through L to give a linear functional $E': L \to R$ with

$$(12) \qquad\qquad E(\xi) = E'(\lambda(\xi)) \quad (\xi \in A).$$

Since J is an ideal, L is an A-module; that is, each $\xi \in A$ acts as a linear operator on L by the rule

$$(13) \qquad\qquad \xi\lambda(\eta) = \lambda(\xi\eta).$$

Thus we have a representation of A in the algebra of linear operators on L. It is an important fact that, if the process has sufficiently simple structure, then J can be so large that L has finite dimension, in which case (A, E) can be replaced, without loss of information, by an algebra of matrices with a linear functional E'.

4.2.3 FINITE MARKOV CHAINS

As an example of the construction of the last section, suppose that X is a Markov chain with initial distribution $(p_i; \ i \in S)$ and transition matrix $P = (p_{ij}; \ i,j \in S)$, where S is finite. Then, by equation (4),

$$(14) \qquad\qquad E(x_1 x_2 \ldots x_n) = p_{x_1} \prod_{r=2}^{n} p_{x_{r-1}x_r}.$$

If therefore ξ is a basis element of the form (2), and $x, y \in S$, then

$$E(\xi xy) = p_{x_1} \prod_{r=2}^{n} p_{x_{r-1}x_r} p_{x_n x} p_{xy}$$

$$= E(\xi x) p_{xy}.$$

Thus by linearity

$$(15) \qquad\qquad E(\xi xy) = E(\xi x) p_{xy}$$

for all $\xi \in A$, so that

(16) $$xy - xp_{xy} \in J$$

or

(17) $$\lambda(xy) = \lambda(x) p_{xy}.$$

From equation (17), it follows by induction on n that, if ξ is of the form (2), then

(18) $$\lambda(x\xi) = \lambda(x) p_{xx_1} \prod_{r=2}^{n} p_{x_{r-1}x_r},$$

so that, for all $\xi \in A$, $\lambda(x\xi)$ is a scalar multiple of $\lambda(x)$. Thus the image

(19) $$xL = \{x\lambda(\xi); \; \xi \in A\} = \{\lambda(x\xi); \; \xi \in A\}$$

of L under the linear operator x is the one-dimensional (or zero-dimensional if $\lambda(x) = 0$) subspace spanned by the vector $\lambda(x)$. The converse of this assertion is easily established, and we therefore have the following theorem [2].

Theorem 2. *The process X is a Markov chain if and only if, for each $x \in S$,*

(20) $$\dim(xL) \leqslant 1.$$

Notice that, if $n \geqslant 1$, the basis element (2) belongs to $x_1 A$, so that its image under λ belongs to $x_1 L$. Moreover

$$\lambda(1) = \lambda(\sigma) = \sum \lambda(x),$$

so that the image under λ of each basis element of A lies in the subspace of L spanned by the subspaces xL. Thus L is spanned by its subspaces xL ($x \in S$). In particular, if X is a Markov chain, L is spanned by the vectors $\lambda(x)$ ($x \in S$), so that

(21) $$\dim L \leqslant |S| < \infty,$$

which justifies the assertion that, in simple but interesting cases, L can be finite-dimensional.

If $\lambda(y) = 0$, then $E(\xi y) = 0$ for all $\xi \in A$. Taking ξ to be of the form (2), we have

$$\mathrm{pr}\{X_1 = x_1, X_2 = x_2, \ldots, X_n = x_n, X_{n+1} = y\} = 0,$$

so that

$$\mathrm{pr}\{X_n = y \text{ for some } n \geqslant 1\} = 0.$$

Thus a state with $\lambda(y) = 0$ is never entered by the process X, and may be removed from S without loss. We may therefore assume that $\lambda(x) \neq 0$ for all $x \in S$.

Suppose that the $\lambda(x)$ are linearly dependent, so that

$$\sum_{y \in S} C_y \lambda(y) = 0$$

for scalars C_y, not all zero. For any $x \in S$, equation (17) shows that $0 = x \sum C_y \lambda(y) = \sum C_y \lambda(xy) = \sum C_y \lambda(x) p_{xy}$, and since $\lambda(x) \neq 0$,

$$\sum_{y \in S} C_y p_{xy} = 0$$

for all x. Thus P must be singular. Indeed, it follows that the inequality (21) may be sharpened to assert that the dimension of L is equal to the rank of P.

In particular, if P is non-singular, the $\lambda(x)$ form a basis for L. Because of equation (17),

$$\sigma \lambda(x) = \sum_{y \in S} y \lambda(x) = \sum_{y \in S} \lambda(yx)$$

$$= \sum_{y \in S} \lambda(y) p_{yx}.$$

In other words, the matrix which represents σ with respect to the basis $(\lambda(x))$ is just P.

Theorem 3. *If X is a Markov chain with non-singular transition matrix P, then P is a matrix representation of σ.*

If, on the other hand, P has rank $r < |S|$, we can choose a subset $(\lambda(x); x \in R)$, where $R \subset S$ and $|R| = r$, as a basis for L, and then σ is represented by the appropriate submatrix of P. In any case, the vectors $\lambda(x)$ have a geometrical significance. Let A_+ be the set of positive elements of A, and $L_+ = \lambda(A_+)$. Then it is easy to see that L_+ is a proper convex cone in L, and that $\sigma L_+ \subseteq L_+$. The vectors $\lambda(x)$ are exactly the extreme rays of L_+.

The impressive feature of the Heller construction is the way in which the algebraic device of taking the quotient of A by the largest left ideal in the null-space of E reduces the process to its bare essentials, and takes the fullest advantage of the simplification expressed, for example, in the Markov property. This appears even more striking when the construction is generalized to processes on arbitrary state-spaces.

4.2.4 GENERAL RANDOM SEQUENCES

At first sight the construction of Section 4.2.2 might seem to be, of its very nature, restricted to processes with only a finite, or perhaps countable, number of possible states, but this is seen not to be so if the construction is expressed in a slightly different form. Observe first that, as a vector space, A is the direct sum of the spaces A_n $(n \geqslant 0)$, where A_n has basis elements

$$x_1 x_2 \dots x_n$$

which are formal products of length n (and A_0 is the one-dimensional space spanned by the identity 1). With respect to this basis, a general element ξ of A_n has a unique expression in the form

$$\xi = \sum_{x_1, x_2, \dots, x_n \in S} f(x_1, x_2, \dots, x_n) x_1 x_2 \dots x_n,$$

where the values of f are real coefficients. Hence A_n is isomorphic to the vector space of real-valued functions defined on the n-fold Cartesian product S^n of S with itself. Moreover,

$$
\begin{aligned}
E(\xi) &= \sum f(x_1, x_2, \dots, x_n)\, E(x_1 x_2 \dots x_n) \\
&= \sum f(x_1, x_2, \dots, x_n)\, \mathrm{pr}\{X_1 = x_1, X_2 = x_2, \dots, X_n = x_n\} \\
&= \mathbf{E}\{f(X_1, X_2, \dots, X_n)\},
\end{aligned}
$$

where $\mathbf{E}$ denotes expectation.

With this in mind, we can generalize Heller's construction as follows. Let

$$(22) \qquad\qquad X = (X_1, X_2, \dots)$$

be a random sequence taking values in an arbitrary state-space S. (Of course, S must be provided with a measure-theoretic structure, but we shall omit explicit reference to it.) Define A_n to be the vector space of bounded functions $f \colon S^n \to R$, and write

$$(23) \qquad\qquad E(f) = \mathbf{E}\{f(X_1, X_2, \dots, X_n)\}.$$

Notice that we do not identify functions which are equal with probability one. Then E is a linear functional on A_n. The vector space A_0 is taken to be the real line, with $E \colon A_0 \to R$ the identity map.

The vector space A is defined as the direct sum

$$A = \bigoplus_{n=0}^{\infty} A_n,$$

so that a typical element ξ of A is a formal sum

$$(24) \qquad\qquad \xi = \sum_{n=0}^{\infty} f_n \quad (f_n \in A_n)$$

in which all but finitely many of the terms are zero. Then E extends by linearity to a linear functional on A;

$$(25) \qquad E(\Sigma f_n) = \Sigma E(f_n).$$

An element ξ of A is positive ($\xi \geqslant 0$) if f_n is a non-negative function for all n, so that

$$(26) \qquad E(\xi) \geqslant 0 \quad if \; \xi \geqslant 0.$$

(It may be remarked parenthetically that there might be advantage for some purposes in enlarging A by allowing it to contain sums of the form (24) in which infinitely many of the f_n are non-zero, subject to the condition

$$(27) \qquad \Sigma E(|f_n|) < \infty.$$

Such an enlargement would not affect the present discussion.)

It remains to make the vector space A an algebra by defining an operation of multiplication. To do this it is sufficient to define the product fg ($f \in A_m, g \in A_n$), since then the product of two elements

$$\xi = \sum_{m=0}^{\infty} f_m, \quad \eta = \sum_{n=0}^{\infty} g_n$$

must be given by

$$(28) \qquad \xi \eta = \sum_{m=0}^{\infty} \sum_{n=0}^{\infty} f_m g_n.$$

If then $f \in A_m$ and $g \in A_m$, their product fg is the element of A_{m+n} defined by

$$(29) \qquad (fg)(x_1, x_2, ..., x_{m+n}) = f(x_1, x_2, ..., x_m) g(x_{m+1}, ..., x_{m+n}).$$

It is easy to check that A is a (real associative) algebra, and that in the particular case when S is finite it is isomorphic to the algebra $A(S)$. As in Section 4.2.2, the pair (A, E) contains all the information necessary to reconstruct the finite-dimensional distributions of X via the sequence (22).

An important element of A is the function σ in A_1 which is identically equal to 1 (note that σ^n is the function in A_n which is identically 1). The Kolmogorov consistency conditions reduce as before to the conditions

$$(30) \qquad E(1) = 1, \quad E(\xi) \geqslant 0 \;\; (\xi \geqslant 0), \quad E(\xi \sigma) = E(\xi),$$

which show in particular that E annihilates the left ideal $A(\sigma - 1)$.

We now proceed exactly as in Section 4.2.2. Let J be the maximal ideal

$$(31) \qquad J = \{\eta \in A; \; E(\xi \eta) = 0 \text{ for all } \xi \in A\}$$

contained in the null-space of E, $L = A/J$ the quotient A-module, and $\lambda: A \to L$ the natural map. Then $E: A \to R$ factors through L, to give a map which will without risk of confusion be denoted also by $E: L \to R$, so that

(32) $$E(\lambda(\xi)) = E(\xi).$$

If A_+ is the set of non-negative elements of A, and $L_+ = \lambda(A_+)$, then L_+ is a convex cone in L, and E is non-negative on L_+. An important element of L_+ is the vector $l_0 = \lambda(1)$.

Theorem 4. *The A-module L has the property that, for every bounded real-valued function f on S^n,*

(33) $$\mathbf{E}\{f(X_1, X_2, ..., X_n)\} = E(fl_0).$$

The cone L_+ is proper in the sense that

  (i) *every element of L is of the form $l_1 - l_2$ ($l_1, l_2 \in L_+$), and*

  (ii) *if l and $-l$ are both in L_+, then $l = 0$.*

Any non-negative element of A maps L_+ into L_+. In particular, σ is a linear operator on L, leaving the cone L_+ and the vector l_0 invariant.

Proof. Equation (33) follows at once from formulae (23) and (32). The property (i) is trivial; any $f_n \in A_n$ is of the form $f_n^+ - f_n^-$ with $f_n^+, f_n^- \geqslant 0$, so that any $\xi \in A$ is of the form $\xi^+ - \xi^-$ with $\xi^+, \xi^- \geqslant 0$, and

$$\lambda(\xi) = \lambda(\xi^+) - \lambda(\xi^-).$$

Suppose that l has the property that both l and $-l$ belong to L. Then there are non-negative elements

$$\xi = \Sigma f_n, \quad \eta = \Sigma g_n$$

of A with

$$\lambda(\xi) = l, \quad \lambda(\eta) = -l,$$

so that $\xi + \eta \in J$. Thus in particular,

$$0 = E\{\sigma^m(\xi + \eta)\} = \sum_{n=0}^{\infty} E\{\sigma^m(f_n + g_n)\},$$

and since all the terms in the sum are non-negative,

$$0 = E\{\sigma^m(f_n + g_n)\}$$
$$= \mathbf{E}\{f_n(X_{m+1}, ..., X_{m+n}) + g_n(X_{m+1}, ..., X_{m+n})\}.$$

Hence, for every $m \geqslant 0$,

$$\mathrm{pr}\{f_n(X_{m+1}, ..., X_{m+n}) = g_n(X_{m+1}, ..., X_{m+n}) = 0\} = 1,$$

whence for every $h_m \in A_m$,

$$E(h_m f_n) = E(h_m g_n) = 0,$$

and so

$$E(h_m \, \xi) = E(h_m \, \eta) = 0.$$

Thus, for any $\zeta \in A$,

$$E(\zeta \xi) = E(\zeta \eta) = 0,$$

showing that ξ and η belong to J, so that $l = \lambda(\xi) = 0$.

Since the product of two positive elements of A is positive,

$$\xi A_+ \subseteq A_+$$

for $\xi \geq 0$, so that

$$\xi L_+ \subseteq L_+.$$

In particular,

$$\sigma L_+ \subseteq L_+.$$

Moreover,

$$\sigma l_0 = \sigma \lambda(1) = \lambda(\sigma 1) = \lambda(\sigma) = \lambda(1) = l_0,$$

and the theorem is proved.

To every random sequence there is therefore associated a vector space L partially ordered by a proper convex cone L_+, a positive linear functional E on L, a vector l_0 in L_+, and a representation of the bounded functions on S^n as linear operators on L satisfying equations (29) and (33). Such a representation is very reminiscent of the formalism of quantum theory, and the present discussion has affinities, which I do not profess fully to understand, with the work for instance of Davies and Lewis [1].

The reader will have no difficulty in seeing that the results of [2], [3] and [4], insofar as they make sense in the general context, carry over in an obvious way. For instance, the sequence is stationary if and only if

$$(34) \qquad\qquad E(\sigma l) = E(l)$$

for all $l \in L$, an equation which makes very clear the role of σ as a shift operator.

Another example not to be found in the work of earlier authors is the following: if $\varphi \in A_1$, then $(\varphi(X_n))$ is a martingale if and only if

$$(35) \qquad\qquad E(\xi \varphi) = E(M_\varphi \, \xi)$$

for all $\xi \in A$, when $M_\varphi : A \to A$ is defined by linearity and the rule

$$(36) \qquad M_\varphi f(x_1, x_2, \ldots, x_n) = f(x_1, x_2, \ldots, x_n) \, \varphi(x_n).$$

It is not my object to set out in detail the possible translations of probabilistic ideas into algebraic form. Instead, the next section will be devoted to showing that there are interesting sequences, with genuinely infinite state-spaces, for which L has finite dimension.

4.2.5   MARKOV PROCESSES OF FINITE RANK

Suppose now that X is a Markov process, with initial distribution

$$(37) \qquad\qquad p(B) = \text{pr}\,(X_1 \in B)$$

and transition function

$$(38) \qquad\qquad P(x, B) = \text{pr}\,\{X_{n+1} \in B \,|\, X_n = x\}.$$

Then the n-step transition probabilities may be computed from P by the recurrence relation

$$(39) \qquad \begin{cases} P_1(x, B) = P(x, B), \\[2mm] P_{n+1}(x, B) = \displaystyle\int_S P_n(x, dy)\, P(y, B), \end{cases}$$

and the distribution

$$(40) \qquad\qquad p_n(B) = \text{pr}\,\{X_n \in B\}$$

of X_n is then given by

$$(41) \qquad\qquad p_n(B) = \int_S p(dx)\, P_{n-1}(x, B).$$

In practical problems it will usually be difficult to compute P_n explicitly and it is important to consider classes of transition functions which, while sufficiently flexible to approximate reasonably general chains, are simple enough to admit explicit analysis. With this in mind, Runnenburg, in a lecture to the 1962 Dublin meeting of the Institute of Mathematical Statistics describing joint work with Steutel [8], suggested the study of transition functions of the form

$$(42) \qquad\qquad P(x, B) = \sum_{j=1}^{r} a_j(x)\, b_j(B),$$

where the a_j are measurable functions on S, and the b_j are finite signed measures on S. Clearly equation (42) defines a transition function if and only if

$$(43) \qquad\qquad \sum_{j=1}^{r} a_j(x)\, b_j(B) \geqslant 0$$

for all x, B, and

$$(44) \qquad \sum_{j=1}^{r} a_j(x) b_j(S) = 1.$$

It is important to note that there is no need to impose the restriction that the a_j or the b_j should individually be positive.

A null set is a subset B of S with $p_n(B) = 0$ for all $n \geqslant 1$. Clearly if some linear relation

$$\sum_j c_j a_j(x) = 0$$

holds except on a null set, where the c_j are not all zero, then by substituting for one of the a_j in terms of the rest, we can throw P into the form (42) for a smaller value of r. Thus there is no loss of generality in supposing that no such linear relation holds. For the same reason, there is no loss of generality in supposing the b_j linearly independent. When these two conditions are satisfied, the transition function (42) will be said to be of rank r.

Runnenburg pointed out that, for a transition function of finite rank, the computation of P_n is very simple; an easy induction gives the formula

$$(45) \qquad P_n(x, B) = \sum_{j,k=1}^{r} c_{jk}^{(n-1)} a_j(x) b_k(B),$$

where $c_{jk}^{(m)}$ is the (j, k)th element of C^m, the mth power of $C = (c_{jk})$, and

$$(46) \qquad c_{jk} = \int_S a_k(x) b_j(dx).$$

The n-step transition probabilities are therefore governed by the powers of the matrix C.

It is sensible to regard the b_j as the coordinates of a vector measure b on S, taking values in a vector space V of dimension r, and the a_j as being the coordinates of a function a from S into the dual space V^*. Then equation (42) becomes

$$(47) \qquad P(x, B) = \langle a(x), b(B) \rangle,$$

which shows that the choice of basis in V is not important. Equation (45) becomes

$$(48) \qquad P_n(x, B) = \langle a(x), C^{n-1} b(B) \rangle,$$

where $C: V \to V$ is the linear operator defined by

$$(49) \qquad C = \int_S b(dx) \otimes a(x)$$

Let V_+ be the smallest closed convex cone containing the range of b. Then formulae (43) and (44) are equivalent to the assertion that the values of a are contained in

$$\text{(50)} \qquad\qquad \Delta = V_+^* \cap H,$$

where V_+^* is the dual cone of V_+, and H is the hyperplane

$$\text{(51)} \qquad\qquad H = \{\varphi \in V^*; \langle \varphi, b(S) \rangle = 1\}.$$

It follows from equation (49) that

$$\text{(52)} \qquad\qquad CV_+ \subseteq V_+, \quad Cb(S) = b(S).$$

Hence C is a positive operator in the sense that it maps the proper cone V_+ into itself, and since $b(S) \in V_+$ formula (52) shows that the largest eigenvalue of C is 1. Thus, although (as Runnenburg remarked) there is no reason for C to be stochastic for any choice of basis for V, the more general theory of positive operators [6] can be brought to bear upon C.

The adjoint

$$\text{(53)} \qquad\qquad C^* = \int_S a(x) \otimes b(dx)$$

of C has $C^* V_+^* \subseteq V_+^*$ and $C^* H \subseteq H$, so that it maps the convex compact set Δ into itself. Thus Brouwer's fixed point theorem shows that there exists $\alpha \in \Delta$ such that

$$\text{(54)} \qquad\qquad C^* \alpha = \alpha.$$

For any $\alpha \in \Delta$,

$$\text{(55)} \qquad\qquad p(B) = \langle \alpha, b(B) \rangle$$

defines a probability measure p, and if equation (54) holds then

$$\text{(56)} \qquad\qquad p(B) = \int_S p(dx) P(x, B).$$

If this particular measure is taken to be the distribution of X_1, then X is stationary.

It will be noted that the properties of $C: V_+ \to V_+$ are very similar to those of $\sigma: L_+ \to L_+$ in the general context of Section 4.2.4. In view of Theorem 3, it is not surprising that they are virtually identical.

Theorem 5. *If a Markov chain of rank r has initial distribution of the form* (55) *for some α in Δ (not necessarily satisfying equation* (54)), *then there* *is an isomorphism* $\theta: L \to V$ *such that the diagram*

(57)

$$
\begin{array}{ccc}
L & \xrightarrow{\ \sigma\ } & L \\
{\scriptstyle\theta}\big\downarrow & & \big\downarrow{\scriptstyle\theta} \\
V & \xrightarrow{\ C\ } & V
\end{array}
$$

commutes.

Proof. It is convenient to select a basis for V, so that P is of the form (42). For $f \in A_n$, $n \geqslant 1$ and $1 \leqslant j \leqslant r$, write

$$
E^j(f) = \int_S \cdots \int_S f(x_1, \ldots, x_n)\, a_j(x_n)\, p(dx_1)\, P(x_1, dx_2) \ldots P(x_{n-1}, dx_n),
$$

and

$$
E_j(f) = \int_S \cdots \int_S f(x_1, \ldots, x_n)\, b_j(dx_1)\, P(x_1, dx_2) \ldots P(x_{n-1}, dx_n).
$$

To complete the definition of E^j and E_j as linear functionals on A, write

$$
E^j(1) = \alpha_j, \quad E_j(1) = b_j(S).
$$

Then, for $m, n \geqslant 1$, $f \in A_m$, $g \in A_n$,

$$
E(fg) = \mathbf{E}\{f(X_1, \ldots, X_m)\, g(X_{m+1}, \ldots, X_{m+n})\}
$$

$$
= \int \cdots \int f(x_1, \ldots, x_m)\, g(x_{m+1}, \ldots, x_{m+n})\, p(x_1)\, P(x_1, dx_2)\ldots
$$

$$
\ldots P(x_{m-1}, dx_m) \sum_j a_j(x_m)\, b_j(dx_{m+1})\, P(x_{m+1}, dx_{m+2})\ldots
$$

$$
\ldots P(x_{m+n-1}, dx_{m+n})
$$

$$
= \sum_j \int \cdots \int f(x_1, \ldots, x_m)\, p(x_1)\, P(x_1, dx_2) \ldots P(x_{m-1}, dx_m)\, a_j(x_m)
$$

$$
\int \cdots \int g(y_1, \ldots, y_n)\, b_j(dy_1)\, P(y_1, dy_2) \ldots P(y_{n-1}, dy_n)
$$

$$
= \sum_j E^j(f)\, E_j(g).
$$

Similar arguments establish this identity when $f \in A_0$ or $g \in A_0$, and linearity shows that, for all $\xi, \eta \in A$,

(58)

$$
E(\xi\eta) = \sum_{j=1}^r E^j(\xi)\, E_j(\eta).
$$

Suppose there are numbers c_j, not all zero, such that

$$\sum_j c_j E^j(\xi) = 0$$

for all $\xi \in A$. Taking $\xi = \sigma^m f$ for $m \geq 0$ and $f \in A_1$, this implies

$$\sum_j c_j \int f(x) a_j(x) p_m(dx) = 0.$$

Since f is arbitrary,

$$\sum_j c_j a_j(x) = 0$$

except perhaps on a null set, and this contradicts the assumption that P is of rank r.

This implies, by equation (58), that $\eta \in J$ if and only if $E_j(\eta) = 0$ for all j. Thus J is the kernel of the linear mapping $\mathcal{E} \colon A \to V$ defined by

(59) $$\mathcal{E}\xi = (E_1(\xi), E_2(\xi), ..., E_r(\xi)).$$

The range of $\mathcal{E}$ is a linear subspace of V; if it is a proper subspace then there are numbers c_j, not all zero, such that

$$\sum_j c_j E_j(\xi) = 0$$

for all $\xi \in A$, and by a similar argument to that of the last paragraph this contradicts the assumption that P has rank r. Hence $\mathcal{E}$ maps A onto V and has kernel J. Therefore $\mathcal{E}$ factors through $L = A/J$:

$$A \xrightarrow{\quad \lambda \quad} L \xrightarrow{\quad \theta \quad} V,$$

where θ is an isomorphism.

For $f \in A_n$,

$$E_j(\sigma f) = \int ... \int f(x_2, ..., x_{n+1}) \, b_j(dx_1) \sum_k a_k(x_1) \, b_k(dx_2) \, P(x_2, dx_3)...$$

$$... P(x_n, dx_{n+1})$$

$$= \sum_k c_{jk} \int ... \int f(x_2, ..., x_{n+1}) \, b_k(dx_2) \, P(x_2, dx_3) ... P(x_n, dx_{n+1})$$

$$= \sum_k c_{jk} E_k(f).$$

Hence, for all $\xi \in A$,

(60) $$E_j(\sigma \xi) = \sum_k c_{jk} E_k(\xi).$$

This means that

$$\mathscr{E}(\sigma\xi) = C\mathscr{E}(\xi),$$

so that

$$\theta(\sigma\lambda(\xi)) = \theta(\lambda(\sigma\xi)) = \mathscr{E}(\sigma\xi) = C\mathscr{E}(\xi) = C\theta(\lambda(\xi)),$$

and

$$\theta\sigma = C\theta$$

on L. This proves that the diagram (57) commutes, and completes the proof.

4.2.6 SOME OPEN PROBLEMS

This paper is not intended to be a final account of a complete theory. Its object is rather to indicate an area in which further research might produce interesting results. With this in mind, a few open problems may be noted.

(*a*) If (X_n) is a Markov process of rank r, and φ is a function defined on S, then the process $(\varphi(X_n))$, which is in general non-Markovian, has the property that the corresponding space L has dimension $\leqslant r$. A possible converse might be that every process for which L has finite dimension is expressible in the form $(\varphi(X_n))$ for some Markov process (X_n) of finite rank.

(*b*) Given a proper cone V_+ in a vector space V of finite dimension r, and a vector $b \in V_+$, under what conditions does there exist a Markov process of rank r whose C-matrix satisfies

$$(61) \qquad CV_+ \subseteq V_+, \quad Cb = b?$$

How can the possible matrices C be characterized? More generally, what operators σ can arise from arbitrary processes as in Section 4.2.4, when L has given finite dimension?

(*c*) The sequence (X_n) is exchangeable [7] if and only if the elements of A commute as operators on L. Does this lead to a proof of de Finetti's theorem?

(*d*) A sequence $(u_n; n \geqslant 0)$ is a renewal sequence [5] if and only if there is a Markov process X with countable state space S, such that

$$(62) \qquad u_n = P_n(x, \{x\})$$

for some $x \in S$. For which renewal sequences may X be chosen to have finite rank?

(e) The seminorm

(63) $\|\eta\| = \sup\{|E(\xi\eta)|; \xi \in A_+, E(\xi) = 1\}$

on A vanishes exactly on J, and so induces a norm on L [4]. With respect
to this σ is a positive contraction, and it would be interesting to study
the properties of σ^n when L is infinite-dimensional. It is suggestive that
various mixing conditions for X can very simply be expressed in terms of
the norm (63).

(f) Is there a useful continuous-time analogue of the theory described
in this paper?

REFERENCES

1. E. B. Davies and J. T. Lewis, "An operational approach to quantum
   probability", *Comm. Math. Phys.* **17** (1970), 239–260.
2. A. Heller, "On stochastic processes derived from Markov chains", *Ann.
   Math. Statist.* **36** (1965), 1286–1291.
3. ——, "Probabilistic automata and stochastic transformations", *Math.
   Systems Theory*, **1** (1967), 197–208.
4. P. W. Holland, "Some properties of an algebraic representation of stochastic
   processes", *Ann. Math. Statist.* **39** (1968), 164–170.
5. J. F. C. Kingman, "An approach to the study of Markov processes", *J.
   R. statist. Soc. B*, **28** (1966), 417–477.
6. M. Krein and M. A. Rutman, "Linear operators leaving invariant a cone in
   a Banach space", *Uspehi Matem. Nauk* **3**, No. 1 (23) (1948), 3–95. (*A.M.S.
   translation*, No. 26.)
7. M. Loève, *Probability Theory*, van Nostrand, Princeton (1955).
8. J. Th. Runnenburg and F. W. Steutel, "On Markov chains the transition
   function of which is a finite sum of products of functions of one variable",
   Ann. Math. Statist. **33** (1962), 1483–1484.

4.3

Partial Sums of Matrix Arrays, and Brownian Sheets

R. PYKE

4.3.1 INTRODUCTION

In this paper, a survey of results and open problems is presented about partial sums formed from lattices of independent and identically distributed (IID) random variables. The basic difference between this situation and the classical one involving a sequence of IID random variables is the partial, rather than linear, ordering of the index set. The topics surveyed below include fluctuation theory, strong laws of large numbers, inequalities, central limit theorems and the law of the iterated logarithm.

I first became acquainted with Rollo Davidson during my sabbatical leave at the University of Cambridge in 1964–65 where he was an exceptionally promising research student. It was not until the summer of 1969 however that I was able to observe and enjoy his unusual aptitude for and enjoyment of mathematical research. In August 1969 he participated in the Twelfth Biennial Seminar of the Canadian Mathematical Society which was held at the University of British Columbia, Vancouver, Canada. During this three-week seminar, Rollo's contributions were considerable, as indicated in part by his involvement in the informal *ad hoc* seminars, as well as the special lectures, and his contributions to the 'Open Problems Book' maintained at the seminar. His great love of the outdoors was also evident at the seminar where on weekends he led several other members on expeditions, including a climb to within a few hundred feet of the summit of The Lions, a well-known Vancouver landmark.

It was Rollo Davidson's strong interest in open problems which is primarily responsible for my choice of topic and format for this paper. Rollo's contribution of solutions, partial or complete, to the seminar's 'Open Problem Book' was greater than anyone else's. One such contribution, discussed in the next section, is related to this paper's topic. Rollo was also in frequent discussion at the seminar with the late Professor Rényi about open problems raised by Rényi in his lectures. (Cf. the footnote on p. 180 of [24] for one result of these discussions.)

The context for this paper is as follows: we consider throughout partial sums of IID random variables which are indexed by 'tuples' of integers. For example, in the two-dimensional case, let $\{X_{ij}: 1 \leqslant i, j < \infty\}$ be IID, and define partial sums

$$(1) \qquad S_{mn} = \sum_{\substack{i \leqslant m \\ j \leqslant n}} X_{ij}.$$

It will be convenient to set $S_{m0} = S_{0n} = 0$ for any $m, n \geqslant 0$. The main difference between these matrix arrays of partial sums and the classical sequences of partial sums $\{S_n\}$ is the non-linear ordering of the index set.

In general, let K_r be the set of r-tuples $\mathbf{k} = (k_1, k_2, ..., k_r)$ with positive integers for coordinates. If $\leqslant$ denotes the coordinate-wise partial ordering on K_r, and if $\{X_\mathbf{k}: \mathbf{k} \in K_r\}$ is a set of IID random variables, define partial sums

$$(2) \qquad S_\mathbf{k} = \sum_{\mathbf{j} \leqslant \mathbf{k}} X, \quad \mathbf{k} \in K_r.$$

Whenever convenient we will extend the domain of $S_\mathbf{k}$ to include indices some of whose coordinates may be zero. In such cases we define $S_\mathbf{k}$ to be zero. Also set $\mathbf{1} = (1, 1, ..., 1)$ and $|\mathbf{k}| = k_1 k_2 ... k_r$.

4.3.2 FLUCTUATION THEORY

The partial sums described by equations (1) or (2) represent a stochastic process whose index set is an integer lattice in a multi-dimensional space. What analogues, if any, of the standard fluctuation theory results for one-dimensional random walks (cf. Chapter 3 of Feller [15]) are valid for these random walks over higher dimensional 'time'? It seems that the probable answer is, none. Although counter-examples to specific conjectures will not be included here (they are easily constructed), it is clear to this writer that analogues to Spitzer's lemma and the invariance results of Sparre Andersen, relating location of maxima and number of positive partial sums, do not obtain. The pertinent random variables (r.v.s) in two and higher dimensions include

$$(3) \quad \begin{cases} N_{mn} = \text{card}\{(i, j) \leqslant (m, n): S_{ij} > 0\}, \quad N_\mathbf{n} = \text{card}\{\mathbf{k} \leqslant \mathbf{n}: S_\mathbf{k} > 0\}; \\ M_{mn} = \max\{S_{ij}: (i, j) \leqslant (m, n)\}, \quad M_\mathbf{n} = \max\{S_\mathbf{k}: \mathbf{k} \leqslant \mathbf{n}\}; \\ L_{mn} = \min\{ij: M_{mn} = S_{ij}\}, \quad L_\mathbf{n} = \min\{|\mathbf{k}|: M_\mathbf{n} = S_\mathbf{k}\}. \end{cases}$$

One possible analogue of Spitzer's lemma would be that the conditional distribution of N_{mn} given $[S_{mn} = 0]$ would be invariant under changes in the common distribution of X_{ij}, provided only that with probability one ties do not occur. However, this is easily disproved. An alternative conjecture might involve the conditional distribution given $[S_{in} = 0 = S_{mj}$ for all $1 \leqslant i \leqslant m,\ 1 \leqslant j \leqslant n]$. This also does not yield an invariance result. A central difficulty here is the absence of a reflection principle, due to the non-linear ordering of K_r.

The basic fluctuation theory result of Sparre Andersen for $r = 1$ is that

$$N_n \overset{L}{=} L_n,$$

where what is asserted here is identity of distribution. Again, it is easily shown that no such result links N_{mn} and L_{mn} in the case $r = 2$.

Consider now the possibility of computing exact distributions of the r.v.s in equations (3) for specific distributions of the X_{ij}. One case of interest, particularly in view of the approximation of partial sums to a Brownian sheet (see Section 4.3.5 below), would be that in which X_{11} is $N(0, 1)$. Another case of interest is the Bernoulli one in which X_{ij} takes values ± 1 with probability $\frac{1}{2}$. At the Twelfth Biennial Seminar referred to in Section 4.3.1, I included in the 'Open Problem Book' the problem of finding the distribution of M_{mn} in this Bernoulli case. (A prize of one Canadian dollar was offered for a solution when $m = n = 4$.) Shortly after the seminar I received the following in a letter from Rollo Davidson, dated 31 August 1969: 'On the way back I wrote a program to evaluate the distribution of the maximum for the 4×4 case. For the 3×3 case, I got (by hand)

k:	0	1	2	3	4	5	6	7	8	9
$v_{33}(k)$:	116	112	103	87	46	30	8	9	0	1

and for the 4×4 case, by $1\frac{1}{2}$ minutes on TITAN,

k	$v_{44}(k)$	k	$v_{44}(k)$	k	$v_{44}(k)$	k	$v_{44}(k)$
0	12203	4	14186	8	2461	12	130
1	5916	5	960	9	32	13	0
2	18552	6	6620	10	672	14	16
3	3499	7	288	11	0	15	0
						16	1

(where $v_{mn}(k)$ is the number of the 2^{mn} basic outcomes for which $M_{mn} = k$). There seems to be here a great distinction between odd and even. (N.B. I included the null rectangle, so M_{mn} is always $\geqslant 0$.)'

The large oscillation shown in the 4×4 case is due in great part to the fact that 12 of the 17 partial sums will be even, whereas only 6 of 11 in the 3×3 case are even. It is difficult to see how exact distributions can be obtained for higher orders. Clearly fluctuation theory for matrix arrays provides a large source of open problems, of which this is but one.

4.3.3 THE STRONG LAW OF LARGE NUMBERS

Under what conditions does

$$(mn)^{-1} S_{mn} \xrightarrow{\text{a.s.}} ?$$

In view of Kolmogorov's SLLN for the classical one-dimensional case, we assume throughout this section that each X_{ij} is integrable and $\mathbf{E}(X_{ij}) = 0$. The problem then becomes that of finding (possibly stronger) necessary and sufficient conditions to ensure that

$$(mn)^{-1} S_{mn} \xrightarrow{\text{a.s.}} 0.$$

(By this convergence we mean that except for a null set, the net

$$\{(mn)^{-1} S_{mn}, \leqslant\}$$

converges in the usual sense. This is equivalent in this case to stating that for every $\varepsilon, \delta > 0$, there exists (M, N) such that

$$\text{pr}\,[\sup_{(m,n) \nleqslant (M,N)} |S_{mn}/mn| > \varepsilon] < \delta.$$

This problem represents a non-trivial extension of the usual SLLN that has recently been solved by Robert Smythe [27], during the preparation of this paper. Smythe's result stated for arbitrary r, is as follows.

Theorem 1 (Smythe). *The SLLN obtains for $\{S_k/|\mathbf{k}|: \mathbf{k} \in K_r\}$ if and only if*

$$(4) \qquad \mathbf{E}\,|X_1(\log^+|X_1|)^{r-1}| < \infty.$$

Smythe's proof of this theorem obtains a convergence theorem for the reversed martingale $\{S_k/|\mathbf{k}|: \mathbf{k} \in K_r\}$ with respect to the directed set $(K_r, \leqslant)$. It utilizes methods of Cairoli [5] who derives a convergence theorem for a particular class of partially ordered (forward) martingales.

The sufficiency part of the proof is the difficult one.† The necessity of condition (4) for the SLLN follows easily from the fact that the SLLN implies

$$X_k/|\mathbf{k}| \xrightarrow{\text{a.s.}} 0.$$

† (Added in proof.) Professor Smythe has pointed out that the sufficiency of $\mathbf{E}\,|X_1|\,(\log^+|X_1|)^{r-1} < \infty$ for the SLLN may be obtained as an application of ergodic theory involving r transformations. The appropriate reference is A. Zygmund, "An individual ergodic theorem for non-commutative transformations", *Acta Sci. Math. Szeged* **14** (1951), 103–110.

From this a two-dimensional application of the Borel–Cantelli lemma implies that

$$\sum_{\mathbf{k} \in K_r} \text{pr}\left[\,|X_{\mathbf{k}}| > \varepsilon\,|\mathbf{k}|\,\right] < \infty$$

for all $\varepsilon > 0$. The finiteness of this sum is equivalent to

$$\int_1^\infty \cdots \int_1^\infty \text{pr}\left[\,|X_1| > s_1 s_2 \dots s_r\,\right] ds_1 \dots ds_r < \infty.$$

Direct evaluation of this integral shows it to be equal to

$$\sum_{j=1}^{r-1} (-1)^{r-j-1} \mathbf{E}\,|X_1| (\log^+|X_1|)^j/j! - (-1)^r \mathbf{E}(|X_1| - 1)^+,$$

from which the necessity of condition (4) follows. This condition represents an interesting discontinuity between the classical case of $r = 1$ and the higher dimensional situations, $r > 1$. This difference is in contrast with the weak law of large numbers (WLLN) which holds in any dimension if and only if X_1 satisfies the classical WLLN for sequences. This follows since $\text{pr}\left[\,|S_{\mathbf{k}}|/|\mathbf{k}| > \varepsilon\right]$ depends only upon $|\mathbf{k}|$ and not on the ordering placed on K_r. Similarly, results about convergence in p-mean are the same in the matrix case as for sequences.

Sufficient conditions for the SLLN to hold for matrix arrays of independent, but not identically distributed, r.v.s are unknown, but some should not be too difficult to obtain. Questions concerning the convergence of multiple series $\sum_{\mathbf{k}} X_{\mathbf{k}}$ may be more difficult; for example, is there an analogue to Kolmogorov's Three Series Theorem?

4.3.4 INEQUALITIES

The proofs of most inequalities for sequences of partial sums involve stopping times; for example, the upcrossing inequalities for martingales or the direct proofs of the Kolmogorov and Hájek–Rényi inequalities. The absence of linearly ordered time, and hence of stopping times, in the case of matrix arrays poses a serious problem. In most cases, however, this problem can be circumvented to obtain analogues to certain of the standard inequalities.

The first inequalities for arrays of partial sums were obtained in 1969 by Wichura [30]. These include the following:

Theorem 2 (Wichura). *If* $\{X_{\mathbf{k}} : \mathbf{k} \in K_r\}$ *are independent r.v.s of mean zero and finite variance, then*

(5) $$\mathbf{E}(M_{\mathbf{n}}^2) \leqslant 4^r \sigma^2,$$

(6) $$\text{pr}\left[\,|M_{\mathbf{n}}| > 2^r c\right] \leqslant [1 - \sigma^2/c^2]^{-1} [1 - (r-1)\,\sigma^2/4c^2]^{-1} \text{pr}\left[\,|S_{\mathbf{n}}| > c\right]$$

and

(7) $$\mathrm{pr}\,[|M_\mathrm{n}| > 2^r\,c] \leqslant [1 - (r-1)\,\sigma^2/4c^2]^{-1}\,\sigma^2/4c^2$$

for $c^2 > (r-1)\,\sigma^2/4$, where $\sigma^2 = \mathbf{E}(S_\mathrm{n}^2)$.

[For $r = 1$, these inequalities reduce to those of Doob ([9], p. 317), Skorokhod ([26], cf. [4], p. 45 with $\alpha = \sigma^2/c^2$) and Kolmogorov ([4], p. 65), respectively.]

The proofs of all of these inequalities for general r are by induction on r. The induction possibilities are made clear by the following discussion. Write $i\mathbf{k} = (i, k_1, \ldots, k_{r-1}) \in K_r$ when $\mathbf{k} = (k_1, \ldots, k_{r-1}) \in K_{r-1}$. Set

$$M_i(\mathbf{m}) = \max\{|S_{i\mathbf{k}}| : \mathbf{k} \leqslant \mathbf{m} \in K_{r-1}\},$$
$$M_i^*(\mathbf{m}) = \max\{|S_{n\mathbf{k}} - S_{i\mathbf{k}}| : \mathbf{k} \leqslant \mathbf{m} \in K_{r-1}\};$$

thus

$$M_\mathrm{n} = \max_{1 \leqslant i \leqslant n} M_i(\mathbf{m})$$

if $\mathbf{n} = n\mathbf{m}$. Let I be the least i, if any, for which $|M_i(\mathbf{m})| > 2c$. Then by the usual argument

$$\mathrm{pr}\,[|M_n(\mathbf{m})| > c] \geqslant \mathrm{pr}\,[I \leqslant n, \text{ and } |M_I^*(\mathbf{m})| < c]$$

$$= \sum_{i=1}^{n} \mathrm{pr}\,[I = n]\,\mathrm{pr}\,[|M_i^*(\mathbf{m})| < c]$$

$$\geqslant \mathrm{pr}\,[I \leqslant n]\min_{1 \leqslant i \leqslant n} \mathrm{pr}\,[|M_i^*(\mathbf{m})| < c].$$

Thus if the second factor is not zero, one concludes that

$$\mathrm{pr}\,[|M_\mathrm{n}| > 2c] \leqslant \mathrm{pr}\,[|M_n(\mathbf{m})| > c]\,(1 - \max_{i \leqslant n} \mathrm{pr}\,[|M_i^*(\mathbf{m})| > c])^{-1}.$$

The r.v.s $M_n(\mathbf{m})$ and $M_i^*(\mathbf{m})$ on the right-hand side are maxima over an $(r-1)$-dimensional array, so that the appropriate induction hypotheses may be invoked.

In 1970, Fernandez [16] obtained an inequality for the maximum partial sum of $D[0, 1]$-valued r.v.s. If for $r = 2$ in our context, one treats the rows $(X_{i1}, \ldots, X_{im})$, $1 \leqslant i \leqslant n$, as vector-valued r.v.s under the supremum norm, one obtains as in Fernandez [16] the inequality,

(8) $$\mathrm{pr}\,[|M_\mathrm{n}| > 4c] \leqslant b^{-1}\,\mathrm{pr}\,[\max_{1 \leqslant j \leqslant m} |S_{nj}| > 2c],$$

where

$$b = 1 - \max_{1 \leqslant i \leqslant n} \mathrm{pr}\,[\max_{1 \leqslant j \leqslant m} |S_{nj} - S_{ij}| > 2c].$$

Since all probabilities on the right-hand side of inequality (8) involve maxima of sequences, one may, as above, apply the Kolmogorov and Skorokhod inequalities to denominator and numerator respectively to obtain inequality (6) from inequality (8).

In [5], Cairoli derives inequalities for a special class of martingales $\{Z_k : k \in K_r\}$ which are defined on product-spaces in the sense that $Z_k = g_k(W_{1k_1}, W_{2k_2}, ..., W_{rk_r})$ for independent real processes $\{W_{in} : n \geq 1\}$ for $i = 1, ..., r$. It has been observed, however, by Cairoli [6] and by Smythe (a personal communication) that the results of [5] apply also to our present situation. In particular then, Cairoli's martingale inequalities yield

$$c \, \mathrm{pr} \, [\,|M_n| > c] \leqslant A_r \sup_{k \leqslant n} \mathbf{E}[\,|S_k| (\log^+ |S_k|)^{r-1}] + A'_r,$$

and

(9) $$\mathbf{E}\,|M_n|^p \leqslant B_p^r \sup_{k \leqslant n} \mathbf{E}\,|S_k|^p \quad (p > 1),$$

where

$$A_r = (r-1)! \, a^{r-1}, \quad A'_r = a \sum_{j=1}^{r-2} j! \, a^j,$$

$$a = (1 - \mathrm{e}^{-1})^{-1} \quad \text{and} \quad B_p = p^p (p-1)^{-p}.$$

Note that for $p = 2$, inequality (9) is equivalent to inequality (5).

For the proof by Smythe [27] of the SLLN stated in the previous section the following analogous inequalities for the averaged sums, $S_k/|k|$, were derived;

$$c \, \mathrm{pr} \left[\max_{k \leqslant n} |S_k|/|k| > c \right] \leqslant A_r \sup_{k \leqslant n} \mathbf{E} \left[\frac{|S_k|}{|k|} \log^+ \frac{|S_k|}{|k|} \right] + A'_r$$

and

$$\mathbf{E}[\max_{k \leqslant n} |S_k|/|k|] \leqslant B_p^r \sup_{k \leqslant n} \mathbf{E}[\,|S_k|^p/|k|^p] \quad (p > 1),$$

where A_r, A'_r and B_p are as in inequality (9).

These latter inequalities involving $S_k/|k|$ are instances of a Hájek–Rényi type of inequality, the general form of which might be described as follows. Let $\mathbf{b} = \{b_k : k \leqslant n\}$ be an array of positive real numbers. Then under suitable monotonicity conditions on $\mathbf{b}$ one should have an inequality of the form

(10) $$\mathrm{pr}\,[\max_{k \leqslant n} |S_k|/b_k > c] \leqslant B/c^2,$$

where B is a universal constant depending on $r, \mathbf{b}$ and the variances of S_k.

For $r = 1$, Hájek and Rényi [17] obtained the bound

$$B = \sum_{i=1}^{n} \operatorname{var}(X_i)/b_i^2,$$

provided only that $0 < b_1 \leqslant b_2 \leqslant \ldots \leqslant b_n$. Using the approach of Wichura, Shorack obtained the bound $B = B'/(1 - B')$, where

$$B' = (c^2/4) \sum_{i,j=1}^{n} b_i^{-2} \operatorname{var}(X_{ij})$$

in the case $r = 2$ and $b_{ij} = b_{i1}$ for all i,j (cf. [30], p. 682). In general however the truth of inequality (10) is still an open question. I conjecture that the proper monotonicity condition on $\mathbf{b}$ is that all rth-order finite differences of $\mathbf{b}$ are non-negative; equivalently, $\mathbf{b}$ is the mass function associated with a measure on K_r.

Further inequalities for matrix arrays were derived in 1971 by Bickel and Wichura [2]. These inequalities, generalizations to the case $r > 1$ of the fluctuations inequalities presented by Billingsley [3], are applied to obtain weak convergence results for processes with a multidimensional index set. A brief review of the weak convergence for partial sums of IID random variables is given in the following section.

4.3.5 THE CENTRAL LIMIT PROBLEM

Let $\mathbf{1} = (1, 1, \ldots, 1)$ and $I = [0, 1] \subset R_r$. Fix $\mathbf{n} = (n_1, \ldots, n_r) \in K_r$. Set $Z_{\mathbf{n}}(\mathbf{t}) = |\mathbf{n}|^{-\frac{1}{2}} S_{[\mathbf{n} \cdot \mathbf{t}]}$, where $\mathbf{t} \in I$, $\mathbf{n} \cdot \mathbf{t} = (n_1 t_1, \ldots, n_r t_r)$ and $[\cdot]$ is the 'greatest-integer-contained-in' function defined coordinate-wise. The stochastic process $Z_{\mathbf{n}}$ is a $D(I)$-valued process. If one preferred to work with $C(I)$-valued processes one could make $Z_{\mathbf{n}}$ into a continuous function by setting

$$X(\mathbf{s}) = \lambda([\mathbf{s}], \mathbf{s}]) X_{[\mathbf{s}]+1}$$

and

$$Z_{\mathbf{n}}^*(\mathbf{t}) = |\mathbf{n}|^{-\frac{1}{2}} \int_{[0, \mathbf{n} \cdot \mathbf{t}]} dX(\mathbf{s}),$$

where λ is Lebesgue measure. We will stick to the more natural process $Z_{\mathbf{n}}$, however, and apply $\xrightarrow{L}$ to processes whose image laws converge weakly in the sense of Dudley–Wichura.

The Central Limit Theorem. *For IID $\{X_k\}$ with means zero and variances one,*

$$Z_{\mathbf{n}} \xrightarrow{L} Z,$$

where Z is a Brownian sheet; *that is, a Gaussian process with mean zero and* $\operatorname{cov}(Z(\mathbf{s}), Z(\mathbf{t})) = |\mathbf{s} \wedge \mathbf{t}|$, *where* $\mathbf{s} \wedge \mathbf{t} = (s_1 \wedge t_1)(s_2 \wedge t_2) \ldots (s_r \wedge t_r)$.

This theorem is proved in Wichura [30], as an application of his inequalities discussed in the previous section. A year earlier, Kuelbs [20] proved the weak convergence of the partial-sum processes Z_n^* for $r = 2$ under additional restrictions on the tails of the distribution of X_{11}. It is interesting to point out that Kuelbs mentions that these restrictions were necessitated by the lack of an analogue to Kolmogorov's inequality.

Although I have mentioned the Central Limit Theorem only for the IID case, the general theory is quite complete. Recent references such as Neuhaus [21], Straf [28] and Bickel and Wichura [2] provide the necessary tools for handling the general Central Limit problem as well as some cases involving dependent random variables. Possibly the only major open problems which remain involve (*i*) rates of convergence and (*ii*) the distribution of specific functions of Brownian sheets. For example, in the latter case, the limiting distribution of M_n, properly normalized, is unknown. Unlike the classical situation of $r = 1$ in which the reflection principle plays a key role in making direct computations feasible, the general case of $r > 1$ appears to be intractable at the present time. This difficulty is related to those of Section 4.3.2, where exact distributions for Bernoulli arrays were discussed.

Analogous problems arise in the study of the tied-down Brownian sheet,

$$W(\mathbf{t}) = Z(\mathbf{t}) - |\mathbf{t}| Z(\mathbf{1}) \quad (\mathbf{t} \in I),$$

which has mean 0 and $\mathrm{cov}(W(\mathbf{s}), W(\mathbf{t})) = |\mathbf{t} \wedge \mathbf{s}| - |\mathbf{t}||\mathbf{s}|$. (For $r = 2$, visualize a roughly (!) shaken bedsheet which is fastened down at all corners and along two adjacent sides.) This is shown by Dudley [11] to be the limiting process for a sequence of empirical processes determined by independent uniform r.v.s on I. One application of this result was by the author in [23]. For statistical purposes it would be useful to know the limiting distribution of specific functionals of the empirical processes, such as analogues to the Kolmogorov–Smirnov supremum and the Cramér–von Mises integral statistics. This leads one to the difficult questions of computing distributions for functionals of W. Some results in this case have been obtained by Durbin [14] and Dugué [13]. For a further discussion on problems connected with empirical processes, see [25].

4.3.6 BROWNIAN SHEETS

An important consequence of the Central Limit Theorem of the preceding section is the *existence* of the continuous process Z on $C(I)$ which we called a Brownian sheet. The first existence proof of this continuous

Gaussian process for $r = 2$ was given by Chentsov [7] in 1956. Also for $r = 2$, Yeh [32] showed in 1960 that the related Wiener measure on $C(I)$ existed. His approach was to prove an extension theorem for the natural additive set function defined on the semi-ring generated by 'intervals'. For general r, the existence of this continuous process may be deduced also using sufficient conditions of Dudley [10] in 1965 and Berman [1] in 1970, both of which involve rates of convergence to zero of

$$\mathbf{E}[Z(\mathbf{t}+\mathbf{h})-Z(\mathbf{t})]^2.$$

Another approach to the existence of Z was given by Park [22], in 1970, who showed that the classical construction of Brownian motion using Haar functions extends straightforwardly to the case of $r > 1$. Other references include Hudson [18], De Hoyos [8] and Dudley [12]. If $J = (\mathbf{s}, \mathbf{t}]$ is an r-dimensional interval in I, we write $Z(J)$ to denote the (random) measure of J determined in the usual way by the values of Z at the corners of J. We shall use Z interchangeably to denote either the point function of the previous paragraph or this associated finitely additive set function. Viewed as a set function, a Brownian sheet is a process $\{Z(J):$ intervals $J \subset I\}$ for which (*i*) $Z(J)$ is $N(0, \lambda(J))$ for each J, where λ is Lebesgue measure, and (*ii*) $Z(J_1)$ and $Z(J_2)$ are independent for disjoint J_1 and J_2.

As a set function it is clear that $Z(B)$ may be defined *almost surely* for each Borel set B of $C(I)$. (Cf. Park [22].) There are, however, too many null sets to permit Z to be defined as a signed measure over all Borel sets. Nevertheless, it is possible to define a Brownian sheet whose domain is much larger than the set of all rectangles. For example, the domain may include all convex subsets with suitably smooth boundaries. This result is due to Dudley [12]. Dudley had pointed out to me earlier that the extension of Z to include all convex polytopes with a bounded number of vertices follows from his 1965 paper [10].

In the following section I derive a Strassen-type law of the iterated logarithm (LIL) for Z. A Hölder condition for Z has also been obtained in unpublished work of G. E. H. Reuter and myself, namely

Theorem 3. *If Z is a Brownian sheet on I,*

$$(11) \qquad \limsup_{\substack{\text{rect } J \subset I \\ \lambda(J) \searrow 0}} \frac{|Z(J)|}{h(\lambda(J))} = 1 \quad \text{a.s.,}$$

where $h(u) = (-2u \log u)^{\frac{1}{2}}$.

In a personal communication from W. E. Pruitt this result and related local results and integral tests have been obtained by him and Steven Orey.

Many open problems still remain, however, about the sample function behaviour of Brownian sheets; some may be answered analogously to the one-dimensional theory, while others will require new methods. For example.

(a) *Does equation* (11) *hold when Z is defined on a larger class of sets than rectangles*?

(b) *If* $A_x(\mathbf{t}, Z) = \lambda\{\mathbf{s} \leqslant \mathbf{t} : Z(\mathbf{s}) \leqslant x\}$, *does* $A_x(\mathbf{t}, Z)$ *have almost surely a smooth density function*? The suggestion here is to find an analogue to local times and possibly additive functionals.

(c) *What topological or dimensional properties are possessed almost surely by the zero sets,* $\{\mathbf{t} \leqslant \mathbf{1} : Z(\mathbf{t}) = 0\}$?

4.3.7 A LIL FOR BROWNIAN SHEETS

The discussion of the previous section implies the existence of a real-valued Brownian sheet Z defined over R_r^+, the closed positive orthant of Euclidean r-space. We assume that Z is defined in $(C_0, \mathscr{B})$, the space of continuous functions over R_r^+ which vanish along the axes, with its Borel field $\mathscr{B}$ as determined by the supremum metric. Let $\mathbf{1} \in R_r^+$ be the vector of unit coordinates and set $I = [0, 1]$, the unit cube. Let $\mathscr{G}$ denote the class of continuous functions g defined on I which vanish along the axes, are absolutely continuous with respect to Lebesgue measure λ over I and whose Radon–Nikodym derivatives $\dot{g}$ satisfy

$$(12) \qquad \int_I \dot{g}^2 \, d\lambda \leqslant 1.$$

It follows from Hölder's inequality that for any Borel set $J \subset I$ and $g \in \mathscr{G}$,

$$(13) \qquad |g(J)| \equiv \left| \int_J \dot{g} \, d\lambda \right| \leqslant \left| \int_J 1 \, d\lambda \right|^{\frac{1}{2}} \left| \int_J \dot{g}^2 \, d\lambda \right|^{\frac{1}{2}} \leqslant [\lambda(J)]^{\frac{1}{2}}.$$

For a fixed $\mathbf{c} \in R_r^+$ with $|\mathbf{c}| = 1$ define for $n \geqslant 3$ and $\mathbf{u} \in I$,

$$(14) \qquad Z_n(\mathbf{u}) = b_n^{-\frac{1}{2}} n^{-r/2} Z(n(c_1 u_1, \dots, c_r u_r)),$$

where

$$b_n = 2 \log \log n.$$

Theorem 4. *With probability* 1, $\{Z_n : n \geqslant 3\}$ *is relatively compact with limit points exactly equal to* $\mathscr{G}$.

Proof. The reader familiar with [29] will realize the great dependence of this proof upon that of Strassen for $r = 1$. The differences which arise are due to the lack of a linear ordering on I when $r \geqslant 2$.

12

Fix $\varepsilon > 0$ and let m be a positive integer. Let π_m be the partition of I into m^r equal cubes of λ-measure m^{-r}. Let $\mathscr{G}_\varepsilon$ denote the open ε-sphere about $\mathscr{G}$ taken with respect to the supremum norm $\|\cdot\|$.

Then for any $s > 1$

$$(15) \qquad \mathrm{pr}\,[Z_n \notin \mathscr{G}_\varepsilon] \leqslant \mathrm{pr}\,[m^r \sum_{J \in \pi_m} |Z_n(J)|^2 > s^2]$$

$$+ \mathrm{pr}\,[m^r \sum_{J \in \pi_m} |Z_n(J)|^2 \leqslant s^2, Z_n \notin \mathscr{G}_\varepsilon] \equiv A + B.$$

Since the elements of π_m are disjoint, the terms $Z_n(J)$ for $J \in \pi_m$ are independent $N(0, m^{-r} b_n^{-1})$ r.v.s. The sum of their squares is then a chi-square r.v. yielding, for any $1 < \rho < s^2$,

$$(16) \qquad A = \int_{s^2 b_n/2} x^{\frac{1}{2}m^r - 1} \exp\,(-x)\,dx/\Gamma(m^r/2)$$

$$\leqslant \exp\,(-\rho \log\log n)$$

for all n sufficiently large.

To study B in formula (15) let $Z_{n,m}$ be the absolutely continuous function on I defined to equal zero along the axes whose derivative satisfies

$$\dot{Z}_{n,m}(u) = Z_n(J) m^r \quad \text{for } \mathbf{u} \in J \in \pi_m.$$

(Thus in two dimensions, for example, $Z_{n,m}$ is derived by quadratically interpolating between the values of Z_n at the points $(i/m, j/m), i, j = 1, \dots, m$.) If

$$m^r \sum_{\pi_m} |Z_n(J)|^2 \leqslant s^2,$$

then $s^{-1} Z_{nm} \in \mathscr{G}$ by inequality (12) since Z_n and Z_{nm} agree at the vertices of elements in π_m. Hence

$$(17) \qquad B = \mathrm{pr}\,[s^{-1} Z_{nm} \in \mathscr{G}, \text{ and } Z_n \notin \mathscr{G}_\varepsilon]$$

$$\leqslant \mathrm{pr}\,[s^{-1} Z_{nm} \in \mathscr{G}, \text{ and } \|s^{-1} Z_{nm} - Z_n\| \geqslant \varepsilon].$$

If s is chosen sufficiently close to 1 so that $s - 1 < \varepsilon/2$, then $s^{-1} Z_{nm} \in \mathscr{G}$ and $\|s^{-1} Z_{nm} - Z_n\| \geqslant \varepsilon$ implies that $\|Z_{nm} - Z_n\| \geqslant \varepsilon/2$, since

$$\|s^{-1} Z_{nm} - Z_{nm}\| = (1 - s^{-1})\|Z_{nm}\| \leqslant s - 1$$

and $\|g\| \leqslant 1$ for $g \in \mathscr{G}$. Thus from equation (17)

$$B \leqslant \mathrm{pr}\,[s^{-1} Z_{nm} \in \mathscr{G}, \text{ and } \|Z_{nm} - A_n\| \geqslant \varepsilon/2].$$

$$\leqslant \mathrm{pr}\,[\|Z_{nm} - Z_n\| \geqslant \varepsilon/2].$$

Set $D_{nm} = Z_n - Z_{nm}$. By construction of Z_{nm}, D_{nm} is zero at all lattice points determining π_m. For $\mathbf{u} \in I$, let $p(\mathbf{u}) = m^{-1} ([mu_1], \dots, [mu_r])$ be that lattice point of the partition π_m which is closest to $\mathbf{u}$ and is in $[\mathbf{0}, \mathbf{u}]$.

Let $\mathscr{P}_m = \{[0, 1/m], (1/m, 2/m], ..., (1-1/m, 1]\}$ be the partition of $[0, 1]$ into intervals of equal length. Thus π_m is the r-fold Cartesian product of $\mathscr{P}_m$ with itself. Let

$$\mathscr{L}_m = (\mathscr{P}_m \times [0, 1]^{r-1}) \cup ([0, 1] \times \mathscr{P}_m \times [0, 1]^{r-2}) \cup ... \cup ([0, 1]^{r-1} \times \mathscr{P}_m).$$

Thus $\mathscr{L}_m$ is the set of all subsets of I determined by restricting one coordinate to fall in one of the members of $\mathscr{P}_m$. For any $L \in \mathscr{L}_m$ set

$$d(L) = \sup\{|D_{nm}(L \cap (0, \mathbf{u}])| : \mathbf{u} \in L\}.$$

The observation to make now is that since $p(\mathbf{u})$ is a lattice point of π_m, we have $D_{nm}(p(\mathbf{u})) = 0$. Consequently, $|D_{nm}(\mathbf{u})| \geq \varepsilon/2$ if and only if

$$|D_{nm}(\mathbf{u}) - D_{nm}(p(\mathbf{u}))| \geq \varepsilon/2.$$

This in turn implies that whenever $\|Z_{nm} - Z_n\| \geq \varepsilon/2$, then $d(L) \geq \varepsilon/2r$ for at least one of the r members of $\mathscr{L}_m$ which contain u. (A picture for the case $r = 2$ would help. However one simply uses the fact that the region $[0, \mathbf{u}]\backslash[0, p(\mathbf{u})]$ is the union of r of the strips in $\mathscr{L}_m$, each intersected with $[0, \mathbf{u}]$.) Thus

$$(18) \quad B \leq \mathrm{pr}\,[\|Z_n - Z_{nm}\| \geq \varepsilon/2] \leq \mathrm{pr}\,[d(L) \geq \varepsilon/2r \text{ for some } L \in \mathscr{L}_m]$$

$$\leq \sum_{L \in \mathscr{L}_m} \mathrm{pr}\,[d(L) \geq \varepsilon/2r].$$

But $d(L)$ has the same distribution for every L by the stationarity of a Brownian sheet's increments. Thus in particular, for each L,

$$\mathrm{pr}\,[d(L) \geq x] = \mathrm{pr}\,[\sup\{|Z_n(\mathbf{u})| : \mathbf{u} \leq (m^{-1}, 1, 1, ..., 1)\} \geq x].$$

By applying Wichura's inequality (6) to a sequence of arrays approximating Z_n in the natural way, this latter probability may be shown to be bounded above by

$$c\,\mathrm{pr}\,[|D_{nm}((m^{-1}, 1, ..., 1))| \geq x2^{-r}],$$

where c depends on x and m but not on n. Thus by Mill's ratio and inequality (18) this implies that

$$(19) \quad B \leq c'\,b_n^{-\frac{1}{2}} \exp\{-m\,\varepsilon^2(\log\log n)/4^{r+1}\,r^2\}$$

$$\leq c' \exp(-\rho \log\log n)$$

for any $1 < \rho < m\,\varepsilon^2/4^{r+1}\,r^2$ and $m > 4^{r+1}\,r\varepsilon^{-2}$, where again c' is a constant depending on r, m and ε but not on n.

Upon combining formulae (16) and (18) into (15) it follows that for m sufficiently large but fixed, there exists $\rho > 1$ so that

$$\mathrm{pr}\,[Z_n \notin \mathscr{G}_\varepsilon] \leq \exp(-\rho \log\log n) = (\log n)^{-\rho}.$$

Thus for any subsequence of the form $n_j = [\alpha^j]$ for $\alpha > 1$, $\sum \mathrm{pr}\,[Z_{n_j} \notin \mathscr{G}_\varepsilon] < \infty$ and so $Z_{n_j} \in \mathscr{G}_\varepsilon$ eventually with probability one. If $n_{j-1} < n \leqslant n_j$ and $n_j^r b_n / n^r b_n = a_{n,j}^2$, then

$$Z_n(\mathbf{u}) = Z_{n_j}\left(\frac{n}{n_j}\mathbf{u}\right) a_{n,j}.$$

Hence for $g \in \mathscr{G}$,

$$
\begin{aligned}
|Z_n(\mathbf{u}) - g(\mathbf{u})| = \Bigg\| &\left[Z_{n_j}\left(\frac{n}{n_j}\mathbf{u}\right) - g\left(\frac{n}{n_j}\mathbf{u}\right)\right] a_{n,j} + g(\mathbf{u})(a_{n,j} - 1) \\
&+ \left[g\left(\frac{n}{n_j}\mathbf{u}\right) - g(\mathbf{u})\right] a_{n,j} \Bigg\| \\
\leqslant \|Z_{n_j} - g\|\,a_{n,j} &+ \|g\|(a_{n,j} - 1) + [1 - (n/n_j)^r]^{\frac{1}{2}}\,a_{n,j},
\end{aligned}
$$

where formula (13) has been used to obtain the last term. For α sufficiently close to 1, $a_{n,j}$ can be made arbitrarily close to 1 to ensure that $\|Z_n - g\| < 2\varepsilon$ whenever $\|Z_{n_j} - g\| < \varepsilon$. Thus $\mathrm{pr}\,[Z_n \notin \mathscr{G}_{2\varepsilon}$ at most finitely often$] = 1$ as desired.

The above argument establishes as in Strassen [29] that with probability 1, $\{Z_n\}$ is relatively compact and $\mathscr{G}$ contains all of its limit points. We now show that with probability 1, $\mathscr{G}$ coincides with the set of limit points. Because of the separability of $\mathscr{G}$ it suffices to show that for any $\varepsilon > 0$,

$$\mathrm{pr}\,[\|Z_n - g\| < \varepsilon \text{ i.o.}] = 1$$

for each $g \in \mathscr{G}$.

In order to show that $\|Z_n - g\| < \varepsilon$ it suffices to show that for all $J \in \pi_m$,

(i) $|Z_n(J \cap (\mathbf{0}, \mathbf{u}])| \leqslant [\lambda(J)]^{\frac{1}{2}} + \delta$ for all $\mathbf{u} \in I$, and

(ii) $|Z_n(J) - g(J)| < \delta$.

provided that δ and m are chosen, as is possible, so that

(20) $$m^r \delta + 2m^{-r/2} + \delta < \varepsilon.$$

This follows since

$$
\begin{aligned}
\|Z_n - g\| \leqslant \sum_{J \in \pi_m} |Z_n(J) - g(J)| &+ \sup\{|Z_n(J \cap (\mathbf{0}, \mathbf{u}])| \\
&+ |g(J \cap (\mathbf{0}, \mathbf{u}])| : J \in \pi_m, \mathbf{u} \in I\}.
\end{aligned}
$$

That (i) holds with probability 1 for each $\delta > 0$ and n sufficiently large was proved above. To establish (ii), let $g \in \mathscr{G}$ be fixed and let m and δ satisfy inequality (20). Set $\pi'_m = \pi_m \backslash \{[0, (m^{-1}, ..., m^{-1})]\}$ and

$$A_n = [\,|Z_n(J) - g(J)| < \delta; J \in \pi'_m].$$

Then by independence

$$(21) \qquad \mathrm{pr}\,(A_n) = \prod_{J \in \pi_{m'}} \int_{(|g(J)|-\delta)\sqrt{(m^r b_n)}}^{(|g(J)|+\delta)\sqrt{(m^r b_n)}} (2\pi)^{-\frac{1}{2}}\,e^{-x^2/2}\,dx.$$

If the lower limit of any of these factors does not exceed zero (i.e. $|g(J)| \leqslant \delta$) then that factor is bounded below by

$$\int_0^{\delta\sqrt{(m^r b_n)}} (2\pi)^{\frac{1}{2}}\,e^{-x^2/2}\,dx \geqslant \tfrac{1}{4}$$

for n sufficiently large. Otherwise, if $|g(J)| > \delta$ then by Mill's ratio this factor is bounded below by

$$Cb_n^{-\frac{1}{2}}\exp\{-m^r b_n(|g(J)|-\delta)^2/2\} \geqslant Cb_n^{-\frac{1}{2}}\exp\{-m^r b_n(|g(J)|^2 - \delta^2)/2\}.$$

Substitution into equation (21), using formulae (12) and (13) (and hence $m^r \sum |g(J)|^2 \leqslant 1$), yields

$$\mathrm{pr}\,(A_n) \geqslant C\frac{1}{\log n}\left\{\frac{\exp\,(m^r\,\delta^2\,b_n/2)}{b_n^{\frac{1}{2}}}\right\}^{N(g,m,\delta)},$$

where $N(g, m, \delta)$ is the number of times $|g(J)| > \delta$ for $J \in \pi_m'$. But the quantity in brackets diverges to ∞ as $n \to \infty$. Thus for n sufficiently large

$$\mathrm{pr}\,(A_n) \geqslant c/\log n.$$

For the subsequence $n_j = m^j$,

$$\sum_j \mathrm{pr}\,(A_{n_j})$$

diverges. But by our use of π_m', the events $\{A_{n_j}\}$ for this particular subsequence are independent, thereby establishing that the events $\{A_{n_j}\}$ occur infinitely often with probability 1.

Although the above pertains to a real-valued Z, there would be no difficulty in generalizing to a vector-valued Brownian sheet.

4.3.8 A LIL FOR PARTIAL SUMS AND EMBEDDINGS

The following Strassen-type LIL for matrix arrays is a special case of a recent result by Wichura [30].

Theorem 5. *Let* $Z_\mathbf{n}$ *be the normalized partial sums of Section* 4.3.5 *for* $\mathbf{n} = (n, \dots, n)$. *Then with probability* 1, $\{b_n^{-\frac{1}{2}}Z_\mathbf{n}: n \geqslant 3\}$ *is relatively compact in the uniform topology on* $C(I)$ *and its limit points coincides with* $\mathscr{G}$.

In [31], Wichura's theorem applies to quite general processes with independent increments, and not just the IID case of this paper.

An interesting feature of Wichura's paper is its use of the classical techniques of Kolmogorov and Hartman and Wintner, rather than the newer embedding techniques of Skorokhod and others used by Strassen. Although the LIL for Brownian sheets described in the previous section is of interest in its own right, my main purpose in deriving it was as a first step in obtaining a LIL for partial sums. The second step was to embed the partial sums into Z in a way which would show that the limit behaviour of the partial sum process would be the same as for Z. This turned out to be very difficult. In fact it now seems possible that embedding methods will not yield this result when $r > 1$.

To be more precise, an embedding approach would seek r.v.s $\{\mathbf{T}(\mathbf{k})\}$ defined on Z such that

$$\{Z(\mathbf{T}(\mathbf{k}))\} \stackrel{L}{=} \{S_{\mathbf{k}}\}$$

and such that $|\mathbf{T}(\mathbf{k}) - \mathbf{k}|$ would be sufficiently small. The r.v.s $\mathbf{T}$ must also be 'stopping times' in some sense. The existence of usable $\mathbf{T}$ seems to be difficult to establish. I have studied a slightly different type of embedding which uses Skorokhod's method on disjoint strips of Z to generate the rows $(X_{i1}, X_{i2}, \ldots)$ for $r = 2$. This may well work satisfactorily for arrays with finite fourth moments, but the general question of whether or not the LIL for matrix arrays of r.v.s with finite variance can be proved using embedding techniques is still open. The reader should note that a successful embedding in a two-dimensional Gaussian process has recently been given by Kiefer [19] for the purpose of studying rates of convergence of empirical processes. Kiefer's construction may well be applicable to our problem of embedding partial sums into a Brownian sheet. It might be noted that for any embedding in which the 'times' $\{\mathbf{T}(\mathbf{k})\}$ were themselves partial sums (as is the case when $r = 1$) the SLLN of Smythe in Section 4.3.3 implies that this would impose a moment restriction on $X_{\mathbf{k}}$ greater than the finiteness of the second moment (perhaps the finiteness of $\mathbf{E}|X_{\mathbf{k}}|^2 (\log^+ |X_{\mathbf{k}}|)^2$ when $r = 2$?).

4.3.9 SUMMARY AND ACKNOWLEDGEMENT

In this paper I have surveyed recent results and emphasized some open problems concerning partial sums of matrix arrays of IID random variables. Topics include fluctuation theory, laws of large numbers, inequalities, weak convergence and laws of the iterated logarithm. The limiting Brownian sheet is also discussed and a Strassen-type law of the iterated logarithm is derived for it.

I wish to express my appreciation to G. E. H. Reuter for many helpful discussions about the material of this paper, particularly concerning the proof in Section 4.3.7, as well as to R. M. Dudley for his interest and comments concerning extensions of the definition of Brownian sheets.

REFERENCES

1. S. Berman, "Gaussian processes with stationary increments: local times and sample function properties", *Ann. Math. Statist.* **41** (1970), 1260–1272.
2. P. J. Bickel and M. J. Wichura, "Convergence criteria for multiparameter stochastic processes and some applications", *Ann. Math. Statist.* **42** (1971), 1656–1670.
3. P. Billingsley, *Convergence of Probability Measures*, John Wiley and Sons, New York (1968).
4. L. Breiman, *Probability*, Addison-Wesley, Reading, Mass. (1968).
5. R. Cairoli, "Une inégalité pour martingales à indices multiples et ses applications", *Séminaire de Probabilités IV. Lecture Notes in Mathematics* **124**, 1–27, Springer-Verlag, Berlin (1970).
6. ——, "Décomposition de processus à indices doubles", *Séminaire de Probabilités V. Lecture Notes in Mathematics* **191**, 37–57, Springer-Verlag, Berlin (1971).
7. N. N. Chentsov, "Wiener random fields depending on several parameters", *Dokl. Akad, Nauk SSSR* (N.S.) **106** (1956), 607–609.
8. A. G. DeHoyos, *Continuity of some Gaussian processes parametrized by the compact convex sets in R^s"*, unpublished dissertation, University of California, Berkeley (1970).
9. J. L. Doob, *Stochastic Processes*, John Wiley and Sons, New York (1953).
10. R. M. Dudley, "Gaussian processes on several parameters", *Ann. Math. Statist.* **36** (1965), 771–788.
11. ——, "Weak convergence of probabilities on non-separable metric spaces and empirical measures on Euclidean spaces", *Ill. J. Math.* **10** (1966), 109–126.
12. ——, "Sample functions of the Gaussian process", to appear in *Ann. Prob.* **1** (1973).
13. D. Dugué, "Characteristic functions of random variables connected with Brownian motion and of the von Mises multidimensional ω_n^2", *Multivariate Analysis*, *Vol.* II (Ed. by P. R. Krishnaiah), pp. 289–301. Academic Press, New York (1969).
14. J. Durbin, "Asymptotic distributions of some statistics based on the bivariate sample distribution function", *Nonparametric Techniques in Statistical Inference* (Ed. by M. L. Puri), pp. 435–451. Cambridge University Press, Cambridge (1970).
15. W. Feller, *An Introduction to Probability Theory and its Applications*, vol. 1, 3rd ed., John Wiley and Sons, New York (1968).
16. P. J. Fernandez, "A weak convergence theorem for random sums of independent random variables", *Ann. Math. Statist.* **41** (1970), 710–712.
17. J. Hájek and A. Rényi, "Generalization of an inequality of Kolmogorov", *Acta Math. Acad. Sci. Hungar.*, **6** (1955), 281–283.

18. W. N. Hudson, *An investigation into Brownian motion processes with multidimensional time parameters*, unpublished dissertation, University of California, Irvine (1970).
19. J. Kiefer, "Skorokhod embedding of multivariate rv's, and the sample df", *Ztschr. Wahrsch'theorie & verw. Geb.* **24** (1972), 1–36.
20. J. Kuelbs, "The invariance principle for a lattice of random variables", *Ann. Math. Statist.* **39** (1968), 382–389.
21. G. Neuhaus, "On weak convergence of stochastic processes with multidimensional time parameters", *Ann. Math. Statist.* **42** (1971), 1285–1295.
22. W. J. Park, "A multi-parameter Gaussian process", *Ann. Math. Statist.* **41** (1970), 1582–1595.
23. R. Pyke, "The weak convergence of the empirical process with random sample size", *Proc. Cam. Phil. Soc.* **64** (1968), 155–160.
24. —— (Editor), *Proceedings of the Twelfth Biennial Seminar, Canadian Mathematical Congress, Montreal* (1970).
25. ——, "Empirical processes", *Jeffery-Williams Lectures: 1968–1972*, Canadian Mathematical Congress, Montreal (1972).
26. A. V. Skorokhod, "Limit theorems for stochastic processes", *Th. Prob. and Applic.* (English translation by SIAM) **2** (1957), 138–171.
27. R. Smythe, "A strong law of large numbers for *n*-dimensional arrays of iid random variables", to appear in *Ann. Prob.* **1** (1973).
28. M. L. Straf, "Weak convergence of stochastic processes with several parameters", *Proc. Sixth Berkeley Symp. Math. Statist. and Prob.* **2** (1972), 187–221, University of California Press, Berkeley.
29. V. Strassen, "An invariance principle for the law of the iterated logarithm", *Ztschr. Wahrsch'theorie & verw. Geb.* **3** (1964), 211–266.
30. M. J. Wichura, "Inequalities with applications to the weak convergence of random processes with multidimensional time parameters", *Ann. Math. Statist.* **40** (1969), 681–687.
31. ——, "Some Strassen-type laws of the iterated logarithm for multiparameter stochastic processes with independent increments", to appear in *Ann. Prob.* **1** (1973).
32. J. Yeh, "Wiener measure in a space of functions of two variables", *Trans. Am. Math. Soc.* **95** (1960), 433–450.

4.4

Some Basic Theorems on Harnesses

DAVID WILLIAMS

4.4.1 INTRODUCTION AND SUMMARY

Hammersley's concept [3] of a *harness* as a model for spatial variation
has a strong intuitive appeal. Here, we take up the study of *Q-harnesses*
which are motivated by the *central harnesses* of Hammersley's paper. The
relation between *Q*-harnesses and central harnesses is significant and
somewhat surprising.

Q-harnesses and the more general *Q-increment-harnesses* exhibit exactly
those properties which Hammersley required for models of long-range
misorientation in the crystalline structure of metals. Their continuous
parameter analogues (*X*-harnesses) include models of spatial variation
investigated by Whittle [9] and Wong [10]. Whittle showed that his model
for soil fertility agreed well with practical observation.

A major advantage of *Q*-harnesses over the central and serial harnesses
of [3] is that it is possible simply to read off basic *Q*-harness theory from
standard results about martingales and random walks. Important problems
on the possible distributional structure of harnesses remain. M. Jacobsen
has shown me some interesting examples of non-Gaussian *Q*-harnesses.

Some general remarks

To each member *X* of an important class of Markov processes (including
all simple random walks and the Brownian motions in R^n ($n \neq 2$)) corresponds a *canonical X*-harness (or *X*-increment-harness) which is Gaussian.
Thus, for example, one-dimensional Brownian motion is its own canonical
increment-harness.

If *f* is an *X*-harness, then the *parameter set* of the stochastic process *f* is
the *state-space A* of the Markov process *X*. It is therefore of interest that

349

the canonical X-harness f is itself *Markovian*; its Markov property is described in terms of the hitting distributions of X.

McKean [7] found the best definition of the Markov property for processes with multi-dimensional parameter set and gave a powerful technique for checking its validity. Wong [10] extended McKean's ideas and methods to obtain an important result characterizing all homogeneous Gaussian Markov processes with n-space R^n as parameter set.

The study of continuous parameter harnesses would involve a considerable amount of technical mathematics of the type found in Wong's paper. Thus, for example, the canonical harness associated with Brownian motion in R^n ($n \geqslant 3$) is a *generalized process*.

Here, we consider in detail only the simplest possible case, namely when X is a discrete parameter Markov chain with countable state-space A. There are some results and some interesting unsolved problems in this setting which have no continuous parameter analogues. However, a brief heuristic look at the 'Brownian' harness in R^n in Section 4.4.3 is necessary to illuminate the 'random walk' results.

4.4.2 HARNESSES AND RANDOM WALKS

We now illustrate our theorems by describing what happens when X is a simple random walk.

Let A be the lattice of points with integer coordinates in d-dimensional Euclidean space R^d with the usual Pythagorean distance $|x-y|$ between points x and y. Let Q be the one-step transition matrix of simple random walk X on A so that

$$Q(x,y) = \frac{1}{2d} \quad \text{if } |x-y| = 1,$$

$$= 0 \quad \text{otherwise.}$$

Suppose that $f = \{f(x) \colon x \in A\}$ is a stochastic process parametrized by A, or random function, or random column vector, on A. We say that f is a *strict sense Q-harness*, or $(Q, 1)$-*harness*, if each $f(x)$ ($x \in A$) has finite expectation and if, for each x in A,

(1) $$\mathbf{E}[f(x)|f(y), y \neq x] = [Qf](x) = \sum_{y \in A} Q(x,y)f(y).$$

If $d = 1$ or 2, in which case X is recurrent, then we shall see that every $(Q, 1)$-harness is Q-*harmonic*: $f = Qf$ on A.

Note. We shall drop qualifying 'with probability one' phrases except when there is some real point to them.

The results for $d \geqslant 3$ are most conveniently sketched if we make the assumption that f is a *finite variance* $(Q, 1)$-harness, i.e. that f is a $(Q, 1)$-harness in which each $f(x)$ $(x \in A)$ has finite variance. Then f has a *unique* 'Riesz' decomposition:

$$f = h + p \quad \text{on } A,$$

where h is Q-harmonic: $h = Qh$ on A, and where p satisfies

$$(2) \qquad \mathbf{E}[p(x)] = 0, \quad \mathbf{E}[p(x)p(y)] = KG(x, y) \quad (x, y \in A),$$

K being a non-negative constant and G the Green function of the random walk:

$$G(x, y) = \sum_{n=0}^{\infty} Q^n(x, y) < \infty \quad (x, y \in A).$$

The sense in which p is a *potential* $(Q, 1)$-*harness* will be explained later.

Note. Equation (2) does not force p to be a *strict* sense, or $(Q, 1)$, harness. However, p inherits the $(Q, 1)$ harness property from f. What does follow from the equation

$$(3) \qquad \mathbf{E}[p(x)p(y)] = KG(x, y) \quad (x, y \in A)$$

is that p is a *wide sense*, or $(Q, 2)$, harness. See Sections II.3 and IV.3 of Doob [2], or later in this paper, for 'wide sense' properties.

One consequence of the Riesz decomposition is that, since

$$\mathbf{E}[p(x)^2] = KG(0, 0) < \infty \quad (x \in A),$$

f cannot differ 'much' from a Q-harmonic function h. Hammersley explained that harness models of long-range crystalline misorientation should have this 'strait-jacket' property.

Another consequence of the (fully and properly stated) Riesz decomposition is that, as far as finite variance $(Q, 1)$-harnesses are concerned, we may restrict our attention to the potential $(Q, 1)$-harnesses p which satisfy equation (2). Now, as $|x - y| \to \infty$,

$$(4) \qquad G(x, y) \sim \frac{C_1(d)}{|x - y|^{d-2}},$$

where, and from now on this type of convention is taken as self-explanatory, $C_1(d)$ is a constant depending on the dimension $d \geqslant 3$. Thus, asymptotically, p has the power-law covariance function of the generalized process obtained at equation (6) of Whittle's paper [9]. Whittle arrived at his process, which is, of course, a Brownian harness, by using a serial harness approach (see Section 7 of Hammersley's paper).

Construction of the Gaussian canonical harness

It is easily shown that a *Gaussian process* p satisfying equation (2) is a strict sense Q harness. We now show that a Gaussian process p satisfying equation (2) with $K = 1$ exists and call such a process a *canonical Q-harness*. That G is symmetric and positive-definite, which is all we need show, is, of course, a standard fact. However, the following argument has independent interest.

Let $\{\tilde{\varepsilon}(x) : x \in A\}$ be a family of independent random variables each Gaussian with mean 0 and variance 1, i.e. each with the $N(0, 1)$ distribution. With I denoting the identity matrix on A, let

$$\tilde{Q} = I - (I - Q)^{\frac{1}{2}} = \sum_{n \geqslant 1} \frac{1}{n \cdot 2^{2n-1}} \binom{2n-2}{n-1} Q^n.$$

Then the stochastic matrix $\tilde{Q}$ is the one-step transition matrix of a random walk which behaves like a Cauchy process. See Section 4.4.3. Define the Green function $\tilde{G}$ of $\tilde{Q}$:

$$\tilde{G} = I + \tilde{Q} + \tilde{Q}^2 + \dots;$$

then $\tilde{G}^2 = G$. For $x \in A$, the series

$$(5) \qquad\qquad\qquad p(x) = \sum_y \tilde{G}(x, y)\, \tilde{\varepsilon}(y)$$

converges in quadratic mean (and indeed with probability 1). The process $\{p(x) : x \in A\}$ so defined is Gaussian and, since $\tilde{G}$ is *symmetric* and $\tilde{G}^2 = G$, satisfies equation (2).

For each x in A, the series defining $[\tilde{Q}p](x)$ converges in quadratic mean and we find that

$$(6) \qquad\qquad\qquad\qquad p = \tilde{Q}p + \tilde{\varepsilon}.$$

By showing up the linear model structure of p with independent (and so uncorrelated) errors, equation (6) exhibits p as a *central* harness, but one associated with the 'Cauchy' matrix $\tilde{Q}$ instead of Q. The Green function $\tilde{G}$ of $\tilde{Q}$ satisfies

$$(7) \qquad\qquad\qquad \tilde{G}(x, y) \sim \frac{C_2(d)}{|x - y|^{d-1}} \quad (|x - y| \to \infty).$$

See Section 4.4.3. Comparison of relation (7) with relation (4) explains why some of Hammersley's results are 'one dimension out' compared with ours and the serial harness results.

Define the 'true' errors $\varepsilon(x)$ $(x \in A)$ by the Poisson equation

$$(8) \qquad\qquad\qquad\qquad \varepsilon = p - Qp.$$

Then it follows from equation (2) that

(9) $\qquad \mathbf{E}[\varepsilon(x)] = 0, \quad \mathbf{E}[\varepsilon(x)\,\varepsilon(y)] = [I-Q]\,(x,y)$

for $x, y \in A$. Further, p is obtained from ε by the equation

(10) $\qquad p(x) = \sum_y G(x,y)\,\varepsilon(y) \quad (x \in A),$

the series converging in quadratic mean. Equation (10) explains one sense in which p is a 'potential', namely the potential due to the 'charge' ε, but the later explanation is more significant.

A characterization of the normal distribution

Only Gaussian $(Q, 1)$-harnesses may be constructed from independent errors $\tilde{\varepsilon}$ by the procedure described at equation (5). The proof of this (not altogether surprising) fact gives some inkling of the kind of mixture of potential theory and characteristic function theory which is needed if any harness-type theory is to be got off the ground. I feel sure that much better results than Theorem A may be obtained by extending these techniques.

Theorem A. *Suppose that* $\{\tilde{\varepsilon}(x) \colon x \in A\}$ *is a family of independent, identically distributed, random variables with common distribution F of mean* 0 *and variance* 1. *Then we may define* $p(x)$ $(x \in A)$ *by equation* (5), *the series converging in quadratic mean and with probability* 1. *The resulting process* p *satisfies equation* (2) *and is therefore a wide-sense Q-harness.*

However, p *is a strict sense Q-harness if and only if*

$$F = N(0, 1).$$

Proof. Throughout the entire proof of the theorem, b denotes some fixed point of A. For typographical convenience, we write $\tilde{G}_{xy}$ for $\tilde{G}(x,y)$, γ_x for $\gamma(x)$, etc.

We need the following

Lemma. *Let* γ *be a bounded complex function on A satisfying*

$$\sum_x (I-\tilde{Q})_{ax}\gamma_x = 0 \quad (a \neq b).$$

Then, for some constants K and C,

$$\gamma_x = K\tilde{G}_{xb} + C.$$

Proof of lemma. It suffices to consider the case when γ is real. Assume therefore that γ is real-valued and set $M = \sup|\gamma(x)|$. Put

$$\delta_x = M + \gamma_x \quad (\text{respectively } M - \gamma_x)$$

according as

$$K = ([I - \tilde{Q}]\gamma)_b \geqslant 0 \quad (\text{respectively } \leqslant 0).$$

Then δ is a bounded $\tilde{Q}$-excessive function on A:

$$2M \geqslant \delta \geqslant \tilde{Q}\delta \geqslant 0.$$

According to the classical Riesz decomposition theorem (T.27.1 of Spitzer [8]), δ may be decomposed into a harmonic part β and a potential part which is simply $|K| G_{xb}$. The expression

$$\beta(x) = \lim_{n \to \infty} [\tilde{Q}^n \delta](x)$$

shows that β is a bounded $\tilde{Q}$-harmonic function:

$$2M \geqslant \beta = \tilde{Q}\beta \geqslant 0.$$

Hence, by the theorem of Choquet and Deny (T.24.1 of Spitzer [8]), β is constant on A. The lemma follows.

Now let

$$\varphi(\theta) = \int_R e^{i\theta t} F(dt)$$

be the characteristic function of F. Then, because F has zero mean and unit variance,

$$|\varphi'(\theta)| = \left| \int it(e^{i\theta t} - 1) F(dt) \right| \leqslant |\theta| \int t^2 F(dt) = |\theta|.$$

Further $\varphi''(\theta)$ exists for all θ.

Since $|1 - \varphi(\theta)| < \theta^2/2$, we may, for $|\theta| < 1$, define

$$\psi(\theta) = \log \varphi(\theta) = \sum_{n \geqslant 1} [\varphi(\theta) - 1]^n/n.$$

It is important now that $\tilde{G}_{xb} \leqslant \tilde{G}_{bb}$ $(x \in A)$. Choose θ_0 with

$$0 < \theta_0 < 1/(2\tilde{G}_{bb}).$$

Then, for $|\theta| < \theta_0$ and $x \in A$,

$$\psi'(\theta \tilde{G}_{xb}) = \frac{\varphi'(\theta \tilde{G}_{xb})}{\varphi(\theta \tilde{G}_{xb})} < \tilde{G}_{xb}.$$

Since

$$\sum_x \tilde{G}_{ax} \tilde{G}_{xb} = G_{ab} < \infty,$$

the series

$$(11) \qquad \sum_{x} \tilde{G}_{ax} \psi'(\theta \tilde{G}_{xb})$$

converges uniformly if $|\theta| < \theta_0$.

Since

$$p_a = \sum \tilde{G}_{ax} \tilde{\varepsilon}_x,$$

we have

$$(12) \qquad \log \mathbf{E}\{\exp[i\alpha p_a + i\theta p_b]\} = \sum \psi(\alpha \tilde{G}_{ax} + \theta \tilde{G}_{xb})$$

for $|\alpha| < \theta_0$ and $|\theta| < \theta_0$. (Recall that $\tilde{G}$ is symmetric so that $\tilde{G}_{ax} = \tilde{G}_{xa}$.) The question of ambiguity of the logarithm on the left-hand side of equation (12) is irrelevant because we now differentiate equation (12) with respect to α and then set $\alpha = 0$:

$$(13) \qquad \frac{i\,\mathbf{E}[p_a \exp(i\theta p_b)]}{\mathbf{E}[\exp(i\theta p_b)]} = \sum_{x} \tilde{G}_{ax} \psi'(\theta \tilde{G}_{xb}).$$

Justification of these last steps is provided by the fact that the series obtained by differentiating equation (12) converges uniformly. Compare with the series (11).

By the harness property,

$$(14) \qquad \mathbf{E}[(p - Qp)_a \exp(i\theta p_b)] = 0 \quad (a \neq b).$$

Substitute equation (13) into equation (14) and use

$$(\tilde{G} - Q\tilde{G})_{ax} = (I - \tilde{Q})_{ax}$$

to find that

$$(15) \qquad \sum_{x} (I - \tilde{Q})_{ax} \psi'(\theta \tilde{G}_{xb}) = 0 \quad (a \neq b).$$

Note. There are no analytic difficulties in the step from equation (13) to equation (15) because, for fixed a, $Q_{ax} = 0$ except for finitely many x.

According to the lemma, we may deduce from equation (15) that, for $|\theta| < \theta_0$ and $x \in A$,

$$\psi'(\theta \tilde{G}_{xb}) = K_b(\theta)\, \tilde{G}_{xb} + C_b(\theta).$$

Let $|x| \to \infty$ and find that $C_b(\theta) = 0$ for $|\theta| < \theta_0$:

$$\psi'(\theta \tilde{G}_{xb}) = K_b(\theta)\, \tilde{G}_{xb}.$$

We now differentiate with respect to θ to get

$$\psi''(\theta \tilde{G}_{xb}) = K_b'(\theta)$$

and let $|x| \to \infty$ to find that $K_b'(\theta) = -1$. Hence, for θ near 0,

$$\psi''(\theta) = -1 \quad \text{and} \quad \psi(\theta) = -\tfrac{1}{2}\theta^2.$$

The rest is standard. The moments of F exist (Section 4.27 of Kendall and Stuart [6]), are the moments of $N(0, 1)$, so determine that $F = N(0, 1)$ (Example 4.7 of [6]).

Q-increment-harnesses

We have seen that no non-Q-harmonic $(Q, 1)$-harness exists if $d = 1$ or $d = 2$; the $(Q, 1)$-harness property is too restrictive. Q-increment-harnesses provide the appropriate generalization.

For $d = 1, 2, 3, ...$, introduce the *potential kernel* $A(.,.)$ corresponding to Q as in Spitzer [8]. Recall that

$$\text{if } d = 1, \quad \text{then } A(x, y) = |x - y|;$$

$$\text{if } d = 2, \quad \text{then } A(x, y) \sim \frac{2}{\pi} \log|x - y| \quad \text{as } |x - y| \to \infty;$$

$$\text{if } d = 3, \quad \text{then } A(x, y) = G(0, 0) - G(x, y).$$

When we speak of an *increment-process* $\{p(x) - p(y): x, y \in A\}$ we understand that a single 'observation' $p(x)$ need not be a random variable, i.e. need not be measurable on the underlying triple $(\Omega, \mathscr{F}, \text{pr})$. The increments $\{p(x) - p(y): x, y \in A\}$ are true (measurable) random variables.

Theorem B. *For $d = 1, 2, 3, ...$, there exists a zero-mean Gaussian increment-process*

$$p = \{p(x) - p(y): x, y \in A\}$$

specified in law by the equation

(16) $\text{var}[p(y) - p(x)] = 2A(x, y) \quad (x, y \in A).$

This process p is a (canonical) $(Q, 1)$-increment-harness in the sense that, for $b \in A$ and $x \neq b$,

$$\mathbf{E}[p(x) - p(b) \,|\, p(y) - p(b), y \neq b] = \sum_y Q(x, y)[p(y) - p(b)].$$

A more general result is given later. For $d = 1$ or $d = 2$, it is impossible to make the individual $p(x)$ $(x \in A)$ random variables. For $d \geqslant 3$, the increments $\{p(x) - p(y): x, y \in A\}$ of the canonical Q-harness $\{p(x): x \in A\}$ form a canonical Q-increment-harness. (Incidentally, this explains why the factor 2 is included in equation (16).)

4.4.3 SOME CONTINUOUS-PARAMETER ANALOGUES

The main point of this extremely brief section, which makes no pretence at rigour, is to clarify our earlier remarks on the 'Cauchy' character of $\tilde{Q}$. Wong does the 'generalized process' stuff rigorously.

Throughout this section, $d \geqslant 3$. Brownian motion in R^d has $\frac{1}{2}\Delta$, where Δ is Laplace's operator, as its infinitesimal generator. It is clear that $(-\frac{1}{2}\Delta)$ is the continuous parameter analogue of $I - Q$.

So (see equation (9)) let us introduce a zero-mean Gaussian generalized process $\varepsilon = \{\varepsilon(x) : x \in R^d\}$ with covariance operator $(-\frac{1}{2}\Delta)$. Thus ε associates with each 'well-behaved' (Schwartz) function α on R^d a real Gaussian variable

$$\int_{R^d} \alpha(x)\, \varepsilon(x)\, dx.$$

For two Schwartz functions α and β,

(9c) $\quad \mathbf{E}\left\{\left[\int_{R^d}\alpha(x)\,\varepsilon(x)\,dx\right]\left[\int_{R^d}\beta(y)\,\varepsilon(y)\,dy\right]\right\} = -\frac{1}{2}\int_{R^d}(\alpha\Delta\beta)(z)\,dz.$

Note. Equation (9c) denotes the continuous parameter analogue of equation (9), and so on.

It is less cumbersome to write equation (9c) as

$$\mathbf{E}\left\{\left[\int\alpha\varepsilon\right]\left[\int\beta\varepsilon\right]\right\} = -\frac{1}{2}\int\alpha\Delta\beta.$$

The covariance (symmetric, positive definite) character of $(-\frac{1}{2}\Delta)$ is brought out by using the alternative expression:

$$-\frac{1}{2}\int\alpha\Delta\beta = \frac{1}{2}\int\operatorname{grad}\alpha\,.\,\operatorname{grad}\beta.$$

Now put

(10c) $\qquad p(x) = \int G(x,y)\,\varepsilon(y)\,dy,$

where G is the Green function of the operator $A = \frac{1}{2}\Delta$:

(4c) $\qquad G(x,y) = \dfrac{C_3(d)}{|x-y|^{d-2}}.$

Then $\{p(x) : x \in R^d\}$ is a zero-mean Gaussian generalized process with

(2c) $\qquad \mathbf{E}\left[\int\alpha p\right]\left[\int\beta p\right] = \int\int G(x,y)\,\alpha(x)\,\beta(y)\,dx\,dy.$

Thus, p is Whittle's process.

Let $\tilde{A}$ be the operator $(-\frac{1}{2}\Delta)^{\frac{1}{2}}$. It is known that $\tilde{A}$ is the generator of an isotropic Cauchy process with Green function

$$(7c) \qquad\qquad \tilde{G}(x,y) = \frac{C_d(d)}{|x-y|^{d-1}}.$$

See Section 7.21 of Ito and McKean [4]. If we define

$$(6c) \qquad\qquad \tilde{\varepsilon} = \tilde{A}p$$

we find that $\tilde{\varepsilon}$ is pure white noise with Dirac delta covariance function. Then

$$(5c) \qquad\qquad p = \int \tilde{G}(x,y)\,\tilde{\varepsilon}(y)\,dy$$

expresses p as a Riesz potential.

4.4.4 RIESZ DECOMPOSITION OF THE GENERAL Q-HARNESS

Let Q be the substochastic one-step transition matrix of a Markov chain $X = \{X(n): n = 0, 1, 2, \ldots\}$, perhaps of finite lifetime, with countable state-space A. We assume that, *for each x in A, $Q(x,x) = 0$ and $Q(x,y) = 0$ for all but finitely many y*. We also assume that Q is *irreducible* in that its Green function G is strictly positive on $A \times A$:

$$0 < G = I + Q + Q^2 + \ldots \leqslant \infty,$$

I denoting the identity matrix on A. Irreducibility implies that either $G < \infty$ on $A \times A$, in which case Q is called *transient*, or $G = \infty$ on $A \times A$ and Q is *recurrent*. We call Q *symmetrizible* if there is a non-zero measure m on A such that

$$(17) \qquad\qquad m(x)\,Q(x,y) = m(y)\,Q(y,x) \quad (x, y \in A).$$

Then $m(x) > 0$ $(x \in A)$ and m is unique save for a positive constant factor. See Kendall [5] and self-dual processes in Chapter VI of Blumenthal and Getoor [1] for the significance of equation (17).

Conventions and notation

(i) The symbol D will denote a subset of A with *finite* complement $D^c = A\backslash D$. The symbol B will denote a proper non-empty subset of A such that *either B or B^c is finite*.

(*ii*) The symbol P^x $(x \in A)$ denotes the law of X when $X(0) = x$. For x in A and b in B, $H_B^+(x,b)$ denotes the following *entrance* probability:

$$H_B^+(x,b) = I(x,b) \quad \text{if } x \in B,$$

$$= \sum_{n=1}^{\infty} P^x[X(n) = b; \; X(m) \notin B, 1 \leqslant m < n] \quad \text{if } x \in B^c.$$

For $y \in B^c$, $H_B^+(x,y) = 0$ for every x.

(*iii*) Write Q_B for the functional restriction of Q to $B^c \times B^c$ and write G_B for the associated Green function.

(*iv*) Suppose that $f = \{f(x): x \in A\}$ is a family of random variables defined on some triple $(\Omega, \mathscr{F}, \mathrm{pr})$. For $\alpha = 1$ or 2, write $f \in \mathscr{L}^\alpha$ if, for each x in A, $f(x) \in \mathscr{L}^\alpha(\Omega, \mathscr{F}, \mathrm{pr})$.

(*v*) $\mathbf{E}_1$ (or $\mathbf{E}$) is used to denote strict sense, and $\mathbf{E}_2$ (or $\hat{\mathbf{E}}$) *wide* sense, conditional expectation. See Sections II.3 and IV.3 of Doob [2]. In particular,

$$(18) \qquad\qquad \mathbf{E}_2[f(x)|f(y), \, y \neq x]$$

denotes that linear combination of the $f(y)$ $(y \neq x)$ which best approximates $f(x)$ in the sense of least-squares approximation. A better explanation is that the conditional expectation at (18) is the orthogonal projection of $f(x)$ onto the closed linear subspace of $\mathscr{L}^2(\Omega, \mathscr{F}, \mathrm{pr})$ spanned by

$$\{f(y): \, y \neq x\}.$$

(*vi*) Two random variables which agree almost everywhere are normally identified.

Definition 1. For $\alpha = 1$ or 2, f is called a (Q, α)-*harness* if $f \in \mathscr{L}^\alpha$ and if, for each x in A,

$$\mathbf{E}_\alpha[f(x)|f(y), y \neq x] = [Qf](x).$$

Definition 2. f is called Q-*harmonic* on A if $f = Qf$ on A.

Definition 3. For $\alpha = 1$ or 2, a *potential* (Q, α)-harness is a (Q, α)-harness p such that, for $x \in A$,

$$\| [H_D^+ p](x) \|_\alpha \to 0$$

as $D \downarrow \varnothing$ (convergence of directed sets).

Notes. (*i*) $\| . \|_\alpha$ is the usual (pseudo-) norm in $\mathscr{L}^\alpha(\Omega, \mathscr{F}, \mathrm{pr})$.

(*ii*) See Definition IV.5.1 of Blumenthal and Getoor [1] for our use of 'potential'.

(*iii*) We use directed sets for a 'coordinate-free' description.

(*iv*) If f is a $(Q, 1)$-harness and $f \in \mathcal{L}^2$, then f is a $(Q, 2)$-harness. If a non-Q-harmonic $(Q, 2)$-harness exists, then there exists a non-Q-harmonic zero-mean Gaussian process which is both a $(Q, 1)$-harness and a $(Q, 2)$-harness.

Theorem 1. *Let* α *be either* 1 *or* 2. *Then every* (Q, α)-*harness has a unique Riesz decomposition*

$$f(x) = h(x) + p(x) \quad (h, p \in \mathcal{L}^\alpha; \ x \in A)$$

where h *is* Q-*harmonic and where* p *is a potential* (Q, α)-*harness. Further, for* $x \in A$,

(19) $\mathbf{E}_\alpha[p(x) | p \text{ on } B, h \text{ on } A] = [H_B^+ p](x),$

(20) $\mathbf{E}_\alpha[p(x) | h \text{ on } A] = 0.$

Corollary 1. *If* Q *is recurrent, then every* $(Q, 1)$-*harness is* Q-*harmonic.*

Corollary 2. *A non-Q-harmonic* $(Q, 2)$-*harness exists if and only if* Q *is both transient and symmetrizible. Suppose that* Q *is both transient and symmetrizible. Then* p *is a potential* $(Q, 2)$-*harness if and only if*

$$\mathbf{E}[p(x) p(y)] = K G(x, y)/m(y) \quad (x, y \in A)$$

for some non-negative constant K. *Further, the* Q-*harmonic part* h *of an arbitrary* $(Q, 2)$-*harness* f *may be identified as follows:*

$$h(x) = \text{l.i.m.} \, [Q^n f](x) \quad (x \in A),$$

l.i.m. denoting limit in quadratic mean.

Problem. *Find a necessary and sufficient condition on* Q *for the existence of a non-Q-harmonic* $(Q, 1)$-*harness.*

Proof of Theorem 1. We consider only the harder case when $\alpha = 1$, so let f be a $(Q, 1)$-harness. Take a set D (of finite complement D^c). Then, for each x,

(21) $\mathbf{E}[f(x) | f \text{ on } D] = [H_D^+ f](x) \quad (x \in A).$

To prove equation (21), note that the $(Q, 1)$-harness property of f implies that the left-hand side of equation (21) defines a function of x which is Q-harmonic on D^c with 'boundary' values f on D. That the right-hand side of equation (21) is the unique solution of this Dirichlet problem is well known—and trivial in this setting.

Let $\sigma\{f(d) : d \in D\}$ denote the smallest σ-algebra of elements of $\mathcal{F}$ relative to which each $f(d)$ $(d \in D)$ is measurable. Introduce the σ-algebra

$\mathscr{T}$ of events which depend on the *infinitely remote* behaviour of f:

$$\mathscr{T} = \bigcap_D \sigma\{f(d)\colon d\in D\},$$

the intersection being taken over all sets D of finite complement. For any particular sequence $\{D(n)\colon n=1,2,3,\ldots\}$ of subsets of A of finite complement with $D(1)\supseteq D(2)\supseteq \ldots$ and $\bigcap D(n)=\varnothing$, it is clear that

$$\mathscr{T} = \bigcap_n \sigma\{f(d)\colon d\in D(n)\}.$$

From equation (21), it follows that the sequence

$$\{H^+_{D(n)}f\colon n=1,2,3,\ldots\}$$

is a martingale. By Lévy's Martingale Convergence Theorem (Theorem VII.4.2 of Doob [2]),

$$\lim_{n\to\infty} H^+_{D(n)}f](x) = \mathbf{E}[f(x)|\mathscr{T}]$$

the limit existing with probability 1, but, more importantly, also *in the mean of order* 1. It is immediate that

$$(22) \qquad \mathbf{E}[f(x)|\mathscr{T}] = \lim_{D\downarrow\varnothing}[H^+_D f](x)$$

in $\mathscr{L}^1$ norm, but it does *not* follow that we get directed-set convergence with probability 1. Put

$$(23) \qquad h(x)=\mathbf{E}[f(x)|\mathscr{T}], \quad p(x)=f(x)-h(x).$$

Because $H^+_D f$ is Q-harmonic on D^c, it is obvious from equation (22) that h is Q-harmonic on A.

When B^c is finite the proof of equation (19) is trivial. So *suppose that B is finite*. Equation (21) implies that, if B and D are disjoint, then for every x,

$$(24) \quad \mathbf{E}[(f-H^+_D f)(x)|f \text{ on } B, f \text{ on } D] = \sum_{b\in B} H^+_{B\cup D}(x,b)[f-H^+_D f](b).$$

We have used here the probabilistically obvious identity:

$$H^+_D(x,d) = H^+_{B\cup D}(x,d) + \sum_{b\in B} H^+_{B\cup D}(x,b)H^+_D(b,d)$$

for $x\in A$, $b\in B$, $d\in D$. Consider what happens when $D\downarrow\varnothing$ in equation (24).

The left-hand side of equation (24) splits as

$$(25) \quad \mathbf{E}[(f-H^+_D f-p)(x)|f \text{ on } B, f \text{ on } D]+\mathbf{E}[p(x)|f \text{ on } B, f \text{ on } D].$$

The $\mathscr{L}^1$ norm of the first term at the expression (25) is dominated by

$$(26) \qquad \|(f-H^+_D f-p)(x)\|_1$$

because of the contraction mapping property of conditional expectations. By Lévy's Martingale Convergence Theorem, the second term at the expression (25) converges in $\mathscr{L}^1$ norm to

$$\mathbf{E}[p(x)\,|\,f \text{ on } B, \mathscr{T}].$$

The right-hand side of equation (24) is a finite sum. In $\mathscr{L}^1$ norm,

$$[f - H_D^+ f](b) \to p(b) \quad (b \in B).$$

Further,

$$H_{B\cup D}^+(x, b) \to H_B^+(x, b) \quad (b \in B)$$

as $D \downarrow \varnothing$ because

$$P^x[X \text{ enters } D \text{ before } B] \downarrow 0.$$

Thus, on letting $D \downarrow \varnothing$ in equation (24), we find that

$$\mathbf{E}[p(x)\,|\,f \text{ on } B, \mathscr{T}] = [H_B^+ p](x).$$

Equation (19) follows and implies that p is a $(Q, 1)$-harness. That p is a potential $(Q, 1)$-harness follows from equation (22) and the fact that $H_D^+ h = h$ on D^c. Finally, equation (20) is implied by equation (23).

Proof of Corollary 1. If Q is recurrent, then, for $x \neq y$,

$$H_y^+(x, y) = H_x^+(y, x) = 1.$$

Hence

$$\mathbf{E}[p(x)\,|\,p(y)] = p(y), \quad \mathbf{E}[p(y)\,|\,p(x)] = p(x).$$

It follows (Section VII.3 of Doob [2]) that

$$p(x) = p(y) \quad (x, y \in A)$$

and that $p = 0$ from the potential property.

Proofs of Theorem 1 for $\alpha = 2$ and of Corollary 2 are simple exercises in Hilbert space theory.

Markov property of canonical Q-harness

Let Q be both transient and symmetrizable relative to some measure m. We assume that some fixed normalization of m is chosen. Then a Gaussian process p with

$$\mathbf{E}[p(x)] = 0, \quad \mathbf{E}[p(x)p(y)] = G(x, y)/m(y),$$

for x and y in A, is simultaneously a $(Q, 1)$ and a $(Q, 2)$-harness. Such a Gaussian process is called a *canonical Q-harness*.

Theorem 2. *Let p be a canonical Q-harness associated with a transient symmetrizable substochastic matrix Q. Let B be a subset of A such that either B or B^c is finite. Then, conditionally on a knowledge of $\{p(b): b \in B\}$,* *the stochastic process $\{p(x): x \in B^c\}$ is a Gaussian Q_B-harness with Riesz decomposition*

$$p(x) = \sum_B H_B^+(x, b) p(b) + p_B(x) \quad (x \in B^c),$$

where $\{p_B(x): x \in B^c\}$ is a canonical Q_B-harness independent of $\{p(b): b \in B\}$.

The reader will be able to prove this quickly with the aid of the identities of Section 10 of Spitzer [8].

In conclusion, we state a generalization of Theorem B. It is curious that recurrent Markov chains conspire to allow its truth.

Theorem 3. *If Q is recurrent, there exists a zero-mean Gaussian increment-process $p = \{p(x) - p(y): x, y \in A\}$ completely specified in law by the equation*

$$\mathrm{var}\,[p(x) - p(y)] = G_x(y, y)/m(y) \quad (x, y \in A),$$

where m is a positive invariant measure for Q. If Q is symmetrizible (necessarily relative to m), then p is a Q-increment-harness: for $x, z \in A$ and $x \neq z$,

$$\mathbf{E}_\alpha[p(x) - p(z) | p(y) - p(z), y \neq x] = \sum_y Q(x, y)\,[p(y) - p(z)],$$

where α is either 1 or 2.

REFERENCES

1. R. M. Blumenthal and R. K. Getoor, *Markov Processes and Potential Theory*, Academic Press, New York and London (1968).
2. J. L. Doob, *Stochastic Processes*, J. Wiley and Sons, New York (1953).
3. J. M. Hammersley, "Harnesses", *Proc. Fifth Berkeley Symp. Math. Statist. and Prob.*, vol. III, pp. 89–117, University of California Press (1966).
4. K. Ito and H. P. McKean, Jr., *Diffusion Processes and their Sample Paths*, Springer-Verlag, Berlin, Heidelberg and New York (1965).
5. D. G. Kendall, "Unitary dilations of one-parameter semigroups of Markov transition operators, and the corresponding integral representations for Markov processes with a countable infinity of states", *Proc. Lond. math. Soc.* (3), **9** (1959), 417–431.
6. M. G. Kendall and A. Stuart, *The Advanced Theory of Statistics*, 2nd ed., vol. 1, Griffin, London (1963).
7. H. P. McKean, Jr., "Brownian motion with a several-dimensional time", *Teor. Veroyatnost.* **4** (4) (1963), 357–378.
8. F. Spitzer, *Principles of Random Walk*, Van Nostrand, Princeton (1964).
9. P. Whittle, "Topographic correlation, power-law covariance functions, and diffusion", *Biometrika* **49** (1962), 305–14.
10. E. Wong, "Homogeneous random fields", *Ann. Math. Statist.* **40** (1969), 1625–1634.

5

SURVEYS OF SPECIAL TOPICS

5.1

Determination of Confounding

R. DAVIDSON

5.1.1 THE PROBLEM

We have a general complete factorial experiment: let there be n factors $F_1, ..., F_n$, where F_j is at p_j levels ($1 \leqslant j \leqslant n$).

We may be presented with an arrangement of the experiment in blocks, and have to find out what interactions are confounded; or we may have to produce an experimental design with blocks not exceeding some size, when we usually wish to confound interactions only of high degree. Again it will be necessary to find out, for any system of blocking, what is confounded. There is here rather a gap between the general theory (see Mann [2]) and the practical applications (see Yates [3] and Kempthorne [1]). Our aim here is to make concrete the notion of an interaction, and then to give a general method of seeing whether it is confounded.

5.1.2 INTERACTIONS AND VECTOR SPACES

We shall let $\mathbf{r}$ be a general subset of $\mathbf{n} = \{1, ..., n\}$. Let $|\mathbf{r}|$ be the number of elements of $\mathbf{r}$, and let $p(\mathbf{r}) = \prod \{p_j : j \in \mathbf{r}\}$—the total number of combinations of levels of the factors F_j as j runs through $\mathbf{r}$. A typical combination of levels will be written $\mathbf{i} = \{i_1, ..., i_n\}$, i_j being the level of F_j. The theoretical mean yield of the plot (experimental unit) bearing the combination of levels $\mathbf{i}$ will be written $y_\mathbf{i}$, and the actual yield $Y_\mathbf{i}$.

A typical experimental treatment comparison is then

$$\sum_\mathbf{i} \xi_\mathbf{i} Y_\mathbf{i}.$$

Without doing the experiment, we may consider the treatment comparison as a (row) vector with $p(\mathbf{n})$ entries, $\boldsymbol{\xi} = \{\xi_\mathbf{i}\}$; which, when multiplied by the (column) vector $\mathbf{Y} = \{Y_\mathbf{i}\}$ of plot yields, gives us the experimental

367

treatment comparison. The ξ all of whose components are 1 will be known as the mean vector. (Strictly speaking, to obtain an estimate of the grand mean yield, we ought to normalize this ξ by dividing all its entries by $p(\mathbf{n})$; but for our purposes the normalization is immaterial.)

The whole space of treatment comparisons is, of course, a vector space, which we call $E_\mathbf{n}$. The $\mathbf{r}$-effect subspace $E_\mathbf{r}$ is the space of those ξ whose entries ξ_i, $\xi_{i'}$ are equal if $i_j = i'_j$ for all $j \in \mathbf{r}$. $E_\mathbf{r}$ is the space of ξ expressing effects due to combinations of F_j indexed by j in $\mathbf{r}$, averaged over the F_j indexed by those j not in $\mathbf{r}$. If $|\mathbf{r}| = 1$, so that there is only one factor—F_k, say—involved, $E_\mathbf{r}$ is called a main-effect subspace. It then has dimension p_k and consists of the mean vector, together with those vectors ξ such that $\xi_i = \xi_{i'}$ for any $\mathbf{i}$, $\mathbf{i}'$ such that $i_k = i'_k$. Thus two components of this ξ are equal if they are taken at the same level of F_k.

We shall from now on distinguish one level of each factor as the zero level.

Proposition 1. (*i*) *If* $\mathbf{r}'$ *is a subset of* $\mathbf{r}$, *then* $E_{\mathbf{r}'}$ *is a subspace of* $E_\mathbf{r}$. (*ii*) $\dim E_\mathbf{r} = p(\mathbf{r})$.

Proof. Any ξ in $E_{\mathbf{r}'}$ satisfies the condition for a ξ in $E_\mathbf{r}$, so (*i*) is proved. As for (*ii*), the conditions on a ξ in $E_\mathbf{r}$ mean that we know ξ when we are given ξ_i for each $\mathbf{i} = \{i_1, ..., i_n\}$ such that $i_j = 0$ for j not in $\mathbf{r}$. On the other hand, the choice of these ξ_i is completely free, and there are $p(\mathbf{r})$ of them, whence the proposition follows.

The interaction $I_\mathbf{r}$ of $\{F_j : j \in \mathbf{r}\}$ is the subspace of vectors in $E_\mathbf{r}$ which are orthogonal to all of its subspaces $E_{\mathbf{r}'}$ (where $\mathbf{r}'$ is a proper subset of $\mathbf{r}$). That is, $I_\mathbf{r}$ is the set of comparisons ξ between the factors indexed by $\mathbf{r}$ that do not contain any information about comparisons between some smaller set of factors.

5.1.3   GENERATION OF $E_\mathbf{r}$ AND $I_\mathbf{r}$

Our aim here is to give useful bases for $E_\mathbf{r}$ and $I_\mathbf{r}$; we shall model our approach on the 2^n-experiment, where we are in no doubt as to the concrete meaning of an interaction. In a 2^n experiment, the (n-factor) interaction is just a single vector ξ, which has entries $+1$ for each combination of levels an even number of which are the zero level, and -1 for all other entries. We shall reduce the general case to a collection of 2^n experiments.

We define $G_\mathbf{r}$ as the subspace of $E_\mathbf{n}$ generated by

(*a*) the mean vector, and

(b) all vectors of the following form. Let $\mathbf{r}'$ be any (arbitrary but not empty) set of indices contained in $\mathbf{r}$. Take the factors indexed by $\mathbf{r}'$ at two levels each, one of which must be the zero level and the other of which is arbitrary. This defines a $2^{|\mathbf{r}'|}$ experiment. Our vector has entries

+1, for each combination of levels in the whole experiment, whose levels of factors indexed by $\mathbf{r}'$ are those of our $2^{|\mathbf{r}'|}$ experiment and which has an even number of these ($\mathbf{r}'$-factors) at the zero level;

−1, in similar circumstances save that for even we read odd;

0, otherwise.

(Here we are taking any $\mathbf{r}'$ included in $\mathbf{r}$, and, for each factor F_j ($j \in \mathbf{r}'$), selecting the zero level and one other level. We then discard all combinations of levels in the big experiment whose $\mathbf{r}'$-levels are not those selected. So we have a $2^{|\mathbf{r}'|}$ experiment, replicated ($p(\mathbf{n} \backslash \mathbf{r}')$ times) over all the factors F_j with j not in $\mathbf{r}'$. We then evaluate the $|\mathbf{r}'|$-factor interaction of this $2^{|\mathbf{r}'|}$ experiment.)

Proposition 2. (i) $G_{\mathbf{r}}$ is a subspace of $E_{\mathbf{r}}$.

(ii) *All type (b) vectors are orthogonal to the mean vector.*

(iii) *Any two type (b) vectors belonging to different $\mathbf{r}'$'s are orthogonal.*

(iv) *The type (b) vectors belonging to any one $\mathbf{r}'$ are linearly independent.*

(v) $\dim G_{\mathbf{r}} = p(\mathbf{r})$.

(vi) $G_{\mathbf{r}} = E_{\mathbf{r}}$.

Proof. Ad (vi): this follows at once from (v), (i) and Proposition 1.

Ad (v): this is a consequence of (ii), (iii) and (iv). For taken together, these say that the mean vector and type (b) vectors of $G_{\mathbf{r}}$ are linearly independent. So we must find out how many type (b) vectors there are. We have

$$1, \qquad \text{mean vector,}$$
$$\sum_{j \in \mathbf{r}} (p_j - 1), \qquad \text{type (b) vectors with } |\mathbf{r}'| = 1,$$
$$\sum_{j \neq j' \in \mathbf{r}} (p_j - 1)(p_{j'} - 1), \quad \text{type (b) vectors with } |\mathbf{r}'| = 2, \quad \text{and so on up to}$$
$$\prod_{j \in \mathbf{r}} (p_j - 1), \qquad \text{type (b) vectors with } \mathbf{r}' = \mathbf{r}.$$

But the sum of all these is just the product

$$\prod_{j \in \mathbf{r}} \{(p_j - 1) + 1\} = p(\mathbf{r}).$$

Ad (i): it is clear that every type (b) vector is in $E_{\mathbf{r}}$, because it is in $E_{\mathbf{r}'}$, which by Proposition 1 is a subspace of $E_{\mathbf{r}}$.

Ad (*ii*): in the $2^{|\mathbf{r}|}$ experiment, the numbers of combinations of treatments with even, and with odd, numbers of their levels the zero level, are equal. It follows then from the definition of a type (*b*) vector that it has equal numbers of $+1$s and -1s among its entries, so it is indeed orthogonal to the mean vector (their scalar product vanishing).

Ad (*iv*): we are considering the $2^{|\mathbf{r}|}$ experiments contained in the complete factorial experiment with factors $F_j(j \in \mathbf{r}')$ at p_j levels ($j \in \mathbf{r}'$); averaged, of course, over all the non-$\mathbf{r}'$ factors. For each of the $2^{|\mathbf{r}|}$ experiments, we consider that entry of the associated type (*b*) vector (treatment comparison of the big experiment) corresponding to the combination of $\mathbf{r}'$-levels none of which is the zero level. This combination of levels assigns a non-zero entry to our type (*b*) vector, and, because it does not appear in any of the other $2^{|\mathbf{r}|}$ experiments, it assigns zero to the corresponding entry in each of their type (*b*) vectors. Linear independence follows.

Ad (*iii*): let the two $\mathbf{r}'$'s referred to be $\mathbf{r}_1$, $\mathbf{r}_2$. They are different, so there exists at least one factor indexed by one and not the other. Let it be F_1, in $\mathbf{r}_1$, and let its levels be $0, 1, ..., L$; with 0 and 1 being the levels selected for our type (*b*) vector belonging to $\mathbf{r}_1$. Then if we consider the type (*b*) vector $\boldsymbol{\eta}$ belonging to $\mathbf{r}_2$, we have that

(1) $$\eta_{\mathbf{i}} = \eta_{\mathbf{i}'}, \quad \text{provided } i_j = i'_j \text{ for all } j \in \mathbf{r}_2.$$

Let the type (*b*) vector belonging to $\mathbf{r}_1$ be $\boldsymbol{\xi}$. Then $\xi_{\mathbf{i}} = 0$ unless $i_1 = 0$ or 1. Further, for all $i_2, ..., i_n$ we have $\xi_{0,i_2,...,i_n} = -\xi_{1,i_2,...,i_n}$, because of the definition of a type (*b*) vector and the change in the parity of the number of factors at the zero level. On the other hand, we have

$$\eta_{0,i_2,...,i_n} = +\eta_{1,i_2,...,i_n},$$

by equation (1), since F_1 is not one of the factors indexed by $\mathbf{r}_2$. Thus, taking the scalar product of $\boldsymbol{\xi}$ and $\boldsymbol{\eta}$, we do indeed get zero.

We have thus produced a concrete basis for $E_\mathbf{r}$. It is now our business to do the same for the interaction subspace $I_\mathbf{r}$, the subspace of vectors in $E_\mathbf{r}$ orthogonal to those in each $E_{\mathbf{r}'}$, $\mathbf{r}'$ strictly included in $\mathbf{r}$.

Proposition 3. (*i*) $I_\mathbf{r}$ *is generated by the type* (*b*) *vectors of* $E_\mathbf{r}$ *which have* $\mathbf{r}' = \mathbf{r}$. (*These will be called* $i_\mathbf{r}$-*vectors.*)

(*ii*) *If* $\mathbf{r} \neq \mathbf{r}'$, $I_\mathbf{r}$ *and* $I_{\mathbf{r}'}$ *are orthogonal.*

(*iii*) $\dim I_\mathbf{r} = \prod_{j \in \mathbf{r}}(p_j - 1)$, *the number of degrees of freedom of the interaction.*

Proof. (*i*) and (*ii*) are immediate from the orthogonal decomposition of the space E_r implicit in (*ii*) and (*iii*) of Proposition 2. (*iii*) follows by enumeration as in the proof of (*v*) of Proposition 2.

So any statement about I_r is equivalent to a statement about the i_r-vectors.

5.1.4 CONFOUNDING

We now have some division of the experimental material (plots) into blocks. Each block makes an additive contribution to the yields of the plots in it (the same contribution for every plot in the block). If b_i is the block contribution received by the plot assigned the combination of levels **i**, then when we work out our experimental treatment comparison

$$\sum_i \xi_i Y_i,$$

it will be distorted by an amount

$$\sum_i \xi_i b_i.$$

Treatment comparisons ξ such that this distortion is always zero, no matter what the values of the b_i (provided they are constant over blocks), are said to be unconfounded (with blocks). A subspace of treatment comparisons is said to be unconfounded when all its vectors are so. By convention we have

$$\sum_i b_i = 0,$$

so the mean vector is always unconfounded.

Proposition 4. *The interaction* I_r *is unconfounded with blocks, if and only if for each block B and* i_r-*vector* ξ, *the following condition holds:*

(2) *the algebraic sum, over the plots in the block B, of the entries* ξ_i *in* ξ
    corresponding to the combinations of levels **i** *given to those plots, is*
    zero.

Proof. Because of the linearity of the condition of non-confounding, one need only verify it for a basis of I_r, which the i_r-vectors form. Let then ξ be a typical i_r-vector: if condition (2) holds, then certainly

$$\sum_i \xi_i b_i = 0$$

for any choice of block constants b_i. Whereas if the latter is true, we have only to choose $b_i = 1 \ (i \in B), b_i = -k \ (i \notin B)$, so that

$$\sum_i b_i = 0.$$

Then

$$\sum_i \xi_i b_i = 0$$

implies, since

$$\sum_i \xi_i = 0$$

((*ii*) of Proposition 2), that

$$\sum_{i \in B} \xi_i = 0,$$

which is condition (2).

It remains now only to translate this proposition into an effective method of computing whether or no the interaction I_r is confounded. Without loss we may suppose the factors involved in I_r to be $F_1, ..., F_r$.

To decide then whether I_r is confounded, we write down the experimental design; the typical plot, receiving $F_1, ..., F_n$ at levels $i_1, ..., i_n$, being written $\mathbf{i} = (i_1, ..., i_r, i_{r+1}, ..., i_n)$. Then we go through the whole design deleting the (for $\mathbf{r}$) redundant 'tail' $(i_{r+1}, ..., i_n)$ of each $\mathbf{i}$, to get $\mathbf{i_r} = (i_1, ..., i_r)$.

For each block B and each choice $\mathbf{l}$ of alternative levels $l_1, ..., l_r$ of $F_1, ..., F_r$, we go through the block discarding those $\mathbf{i_r}$ that do not have, for every $j \ (1 \leqslant j \leqslant r), i_j = l_j$ or 0. Then we find the number of non-discarded $\mathbf{i_r}$s in the block that have an even number of their component i_j zero, and subtract from this the number of similar $\mathbf{i_r}$s that have an odd number of their component i_j zero. Let the result be $N(B, \mathbf{l})$; then I_r is unconfounded if and only if $N(B, \mathbf{l}) = 0$ for every B and $\mathbf{l}$.

It is easily shown, by reference to the definition of a type (*b*) vector, that this criterion is the same as that of Proposition 4. The value of this formulation is that it lends itself readily to practical use.

5.1.5 NOTES AND COROLLARIES

(*a*) In our formulation, the sizes of the blocks do not matter; they need not, indeed, be all the same.

(*b*) It is not necessarily the case—even when the blocks are all of the same size—that $N(B, \mathbf{l}) = 0$ implies $N(B', \mathbf{l}) = 0$: that is, it really is necessary to examine all the blocks. (We have a counter-example, in a $3 \times 2 \times 2 \times 2$ design in blocks of 6 with main effects unconfounded.)

(*c*) If we are doing the computations by hand, it is convenient to write the blocks down as vertical columns (of horizontal treatment combination vectors), and use strips of paper to cover up the factors not of current interest and the plots with currently non-selected alternative levels: the procedure at the end of the last section may then be carried out at a glance. The procedure, which is an enumeration, may be done on an electronic computer; but the only programme I have written—which will handle experiments up to 8^8—is rather slow.

(*d*) **Proposition 5.** *If the experiment is arranged in blocks of size s, then the main effect of the factor F_j, at p_j levels, is confounded unless p_j divides s.*

Proof. We carry out the procedure at the end of the last section for the main effect subspace I_j associated with F_j. It is then clear that F_j is unconfounded if and only if, in each block, the number of plots receiving F_j at the zero level is the same as the number receiving it at any other given level. This is impossible unless p_j divides s.

REFERENCES

1. O. Kempthorne, *Design and Analysis of Experiments*, Wiley, New York (1952).
2. H. B. Mann, *Analysis and Design of Experiments*, Dover, New York (1949).
3. F. Yates, *The Design and Analysis of Factorial Experiments*, Imperial Bureau of Soil Science Technical Communication No. 35 (1937).

5.2

On the Numerical Range of an Operator

B. BOLLOBÁS

5.2.1 INTRODUCTION

In 1970 Rollo Davidson became more and more interested in the theory of numerical ranges. He found it rather astonishing that the constants of the power inequality (see Theorem 9), $(e/n)^n n!$, cropped up in his calculations as well. Naturally this can be coincidence since the power inequality is a simple consequence of Cauchy's inequality.

I think Davidson believed that the methods of the theory of numerical ranges will be useful in tackling the problems of the (Γ, γ) and (M, m) diagrams. I am writing this brief survey in the hope that he will be proved right.

The numerical range of an operator on a Hilbert space was defined by Toeplitz [23] in 1918 in connection with a theorem of Fejér and quadratic forms. A number of interesting properties of this numerical range were proved by Hausdorff [16] and Stone [22]. However, the numerical range of an operator was never of great interest in Hilbert space theory.

The birth of the more important general concept is much more recent and is due to Lumer [17] and Bauer [1]. In 1962 Lumer [17] defined the so-called semi-inner-product on a linear space and with the aid of these semi-inner-products extended the definition of numerical range used for Hilbert spaces. Bauer [1] introduced the concept of numerical range for finite dimensional spaces but his definition is readily extendable to arbitrary normed spaces.

The numerical range of an operator on a normed space X is a bounded subset of the complex plane. It depends on the shape of the unit ball with respect to the action of the operator. Its closure contains the spectrum, which depends only on the algebraic structure of $\mathcal{B}(X)$, the normed algebra of continuous linear operators on X. The numerical range is easier to handle than the spectrum and often is more informative.

374

Among several other important results the theory of numerical ranges leads to the metric characterization of B^*-algebras (the so-called Vidav–Palmer theorem), given by Vidav, [24], Berkson [2], Glickfeld [14] and Palmer [19], and to a simple and natural proof of the classical Gelfand–Naimark theorem (see [15]), stating that a B^*-algebra is an abstract C^*-algebra.

The aim of this note is to give a concise account of those results and simple methods in the theory of numerical ranges which are likely to be useful in probability theory as well. An excellent and more comprehensive account of the theory of numerical ranges can be found in the recent book by Bonsall and Duncan [10]. However, the results of Sections 5.2.2 and 5.2.3 will be presented in a considerably simpler way than in [10], and most of the results of the last two paragraphs have been proved after [10] had been written.†

To avoid inessential inconveniences we shall always work with complex Banach spaces and algebras instead of normed spaces and algebras. We write **C** for the complex plane and **R** for the real line.

5.2.2 THE SPATIAL NUMERICAL RANGE

Let X be a complex Banach space with dual X'. Let $X'_\mathbf{R}$ be the space of continuous real functionals on X. Denote by $S(X)$ the unit sphere of X and by $\mathscr{B}(X)$ the algebra of continuous linear operators on X. The same notation, $\|\cdot\|$, will be used for the norms in X, X' and $\mathscr{B}(X)$. If $T\in\mathscr{B}(X)$, denote by $T^*\ (\in\mathscr{B}(X'))$ the adjoint of T. For $x\in X$ put

$$D(X,x) = \{f: f\in X', \|f\| = f(x) = 1\}.$$

Naturally $D(X,x)$ is not empty when $x\in S(X)$.

If $Q\subset\mathbf{C}$, denote by $\bar{Q}$, ∂Q and co Q the closure, the boundary and the convex hull of Q, respectively. We shall also write

$$\sup \mathrm{Re}\, Q = \sup\{\mathrm{Re}\,\lambda:\ \lambda\in Q\}.$$

Definition. The *spatial numerical range* of $T\in\mathscr{B}(X)$ is the following subset of the complex plane:

$$V(T) = \{f(Tx):\ x\in S(X), f\in S(X'), f(x) = 1\}.$$

The *numerical radius* of T is

$$v(T) = \sup\{|\lambda|:\ \lambda\in V(T)\}.$$

If for $x\in S(X)$ we put

$$V_x(T) = \{f(Tx):\ f\in D(X,x)\},$$

† The second volume of [10], containing recent results, is to appear shortly.

then

$$V(T) = \bigcup_{x \in S(X)} V_x(T).$$

Clearly $V(T) \subset \{z \colon z \in \mathbf{C}, |z| \leqslant \|T\|\}$, i.e. $v(T) \leqslant \|T\|$. $V(T)$ might not be closed or convex. It was proved by Bonsall, Cain and Schneider [9] that $V(T)$ is connected. However, it is not known whether $V(T)$ (or even $\overline{V(T)}$) is also arcwise connected for all T.

Theorem 1. $\sup \operatorname{Re} V(T) = \sup\{c \colon c \in \mathbf{R}$, there exists $x = x(c) \in X, x \neq 0$, such that

$$\|(1 - rc + rT)x\| \geqslant \|x\| \text{ for all } r \geqslant 0\}.$$

Proof. It suffices to show that, for $x \in S(X)$,

(1) $\sup \operatorname{Re} V_x(T) = \max\{c \colon c \in \mathbf{R}, \|(1 - rc + rT)x\| \geqslant \|x\| \text{ for all } r \geqslant 0\}$.

Let P be the real normed linear subspace of X spanned by x and Tx. Then, by the Hahn–Banach theorem and elementary geometrical considerations, both sides are equal to

(2) $\max\{g(Tx) \colon g \in P'_{\mathbf{R}}, \|g\| = g(x) = 1\}$.

Remark. By considering again the support lines of a two-dimensional convex set, this last value is easily seen to be equal to

(3) $\inf\{c \colon c \in \mathbf{R}$, there is an $\varepsilon > 0$ such that $\|(1 - \varepsilon c + \varepsilon T)x\| \leqslant \|x\|\}$.

Evidently $V(T) \subset V(T^*)$ and it is known that this inclusion might be strict. However, the following extension of the Bishop–Phelps theorem [4] can be applied to show that these two sets do not differ too much.

Theorem 2 ([5]). *Suppose* $x \in S(X), f \in S(X')$ *and* $|f(x) - 1| \leqslant \varepsilon^2/2 \ (0 < \varepsilon < \tfrac{1}{2})$. *Then there exist* $y \in S(X)$ *and* $g \in S(X')$ *such that*

$$g(y) = 1, \quad \|f - g\| \leqslant \varepsilon \quad and \quad \|x - y\| < \varepsilon + \varepsilon^2.$$

Theorem 3 ([5]). $\overline{V(T)} = \overline{V(T^*)}$. *In particular,* $v(T) = v(T^*)$.

Proof. If $\eta \in V(T^*)$, there exist $f \in S(X')$ and $\varphi \in S(X'')$ for which $\varphi(f) = 1$ and $\varphi(T^*f) = \eta$. Let $0 < \varepsilon < \tfrac{1}{2}$. The set $\{\hat{x} \colon x \in X, \|x\| \leqslant 1\} \subset X''$ is dense in $\{\psi \colon \psi \in X'', \|\psi\| \leqslant 1\}$, so choose $x \in X, \|x\| \leqslant 1$ such that

$$|\varphi(f) - \hat{x}(f)| < \varepsilon^2/2 \quad and \quad |\varphi(T^*f) - \hat{x}(T^*f)| < \varepsilon.$$

By Theorem 2 there exist $y \in S(X)$, $g \in S(X')$ for which $g(y) = 1$, $\|x - y\| < \varepsilon + \varepsilon^2$ and $\|f - g\| \leqslant \varepsilon$; consequently $|f(T(x - y))| < \|T\|(\varepsilon + \varepsilon^2)$ and

$$|f(Ty) - g(Ty)| \leqslant \|T\|\varepsilon.$$

These inequalities imply

$$|g(Ty) - \eta| < |g(Ty) - f(Tx)| + \varepsilon < \|T\|(2\varepsilon + \varepsilon^2) + \varepsilon.$$

As $g(Ty) \in V(T)$, this gives $\eta \in \overline{V(T)}$ and thus $\overline{V(T^*)} = \overline{V(T)}$.

An important property of the spatial numerical range is that its closure contains the spectrum of the operator. Clearly every eigenvalue of an operator is actually in the numerical range.

Theorem 4 (Williams [25]). $\operatorname{Sp} T \subset \overline{V(T)}$.

Proof. Suppose $\lambda \in \mathbf{C}$ and $\inf\{|\lambda - \eta| : \eta \in V(T)\} = \varepsilon > 0$.
    Let $x \in S(X)$, $f \in D(X, x)$. Then

$$\|(\lambda I - T) x\| \geq \|f(\lambda I - T) x\| = |\lambda - f(Tx)| \geq \varepsilon,$$

where I denotes the identity operator. So

$$(4) \qquad\qquad \|(\lambda I - T) x\| \geq \varepsilon \|x\|$$

for all $x \in X$. As $\overline{V(T)} = \overline{V(T^*)}$, we have similarly

$$(5) \qquad\qquad \|(\lambda I - T)^* f\| \geq \varepsilon \|f\|$$

for all $f \in X'$.

Inequality (4) implies that $\lambda I - T$ is a one-to-one mapping of X onto a closed subspace Y and inequality (5) shows that this subspace is dense in X. Thus $Y = X$ and it follows from Banach's isomorphism theorem that $\lambda I - T$ has a continuous inverse. Consequently $\lambda \notin \operatorname{Sp} T$, completing the proof.

Definition. $T \in \mathscr{B}(X)$ is said to be *dissipative* if

$$\sup \operatorname{Re} V(T) \leq 0.$$

Proposition 1. *If $T \in \mathscr{B}(X)$ is dissipative then*

$$\|x - rTx\| \geq \|x\| \quad \text{for all } r \geq 0.$$

Proof. Suppose $\sup \operatorname{Re} V(T) < 0$. Then, given $x \in S(X)$, Theorem 1 implies that there exists an $r_0 > 0$ such that

$$\|x + r_0 Tx\| < \|x\|.$$

The convexity of the unit ball implies that

$$\|x - rTx\| > \|x\| \quad \text{for all } r > 0.$$

The next theorem was proved by Sinclair [21], by applying the Markov–Kakutani fixed point theorem. Here we give a trivial proof, based on Proposition 1.

Theorem 5. *If* $\lambda \in \partial\, \overline{\mathrm{co}\, V(T)}$ *and* $(\lambda I - T)x = 0$ *then*

$$\|x + (\lambda I - T)y\| \geqslant \|x\|$$

for all $y \in X$.

(That is, the kernel of $\lambda I - T$ is orthogonal to its range, in the sense of Birkhoff.)

Proof. It suffices to show that if T is dissipative and $Tx = 0$ then

$$\|x + Ty\| \geqslant \|x\|, \quad \text{for all } y \in X.$$

By Proposition 1 we have $(N > 0)$

$$\left\| x - \frac{y}{N} - NT\left(x - \frac{y}{N}\right) \right\| = \left\| x - \frac{y}{N} + Ty \right\| \geqslant \left\| x - \frac{y}{N} \right\|.$$

The result follows by letting N tend to infinity.

Corollary 1. (Nirschl–Schneider [18]). *Let* $\lambda \in \partial\, \overline{\mathrm{co}\, V(T)}$. *If* $(\lambda I - T)^2 u = 0$ *then* $(\lambda I - T)u = 0$.

Proof. Apply Theorem 5 with $x = (\lambda I - T)u$ and $y = -u$.

Corollary 2. *The eigenvalues* λ *with* $|\lambda| = \|T\|$ *have index* 1.

In fact Crabb and Sinclair [12] recently extended Theorem 5 for the values $\lambda \in \partial\, \overline{V(T)}$.

5.2.3 THE NUMERICAL RANGE IN A BANACH ALGEBRA

Let A be a complex unital Banach algebra (i.e. A is a complete normed algebra, with identity 1 and $\|1\| = 1$) with dual A'. If $a \in A$, let L_a be the operator of left multiplication on A: $L_a(x) = ax$.

Definition. The *numerical range* of $a \in A$ is

$$V(A, a) = V(L_a)$$

and the *numerical radius* is

$$v(A, a) = \sup\{|\lambda|: \lambda \in V(A, a)\}.$$

Theorem 6 ([8]). *If* $y, y^{-1} \in S(A)$ *then*

$$V(A, a) = V_y(L_a) = \{f(ay): f \in D(A, y)\}.$$

Proof. As

$$V(A, a) = \bigcup_{x \in S(A)} V_x(L_a)$$

and $y \in S(A)$, one has $V(A, a) \supset V_y(L_a)$.

Given $x \in S(A)$ and $f \in D(A, x)$, define $g \in D(A, y)$ by

$$g(z) = f(zy^{-1}x), \quad z \in A.$$

Then $f(ax) = g(ay) \in V_y(L_a)$.

Let us deduce now a number of corollaries of this simple result. The first one of them is the usual and convenient definition of the numerical range of an element of a Banach algebra.

Corollary 1. $V(A, a) = V_1(L_a) = \{f(a): f \in D(A, 1)\}$.

Corollary 2. $V(A, a)$ *is a compact convex subset of the complex plane.*

Proof. $D(A, 1)$ is a convex weak* compact subset of A' and the linear mapping $D(A, 1) \to \mathbf{C}$, given by $f \to f(a)$, is weak* continuous.

Corollary 3. *If R_a is the operator of right multiplication by a on A, i.e. $R_a(x) = xa \ (x \in A)$ then*

$$V(A, a) = V(L_a) = V(R_a) = V_1(L_a) = V_1(R_a).$$

Equality (1), applied in the case of $x = 1$ and $T = L_a$, implies the following result.

Theorem 7. $\sup \operatorname{Re} V(A, a) = \max \{c: c \in \mathbf{R}, \| 1 - rc + ra \| \geqslant 1 \text{ for all } r \geqslant 0\}$.

As, by Corollary 2 to Theorem 6, $V(A, a)$ is compact and convex, the above theorem implies that $V(A, a)$ is determined by the norms of linear polynomials of a. So, in particular, one has the following useful result.

Corollary. *Let B be a subalgebra of A containing the unit element. Then*

$$V(B, b) = V(A, b)$$

for each $b \in B$.

If X is a Banach space, an operator $T \in \mathscr{B}(X)$ can be considered as an element of the unital Banach algebra $\mathscr{B}(X)$. Thus T has two natural numerical ranges: a spatial numerical range, $V(T)$, and a numerical range as an element of the algebra $\mathscr{B}(X)$, $V(\mathscr{B}(X), T)$. As $V(\mathscr{B}(X), T)$ is convex and compact and $V(T)$ might be neither, the following relation between these numerical ranges, which is easily obtained from Theorems 1 and 7, is about the closest possible.

Theorem 8. $\overline{\operatorname{co} V(T)} = V(\mathscr{B}(X), T)$. *In particular,* $V(T) \subset V(\mathscr{B}(X), T)$ *and* $v(T) = v(\mathscr{B}(X), T)$.

5.2.4 THE NORM AND THE NUMERICAL RANGE

The norm of an operator T on a Banach space X, whose spectral radius is zero, can be arbitrarily large. On the other hand, the numerical range of T depends on the norm structure of X (or $\mathscr{B}(X)$) as well, and it was proved by Glickfeld [13] that $\|T\|$ is always bounded by $ev(T)$ (see Theorem 9). This and other similar properties of the numerical range make it sometimes more useful than the spectrum. In this section we shall show some of the relations among norms and numerical ranges, in particular among numerical radii. For the sake of convenience we shall work with algebras, though the same results hold for operators and spatial numerical ranges as well.

Let A be a complex unital Banach algebra and let $a \in A$. The first proposition of this section is a consequence of the equality of the values (1) and (3).

Proposition 2.

$$\sup \operatorname{Re} V(A, a) = \inf\{c \colon c \in \mathbf{R}, \text{there is an } \varepsilon > 0 \text{ such that}$$
$$\|1 - \varepsilon c + \varepsilon a\| \leqslant 1\}.$$

Now for $b \in A$ define, as usual,

$$\exp(b) = \sum_{k=0}^{\infty} \frac{b^k}{k!} = \lim_{n \to \infty} \left(1 + \frac{b}{n}\right)^n.$$

If $b_1, b_2 \in A$ commute, we have $\exp(b_1 + b_2) = \exp(b_1)\exp(b_2)$.

Proposition 3.

$\sup \operatorname{Re} V(A, a) \leqslant 1$ *if and only if* $\|\exp(ra)\| \leqslant \exp(r)$ *for all* $r \geqslant 0$.

Proof (a). If $\|\exp(ra)\| \leqslant \exp(r)$ then

$$\|1 + ra\| \leqslant \exp(r) + (\exp(r\|a\|) - 1 - r\|a\|).$$

So for any $\eta > 0$, if r is small enough, then the right-hand side is less than $(1 - (1 + \eta)r)^{-1}$. Consequently

$$\|1 - (1 + \eta)r + (1 - (1 + \eta)r)ra\| \leqslant 1.$$

By Proposition 2 we have $\sup \operatorname{Re} V(A, a) \leqslant 1$.

(b). Suppose $\sup \operatorname{Re} V(A, a) \leqslant 1$. Proposition 2 implies that there is $\varepsilon > 0$ such that $1/n < \varepsilon$ implies $\|1 - 1/n + a/n\| \leqslant 1$.

Let t be a positive real number and choose the integer m such that $m \leqslant tn < m + 1$. As $\|(1 - 1/n + a/n)^m\| \leqslant 1$, we have

$$\left\|\sum_{k=0}^{m} \binom{m}{k} n^{-k}(1 - 1/n)^{m-k} a^k\right\| \leqslant 1.$$

If we keep t fixed and let $n \to \infty$, it is trivial that the series on the left-hand side tends to $\exp(ta - t)$, and this completes the proof.

Let K be a compact convex set of the complex plane. The following two functions are defined in terms of K:

$$k(\varphi) = \sup\{\operatorname{Re} \lambda \exp(-i\varphi): \lambda \in K\}$$

and

$$\omega(z) = \exp(rk(\varphi)),$$

where $z \in \mathbf{C}$, $z = r \exp(i\varphi)$, $r \geq 0$ and φ is real.

Put

$$k_a(\varphi) = k(a, \varphi) = \sup\{\operatorname{Re} \lambda: \lambda \in V(A, a \exp(i\varphi))\},$$

where $-\infty < \varphi < \infty$. The functions $k(\varphi)$ and $k_a(\varphi)$ will be called the *indicator functions* of the set K and element a, respectively. We call the function $\omega(z)$ the *majorant* belonging to K. Clearly $\omega(rz) = (\omega(z))^r$ for $r \geq 0$.

As K is uniquely determined by its indicator function, the next result is an immediate consequence of Proposition 3.

Proposition 4. $k(a, \varphi) \leq k(\varphi)$ *for all* φ *if and only if* $\|\exp(za)\| \leq \omega(z)$ *for all* $z \in \mathbf{C}$, *i.e.* $V(A, a) \subset K$ *if and only if* $\|\exp(za)\| \leq \omega(z)$ *for all* $z \in \mathbf{C}$.

Corollary. $T \in \mathcal{B}(x)$ *is dissipative if and only if* $\|\exp(rT)\| \leq 1$ *for all* $r \geq 0$.

The following important relation between a^n and $v(a)$ was proved independently in [6] and [11].

Theorem 9.

$$g(n) = \sup\left\{\frac{\|a^n\|}{v^n(a)}: A \text{ is a complex unital Banach algebra}, a \in A\right\}$$

$$= \left(\frac{e}{n}\right)^n n! \quad \text{for every positive integer } n.$$

In other words, if $g(n)$ is the least constant for which $\|a^n\| \leq g(n) (v(a))^n$ for any algebra element a then $g(n) = (e/n)^n n!$.

Proof. (a) $g(n) \leq (e/n)^n n!$ If $v(A, a) \leq 1$, then $\|\exp(za - |z|)\| \leq 1$, so

$$1 \geq \left\|\frac{n^n}{2\pi} \int_{|z|=n} \exp(za - n) z^{-n-1} dz\right\| = (n/e)^n \|a^n\|/n!.$$

In other words the power inequality $\|a^n\| \leq (e/n)^n n!$ is a simple consequence of Cauchy's inequality.

(b) Let P be the vector space of formal power series in one variable:

$$P = \left\{ \xi : \xi = \sum_{k=0}^{\infty} \xi_k a^k, \ \xi_k \in \mathbf{C} \right\}.$$

With the usual multiplication (or, in other words, with the convolution of sequences) P becomes an algebra: if $\xi, \eta \in P$, then

$$\xi\eta = \zeta = \sum_{k=0}^{\infty} \zeta_k a^k$$

with

$$\zeta_k = \sum_{h=0}^{\infty} \xi_h \eta_{k-h}.$$

If z is an arbitrary complex number, define $\exp(za) \in P$ by

$$\exp(za) = \sum_{n=0}^{\infty} \frac{z^n}{n!} a^n.$$

Let R be an arbitrary real number, about which, for the time being, we shall only suppose that it is at least e. Let B be the absolutely convex hull of the set

$$\left\{ \exp(za - |z|) \sum_{k=0}^{\infty} \eta_k R^{-k} a^k : \sum_{k=0}^{\infty} |\eta_k| \leqslant 1 \right\}.$$

[The absolutely convex hull of a set is defined by replacing the finite sum $\sum c_i x_i$ ($c_i \geqslant 0$, $\sum c_i = 1$) in the definition of convex hull by the finite sum $\sum c_i x_i$ (c_i complex, $\sum |c_i| \leqslant 1$).]
As

$$\exp(z_1 a - |z_1|) \exp(z_2 a - |z_2|) = r \exp((z_1 + z_2) a - |z_1 + z_2|)$$

with $0 \leqslant r \leqslant 1$, it is trivial that B is closed under multiplication. Consequently,

$$P_0 = \bigcup_{n=1}^{\infty} nB \subset P$$

is a subalgebra of P. It is easily checked that

$$\|\xi\| = \inf\{r : r > 0, \xi/r \in B\}$$

defines a norm on P_0.

Let A_0 be the completion of P_0. Let us now look at the element $a \in A_0$. By construction and Proposition 4, we have $v(A_0, a) \leqslant 1$. On the other hand, the coefficient of a^n in

$$\exp(za - |z|) \sum_{k=0}^{\infty} \eta_k R^{-k} a^k \left(\sum_{h=0}^{\infty} |\eta_h| \leqslant 1 \right)$$

is easily seen to be at most

$$\sup\{|z|^k \exp(-|z|)/k! : z \in \mathbf{C}, k = 0, 1, ..., n\} = n^n \exp(-n)/n!$$

Consequently, by the definition of the norm, $\|a^n\| \geqslant (e/n)^n n!$

It is easily seen that the Banach algebra A_0 constructed in the proof of Theorem 9 is extremal in the following sense: A_0 is generated by a; $v(a) = 1$; and if X is any Banach algebra, $x \in X$ and $v(x) \leqslant 1$, then $h(1) = 1$, $h(a) = x$ defines an algebra homomorphism $h: A_0 \to X$ of norm 1. In fact, it was proved in [7] that a similar extremal algebra can be constructed for any closed convex set K of the complex plane.

Theorem 10. *Let $K \subset \mathbf{C}$ be a non-empty compact convex set. Then there exists a complex unital Banach algebra $A(K)$, generated by an element a, with the following properties.*

(i) $V(A, a) = K$.

(ii) If X is any Banach algebra containing an element x such that $V(X, x) \subset K$ then there is an algebra homomorphism $h: A(K) \to X$ of norm 1, mapping a into x. In particular, if $G(z)$ is any function which is regular on K, we have $V(X, G(x)) \subset V(A, G(a))$.

5.2.5 HERMITIAN OPERATORS

It is an important feature of the theory of numerical ranges that it provides the natural setting for the so-called Hermitian operators in Banach spaces. X, as usual, will denote a Banach space.

Definition. An operator $T \in \mathcal{B}(X)$ is said to be *Hermitian* if any of the following conditions is satisfied.

(i) $V(T)$ *is real.*

(ii) $V(\mathcal{B}(X), T)$ *is real.*

(iii) $\|\exp(irT)\| = 1$ *for all real r.*

The equivalence of these conditions follows from Theorem 8 and Proposition 4. We omit the straightforward proof of the next proposition, which justifies the term Hermitian.

Proposition 5. *An element of a complex unital B*-algebra is Hermitian if and only if $h^* = h$.*

As we are interested mainly in the norms of functions of Hermitian operators, instead of Hermitian operators on Banach spaces we may always consider the Hermitian elements of a normed algebra.

Vidav's famous lemma [24], which is an important step in the Vidav–Palmer characterization theorem of B^*-algebras, states that if $h \in A$ is Hermitian then $\rho(h) = v(h)$, where $\rho(h)$ denotes the spectral radius of h. This result was extended by Sinclair [20] who proved that the spectral radius of a Hermitian element is also equal to its norm.

We deduce this result from a theorem, the first part of which is a simple form of S. N. Bernstein's classical theorem [3]. Part (*ii*) follows from (*i*) and from the simple observation that if

$$|F(z)| = \left| \sum_{k=0}^{\infty} f_k z^k/k! \right| \leqslant \exp(|\operatorname{Re} z|), f_0 = 1 \quad \text{and} \quad f_1 = 1,$$

then $f_2 = 1$.

Theorem 11 (Bernstein). *Let* $F(z)$ *be an entire function, such that* $|F(z)| \leqslant \exp(|\operatorname{Re} z|)$ *for all* $z \in \mathbf{C}$. *Then*

(*i*) $|F'(z)| \leqslant \exp(|\operatorname{Re} z|)$ *for all* $z \in \mathbf{C}$,

(*ii*) *if* $F(0) = F'(0) = 1$ *then* $F(z) = \exp(z)$.

Corollary (Sinclair). *If* $h \in A$ *is Hermitian then*

$$\rho(h) = v(h) = \|h\|.$$

Proof. By replacing h by λh ($\lambda \in \mathbf{R}$) one may assume that

$$1 \in V(A, h) \subset [-1, 1].$$

Then by Proposition 4 we have

(6) $\|\exp(zh)\| \leqslant \exp(|\operatorname{Re} z|), \quad z \in \mathbf{C}.$

Choose any $f \in A'$, $\|f\| \leqslant 1$. Then, by equation (6),

(7) $|f(\exp(zh))| = |F(z)| \leqslant \exp(|\operatorname{Re} z|),$

where F is entire. So, by Theorem 11, $|f(h)| = |F'(0)| \leqslant 1$, implying $\|h\| \leqslant 1$.

As $1 \in V(A, h)$, there is a functional $f \in A'$, $\|f\| = 1$ such that

$$f(1) = f(h) = 1.$$

By equation (6) and Theorem 5.3 (*ii*) we obtain $F(z) = \exp(z)$, i.e. $f(h^n) = 1$, implying $\|h^n\| = 1$ for all n. As for any element x,

$$\rho(x) = \lim_{n \to \infty} (\|x^n\|)^{1/n}$$

and $\rho(x) \leqslant v(x) \leqslant \|x\|$, we obtain $\rho(h) = v(h) = \|h\|$.

Sinclair also extended his result to linear polynomials of Hermitian elements. We only state this theorem; it can be deduced from a number of generalizations of Bernstein's theorem.

Theorem 12. *If T is a hermitian operator, for all $\lambda \in \mathbf{C}$ we have*

$$\rho(T + \lambda I) = \|T + \lambda I\|.$$

This implies the following characterization of Hermitian operators.

Theorem 13. *An operator T is Hermitian if and only if $\operatorname{Sp} T \subset \mathbf{R}$ and for some $\varepsilon > 0$*

$$\|T + ir\| = \rho(T + ir)$$

for all $r \in \mathbf{R}, |r| \leqslant \varepsilon$.

Proof. Put $\overline{\operatorname{co}} \operatorname{sp} T = [t_1, t_2]$. Then if $r \in \mathbf{R}$ and n is large enough, we have

$$\|(1 + irT/n)^n\| = \max\{|1 + irt_1/n|^n, |1 + irt_2/n|^n\}.$$

Letting $n \to \infty$, one obtains

$$\|\exp(irT)\| \leqslant 1,$$

i.e. T *is* Hermitian.

One might expect that if $\|T + z\| = \rho(T + z)$ for all $z \in \mathbf{C}$ then $T = \lambda T_1 + \mu$, where T_1 is Hermitian and $\lambda, \mu \in \mathbf{C}$. However, this is not so.

REFERENCES

1. F. L. Bauer, "On the field of values subordinate to a norm", *Numer. Math.* **4** (1962), 103–111.
2. E. Berkson, "Some characterizations of C^*-algebras", *Ill. J. Math.* **10** (1966), 1–8.
3. S. N. Bernstein, *Propriétés Extremales*, Gauthiers–Villars, Paris (1926).
4. E. Bishop and R. R. Phelps, "A proof that every Banach space is subreflexive", *Bull. Am. math. Soc.* **67** (1961), 97–98.
5. B. Bollobás, "An extension to the theorem of Bishop and Phelps", *Bull. Lond. math. Soc.* **2** (1970), 181–182.
6. ——, "The power inequality on Banach spaces", *Proc. Cam. Phil. Soc.* **69** (1971), 411–415.
7. ——, "The numerical range in Banach algebras and complex functions of exponential type", *Bull. Lond. math. Soc.* **3** (1971), 27–33.
8. F. F. Bonsall, "The numerical range of an element of a normed algebra", *Glasgow Math. J.* **10** (1969), 68–72.
9. F. F. Bonsall, B. E. Cain and H. Schneider, "The numerical range of a continuous mapping of a normed space", *Aequationes Math.* **2** (1968), 86–93.
10. F. F. Bonsall and J. Duncan, "Numerical ranges of operators on normed spaces and of elements of normed algebras", *Lond. math. Soc. Lecture Notes*, 2, Cambridge University Press (1971).

11. M. J. Crabb, "The power inequality on normed spaces", *Proc. Edin. math. Soc.* **17** (1971), 237–240.

12. M. J. Crabb and A. M. Sinclair, "On the boundary of the spatial numerical range", *Bull. Lond. math. Soc.* **4** (1972), 17–19.

13. B. W. Glickfeld, "On an inequality of Banach algebra geometry and semi-inner-product space theory", *Notices of Am. math. Soc.* **15** (1968), 339–340.

14. ——, "A metric characterization of $C(X)$ and its generalization to C^*-algebras", *Ill. J. Math.* **10** (1966), 547–566.

15. I. M. Gelfand and M. A. Naimark, "On the embedding of normed rings into the ring of operators in Hilbert space", *Mat. Sbornik* **12** (1943), 197–213.

16. F. Hausdorff, "Der Wertvorrat einer Bilinearform", *Math. Zeit.* **3** (1919), 314–316.

17. G. Lumer, "Semi-inner-product spaces", *Trans. Am. math. Soc.* **100** (1961), 29–43.

18. N. Nirschl and H. Schneider, "The Bauer fields of values of a matrix", *Numer. Math.* **6** (1964), 355–365.

19. T. W. Palmer, "Characterizations of C^*-algebras", *Bull. Am. math. Soc.* **74** (1968), 538–540.

20. A. M. Sinclair, "The norm of a Hermitian element in a Banach algebra", *Proc. Am. math. Soc.* **28** (1971), 446–450.

21. ——, "Eigenvalues in the boundary of the numerical range", *Pacific J. Math.* **35** (1970), 231–234.

22. M. H. Stone, "Linear transformations in Hilbert space", *Am. math. Soc. Coll. Publ. XV* (1932).

23. O. Toeplitz, "Das algebraische Analogon zu einem Satze von Fejér", *Math. Zeit.* **2** (1918), 187–197.

24. I. Vidav, "Eine metrische Kennzeichnung der selbstadjungierten Operatoren", *Math. Zeit.* **66** (1956), 121–128.

25. J. P. Williams, "Spectra of products and numerical ranges", *J. Math. Anal. and Appl.* **17** (1967), 214–220.

5.3

Sample Path Properties of Processes with Stationary Independent Increments

S. J. TAYLOR†

In memory of Rollo Davidson, who was unstinting in his help to others

5.3.1 INTRODUCTION

Much has been discovered about this special class of Markov processes in the last twenty years so that a complete account is not practicable. We therefore concentrate on a few areas where significant progress has been made in the recent past. The main emphasis will be on properties determined by the local structure of the sample path. Since we have no new results (apart from a new way of looking at local time in Section 5.3.11) we will try to describe concepts and methods rather than give detailed arguments. As well as summarizing the present state of knowledge, we formulate several open problems which are now accessible and for which a solution would be interesting.

While writing this paper I have benefited greatly from conversations with a substantial proportion of the authors whose results are being described. I also discovered that B. E. Fristedt [32] was writing a comprehensive account of independent increment processes which gives detailed arguments and covers the field quite exhaustively. I am indebted to Fristedt for allowing me to see sections of his manuscript before publication, and would urge the reader who is interested in parts of the subject not covered in the present article to look first in [32] for help.

† This survey is an extended version of a lecture presented at the Joint Statistical Meetings held in Fort Collins, Colorado, 23–26 August 1971. While preparing the article, the author was supported in part by the National Science Foundation, under Grant GP-28683.

5.3.2 MATHEMATICAL FRAMEWORK

We will consider only processes taking values in Euclidean space R^d, though recent work by Port and Stone [77] shows that it is possible to study the behaviour of processes taking values in a locally compact abelian group. The finite-dimensional distributions of a process with stationary independent increments are completely determined by the distribution of the increment $\{X(t_0+1)-X(t_0)\}$ in unit time together with the distribution of the starting point $X(0)$. We assume that $X(0) = 0$, so that the defining distribution is that of $X(1)$. The class of possible distributions for $X(1)$ is precisely the class of infinitely divisible laws. This means that the characteristic function is

$$\mathbf{E}\exp\{i(z,X(1))\} = \psi(z),$$

where $\psi(z)$ is called the *exponent* of the distribution and has a unique representation in the form (see Loève [65] or Bochner [11])

$$(1) \qquad \psi(z) = i(z,a) - \tfrac{1}{2}Q(z) + \int\left[\exp(i(z,y)) - 1 - \frac{i(z,y)}{1+|y|^2}\right]\nu(dy).$$

Here a is a constant vector in R^d, (x,y) denotes the scalar product, $Q(z)$ is a non-negative quadratic form, and ν is a Borel measure in R^d (called the Lévy measure) such that $\nu\{0\} = 0$ and

$$\int_{R^d}\min(1,|y|^2)\,\nu(dy)<\infty.$$

The first term in equation (1) can be thought of as deterministic—but note that different forms of the third term are possible and these will lead to different constants a. The term $-\tfrac{1}{2}Q(z)$ is the Gaussian component of the distribution. If ν happens to satisfy the additional condition

$$(2) \qquad \int_{R^d}\min(1,|y|)\,\nu(dy)<\infty$$

then there is a unique representation in the form

$$(3) \qquad \psi(z) = i(z,b) - \tfrac{1}{2}Q(z) + \int[1-\exp(i(z,y))]\nu(dy),$$

where ν is the same (Lévy) measure, but b may be different from a.

We will assume without further comment that we are talking about a standard version of the Markov process whose finite dimensional distributions are determined by equation (1) together with the condition of independent increments (see, for example, Blumenthal and Getoor [10]). In this formulation each point ω of the probability space Ω corresponds to a function $X(t) = X(t,\omega)$ mapping $[0,\infty)$ to R^d and these functions

$X(t)$ are bounded for finite time intervals, and are right continuous with left limits everywhere. Throughout the present paper we are interested in sample path properties—that is, properties of the function $X(t) = X(t, \omega)$ for fixed points ω in Ω. Every statement we make about a sample path should be understood as having the qualifying phrase 'with probability 1'. Thus if we say that the paths of the process have property Q we mean that the subset of Ω given by $\{\omega \in \Omega \colon X(.) = X(., \omega)$ has property $Q\}$ is measurable and has probability 1. Since it is usual to suppress the ω in the notation for a sample path we will usually also suppress the qualifying phrase 'with probability 1'. Of course, the properties of the sample path we obtain depend on the process being considered which in turn is determined by the particular exponent (1).

Many of the results we describe in this paper are most easily proved using the methods of modern generalized potential theory. The connection between potential theory and Markov processes is due to Hunt [50], but the recent work of Port and Stone [78] has formulated the potential theory of independent increment processes in a more useful way. As far as possible, we will omit all details of a potential theoretic nature as they require a fair amount of technical machinery which is at least one step removed from the analysis of path structure.

5.3.3 SPECIAL CASES

It is only recently that much progress has been made in the study of general processes based on equation (1). The study in depth of the properties of special cases has influenced the development at each stage. If in equation (1) we put $a = 0$, $Q(z) = |z|^2$, $\nu(R^d) = 0$, we obtain the characteristic function of the (symmetric) normal distribution and the resulting process is called Brownian motion in R^d. The existence of such a process is essentially due to Wiener [99],† and its properties were extensively studied by Lévy [63]. The normal distribution is a special case of the more general class of limit laws called *stable*. These can be defined as the class of probability distributions τ in R^d such that, if X_1, X_2 are independent with distribution τ and a_1, a_2 are fixed scalars then

$$a_1 X_1 + a_2 X_2 - l = cX$$

for a suitable $l \in R^d$ and scalar c, where X has the same distribution τ. If the distribution of X is stable, then it has a representation (1) in the form

---

† [For the origins of Wiener's work, and hence of stochastic analysis as a whole, see [100], which is added as a compactification point to Professor Taylor's list of references. D.G.K.]

(Bochner [11])

(4) $$\psi(z) = i(a,z) - \lambda |z|^\alpha \int_{S_d} w_\alpha(z,\theta)\,\mu(d\theta);$$

where $\lambda > 0$, $0 < \alpha \leqslant 2$ (α is called the index), μ is a probability distribution on the surface of the unit sphere $S_d(|\theta| = 1)$, and

$$w_\alpha(z,\theta) = [1 - i\operatorname{sgn}(z,\theta)\tan\tfrac{1}{2}\pi\alpha]\left|\left(\frac{z}{|z|},\theta\right)\right|^\alpha, \quad \text{if } \alpha \neq 1,$$

$$w_1(z,\theta) = \left|\left(\frac{z}{|z|},\theta\right)\right| + \frac{2i}{\pi}(z,\theta)\log|(z,\theta)|.$$

Note that, if $\alpha = 2$, equation (4) reduces to Brownian motion after appropriate standardization. If $\alpha = 1$, we have a Cauchy distribution and the resulting process is called a Cauchy process. Although a standard argument shows that all the stable distributions have bounded continuous densities, it is only possible to write down these densities in closed form in a few cases such as the symmetric Cauchy distribution and the Gaussian case, $\alpha = 2$. If μ is a uniform distribution on S_n (and $a = 0$) we obtain the symmetric stable process of index α,

(5) $$\psi(z) = -c|z|^\alpha.$$

It is clear from the form of equation (4) that if $X(1)$ is stable with $a = 0$, $\alpha \neq 1$ (also in some cases if $\alpha = 1$), then $X(t)$ and $r^{-1/\alpha}X(rt)$ have the same distribution for any $r > 0$. This implies that $r^{-1/\alpha}X(rt)$ is a version of the same process as $X(t)$, and this fact is called the 'scaling property' of stable processes (including Brownian motion). When it is valid, the scaling property often provides concise methods of obtaining results.

The real line has a natural order, and this makes it interesting to consider the class of processes in R^1 which are monotone increasing. These are called *subordinators* and we have a Laplace transform for the distribution. Further, in this case $Q \equiv 0$ and the Lévy measure must satisfy inequality (2) so that equation (3) becomes

$$\mathbf{E}\exp(-uX(1)) = \exp(-g(u)),$$

where

(6) $$g(u) = du + \int_0^\infty (1 - \exp(-ur))v(dr),$$

$d \geqslant 0$, and v is the (same) Lévy measure now concentrated on the positive half-line. In equation (4), the only stable distributions which lead to subordinators have index α satisfying $0 < \alpha < 1$ and μ a point mass on $\{1\}$.

These correspond to

(7) $$g(u) = cu^{\alpha}, \quad 0 < \alpha < 1, \quad c > 0.$$

Subordinators are an important tool in the general theory as well as being of independent interest. For a full account of their properties see Fristedt [32]; but note that Fristedt widens the class of processes considered by adjoining the point ∞ to R^1 and allowing the Lévy measure to have a finite mass σ on $\{\infty\}$. This is equivalent to 'killing' our process at time T, an independent random variable distributed with density $(1/\sigma)\exp(-\sigma x)$, $x \geqslant 0$, and then saying that $X(t) = \infty$ for $t \geqslant T$.

5.3.4 PARAMETERS

In the historical development of the subject, sample path properties have usually been obtained first for Brownian motion (stable of index 2), then for symmetric stable processes of index α $(0 < \alpha < 2)$, and then more generally. For stable processes it was found that the structure of the path depends on the relation between the index α and the dimension d of the state-space so it was natural to look for numerical parameters of the distribution (1), or the corresponding process, which would play the role which α plays in the stable case. In equation (1) it is the third term which is difficult to analyse since the growth behaviour of ν near 0 may be very irregular. There is no single numerical parameter which gives all the properties, so several have been defined—for stable processes they are all equal to the index α. We collect the various definitions together here for easy reference.

The parameter β depends on the behaviour of the Lévy measure ν near 0:

(8) $$\beta = \inf\left\{\alpha > 0 : \int_{|y| < 1} |y|^{\alpha} \nu(dy) < \infty\right\}$$
$$= \inf\{\alpha > 0 : r^{\alpha} \nu\{y : |y| > r\} \to \infty \text{ as } r \to 0\}.$$

The parameters β', β'' depend on the behaviour of the real part of the exponent $\psi(z)$:

(9) $$\beta'' = \sup\{\alpha \geqslant 0 : |z|^{-\alpha} \operatorname{Re}\psi(z) \to \infty \text{ as } |z| \to \infty\},$$

(10) $$\beta' = \sup\left\{\alpha \geqslant 0 : \int |z|^{\alpha - d} \frac{1 - \exp(-\operatorname{Re}\psi(z))}{\operatorname{Re}\psi(z)} \, dz < \infty\right\}.$$

Blumenthal and Getoor [6] who first defined these parameters showed that $0 \leqslant \beta'' \leqslant \beta' \leqslant \beta \leqslant 2$ and that all these parameters could be distinct. They also defined a parameter for subordinators determined by the growth of $g(u)$ as $u \to \infty$. Using the definition (6), this is

(11)
$$\sigma = \sup \left\{ \alpha \leqslant 1 : \int_1^\infty \frac{u^{\alpha-1}}{g(u)} \, du < \infty \right\}$$

and

$$\beta' \leqslant \sigma \leqslant \min(1, \beta).$$

A final parameter, due to Pruitt [76], is determined more directly in terms of the sample path by means of the behaviour of the expected time spent in a small sphere:

(12)
$$\gamma = \sup \left\{ \alpha \geqslant 0 : \limsup_{a \to 0} a^{-\alpha} \int_0^1 \mathrm{pr}\{X(t) \leqslant a\} \, dt < \infty \right\}.$$

Pruitt shows that for a subordinator $\gamma = \sigma$, for a symmetric process $\gamma = \min(\beta', d)$, but in general it is possible to have γ different from β'. He also shows that γ is related to the moments of $X(t)$ by

$$\gamma = \sup \left\{ \alpha \geqslant 0 : \int_0^1 \mathbf{E} |X(t)|^{-\alpha} \, dt < \infty \right\}.$$

5.3.5 CONTINUITY PROPERTIES OF SAMPLE PATHS

It is immediate from the form of the characteristic function

$$\mathbf{E} \exp(i(z, X(t))\} = \exp(t\psi(z)),$$

with $\psi(z)$ given by equation (1), that for any $\delta > 0$,

$$\mathrm{pr}\{|X(t)| > \delta\} \to 0 \quad \text{as } t \to 0;$$

and because we are only considering a nice version of the process we can assert that

(13)
$$\lim_{t \to t_0} X(t) = X(t_0), \quad \text{with probability 1.}$$

This means that, at any fixed time instant t_0, $X(t)$ will be continuous at t_0 with probability 1. The question of continuity in an interval is more subtle, but a complete answer is possible in terms of the exponent (1).

If $\nu \equiv 0$, the process reduces to Brownian motion (with a drift if $a \neq 0$, and a lack of symmetry unless Q is I) and the continuity of the paths for this process was essentially obtained by Wiener [99]. If $\nu(R^d) \neq 0$, then $X(t)$ is not continuous for all t: in fact, one method of deriving the form (1) of the exponent leads to the fact that $\nu(B)$, for any Borel $B \subset R^d$, is the expected number of jumps $J(t) = X(t) - X(t-0)$ in unit time, satisfying $J(t) \in B$. This gives further information as follows:

(*a*) The number of discontinuities of $X(t)$ in a bounded interval is finite if, and only if, $\nu(R^d) < \infty$.

(*b*) If $\nu(R^d) = \infty$, the set of discontinuities of $X(t)$ is everywhere dense in $[0, \infty)$, though still countable. Further $\nu(R^\infty) = \infty$ implies that the distribution of $X(t)$ is continuous.

It is worth noting the possibility that ν is supported by a lattice in R^d: this implies that $\nu(R^d) < \infty$ and if in addition $a = 0$, $Q \equiv 0$, in equation (1), then the process takes only values on the lattice of support of ν, and we are really looking at a special kind of Markov chain in continuous time. The relevant questions to ask about the sample paths are very different in this case and we will not consider them here—see, for example, the book by K. L. Chung [16].

If $a = 0$, $Q \equiv 0$ and $0 < \nu(R^d) < \infty$ then the distribution of $X(t)$ has 0 as an atom (it was shown by Blum and Rosenblatt [3] that 0 is the only atom when ν has no atoms). Further, if we know that $X(t_0) = x_0$, there will be a finite holding time δ (with a negative exponential distribution) such that $X(t) = x_0$ for $t_0 \leqslant t < t_0 + \delta$ and the first jump $J(t_0 + \delta) = X(t_0 + \delta) - x_0$ will be distributed in R^d according to the measure ν. We should think of this case as a generalized Poisson process in which jumps distributed according to ν replace the fixed jumps of the Poisson process. The fact that the Poisson process (and its generalization) is included in the family of processes given by equation (1) is important—for it has recently been used as a powerful simplifying tool by Bretagnolle [15] to avoid the analytic difficulties which can arise in certain special cases. However, the sample path properties of this generalized Poisson case are relatively simple, and we will not consider it further in this paper. To reduce the number of special cases for formulating results we therefore assume henceforth that, in equation (1), either

(*a*) $Q \not\equiv 0$; that is, there is a Gaussian component; or

(*b*) $Q \equiv 0$, but $\nu(R^d) = \infty$.

Again in the interests of brevity we will assume that the process is 'honestly d-dimensional' in the sense that in case (*b*) there is no hyperplane H through 0 of dimension $(d-1)$ such that

$$\nu(R^d \backslash H) < \infty.$$

Similarly, if we have a modified Brownian motion with $\nu = 0$, $Q \not\equiv 0$, then we assume that Q has rank d—this ensures that the process requires the whole of R^d for its state space.

5.3.6 VARIATION AND HÖLDER CONDITIONS

Lévy [63] showed that Brownian motion was not of bounded variation, though

$$\sum_{i=1}^{k_n} |X(t_{n,i}) - X(t_{n,i-1})|^2$$

converges to a limit as the mesh, $\max(t_{n,i} - t_{n,i-1}) \to 0$, where

$$0 = t_{n,0} < t_{n,i-1} < \ldots < t_{n,k_n} = 1, \ n = 1, 2, \ldots,$$

is a sequence of nested dissections of $[0, 1]$. If $X(t)$ is a symmetric stable process (5) of index α, then Blumenthal and Getoor [4] showed that

(14) $$\sum |X(t_{n,i}) - X(t_{n,i-1})|^\delta$$

converges to a finite limit if $\delta > \alpha$, and tends to $+\infty$ if $\delta < \alpha$. For general processes the variational structure seems most closely related to the parameter β defined in equation (8). Partial results about the δ-variation

$$W(X, \delta) = \sup_{0 \leqslant t_0 < t_1 < \ldots < t_n = 1} \sum_{i=1}^{n} |X(t_i) - X(t_{i-1})|^\delta$$

were obtained by Blumenthal and Getoor [6] who showed that, for $\beta < 1$, $W(X, \delta)$ is finite for $\beta < \delta$. The corresponding result for $\delta > \beta \geqslant 1$ has not yet been established in general. There has been some recent progress about the convergence in probability of the sums (14) (for suitable functions as well as powers) due to Sharpe [87], Millar [68], Greenwood and Fristedt [39] and Fristedt [30] and there are partial results about stronger forms of convergence.

For Brownian motion, Lévy [63] showed that it was possible to obtain a uniform modulus of continuity in the form

(15) $$\limsup_{\substack{t_2-t_1=\varepsilon\to 0 \\ 0\leqslant t_1 < t_2 \leqslant 1}} \frac{|X(t_2) - X(t_1)|}{(2\varepsilon |\log \varepsilon|)^{\frac{1}{2}}}$$

and Chung, Erdös and Sirao [17] obtained the sharpest possible result of this kind. It is immediate that Brownian motion is not a differentiable function at a fixed point t_0, but Dvoretzky and Erdös [21] showed that the Brownian motion process $X(t)$ is nowhere differentiable; and in fact it has no points of local increase.

For general processes with $Q \equiv 0$, a necessary and sufficient condition for $X(t)$ to be of bounded variation is that

(16) $$\int \min(1, |y|) \nu(dy) < \infty;$$

and if this condition is satisfied we can always think of $X(t)$ as the difference of two independent subordinators. Processes with $Q \equiv 0$ satisfying inequality (16) are therefore differentiable almost everywhere (Lebesgue), but I am not aware of the solution to the following:

Problem A. *Find a necessary and sufficient condition for $X(t)$ to be differentiable nowhere?*

In connection with this problem, it is worth remarking that Berman [1] has recently obtained results about nowhere differentiability for Gaussian processes with continuous local time.

For Brownian motion there are local 'laws of the iterated logarithm' (see Lévy [63]) which can be formulated as follows:

$$(17) \qquad \limsup_{t \to 0} \frac{|X(t)|}{(2t \log |\log t|)^{\frac{1}{2}}} = 1,$$

$$(18) \qquad \liminf_{t \to 0} \frac{M(t)(\log |\log t|)^{\frac{1}{2}}}{t^{\frac{1}{2}}} = C_d,$$

where C_d is a non-zero constant depending on the dimension and

$$M(t) = \sup_{0 \leqslant \tau \leqslant t} |X(\tau)|.$$

The laws corresponding to equations (17) and (18) are known for stable processes—see Takeuchi [88] and Taylor [96]—but it is more difficult to get precise information about more general processes. Blumenthal and Getoor [6] show that, if β is given by equation (8), then

$$\limsup_{t \to 0} t^{-1/\alpha} |X(t)| = \begin{cases} 0 & \text{if } \alpha > \beta, \\ +\infty & \text{if } \alpha < \beta, \end{cases}$$

but there is no known corresponding parameter for the behaviour of $M(t)t^{-\gamma}$ except in the case of subordinators where the index $1/\sigma$ given by equation (11) is the relevant one. In the case of subordinators Breiman [13] obtained very precise results about the stable case, and Fristedt [27] considers the general situation; also Fristedt and Pruitt [33] have recently completely settled the lower asymptotic results for $M(t)$ ($= X(t)$ in this case). A large number of authors have obtained sharp results about envelopes near zero for $X(t)$, including Fristedt ([27] for stable subordinators, [31] for general symmetric processes), Breiman [13], Takeuchi [89] for stable processes, Hendricks [46] for processes with stable components; and there are corresponding problems about the growth behaviour of $|X(t)|$ and $M(t)$ as $t \to \infty$.

The Gaussian processes are the only ones with continuous sample paths so that there cannot be any uniform Hölder condition corresponding to equation (15) for general processes. However, Hawkes [44] obtained a useful uniform lower bound for the rate at which the stable subordinator of index α given by equation (7) grows. This is given by

$$(19) \qquad \liminf_{\substack{\varepsilon \to 0 \\ 0 < h < \varepsilon}} \limits_{0 \leqslant t \leqslant 1} \frac{X(t+h) - X(t)}{\varphi(h)} = c,$$

where $\varphi(h) = h^{1/\alpha} |\log h|^{-(1-\alpha)/\alpha}$ and c is a (known) finite positive constant. It would be interesting to have corresponding results for more general processes.

5.3.7 RECURRENCE

There are two distinct notions of recurrence which should not be confused. We call a process *point recurrent* if

$$(20) \qquad\qquad Z = \{t: X(t) = X(0)\} \quad \text{is unbounded};$$

while we say it is *recurrent* (or neighbourhood recurrent) if, for each open set G containing $X(0)$, the occupation time set

$$(21) \qquad\qquad\qquad \{t: X(t) \in G\} \quad \text{is unbounded}.$$

It is immediate that each of the conditions (20), (21) defines an event of probability zero or one; and that, if condition (21) is false, then

$$(22) \qquad\qquad\qquad |X(t)| \to \infty \quad \text{as } t \to \infty.$$

Processes which are not recurrent are called *transient*.

It was proved by Lévy [63] that Brownian motion is point recurrent for $d = 1$, recurrent but not point recurrent for $d = 2$ and transient for $d \geqslant 3$. Kac [55] showed that the symmetric stable process is point recurrent if and only if $\alpha > d = 1$, and recurrent if and only if $\alpha \geqslant d$. The criterion for the recurrence and point recurrence of the general stable process of index α is precisely the same, except that if $\alpha = 1 = d$, the asymmetric Cauchy process is transient.

The general criterion for recurrence was first obtained by Kingman [61] who showed that an independent increment process $X(t)$ is recurrent if and only if the 'discrete skeleton' $X(nh)$, $n = 1, 2, \ldots$, is recurrent for suitable small positive h. This allowed him to obtain a criterion in terms of $\varphi(z)$ by applying the result of Chung and Fuchs [18] for the recurrence of sums of independent random variables. More recently, Port and Stone [78] showed that the process is recurrent if and only if

$$(23) \qquad \int_{|z|<1} \mathrm{Re}\left[\frac{-1}{\psi(z)}\right] dz = \infty,$$

where $\psi(z)$ is the exponent (1) of the process.

The problem of obtaining a general criterion for point recurrence is much deeper. There is an incomplete result for symmetric processes due to Bretagnolle and Dacunha-Castelle [14], but the complete picture comes from the detailed analysis of $p(r)$—the hitting probability of single point r—recently carried out by Kesten [60]. For $d \geqslant 2$, processes which are honestly d-dimensional satisfy $p(x) \equiv 0$, so there is no possibility of any of these being point recurrent. The classification for $d = 1$ given by Kesten and reformulated by Bretagnolle [15] shows that $p(r) > 0$ for a set of values of r of positive Lebesgue measure, if and only if, for $\lambda > 0$,

$$(24) \qquad \int_{-\infty}^{\infty} \mathrm{Re}\left[\frac{1}{\lambda - \psi(z)}\right] dz < +\infty.$$

Kesten points out that a process is point recurrent if and only if it is recurrent and $p(r) > 0$ on a set of positive measure, so that conditions (23) and (24) together give the criteria for point recurrence on the line.

5.3.8   HITTING PROBABILITIES

For any Markov process the probability of entering a Borel set E in the state space is considered by Blumenthal and Getoor [10], and others. For independent increment processes this probability,

$$(25) \qquad \Phi(x, E) = \mathrm{pr}_x\{X(t) \in E \text{ for some } t > 0\},$$

depends only on $E - x$. Sets E such that $\Phi(x, E) \equiv 0$ are called *polar*. There are no very tractable general criteria for deciding whether or not a set E is polar. For the symmetric stable process of index α in R^d, it turns out that the potential theory of the process is the same as that investigated by Riesz [84], of order $d - \alpha$, because the potential kernel has a density

$$(26) \qquad u(x, y) = c/|x - y|^{d-\alpha}$$

This means that a Borel set E in R^d is polar if and only if it has zero Riesz capacity of order $(d - \alpha)$. For $d = 1$, it turns out that the class of polar sets is the same for all stable processes of index $\alpha < 1$; and it is even possible to obtain explicit bounds for the hitting probability of E in terms of the corresponding hitting probability for the symmetric stable process of the same index (see Hawkes [41]). For $\alpha > 1 = d$ there is no problem as only the empty set is polar, while for $\alpha = 1 = d$ there is an explicit (logarithmic) kernel in the symmetric case and a quite different behaviour in the asymmetric case. Port and Stone [76] showed that the asymmetric Cauchy process on the line hits singletons, so that there are no non-empty polar sets. The easiest question to which the answer is not known is the following (Pruitt and Taylor [81]):

Problem B. *Do all stable processes of index α in R^d ($\alpha \neq 1, d \geqslant 2$) have the same class of polar sets?*

Another question of interest is that of regularity. Given a non-polar set A, we say that x is regular for A if the process starting at x hits A immediately; that is,

(27) $$\mathrm{pr}_x[\inf\{t > 0, X(t) \in A\} = 0] = 1.$$

Because of our restriction to independent increment processes, x is regular for A if and only if 0 is regular for $A - x$, so it would be interesting to ask for a criterion (depending on the process) on a Borel set B which is necessary and sufficient to ensure that 0 is regular for B. Since the potential theory for the Brownian motion process is the classical one, Wiener's test provides such a criterion for Brownian motion. This test was extended to the symmetric stable process by Lamperti [62], to stable subordinators by Hawkes [41] and to the Cauchy process in R^1 by Takeuchi and Watanabe [90], but I do not know of any criterion for a general process.

A related question to which the answer is again known only in special cases is that of classifying in the case of a transient process the (unbounded) Borel sets B for which the occupation time set,

(28) $$Z(B) = \{t > 0: X(t) \in B\},$$

is bounded.

A complete answer is now available to the question of whether 0 is regular for $\{0\}$. This property was proved to hold for the asymmetric Cauchy process by Port and Stone [76], and a complete analysis of the possible types of processes for which 0 is regular for $\{0\}$ is given by Bretagnolle [15]. It is worth noting that 0 regular for $\{0\}$ does not imply recurrence; nor does point recurrence imply that 0 is regular for $\{0\}$.

For a fixed process, a set A is called *thin* if it has no regular points. We can ask the following interesting question, due to Getoor.

Problem C. *For which processes is the class of thin sets the same as the class of polar sets?*

It is known that all stable process in R^1 satisfy this criterion, but it is not even known whether stable process in R^d ($d \geqslant 2$) can have thin sets which are non-polar. This question is related to that of finding a criterion on the process which ensures the validity of a suitably formulated maximum principle—see Orey [69].

5.3.9 MULTIPLE POINTS

For a process which is point recurrent, the set

$$Z = \{t > 0: X(t) = 0\}$$

clearly has infinite cardinality and if 0 is regular for $\{0\}$, Z in fact has cardinal c of the continuum. However, if $h(0) = 0$, the probability that the sample path ever revisits $X(t_0)$ for $t > t_0$ is zero. It is still possible, nevertheless, that *some* points on the path are visited more than once. Such a point is called multiple. For a fixed sample path $X(t)$ we define the maximum multiplicity of the path to be the largest cardinal p such that there exists a point x_0 for which

$$\{t > 0: X(t) = x_0\} \quad \text{has cardinal } p.$$

A path is called *simple* if $p = 1$, that is

$$t_1 \neq t_2 \Rightarrow X(t_1) \neq X(t_2),$$

or no point is visited twice. A zero–one argument shows that for any process there is a unique p such that all sample paths have maximum multiplicity p.

Problem D. *Obtain a criterion for a process to have simple paths. Determine the maximum multiplicity for a general process.*

The solution of this problem was obtained for Brownian motion in a series of papers by Dvoretzky and co-workers [22–25]. Here the path is simple for $d \geqslant 4$, has maximum multiplicity 2 for $d = 3$; while for $d = 1, 2$ points of multiplicity c exist. These results were extended to symmetric stable processes by Takeuchi [89] and Taylor [95] and almost complete results were obtained by Hendricks [47] for processes with stable components of different indices. It is worth noting that for any integer k, there are processes of maximum multiplicity k; the symmetric stable processes of index $\alpha = k/(k+1)$ in R^1 and of index $\alpha = 2k/(k+1)$ in R^2 are examples. I presume, though I do not think even this has been proved, that for $d \geqslant 4$ all honestly d-dimensional processes are simple, while for $d = 3$ there are no processes of multiplicity greater than 2 (attained by symmetric stable process of index $\alpha > 3/2$). Some information, including confirmation of the conjecture for $d \geqslant 4$, was obtained by Orey [69] for symmetric processes.

5.3.10 HAUSDORFF MEASURE OF THE RANGE

If we define

$$R(t) = \{x \in R^d : X(\tau) = x \text{ for some } \tau,\, 0 \leqslant \tau \leqslant t\}$$

to be the image on the path of the time interval $[0, t]$, we can think of $R(t)$ as a random (Borel) set in the state space determined by the sample point ω. If $d > 2$ (or if $d = 1$, and the hitting probability $p(r) = 0$ for all points r), $R(t)$ will have zero Lebesgue measure. This means that measures finer than Lebesgue have to be used to 'measure' the size of the set $R(t)$. A class of measures which is useful for analysing sets of zero Lebesgue measure was defined by Hausdorff (but see Rogers [85] for a full account of their properties).

Given a function $h: [0, \infty) \to [0, \infty)$ which is continuous, monotone increasing and satisfies $h(0) = 0$, the Hausdorff h-measure is a Carathéodory outer measure defined for all subsets of a metric space (in our case R^d) by

$$(29) \qquad h(E) = \lim_{\delta \to 0} \left[\inf_{\substack{E \subset \cup C_i \\ d(C_i) < \delta}} \sum_{i=1}^{\infty} h[d(C_i)] \right],$$

where $d(C_i)$ denotes the diameter of set C_i and the infimum is taken over countable covers of E by sets of diameter less than δ. Since $h(E)$ is a metric outer measure, Borel sets will be h-measurable. For the analysis of subsets of R^d of zero Lebesgue measure it is appropriate to assume that $h(s)/s^d \to \infty$ as $s \to 0+$. For example, the classical Cantor ternary set C on the line satisfies

$$h^\alpha(C) = \begin{cases} 0, & \alpha > \delta, \\ 1, & \alpha = \delta, \quad \text{if } h^\alpha(s) = s^\alpha, \\ +\infty, & \alpha < \infty, \end{cases}$$

where $\delta = \log 2/\log 3$. For a given set A, the interesting question to decide is: 'for which functions h does $h(A)$ turn out to be zero or $+\infty$?' If possible, and it is not always possible, one likes to find a suitable h which makes $0 < h(A) < \infty$.

If we restrict attention to the powers

$$h^\alpha(s) = s^\alpha, \quad \alpha > 0,$$

then all subsets of R^d have a numerical dimension, which is a real number $\gamma \leqslant d$ given by

$$\dim E = \gamma = \inf\{\alpha > 0 : h^\alpha(E) = 0\}.$$

Even if $\dim E = \gamma$ it is possible for $h\gamma(E)$ to be zero, finite and positive, or infinite; but

$$\alpha > \gamma \Rightarrow h^{\alpha}(E) = 0,$$

$$0 \leqslant \alpha < \gamma \Rightarrow h^{\alpha}(E) \quad \text{is non } \sigma\text{-finite}.$$

The first process to have the Hausdorff measure of its range examined was Brownian motion. Taylor [92] showed that, for $d > 2$, it has a range $R(t)$ of dimension 2 and zero h^2-measure. Lévy conjectured [64] that, for $d \geqslant 3$,

$$0 < h(R(t)) < +\infty \quad \text{for} \quad h(s) = s^2 \log |\log s|,$$

and this result was proved by Ciesielski and Taylor [19]. The case of Brownian motion in the plane was settled by Erdös and Taylor [26], Ray [82] and Taylor [94]—here the correct measure function is

$$h(s) = s^2 |\log s| \, \log\log |\log s|.$$

More or less complete results are available for the stable processes. The correct dimension was obtained by McKean [67] for the symmetric case, and by Blumenthal and Getoor [5] for the general stable process, while the exact measure function for the symmetric Cauchy process on the line is due to Ray [83], for stable subordinators to Taylor and Wendel [97], and for all stable processes with the scaling property to Pruitt and Taylor [81]. There is only one remaining unresolved question about the Hausdorff measure of the range for stable processes.

Problem E. *Find a Hausdorff measure function $h(s)$ such that $0 < h(R(t)) < \infty$, where $R(t)$ is the range of a non-symmetric Cauchy process in R^d ($d \geqslant 2$).*

For general processes the dimension of the range is now known. Blumenthal and Getoor [6] showed that (see Section 5.3.4. for definitions of parameters)

$$\beta'' \leqslant \dim R(t) \leqslant \beta,$$

Horowitz [49] showed that for subordinators,

$$\dim R(t) = \sigma,$$

and Pruitt [79] showed that, in general,

$$\dim R(t) = \gamma.$$

This paper [79] also contains an example of a process for which the dimension of the range is smaller than that for its symmetrization.

Fristedt and Pruitt [33] have recently given an explicit method for finding an exact measure function $h(s)$ appropriate for measuring the range of a given subordinator, but as yet there is no way of obtaining the correct h for a general process. The only other type of process for which the correct measure function is known is a process with independent stable components of different indices (Pruitt and Taylor [80]) which includes as a special case the graph of a stable process (Jain and Pruitt [54]).

It is worth making a few remarks about the techniques which have been used to give the results described above. In order to obtain dim $R(t)$ it is good enough to obtain the capacity dimension using Riesz [84] capacities of the general type defined by Frostman [34]. Since there are various equivalent methods of proving that a set has positive capacity and the potential kernel (26) takes a very simple form for the symmetric stable process, it was easy to obtain results in these cases. The technique has been exploited by many authors—including Taylor [92], McKean [66, 67], Blumenthal and Getoor [5] and Orey [69]. An interesting variant of the above argument has been the use of symmetric stable processes of varying indices to see which of them will hit a given set such as $R(t)$. If E is a Borel set in R^d,

$$(30) \qquad \dim E = d - \inf\{\alpha > 0 : E \text{ is not polar for } X_{\alpha,d}(t)\},$$

where $X_{\alpha,d}(t)$ denotes a symmetric stable process of index α in R^d (which is independent of E, if E is random). For a path with multiple points equation (30) was useful for determining the dimension of the set of points of multiplicity k—see Taylor [95] and Fristedt [29].

If one wants to obtain a precise measure function h such that

$$0 < h(R(t)) < \infty$$

then the connection between capacities of Frostman type and Hausdorff measure is not sufficiently tight—see Taylor [91]. Difficulties arise because, in order to obtain economical coverings in equation (29), it is essential to use sets C_i of widely different sizes at different parts of the sample path. Two separate arguments are required, one to show that the h-measure is positive and the other that it is finite, and both arguments are probabilistic in nature. Several authors have successfully used variants of these techniques, but perhaps it is worth while to try to describe the ideas involved.

(a) To obtain the lower bound, it helps to consider the natural random measure

$$F_\omega(E) = |\tau : X(\tau) \in E, \ 0 \leqslant \tau \leqslant t|$$

$$= \int_0^t I_E(X(t)) \, dt,$$

which is the time spent in the Borel set E up to time t. F_ω is supported by the range $R(t)$ and is, in an obvious sense, spread uniformly on $R(t)$. If F is any finite measure on the Borel sets of R^d, h a fixed Hausdorff measure function, Rogers and Taylor [86] showed how to analyse F using the upper density

$$(31) \qquad \limsup_{a \to 0} \frac{F(S(x,a))}{h(2a)} = \bar{D}_h F(x).$$

The set of points x in R^d for which $\bar{D}_h F(x)$ is small cannot have arbitrarily large F-measure. For our particular measure F_ω, equation (31) has a simple meaning, for it becomes

$$(32) \qquad \limsup_{a \to 0} \frac{T(x,a,t)}{h(2a)},$$

where $T(x,a,t)$ is the total time spent in $S(x,a)$ up to time t. Now it is clear that, if x is not in $R(t)$ (or its closure), then the limit (32) gives zero, while for a fixed point x in $R(t)$ the limit (32) may give a finite positive limit for a suitable choice of h. When such an h can be found, it follows that

$$(33) \qquad \bar{D}_h F(x) = c_1$$

for almost all $(F_\omega) x$ on $R(t)$ and the results of [86] then give

$$(34) \qquad h(R(t)) > c_2 > 0.$$

(b) To obtain a finite upper bound, the results of [86] do not help as we have to cover all of $R(t)$, and there will be an exceptional set (of zero F_ω-measure) where (33) fails. The technique in this case is to use a sequence Λ_n of meshes which get fine as $n \to \infty$ and (i) give an economical covering in equation (29) if any covering exists; and (ii) have an 'almost nested' property so that any covering has a subcovering in which no point is covered more than a fixed number of times. For a large integer n_1 one looks at a set S of Λ_{n_1} which is 'hit' by $X(t)$, and computes the probability that there is *no* integer n with $n_1 \geqslant n \geqslant n_2$ such that $X(t)$ spends a large time in a set $S_n \in \Lambda_n$, which satisfies $S_n \supset S$, and

$$\frac{F_\omega(S_n)}{h(d(S_n))} > c_3 > 0.$$

Such sets are called 'bad', and one needs to use a first moment argument to show that bad sets make negligible contribution for at least some large n_1. Under the conditions we have assumed on Λ_n, the 'good' sets of Λ_{n_1} can be covered economically and this leads to

$$(35) \qquad h(R(t)) < c_4 < \infty.$$

If one has both conditions (34) and (35) for the process, then a simple argument shows the existence of a finite c with $c_2 \leqslant c \leqslant c_4$ and

$$(36) \qquad h(R(t)) = ct.$$

This is the sense in which h is right for measuring the range of the process. Note that h is not unique and may not even be comparable with the powers h^α—see Fristedt and Pruitt [33].

5.3.11 LOCAL TIME

If x is a particular point of the state space which is 'hit' by the process, it is interesting to obtain information about the 'size' of the occupation time set

$$(37) \qquad Z_t^x = \{\tau: 0 \leqslant \tau \leqslant t \text{ and } X(\tau) = x\}.$$

We should think of the local time at x as a 'measure' of the size of Z_t^x. This means that it should be a monotone function of t which grows only for times t at which $X(t) = x$. For our independent increment processes the necessary and sufficient condition for the existence of such a function with the right properties is that 0 is regular for $\{0\}$. In the present section we will always assume that this property holds. This means that we are only considering processes taking values on the line, which hit zero with probability 1, though our condition is slightly stronger than that. Our object in the present section is to give several distinct ways of obtaining a local time with the desired properties. Imprecise information about the size of Z_t^x was obtained by Taylor [93] for Brownian motion, by Blumenthal and Getoor [8] for some stable processes and by Hawkes [42] for the asymmetric Cauchy process. See also the related papers by Kesten [59] and Getoor [35].

(i) Density of occupation times

For any Borel set $B \subset R$, we can consider

$$(38) \qquad T_t(B) = \int_0^t I_B(X(\tau)) \, d\tau$$

as a set function in B. Under the conditions we have imposed it is absolutely continuous with respect to Lebesgue measure so that there is a Borel measurable function $l(x, t)$ such that

$$(39) \qquad T_t(B) = \int_B l(x, t) \, dx$$

for all Borel sets B. This is only of interest if we can choose a smooth version of $l(x, t)$—in fact in addition to equation (39) we would like $l(x, t)$ to be monotone in t for each fixed x and jointly continuous in (x, t). The existence of such a function $l(x, t)$ was proved for Brownian motion by Trotter [98] and for a wide class of processes, including stable processes of index $\alpha > 1$, by Boylan [12].

When a jointly continuous $l(x, t)$ exists, it is clear that we can think of equation (39) as a Riemann integral so that

$$l(x, t) = \lim_{h, k \to 0} \frac{T_t[x - h, x + k]}{h + k},$$

and the local time is a pointwise derivative of $T_t(B)$ at x. Griego [40] shows, under very weak conditions, satisfied by independent increment processes with 0 regular for $\{0\}$, but now known to be insufficient to guarantee the existence of a continuous $l(x, t)$ that, for a fixed sequence of neighbourhoods B_n decreasing to $\{x\}$,

(40) $$\lim_{n \to \infty} \frac{T_t(B_n)}{|B_n|} = l(x, t),$$

where $l(x, t)$ is a local time at x in the sense of a continuous additive functional—which we discuss in the next paragraph. In equation (40) the limit is uniform in t for $t \leqslant T$, but it should either be understood as a limit in pr_x-probability or as a pointwise limit on a subsequence.

(ii) Continuous additive functional

This is an abstract formulation based on the general theory of Markov processes (see Blumenthal and Getoor [9] and Getoor [36]). We will not describe this definition precisely as this would require the establishment of a great deal of technical notation. The theory picks a particular level x and considers a monotone (in t) continuous function $A(x, t) = A(x, t, \omega)$ of the process which grows precisely on the set of times t for which $X(t) = x$. The continuous additive functional (CAF) $A(x, t)$ is essentially unique apart from a multiplicative constant. Blumenthal and Getoor show that it is always possible to choose a version of $A(x, t)$ for each x, so that $A(x, t)$ is jointly measurable in (x, t), and equation (39) is satisfied with $l(x, t) = A(x, t)$.

(iii) Jumps across the level

A recent result of Getoor and Millar [38] shows that, under mildly restrictive conditions, the number of small jumps across the level x (suitably standardized) is proportional to the local time at x uniformly in t. To be more precise, suppose $l(x, t)$ exists as a CAF for each level x, satisfies

14

equation (39) and is jointly continuous in (x, t) and that, in addition,

$$\sum F_n^{-1} < \infty, \quad \text{where} \quad F_n = \int_{2^{-n-1}}^{2^{-n}} x \nu(dx),$$

where ν is the Lévy measure of the process. Let $Q_n(t)$ be the number of jumps $J(X, s) = X(s) - X(s-0)$ with $0 < s \leqslant t$, $X(s-0) < x < X(s)$ and $2^{-n-1} < J(X, s) < 2^{-n}$. Then

$$(41) \qquad \qquad \lim_{n \to \infty} \sup_{0 \leqslant t \leqslant T} \left| \frac{Q_n(t)}{F_n} - l(x, t) \right| = 0.$$

It would be of some interest to know if equation (41) is true uniformly in x, under appropriate conditions.

(iv) Hausdorff measure of the level set

The occupation time set Z_t^x defined in equation (37) is a random set determined by the path which has zero Lebesgue measure for the processes we are considering. If we could find an exact Hausdorff measure function f such that

$$(42) \qquad \qquad 0 < f(Z_t^x) < \infty$$

then surely this would be a way of defining the local time at x. This was done in the case of a stable process of index $\alpha > 1$ by Taylor and Wendel [97] who showed that, for

$$f(s) = s^\beta (\log |\log s|)^{1-\beta}, \beta = 1 - \frac{1}{\alpha},$$

and a fixed level x,

$$(43) \qquad \qquad f(Z_t^x) = cl(x, t)$$

for some (unknown) absolute constant $c > 0$. It follows that equation (43) will be valid for almost all (Lebesgue) x so that we have equation (39) satisfied with $l(x, t)$ replaced by $c^{-1} f(Z_t^x)$.

The techniques of [97] can now be extended to all processes in which 0 is regular for $\{0\}$ by the following series of steps:

(a) Consider $A(x, t)$, the local time at x defined as a CAF whose support is $\{x\}$.

(b) Consider the function inverse to $A(x, t)$,

$$\tau(t) = \inf\{u : A(x, u) > t\},$$

where we adopt the usual convention that the infimum of the empty set is $+\infty$. If $X(t)$ is recurrent, then $\tau(t)$ will be a subordinator. However, it is possible for $X(t)$ to be transient and in this case we can think of $\tau(t)$ as a subordinator with an independent 'killing time' T which has a negative exponential distribution

$$\mathrm{pr}\,\{T>k\} = \exp(-\gamma k).$$

The precise formulation of this situation is obtained in Blumenthal and Getoor [10, p. 214]. We now assume that X is recurrent and formulate the rest of the results for this case—it is obvious how to modify the argument for transient X.

(c) The exponent of the subordinator $\tau(t)$ in the form (6) is

$$g(\lambda) = [u^\lambda(0)]^{-1},$$

where $u^\lambda(.)$ is the continuous density of the potential operator U^λ for the process, and our conditions are sufficient to ensure that U^λ has a continuous density. It follows from the smoothness in the situation that

$$u^\lambda(0) = \int_{-\infty}^{\infty} \mathrm{Re}\left(\frac{1}{\lambda+\psi(x)}\right) dx,$$

so we can regard the exponent g of the subordinator as explicitly determined by ψ for the process.

(d) Apart from a countable set, the range of the subordinator $\tau(t)$ corresponds to the points of increase of $A(x,t)$ and is therefore the same as the level set Z_∞^x. Using the result of Fristedt and Pruitt [33] on subordinators, we can find a correct measure function f (in a finite number of explicit steps starting from g) for measuring the range of the subordinator.

(e) Using this measure function f, we obtain equation (43) for the fixed level x. It is worth noting that we do not obtain a result of the form (36) for the zero set because the time interval over which we are measuring the subordinator is not $[0,t]$ but rather the random interval $[0, A(x,t)]$.

The above sequence of arguments (a)–(e) works only for a fixed level x. A Fubini type argument allows one to conclude that equation (43) will be true for almost all (Lebesgue) x. This leads to the obvious question:

Problem F. *Under what conditions is equation (43) true for all x, where Z_t^x is given by equation (37)?*

Note that the graph of a process can itself be thought of as an independent increment process $(X(t),t)$ taking values in R^2 whose range is a subset of the plane. Problem F concerns the intersections of this graph with level lines parallel to the t-axis. A partial result in the direction of our problem was recently obtained by Berman [2] who proved that

$$\dim Z_t^x = \tfrac{1}{2} \text{ for } all \ x$$

in the case of a class of Gaussian processes which includes Brownian motion. A recent paper by Getoor and Kesten [37] obtains necessary conditions and (different) sufficient conditions for the existence of a continuous $l(x,t)$ satisfying equation (39). These lead to the conclusion that the asymmetric Cauchy process, though it has a local time at each fixed point cannot have an $l(x,t)$ which is jointly continuous. An examination of the Hausdorff measure of the level sets for the graph of the asymmetric Cauchy process might give some insight into this curious phenomenon.

There is one other question brought to mind by the argument (b) above.

Problem G. *Which subordinators can arise as the inverse of the local time of an independent increment process?*

5.3.12   SOME RANDOM INTERSECTIONS

In our discussion of hitting sets we have mainly considered singletons $\{x_0\}$. There are many questions to be asked about the hitting of a more general Borel set B by a process. For instance, one can ask for the distribution of $X(T_B)$ where

$$T_B = \inf\{t>0: X(t)\in B\}$$

or for joint distributions of T_B and $X(T_B)$. The general potential theory provides a machinery for tackling this type of question and explicit information has been obtained for many types of process (see, for example, Blumenthal, Getoor and Ray [7], Port [70–75], Ikeda and Watanabe [51]). Instead of describing these results in any detail, I will talk briefly about another class of problems where complete solutions are not yet in sight.

For a set B of positive Lebesgue measure and a Brownian motion process, Darling and Kac [20] gave a technique, using computation of the moments, for obtaining the distribution of

$$|\{\tau\leqslant t: X(\tau)\in B\}| = \int_0^t I_B\{X(s)\}\,ds,$$

which in this case is a proper random variable. These methods do not work if B is a set of zero Lebesgue measure, and one would like information about the occupation time set

(44) $O(B) = \{t>0; \ X(t)\in B\};$

or the intersection with the range

(45) $$B \cap R(t)$$

for Borel sets B of zero Lebesgue measure. One way of attacking this problem is to have a method of relating the size of the image

$$X(A) = \{x \in R^d : X(t) = x \text{ for some } t \in A\}$$

to the size of A. For general processes, Hendricks [48] has an example which shows that the dimension of $X(A)$ is not always a constant multiple of the dimension of A, and this same example shows that the dimension of $O(B)$ is not, in general, determined by the dimension of B. There are some positive results as follows.

For a fixed set A, and a symmetric stable process of index α, we have

(46) $$\dim X(A) = \min(\alpha \dim A, d).$$

This was proved by McKean [67] for $\alpha = 2$ (Brownian motion) and by Blumenthal and Getoor [5] for $0 < \alpha < 2$, with partial results in [6] for more general processes. Recently Kaufman [57] has proved that for Brownian motion in R^d ($d \geqslant 2$) equation (46) is true for all A, and this result was extended by Hawkes [43] to stable subordinators and symmetric stable processes. Hawkes then uses this result to obtain the dimension of the intersection (45) under a strong regularity condition on B.

Another recent result of Kaufman [58] concerns the dimension of the intersection

(47) $$A \cap O(B)$$

for fixed sets A, B on the line and linear Brownian motion. He obtains the dimension in terms of a property of the Cartesian product $A \times B$ and shows that it does not depend only on $\dim A$, $\dim B$. The size of the set (47) could be studied for more general processes.

In [54], Jain and Pruitt consider the collision set of two independent stable processes X_α and X_β in R^1. They obtain incomplete information about the dimension of the set

$$\{t : X_\alpha(t) = X_\beta(t)\}.$$

Another method of attacking the size of (30) was used in [95] to deal with the case where B is itself a random set independent of the process. The idea used here is to have an independent symmetric stable process $X_{\beta,d}(t)$ of index β and to compute the range of β for which $X_{\beta,d}$ hits $B \cap R(t)$ with positive probability. This simple technique can always yield the dimension of the set when $d \leqslant 2$: Fristedt [29] showed how to extend the argument to higher dimensions by using a projection argument. For example, in [95] it was proved that for a transient stable process of index α in R^d, the set E_k of points which are entered at least k times has dimension $\alpha - (k-1)(d-\alpha)$ provided this is positive (otherwise the set E_k is empty).

The techniques used so far are a long way from giving exact measure functions for sets like (44) or (45), or even $X(A)$ (which is likely to be an easier problem). At best, they only give results about the dimension. For example, for Brownian motion in the plane, the sets E_k, $k = 1, 2, \ldots$, and E_c all have dimension 2: it would be interesting to know which measure functions give E_k zero h-measure.

Conjecture. *For Brownian motion in the plane,*

$$h_\alpha(E_k) = \begin{cases} 0, & \text{if } \alpha \leqslant k, \\ \infty, & \text{if } \alpha > k, \end{cases}$$

where

$$h_\alpha(s) = s^2 |\log s|^\alpha.$$

REFERENCES

1. S. M. Berman, "Gaussian processes with stationary increments: local times and sample function properties", *Ann. Math. Statist.* **41** (1970), 1260–1272.
2. ———, "Gaussian sample functions: uniform dimension and Hölder conditions nowhere", *Nagoya Math. Journal* **46** (1972), 63–86.
3. J. R. Blum and M. Rosenblatt, "On the structure of infinitely divisible distributions", *Pacific J. Math.* **9** (1959), 1–7.
4. R. M. Blumenthal and R. K. Getoor, "Some theorems on stable processes", *Trans. Am. math. Soc.* **95** (1960), 263–273.
5. ———, "A dimension theorem for sample functions of stable processes", *Ill. J. Math.* **5** (1960), 370–375.
6. ———, "Sample functions of stochastic processes with stationary independent increments", *J. Math. Mech.* **10** (1961), 493–516.
7. R. M. Blumenthal, R. K. Getoor and D. B. Ray, "On the distribution of first hits for the symmetric stable processes", *Trans. Am. math. Soc.* **99** (1961), 540–554.
8. R. M. Blumenthal and R. K. Getoor, "The dimension of the set of zeros and the graph of a symmetric stable process", *Ill. J. Math.* **6** (1962), 308–316.
9. ———, "Local times for Markov processes", *Ztschr. Wahrsch'theorie & verw. Geb.* **3** (1964), 50–74.
10. ———, *Markov Processes and Potential Theory*, Academic Press, New York (1968).
11. S. Bochner, *Harmonic Analysis and the Theory of Probability*, University of California Press, Berkeley (1955).
12. E. S. Boylan, "Local times for a class of Markov processes", *Ill. J. Math.* **8** (1964), 19–39.
13. L. Breiman, "A delicate law of the iterated logarithm for non-decreasing stable processes", *Ann. Math. Statist.* **39** (1968), 1818–1824.

14. J. Bretagnolle and D. Dacunha-Castelle, "Pointwise recurrence of processes with independent increments", *Institut H. Poincaré* **B3** (1967), 677–682.

15. J. Bretagnolle, "Résultats de Kesten sur les processus à accroissements independants", *Univ. de Strasbourg, Seminaire de Probabilités* (1969), 21–36.

16. K. L. Chung, *Markov Chains with Stationary Transition Probabilities*, Springer, Berlin (1960).

17. K. L. Chung, P. Erdös and T. Sirao, "On the Lipschitz's condition for Brownian motion", *J. math. Soc. Japan* **11** (1959), 263–274.

18. K. L. Chung and W. H. J. Fuchs, "On the distribution of values of sums of random variables", *Mem. Am. math. Soc.* **6** (1951), 1–12.

19. Z. Ciesielski and S. J. Taylor, "First passage times and sojourn times for Brownian motion in space and the exact Hausdorff measure of the sample path", *Trans. Am. math. Soc.* **103** (1962), 434–450.

20. D. A. Darling and M. Kac, "On occupation times for Markov processes", *Trans. Am. math. Soc.* **84** (1957), 444–458.

21. A. Dvoretzky, and P. Erdös and S. Kakutani, "Nonincrease everywhere of the Brownian motion process", *Proc. Fourth Berkeley Symposium*, vol. 2 pp. 103–116 (1960).

22. ——, "Double points of paths of Brownian motion in *n*-space", *Acta Sci. Math.* **12** (1950), 75–81.

23. ——, "Multiple points of paths of Brownian motion in the plane", *Bull. Res. Council Israel* **F3** (1954), 364–371.

24. ——, "Points of multiplicity *c* of plane Brownian paths", *Bull. Res. Council Israel* **F7** (1958), 175–180.

25. A. Dvoretzky, P. Erdös, S. Kakutani and S. J. Taylor, "Triple points of Brownian motion in 3-space", *Proc. Cam. Phil. Soc.* **53** (1957), 856–862.

26. P. Erdös and S. J. Taylor, "On the Hausdorff measure of Brownian paths in the plane", *Proc. Cam. Phil. Soc.* **57** (1961), 209–222.

27. B. E. Fristedt, "The behaviour of increasing stable processes for both small and large times", *J. Math. and Mech.* **13** (1964), 849–856.

28. ——, "Sample function behaviour of increasing processes with stationary independent increments", *Pacific J. Math.* **21** (1967), 21–33.

29. ——, "An extension of a theorem of S. J. Taylor concerning the multiple points of the symmetric process", *Ztschr. Wahrsch'theorie & verw. Geb.* **9** (1967), 62–64.

30. ——, "Variation of symmetric one dimensional stochastic processes with stationary independent increments", *Ill. J. Math.* **13** (1969), 717–721.

31. ——, "Upper functions for symmetric processes with stationary independent increments", *Indiana Math. J.* **21** (1971), 177–185.

32. ——, *Sample Functions of Stochastic Processes with Stationary, Independent Increments. Advances in Probability*, vol. 3, Marcel Dekker, New York (1973).

33. B. E. Fristedt and W. E. Pruitt, "Lower functions for increasing random walks and subordinators", *Ztschr. Wahrsch'theorie & verw. Geb.* **18** (1971), 167–182.

34. O. Frostman, "Potential d'équilibre et capacité des ensembles avec quelques applications à la théorie des fonctions", *Medd. Lund. Univ. Math. Seminar* **3** (1935).

35. R. K. Getoor, "The asymptotic distribution of the number of zero free intervals of a stable process", *Trans. Am. math. Soc.* **106** (1963), 127–138.

36. ——, "Continuous additive functionals of a Markov process with applications to processes with independent increments", *J. Math. Analysis* **13** (1966), 132–153.
37. R. K. Getoor and H. Kesten, "Continuity of local times for Markov processes", *Comp. Math.*, **24** (1972), 277–303.
38. R. K. Getoor and P. W. Millar, "Some limit theorems for local time", *Indiana Math. J.*, to appear.
39. P. Greenwood and B. E. Fristedt, "Variations of processes with stationary independent increments", *Ztschr. Wahrsch'theorie & verw. Geb.* **23** (1972), 171–186.
40. R. J. Griego, "Local time as a derivative of occupation times", *Ill. J. Math.* **11** (1967), 54–63.
41. J. Hawkes, "Polar sets, regular points and recurrent sets for the symmetric and increasing stables processes", *Bull. Lond. math. Soc.* **2** (1970), 53–59.
42. ——, "Measure function properties of the asymmetric Cauchy process", *Mathematika* **17** (1970), 68–78.
43. ——, "Some dimension theorems for the sample functions of stable processes", *Indiana Math. J.* **20** (1971), 733–738.
44. ——, "A lower Lipschitz condition for the stable subordinator", *Ztschr. Wahrsch'theorie & verw. Geb.* **17** (1971), 23–32.
45. ——, "On the Hausdorff dimension of the intersection of the range of a stable process with a Borel set", *Ztschr. Wahrsch'theorie & verw. Geb.* **19** (1971), 90–102.
46. W. J. Hendricks, "Lower envelopes near zero and infinity for processes with stable components", *Ztschr. Wahrsch'theorie & verw. Geb.* **16** (1970), 261–278.
47. ——, Ph.D. thesis, University of Minnesota (1969).
48. ——, "Hausdorff dimension in a process with stable components—an interesting example, *Ann. Math. Statist.* **43** (1972), 690–694.
49. J. Horowitz, "The Hausdorff dimension of the sample path of a subordinator", *Israel J. Math.* **6** (1968), 176–182.
50. G. A. Hunt, "Markov processes and potentials I, II, III", *Ill. J. Math.* **1** (1957), 44–93 and 316–369; **2** (1958), 151–213.
51. N. Ikeda and S. Watanabe, "On some relations between the harmonic measure and the Lévy measure for a certain class of Markov process", *J. Math. Kyoto Univ.* **2** (1962), 79–95.
52. K. Ito and H. P. McKean, *Diffusion Processes and their Sample Paths*, Springer, Berlin (1965).
53. N. Jain and W. E. Pruitt, "The correct measure function for the graph of a transient stable process", *Ztschr. Wahrsch'theorie & verw. Geb.* **9** (1968), 131–138.
54. ——, "Collisions of stable processes", *Ill. J. Math.* **13** (1969), 241–248.
55. M. Kac, "Some remarks on stable processes", *Publ. Inst. Statist. Univ. Paris* **110** (1957), 303–306.
56. R. P. Kaufman, "Brownian motion and dimension of perfect sets", *Can. J. Math.* **22** (1970), 674–680.
57. ——, "Une propriété métrique du mouvement brownien", *C. R. Acad. Sci. Paris* **268** (1969), 727–728.
58. ——, personal communication.

59. H. Kesten, "Positivity intervals of stable processes", *J. Math. Mech.* **12** (1963), 391–410.
60. ——, "Hitting probabilities of single points for processes with stationary independent increments", *Memoir* 93, *Amer. Math. Soc.* (1969).
61. J. F. C. Kingman, "Recurrence properties of processes with stationary independent increments", *J. Australian Math. Soc.* **4** (1964), 223–228.
62. J. Lamperti, "Wiener's test and Markov chains", *J. Math. Analysis and Appl.* **6** (1963), 58–66.
63. P. Lévy, *Processus Stochastiques et Mouvements Browniens*, Gauthier-Villars, Paris (1948).
64. ——, "La mésure de Hausdorff de la courbe du mouvement brownien", *Giorni. 1st. Ital. Attuari* **16** (1953), 1–37.
65. M. Loève, *Probability Theory* 3rd ed., Van Nostrand, Princeton (1963).
66. H. P. McKean, "Hausdorff–Besicovitch dimension of Brownian motion paths", *Duke Math. J.* **22** (1955), 229–234.
67. ——, "Sample functions of stable processes", *Annals Math.* **61** (1955), 564–579.
68. P. W. Millar, "Path behaviour of processes with stationary independent increments", *Ztschr. Wahrsch'theorie & verw. Geb.* **17** (1971), 53–73.
69. S. Orey, "Polar sets for processes with stationary independent increments", *Markov Processes and Potential Theory*, pp. 117–126, Wiley, New York (1967).
70. S. C. Port, "The exit distribution of an interval for completely asymmetric stable processes", *Ann. Math. Statist.* **41** (1970), 39–43.
71. ——, "Hitting times for transient stable processes", *Pacific J. Math.* 21 (1967), 161–165.
72. ——, "On hitting places for stable processes", *Ann. Math. Statist.* **38** (1967), 1021–1026.
73. ——, "Hitting times and potentials for recurrent stable processes", *J. d'Anal. Math.* **20** (1967), 371–395.
74. ——, "Potentials associated with recurrent stable processes", *Markov Processes and Potential Theory*, pp. 135–163, Wiley, New York (1967).
75. ——, "A remark on hitting places for transient stable processes", *Ann. Math. Statist.* **39** (1968), 365–371.
76. S. C. Port and C. J. Stone, "The asymmetric Cauchy processes on the line", *Ann. Math. Statist.* **40** (1969), 137–143.
77. ——, "Potential theory for infinitely divisible processes on Abelian groups", *Bull. Am. math. Soc.* **75** (1969), 848–851. Full account in *Acta Math.* **122** (1969), 19–114.
78. ——, "Infinitely divisible processes and their potential theory I, II", *Ann. Inst. Fourier* **21** (2) (1971), 157–275 and **21** (4) (1971), 179–265.
79. W. E. Pruitt, "The Hausdorff dimension of the range of a process with stationary independent increments", *J. Math. Mech.* **19** (1969), 371–378.
80. W. E. Pruitt and S. J. Taylor, "Sample path properties of processes with stable components", *Ztschr. Wahrsch'theorie & verw. Geb.* **12** (1969), 267–289.
81. ——, "The potential kernel and hitting probabilities for the general stable process in R^N", *Trans. Am. math. Soc.* **146** (1969), 299–321.

82. D. Ray, "Sojourn times and the exact Hausdorff measure of the sample path for planar Brownian motion", *Trans. Am. math. Soc.* **106** (1963), 436–444.

83. ——, "Some local properties of Markov processes", *Proc. Fifth Berkeley Symposium*, vol. 2, Pt. 2, 201–212 (1965).

84. M. Riesz, "Integrales de Riemann–Liouville et potentials", *Acta Sci. Math. Szeged* **9** (1938), 1–42.

85. C. A. Rogers, *Hausdorff Measures*, Cambridge University Press, Cambridge (1970).

86. C. A. Rogers and S. J. Taylor, "Functions continuous and singular with respect to a Hausdorff measure", *Mathematika* **8** (1961), 1–31.

87. M. Sharpe, "Sample path variations of homogeneous processes", *Ann. Math. Statist.* **40** (1969), 399–407.

88. J. Takeuchi, "A local asymptotic law for the transient stable process", *Proc. Japan Acad.* **40** (1964), 141–144.

89. ——, "On the sample paths of the symmetric stable processes in space", *J. Math. Soc. Japan* **16** (1964), 109–127.

90. J. Takeuchi and S. Watanabe, "Spitzer's test for the Cauchy process on the line", *Ztschr. Wahrsch'theorie & verw. Geb.* **3** (1964), 204–210.

91. S. J. Taylor, "On the connection between Hausdorff measures and generalized capacities", *Proc. Cam. Phil. Soc.* **57** (1961), 524–531.

92. ——, "The Hausdorff α-dimensional measure of Brownian paths in n-space", *Proc. Cam. Phil. Soc.* **49** (1953), 31–39.

93. ——, "The α-dimensional measure of the graph and the set of zeros of a Brownian path", *Proc. Cam. Phil. Soc.* **51** (1955), 265–274.

94. ——, "The exact Hausdorff measure of the sample path for planar Brownian motion", *Proc. Cam. Phil. Soc.* **60** (1964), 253–258.

95. ——, "Multiple points for the sample paths of the symmetric stable process", *Ztschr. Wahrsch'theorie & verw. Geb.* **5** (1966), 247–264.

96. ——, "Sample path properties of a transient stable process", *J. Math. Mech.* **16** (1967), 1229–1246.

97. S. J. Taylor and J. G. Wendel, "The exact Hausdorff measure of the zero set of a stable process", *Ztschr. Wahrsch'theorie & verw. Geb.* **6** (1966), 170–180.

98. H. F. Trotter, "A property of Brownian motion paths", *Ill. J. Math.* **2** (1958), 425–433.

99. N. Wiener, "Differential space", *J. Math. Phys.* **2** (1923), 131–174.

100. (i) ——, *I am a Mathematician*, Victor Gollancz, London (1956).

    (ii) P. Lévy, *Quelques Aspects de la Pensée d'un Mathématicien*, Albert Blanchard, Paris (1970).

5.4

Separability and Measurability for Stochastic Processes: A Survey

D. G. KENDALL

1. This expository article is intended to supplement my review elsewhere (in [4] and [5]) of what might be called 'Doob's theory of versions', by an examination of two especially important topics from the point of view adopted in the two papers just mentioned. My own grasp of these topics has been much hampered by a tendency in most of the existing literature to obscure the distinction between the topological and the measure-theoretic/probabilistic strands in the arguments. I have here tried to pull them apart in a systematic way, and hope that this polarization of the theory will be found helpful by others. I have taken the opportunity to set out analogues of some of Doob's theorems in a wider setting than is usual, the various 'real lines' which normally occur being replaced by distinct topological spaces. Technically this has an obvious value, in facilitating wider applications, and didactically also it is not to be despised, for it helps to emphasize the contrast between the topological and probabilistic concepts.

Another source of confusion in the customary presentation is notational. I have tried to remedy this by systematically using f for a generic element of a function-space Z^A, and X_a for the ath coordinate-projection ($a \in A$). In topological contexts we shall write f_a for the value of the function f at a, f_E for the range of the function f when its domain is restricted to the subset E of A, and $f|E$ for the function f itself when so restricted. In probabilistic contexts, when Z^A is the probability-space, we shall think of X_a as a random variable and X_a ($a \in A$) as a stochastic process with parameter-set A. Thus $X_a(f)$ and f_a have the same meaning (each is the value of f at a), but we think of $X_a(f)$ as the realization of the random variable X_a when the random element happens to be f, whereas we think

415

of f_a as the value at a of some prescribed function f. To link up this practice with some other notations, observe that X_a is what is sometimes written as $X_a(\cdot)$, and $X(f)$ is just f itself. We shall be much concerned with the 'evaluation mapping' which we can write as

$$(f, a) \rightarrow f_a$$

in topological contexts, and as

$$(f, a) \rightarrow X_a(f)$$

in probabilistic contexts. Occasionally a formula will have to be regarded from both points of view; the reader who is puzzled should then return to this statement of notational conventions and try writing it out in the alternative way.

I am very conscious of the fact that in notational difficulties there may be no wholly satisfactory solution; thus no one has yet found a really adequate name for the function $(\cdot)$. One has to do one's best, and hope for understanding.

2. To begin with let A and Z be two topological spaces, and let us make the following

Definition. An element f of the function-space Z^A will be called *separable* when its graph,

$$\text{Graph } (f) = \{(a, f_a) : a \in A\}$$

is a separable subspace of the product-space $A \times Z$.

We shall now list a large number of mutually equivalent necessary and sufficient conditions ($=$ nscs) for the separability of such an f. Apart from a few explanatory remarks the proofs of the equivalences will be left to the reader; they are almost self-evident when the nscs are arranged in the order in which they are presented here. It may be helpful to bear in mind that $f \,|\, E$ and f_E both denote the empty set when E itself is empty.

(*i*) Graph $(f \,|\, D)$ *is dense in* Graph (f) *for some denumerable subset* D *of* A.

Evidently the subset D must then be dense in A. Frequently we shall have some specific dense denumerable set D in mind, and then we shall speak of the *D-separability* of f. An element f of Z^A may be separable with respect to one such D, and not with respect to another. We now continue this list of mutually equivalent nscs for *D*-separability.

(*ii*) *If* $a \in U$ (*open*) *and if* $f_a \in G$ (*open*), *then there exists* $d \in D$ *such that*

$$d \in U \quad \text{and} \quad f_d \in G.$$

(iii) *For each open U and G (in A and Z respectively)*

$$f_U \text{ meets } G \text{ implies that } f_{U \cap D} \text{ meets } G.$$

(iv) *For each open U in A and closed F in Z,*

$$f_{U \cap D} \subset F \quad implies \quad f_U \subset F.$$

(v) *For each open U in A,*

$$f_U \subset \overline{f_{U \cap D}}.$$

(vi) *For each open U in A,*

$$\overline{f_U} = \overline{f_{U \cap D}}.$$

3. We shall now make the situation a little less general by making the following assumptions (to hold until further notice).

Assumption (a). *The 'state-space' Z is Hausdorff, second-countable and compact.*

Assumption (b). *The 'parameter-set' A is Hausdorff, second-countable and has D as a dense denumerable subset.*

Let U_n denote a generic member of a countable basis for the open sets in A, and let *the complement of* K_n denote a generic member of a countable basis for the open sets in Z; thus each K_n is closed and so compact (we shall call such a K_n *co-basic*), and we observe that each closed set in Z is a countable intersection of such co-basic closed (and so compact) sets. We can now continue with further equivalent nscs for D-separability.

(vii) *For each m and n,*

$$f_{U_m \cap D} \subset K_n \quad implies \quad f_{U_m} \subset K_n.$$

(viii) *For each m and n,*

$$either \quad f_{U_m} \subset K_n, \quad or \quad f_{U_m \cap D} \text{ meets } K_n^c.$$

Now K_n^c, the (open) complement of K_n, is a Baire set because of Assumption (a), and so it can be represented as a union of a sequence of compact sets, say by

$$K_n^c = \bigcup_r K_{n,r},$$

where each $K_{n,r}$ is compact. We now have another equivalent nsc.

(ix) *For each m and n,*

$$either \quad f_{U_m} \subset K_n, \quad or \quad f_{U_m \cap D} \text{ meets } K_{n,r} \text{ for some } r.$$

Here is yet another nsc.

(x) For each m and n, either

(1) $f_a \in K_n$ for all a in U_m,

or

(2) $f_d \in K_{n,r}$ for some r and some d in $U_m \cap D$.

Now condition (1) determines a compact set of fs, while condition (2) determines a set of fs which is a countable union of compacts. Thus we have

Proposition 1. *The D-separable fs in Z^A constitute a set Γ_D which belongs to the class $\mathscr{K}_{\sigma\delta}$.*

On transforming the nsc (v) we also have

Proposition 2. *The element f is in Γ_D if and only if*

(3) $f_a \in \overline{f_{U \cap D}}$

for each a in A and for each basic open set U containing a.

Notice that, for *fixed a*, condition (3) only restricts f at a countable set of arguments.

4. We now leave *topology*, and turn to *probability*. Let

(4) $(Z^A, \mathscr{B}_0, \mathrm{pr}; Z, A; X_a\,(a \in A))$

be 'the canonical model for a name-class of stochastic processes', in the sense of [4], so that $\mathscr{B}_0$ denotes the σ-algebra of Baire sets in Z^A, pr denotes the Daniell–Kolmogorov probability-measure on $\mathscr{B}_0$ uniquely determined by the finite-dimensional distributions (i.e. by the 'name' of the name-class) and X_a denotes the random variable which is the ath co-ordinate function. Of course A is the parameter-set and Z is the state-space for the process. As was explained in paragraph 1, we write f for a generic element of the probability-space Z^A (instead of the more customary ω, which we reserve for general 'versions'), and thus any set of fs referred to in paragraph 3 is a subset of the probability-space and can be thought of as a random event if it belongs to $\mathscr{B}_0$ (which often it will *not* do). Whether such a set of fs is an element of $\mathscr{B}_0$ or not, we can always assign to it its *outer* measure with respect to pr on $\mathscr{B}_0$, and as usual we shall call the set *thick* when this outer measure is equal to 1. We can now make the

Definition. The name-class will be called *D-separable* when the set Γ_D of all D-separable fs is thick.

We know (from the discussion in [4]) that the thickness of Γ_D is the nsc for the name-class to contain at least one version which is separable, with respect to D, in the sense of Doob; that is, which has D-separable 'sample paths' outside a measurable set-of-paths of probability-measure zero.

We also know (again from [4]) that when this nsc is satisfied then the name-class contains a uniquely important member, *the canonical extension of the canonical model* (4), in which $\mathscr{B}_0$ is extended to the smallest σ-algebra big enough to contain the additional member Γ_D, and in which pr gives measure 1 to Γ_D.

In other words, the D-separability of the name-class is the nsc for the name-class to contain at least one version which is 'nice', i.e. is D-separable in the sense of Doob, and when it is satisfied there is one such 'nice' version *which has a pre-eminent status*.

The D-separability of the name-class is obviously a condition bearing only on the finite-dimensional distributions, but the direct verification that it is satisfied will not necessarily be a simple matter. The next proposition is of considerable assistance in this respect.

Proposition 3. *The name-class is D-separable if and only if, for the Kakutani model in which $\mathscr{B}_0$ is replaced by the (larger) σ-algebra $\mathscr{B}$ of Borel sets, and in which pr is (uniquely) extended to a probability-measure P on $\mathscr{B}$ which is regular, we have*

$$P(\Gamma_D) = 1.$$

Proof. We noted in [4] that each set in the class $\mathscr{K}_{\sigma\delta}$ (necessarily a Borel set) has a P-measure equal to its outer measure with respect to pr on $\mathscr{B}_0$. Now use Proposition 1. (Notice that, as a consequence of this proposition, when the name-class is D-separable then the Kakutani model is an extension of that *canonical extension of the canonical model* which makes Γ_D measurable and gives it measure 1.)

5. From the general arguments which were collected in [4] and [5] we know that if there exists some version in a name-class which is D-separable, as a version, in the sense of Doob, then that name-class must itself be D-separable (because Γ_D must then be (pr,$\mathscr{B}_0$)-thick). We also know that if on the top of some version in the name-class we have been able to build a standard modification which yields a new process that is D-separable in the sense of Doob, then again the name-class must be D-separable.

If we look at this matter from the opposite point of view, we already know that D-separability of the name-class certainly does imply the existence of a D-separable version, for we have only to cite as an example the canonical extension of the canonical model which is defined by its minimality coupled with the requirement that Γ_D is to be measurable and of measure 1. As Meyer [6] has remarked, Doob's further theorem, about the existence of D-separable modifications, lies much deeper; when generalized to the present situation it tells us that if the name-class is D-separable then *every complete* version therein allows a standard modification to be built on top of it in such a way that *all* the sample paths become D-separable in the topological sense.

We propose to give a proof of this result in the (Z, A) case, following Doob's own proof [1] in the $(\bar{R}, R)$ case, but first we must make the

Definition. We shall say that a name-class *satisfies Doob's condition relative to a given dense denumerable subset D of A* when, for each a in A, the Baire subset B_0^a of Z^A defined by the condition

$$(5) \qquad\qquad f_a \in \bigcap_{\substack{a \in U, \\ U \text{ open}}} \overline{f_{U \cap D}}$$

has pr-measure 1.

Notice that the intersection at (5) is in no way altered if we restrict the open sets U to be *basic* open sets, so that condition (5) says essentially that, for each basic U_m containing a, and for each co-basic K_n covering $f_{U_m \cap D}$, the element f_a of Z must lie in K_n. (The notation here is the same as that in paragraph 3.) Thus condition (5) can be replaced by

$$(5') \qquad \text{for each } n, \text{ and for each } m \text{ such that } a \in U_m,$$
$$\text{either } f_a \in K_n, \text{ or } f_d \in K_{n,r} \text{ for some } r \text{ and some}$$
$$d \text{ in } U_m \cap D.$$

This is very like our nsc (x), *but 'a' is here fixed*. It is this fixing of a which makes condition (5) ($= (5')$) determine a Baire set B_0^a, and this is why Doob's condition,

$$\text{pr}(B_0^a) = 1 \quad \text{for each } a \in A,$$

is indeed a condition *on the name-class*. We now prove

Proposition 4. *Doob's condition is necessary and sufficient for the name-class to be D-separable, and when it is satisfied then on top of every complete version in the name-class we can build a standard modification for which each individual sample path is topologically D-separable.*

Proof. It will be enough to show that we can build a standard modification of the required type on top of the *completed canonical model*,

$$(6) \qquad\qquad (Z^A, \mathscr{B}_0^+, \text{pr}; Z, A; X_a \, (a \in A)),$$

whenever Doob's condition is satisfied by the name-class. For if this can be done, and if the standard modification of the model (6) is

$$\mathbf{M}f = \{Y_a(f) \colon a \in A\},$$

then when $(\Omega, \mathscr{F}, \dots, W_a \; (a \in A))$ is a complete version in the same name-class we can modify its generic path $W(\omega)$ to

$$\mathbf{M}W(\omega) = \{Y_a(W(\omega)) \colon a \in A\},$$

which will be topologically D-separable for every ω, and so we get the desired standard modification by putting onto $(\Omega, \mathscr{F})$ the new system of random variables

$$Y_a(W) \quad (a \in A).$$

If we can carry out the above programme then we shall have proved the proposition as a whole, because we shall have the closed cycle of implications,

Doob's condition

implies

there exists a D-separable standard modification

implies

Γ_D is thick with respect to pr on $\mathscr{B}_0$

implies

the name-class is D-separable

implies

$$P(\Gamma_D) = 1$$

implies

$$P(B_0^a) = 1 \quad \text{for each } a$$

implies

$$\mathrm{pr}\,(B_0^a) = 1 \quad \text{for each } a$$

and this is

Doob's condition.

(Note that $\Gamma_D \subset B_0^a$, because of Proposition 2.)

We now set about the construction of the standard modification on top of the model (6), assuming that Doob's condition holds. Let $K_a(f)$ be the *non-empty* compact intersection of the family

$$\overline{f_{U \cap D}} \quad (a \in U)$$

of compact sets having the finite-intersection property. If f is a generic element of Z^A, we define its transform $\mathbf{M}f$ by

$$(\mathbf{M}f)_a = f_a \in K_a(f) \quad \text{whenever } B_0^a \text{ contains } f,$$

$$(\mathbf{M}f)_a = z_a \in K_a(f) \quad \text{for every other } a \text{ in } A.$$

(We can make this slightly less offensively non-constructive in most cases. For example, when $Z = \bar{R} = R \cup \{-\infty, \infty\}$, we can take

$$z_a = \max K_a(f).)$$

Now it is obvious from condition (5) that $B_0^a = Z^A$ when $a \in D$, and so

$$(\mathbf{M}f)_d = f_d \quad \text{for each } d \text{ in } D.$$

This means that the set on the right-hand side of condition (5) is in no way altered when we replace f there by $\mathbf{M}f$, and then it is easily verified that $\mathbf{M}f$ satisfies condition (5) *for all f and a*, whence by Proposition 2 it follows that $\mathbf{M}f$ is topologically D-separable for every f, as required.

Examination of the construction shows that $\mathbf{M}$ is a standard modification; it has the required measurability properties because if B_0 is a generic Baire set in Z, then the sets $\{f : (\mathbf{M}f)_a \in B_0\}$ and $\{f : f_a \in B_0\}$ differ at most by a subset of the measurable null set $(B_0^a)^c$, and by the assumed completeness this is also a measurable (null) set. As for the other requirement, that $\mathrm{pr}(\{f : (\mathbf{M}f)_a \neq f_a\}) = 0$ for each a, this is likewise satisfied in virtue of Doob's condition.

6. We now fix the name-class, and consider the effect of altering D. There are three important matters which come under this heading; first we may ask, given an arbitrary name-class, is it necessarily separable for at least one D, which may depend on the name-class being considered? Secondly we may ask whether by imposing some light restriction on the name-class we can ensure that it will be D-separable for *every* dense denumerable D. Finally we may query the absolute status of probabilities calculated from separable versions. These questions were answered, in the classical $(\bar{R}, R)$ case, by Doob. In carrying out the extensions to the (Z, A) case I have been much influenced by Meyer [6] and Neveu [8].

We shall find that for these questions a convenient starting point is provided by Proposition 3. The *first question* then becomes: can we choose D so that $P(\Gamma_D) = 1$, for the given name-class? From Proposition 1 we know that Γ_D is always in the class $\mathscr{K}_{\sigma\delta}$, and of course it will have P-measure 1 if and only if each of the 'factors' implied by the δs has P-measure 1. Thus the name-class will be D-separable if and only if D can be so chosen that

(7) $P(\{f : f_{U \cap D} \subseteq F \text{ implies } f_U \subseteq F\}) = 1$

for each fixed basic open U and each fixed co-basic closed F (of course such Fs are compact).

In what immediately follows, let J denote a generic finite non-empty subset of A. Now from the regularity of P (see Meyer [6], p. 24) we know that

$$(8) \qquad P(\{f : f_U \subset F\}) = \inf_{J \subset U} P(\{f : f_J \subset F\}),$$

and so for each U and F we can find an increasing sequence $J_1, J_2, \ldots$ of finite non-empty subsets of U such that

$$P(\{f : f_{J_q} \subset F\})$$

exceeds the infimum at equation (8) by less than $1/q$. We shall therefore have, on writing $D(U, F)$ for the (denumerable) union of these J_qs,

$$(9) \qquad P(\{f : f_U \subset F\}) = P(\{f : f_{U \cap D(U,F)} \subset F\}).$$

We now let U and F range respectively through $U_1, U_2, \ldots$ and $K_1, K_2, \ldots$, and form the (denumerable) union of the $D(U, F)$s, calling this D. With this D we shall have equation (7), and as a corollary D must be dense. Thus our first question has been answered affirmatively, and we have

Proposition 5. *Each name-class is D-separable for some dense denumerable D.*

In concrete calculations this result may not be very helpful, because our proof of it does not tell us how D is to be constructed. We therefore turn to the second question, noting first that a D will make a fixed name-class D-separable if and only if the term on the right-hand side of the inequality,

$$P(\{f : f_U \subset F\}) \leqslant P(\{f : f_{U \cap D} \subset F\})$$

(which always holds), cannot for any choice of U and F be reduced by the adjunction to D of *one single point*. This follows (using (8)) by applying Boole's inequality to the P-measure of the difference-set

$$(10) \qquad \{f : f_{U \cap D} \subset F\} \backslash \{f : f_{U \cap D^+} \subset F\},$$

where D^+ is D with any finite set adjoined to it. Thus we have

Proposition 6. *For a denumerable D to be such that a given name-class is D-separable, it is necessary and sufficient that*

$$(11) \qquad \mathrm{pr}(\{f : f_{U_m \cap D} \subset K_n, \text{ and } f_a \in K_n^c\}) = 0$$

for each basic open U_m in A, each co-basic closed K_n in Z and each a in U_m.

The detailed proof consists in the use of equation (8), the Boolean estimation of the P-measure of the set (10), and the fact that P-measure and pr-measure coincide on the Baire set at equation (11).

The importance of a solution to our *second question* is that, when an acceptable restriction on the name-class makes its answer affirmative, then we are in the happy position of being able to choose D to suit our convenience. In the $(\bar{R}, R)$ situation Doob proved that any dense denumerable D will make the name-class D-separable, provided that the name-class is 'continuous in probability'. We shall show that this is true in the (Z, A) case also, but first we must consider how 'continuity in probability' is to be defined in the more abstract situation.

We recall that throughout this whole discussion we have been making Assumption (a). Now this implies that Z is metrizable, and that the Banach space $C(Z)$ of real-valued continuous functions over Z (with the supremum norm) is separable. Let then φ denote a generic member of some countable dense set in $C(Z)$. For any compact K in Z, and any subset E whatsoever in A, we shall then have

$$f_E \subset K \text{ if and only if, for every } \varphi, \ (\varphi \circ f)_E \subset \varphi(K).$$

Here $\varphi \circ f$ denotes the composition 'f followed by φ', and the italicized fact is easily proved on noting that if f_E is not covered by K then (for some a in E) $\{f_a\}$ and K are disjoint, and so we can find a real-valued continuous function which vanishes over K and assumes the value 1 at f_a. On approximating to this sufficiently well by one of the φs we immediately get the implication in one direction, while that in the other direction is trivial.

It is also worth noticing that $\varphi(K)$, as the continuous image of a compact set, is itself compact, and that each φ, being continuous from Z to $\bar{R}$, is necessarily Baire–Baire ($=$ Borel–Borel) measurable.

Thus in testing a name-class for D-separability we can replace the Z-valued stochastic process occurring in the test by an $\bar{R}$-valued stochastic process, via a composition with φ, *provided that we impose the test simultaneously for every φ*. As the set of φs is countable, this last clause will not create any difficulties.

We therefore make the following definition, extending the customary definition of 'continuity in probability' to the abstract case.

Definition. A (Z, A) stochastic process will be called *continuous in probability* when for each φ in a countable dense set in $C(Z)$, for each a in A, and for each positive ε_1 and ε_2,

(12) $$\mathrm{pr}\left(\{\omega : \left|(\varphi \circ W_b)(\omega) - (\varphi \circ W_a)(\omega)\right| > \varepsilon_1\}\right) < \varepsilon_2$$

for every b in some open neighbourhood of a which depends on $\varphi, a, \varepsilon_1$ and ε_2. A name-class will be called continuous in probability if all versions therein enjoy this property.

We need only apply the test in the definition to any one version

$$(\Omega, \mathscr{F}, \text{pr}; Z, A; W_a \, (a \in A))$$

in the name-class, for it is satisfied by all such versions if it is satisfied by any one of them. Notice that inequality (12) *is* meaningful, i.e. $\varphi \circ W_b$ is a random variable for each fixed b, because φ is Baire–Baire measurable. Notice also that if a name-class satisfies the requirements of the definition with respect to any one countable dense set of φs in $C(Z)$, then it does so for any other such set of φs. Thus the φs serve only an auxiliary role, and do not impair the absolute character of the definition.

Now consider a fixed name-class which is known to be continuous in probability, and any fixed denumerable dense set D in A. Take a fixed denumerable dense set of φs in $C(Z)$ and label them as $\varphi_1, \varphi_2, \ldots$. Let a be any fixed point of A, and U_m some fixed basic open set containing a. We shall prove that the condition (11) in Proposition 6 is satisfied, where the probabilities will be calculated using the canonical model in the name-class, (4). After setting

$$\varepsilon_1 = 1/q, \quad \varepsilon_2 = 1/q^3,$$

we write $N(\varphi; q)$ for the open neighbourhood of a within which inequality (12) holds. Without loss of generality we can suppose that $N(\varphi; q)$ shrinks, for fixed φ, as q increases. We now define a new open neighbourhood of a by writing

$$N(q) = N(\varphi_1; q) \cap N(\varphi_2; q) \cap \ldots \cap N(\varphi_q; q) \cap U_m,$$

and we observe that the denumerable dense set D will certainly meet $N(q)$. If we label D in some way by the positive integers, we can define d_q to be the 'first' element of D which lies in $N(q)$.

Now consider the Baire set in Z^A defined by

$$B_0^q = \left\{ f : |\varphi_j(f_{d_q}) - \varphi_j(f_a)| \leqslant \frac{1}{q} \quad \text{for } j = 1, 2, \ldots, q \right\}.$$

The complement of B_0^q is the union of the q Baire sets

$$\left\{ f : |\varphi_j(f_{d_q}) - \varphi_j(f_a)| > \frac{1}{q} \right\},$$

where $j = 1, 2, \ldots, q$, and *each one* of these last Baire sets has a probability of $1/q^3$ at most, because d_q is in $N(q)$ and so in $N(\varphi_j; q)$ for each of the values $j = 1, 2, ..., q$. Thus

$$\sum_{q \geq 1} \mathrm{pr}\left((B_0^q)^c\right) \leqslant \sum q/q^3 < \infty,$$

and therefore by the Borel–Cantelli lemma we can conclude that

$$\mathrm{pr}\,(\liminf B_0^q) = 1.$$

In other words, outside a Baire set of fs of probability zero we shall have

$$|\varphi_j(f_{d_q}) - \varphi_j(f_a)| \leqslant \frac{1}{q}, \quad \text{if } q \geqslant \max\,(j, Q(a, U, f)).$$

Outside the same Baire set of probability zero, therefore, we shall have

$$\lim_{q \to \infty} f_{d_q} = f_a,$$

and so equation (11) must be satisfied, because every $d_q \in U_m \cap D$ (see Figure 1). We summarize our results in

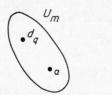

The situation in A   The situation in Z

Figure 1.

Proposition 7. *If a (Z, A) name-class is continuous in probability, then it is D-separable for every choice of the dense denumerable set D in A.*

We have seen (Proposition 5) that there is always at least one dense denumerable D for which a given name-class is D-separable, and of course there will always be more than one, even if the name-class is not continuous in probability; for example, any denumerable set which contains D as a subset will do just as well. (This is evident from Proposition 6.) Suppose then that a given name-class is D_1-separable and also D_2-separable. Let U be open in A, and K compact in Z, and let

$$E(U, K) = \{f : f_U \subset K\}.$$

Then we know that there exists a complete version which for all such U and K makes $E(U, K)$ measurable and gives to it the same measure as the

Baire set

$$E_1(U, K) = \{f : f_{U \cap D_1} \subseteq K\},$$

and that there exists another complete version which for all such U and K makes $E(U, K)$ measurable and gives to it the same measure as the Baire set

$$E_2(U, K) = \{f : f_{U \cap D_2} \subseteq K\}.$$

These measures can be calculated using the Daniell–Kolmogorov model, for they are measures of Baire sets, *but are they equal*? If the answer to this *third question* were negative, the theory of separability might still have some technical use, but it would be irrelevant to the applications. The correct answer is, however, *affirmative*, and it is convenient to formulate this as the important

Remark. For each name-class there is an infinity of ways of choosing the dense denumerable set D in such a way that the name-class is D-separable, and then in several different ways we can set up a complete D-separable version of the process for which $E(U, K)$ (U open in A, K compact in Z) is measurable and has a probability equal to the pr*-measure of the Baire set*

$$\{f : f_{U \cap D} \subseteq K\}.$$

This evaluation of the probability of $E(U, K)$ is independent of the choice of D and of the choice of complete D-separable version, and in fact the probability assigned to $E(U, K)$ is equal to the Kakutani measure of the Borel set $E(U, K)$.

What has been said about $E(U, K)$ is equally true of every set in the Doob σ-algebra (the σ-algebra generated by the Baire sets and the sets $E(U, K)$). All Doob sets are Borel sets, but a Borel set need not be a Doob set.

Some words of proof are of course necessary. Suppose then that the name-class is D_j-separable for $j = 1, 2$; then Γ_{D_j} is a Borel set with Kakutani measure 1 for each value of j (by Proposition 3), and accordingly the intersection $\Gamma = \Gamma_{D_1} \cap \Gamma_{D_2}$ is also a Borel set of Kakutani measure 1 and indeed (in virtue of Proposition 1) we can say what kind of Borel set it is, namely a $\mathscr{K}_{\sigma\delta}$. By an argument given in [4] it follows that Γ is a Borel set which has *outer* measure 1 with respect to pr and the Baire σ-algebra $\mathscr{B}_0$. However, it is clear from our definitions that

$$E_1(U, K) \triangle E_2(U, K)$$

will be a Baire set disjoint with Γ, and so it must have pr-measure zero; that is, we must have

$$\mathrm{pr}\,(E_1(U, K)) = \mathrm{pr}\,(E_2(U, K),$$

as claimed in the remark, and moreover the common value of these two probabilities must be equal to the Kakutani measure $P(E(U, K))$ of the compact Borel set $E(U, K)$. So much for the first paragraph of the remark.

The claims in the second paragraph go a little deeper and involve a new concept, that of the *Doob σ-algebra*. I hope this name will be thought appropriate, though it should be remarked that the problem of attaching probabilities to what we call Doob sets was also examined at an early stage by Pitt [9]. Clearly the Doob σ-algebra $\mathscr{D}$ is that generated by the sets $E(U, K)$ together with all sets of the form

$$F(a; K) = \{f : f_a \in K\} \quad (a \in A)$$

(for these latter generate the Baire sets). Moreover, in choosing a class of sets to generate the Doob σ-algebra we note that we need only consider $E(U, K)$ and $F(a, K)$ for the countable system of basic Us and the countable system of co-basic Ks, though of course we must consider the (in general uncountable) system of *all* as in A. In the special case when $A = R$ and $Z = \bar{R}$, this basis for the Doob σ-algebra will have cardinality c, and so the Doob σ-algebra itself will have cardinality c (Halmos, [3], p. 26, example 9). As the cardinality of $\bar{R}^R$ is 2^c, and as every singleton-set is a Borel set, this example makes it quite clear that *Borel sets need not be Doob sets*. On the other hand it is not difficult to show (see, for example, Nelson, [7], Theorem 3.1) that when Z and A are *both* second-countable compact Hausdorff spaces (i.e. when we add compactness to our standard assumptions about A), then the set of all continuous mappings from A to Z is a Doob set. (Further results of this sort can be extracted from Nelson's paper.)

We now have to substantiate our claim that each Doob set E (which must, of course, be a Borel set) has a unique probability measure, in the sense that whatever dense denumerable D we choose from among those permitted to us (i.e. those which make the name-class D-separable), and whatever version we then construct which is complete and D-separable as a stochastic process in the sense of Doob, the value we get for $\mathrm{pr}(E)$ will be the same, and will thus of course necessarily coincide with the value $P(E)$ assigned to it by the coordinate-process on function-space using Kakutani measure.

We deal with this last question as follows. First we note the topological fact that (for $j = 1, 2$)

$$\Gamma_{D_j} \cap E(U, K) = \Gamma_{D_j} \cap E_j(U, K).$$

Now let $\mathscr{E}$ denote the class of all sets of the form $E(U, K)$ or $F(a, K)$, and let $\mathscr{E}_j$ denote the class of all sets of the form $E_j(U, K)$ or $F(a, K)$; here a

is to run freely through the whole of A. Then evidently we shall have (for $j = 1, 2$)

$$\Gamma_{D_j} \cap \mathscr{E} = \Gamma_{D_j} \cap \mathscr{E}_j,$$

and so by an elementary theorem of measure theory (Halmos [3], p. 25) we can conclude that

$$\Gamma_{D_j} \cap \mathscr{D} = \Gamma_{D_j} \cap \mathscr{B}_0 \quad (\text{for } j = 1, 2),$$

because $\mathscr{E}_j$ for each j generates the σ-algebra of Baire sets, $\mathscr{B}_0$, and because Z^A belongs to each of $\mathscr{E}$ and $\mathscr{E}_j$.

Now let E be any Doob set. Then there must exist Baire sets E_1 and E_2 such that

$$\Gamma_{D_1} \cap E = \Gamma_{D_1} \cap E_1, \quad \text{and} \quad \Gamma_{D_2} \cap E = \Gamma_{D_2} \cap E_2;$$

of course these Baire sets need by no means be unique. Consider the intersection of the symmetric difference $E_1 \triangle E_2$ with Γ. If f is a typical element of it, then say f belongs to E_1 and to Γ, but not to E_2; thus it belongs to Γ_{D_1} and so (by the first of the above equalities) to E; however, it also belongs to Γ_{D_2} and so (by the second of the above equalities) to E_2, giving a contradiction. This and the mirror-image argument show that the Baire set $E_1 \triangle E_2$ is disjoint with the thick set Γ, and so must have pr-measure zero, whence

$$\mathrm{pr}\,(E_1) = \mathrm{pr}\,(E_2).$$

In practice, when we use a complete D-separable version, we replace a Doob set E by a Baire set E_D such that

$$\Gamma_D \cap E = \Gamma_D \cap E_D,$$

and note that our version not only makes E measurable (for it is complete and makes Γ_D measurable and full) but also assigns to it the *calculable* measure

$$\mathrm{pr}\,(E_D).$$

We have shown that this number $\mathrm{pr}\,(E_D)$ depends neither on D nor on the version, so it has an 'absolute' status. Moreover, Proposition 5 tells us that *every* name-class is D-separable for *some* D, so that *separability theory attaches a unique probability* $\mathrm{pr}\,(E)$ *to every Doob set* E, *whatever name-class of stochastic processes we are concerned with.*

In principle (and often in practice) we can evaluate $\mathrm{pr}\,(E)$ by using the fact that it must be the Kakutani value $P(E)$. Thus the Kakutani model provides a short cut to all results which can be obtained by means of separability theory, but the very fact that it does more than this (i.e. fixes the measures of all *non-Doob* Borel sets) implies an inflexibility which can be inappropriate on occasion.

7. We are now about to turn our attention to process-measurability. This is an example of a 'nice property' which is very convenient, when present, and which cannot be expressed by requiring that the set $\{f : f \in \Gamma\}$ shall be 'big' in some suitable sense for an appropriate Γ, because it is not a restriction on the individual sample paths f, but rather on 'the process as a whole'.

Consider first the evaluation-mapping from the Cartesian product $Z^A \times A$ into Z, defined by

$$(13) \qquad\qquad (f, a) \to f_a.$$

We can ask if this is measurable relative to the σ-algebras

$$(14) \qquad\qquad \mathscr{B}_0(Z^A) \otimes \mathscr{B}(A), \quad \text{and} \quad \mathscr{B}_0(Z),$$

where $\mathscr{B}(A)$ denotes the σ-algebra of Borel sets in A and $\otimes$ denotes the operation of forming a product σ-algebra. Notice that as we do not have any *measure* on the measurable subsets of $Z^A \times A$, *it makes no sense* to ask whether, after forming the product σ-algebra, we are to complete under the measure. It is true that, if A happens to be the real line, we might attach some significance to Lebesgue measure on its Borel sets, and then $\mathrm{pr} \times$ Lebesgue would be a (σ-finite) measure on the measurable sets in $Z^A \times A$, so that completion with respect to that would then be possible. Normally, however, even in this very special situation, *completion of the product σ-algebra is not recommended*. The reason for this is that if we eschew completion, and none the less manage to prove process-measurability, we shall then be able to make an extremely important deduction which would otherwise be false. (We shall comment further on this point in a moment.)

We now return to formulae (13) and (14), and observe that the question which we asked about the measurability of the evaluation-mapping is one which *never* has an affirmative answer save in uninteresting trivial cases. For were it to permit an affirmative answer, we could immediately deduce (from the fact that any 'section' of a product-measurable set is measurable) that the mapping

$$a \to f_a$$

would have to be measurable *for every f*, and since 'every f' literally means every possible Z-valued function over A, we clearly cannot expect that to be true.

Suppose, however, that instead of thinking about the canonical model (4), we turn to some version in the same name-class, say

$$(\Omega, \mathscr{F}, \dots, W_a \, (a \in A)).$$

The corresponding question will now take the form, is the mapping

$$(\omega, a) \to W_a(\omega),$$

from $\Omega \times A$ into Z, measurable with respect to the σ-algebras

$$\mathscr{F} \otimes \mathscr{B}(A), \quad \text{and} \quad \mathscr{B}_0(Z)?$$

This time there are situations in which an affirmative answer *can* be given, and when this is so then the 'section' theorem yields the highly significant conclusion (alluded to above) that the mapping

$$a \to W_a(\omega)$$

is measurable, and *the sample path is measurable for every individual ω.*

Even in the classical $(\bar{R}, R)$ case there are still some mysteries about the process-measurability problem; for a recent statement of the position see Dudley [2]. The instances which occur in practice, however, are mostly covered by an important theorem of Doob which says, in the $(\bar{R}, R)$ case, that if the name-class is continuous in probability then on top of *every* version therein (complete or not!) we can build a standard modification such that for the new process the desired measurability property holds— and over and above this we can simultaneously arrange that *every* sample path is D-separable with respect to an arbitrarily assigned dense denumerable set D in A. Obviously to prove this theorem it suffices to prove it first for the (uncompleted) canonical model, and then the measurability and separability properties can be 'carried back' to an arbitrary version.

We shall prove an analogous result in the (Z, A) case, but first we must develop a technical device which will be important for the proof.

8. *For this section of the paper we assume Assumption (a) only*; the parameter-set A, when it occurs at all, will be *quite arbitrary*.

In the proof of the classical $(Z = \bar{R} = R \cup \{-\infty, \infty\}, A = R)$ form of the Doob measurability theorem (for which see, for example, Neveu [8]) a vital role is played by the fact that $\limsup Y^k$ is measurable whenever each Y^k is measurable. As soon as we turn to Z-valued random variables we are held up by the fact that the usual limit-superior operation is no longer available.

On analysing the contribution that the limit-superior operation makes to the proof of the theorem in the real-variable case, we find that it is exactly the following features of the operation which are essential:

(*i*) $\limsup Y^k$ is measurable whenever each Y^k is measurable.

(*ii*) $\limsup y^k$ lies in the cluster-set (and so in the closure of the track) of the sequence $\{y^k : k = 1, 2, \ldots\}$.

Here by the 'track' of a sequence we mean its exact range when it is considered as a mapping from the positive integers into the target-space, while a point of the target-space is said to lie in the 'cluster-set' of the sequence if and only if each neighbourhood of that point is hit by the sequence infinitely often (i.o.). (We shall then call it a cluster-point for the sequence.) We shall need some of the properties of cluster-sets for sequences whose target-space Z satisfies Assumption (*a*). For example, a point z lies outside the cluster-set C of the sequence $\{z^k: k = 1, 2, ...\}$ if and only if it has a neighbourhood which is ultimately avoided by the sequence. Accordingly the cluster-set C is closed, and so compact. Evidently if we pass from the given sequence to a subsequence, then the cluster-set shrinks. From the compactness of Z we see that the cluster-set must be non-empty. We have already parenthetically remarked that the cluster-set is always a part of the closure of the track. Finally we have the very important fact that *the cluster-set reduces to a singleton* $\{z\}$ *when and only when* $\lim z^k = z$ *exists.*

If, therefore, in our extended definition we can preserve (*i*) and (*ii*), then we shall be able to construct a measurable mapping Y which coincides with the limit of the measurable mappings Y^k wherever that limit exists, and which for each fixed argument selects as 'value' a point in the target-space which lies in the cluster-set and so in the closure of the track of the sequence $Y^1, Y^2, ...$ at that argument.

We first note that in the simple case when $Z = \bar{R}$,

$$(15) \qquad \limsup z^k = \sup\{\rho: z^k > \rho \text{ i.o.}\}.$$

Here (and also later) ρ denotes a *rational*. It is in the form (15) that we shall generalize the limit-superior operation.

We now turn to the important case when $Z = \bar{R}^\infty$, so that

$$z^k = \{z_n^k: n = 1, 2, ...\}.$$

We recall that projections are continuous, and so send compact sets into compact sets; thus projections of the cluster-set C of the sequence $\{z^k: k = 1, 2, ...\}$ are compact and non-empty. Accordingly the projection of C onto the first coordinate axis is compact and non-empty, and so has a largest element, which will in fact be the limit superior,

$$(16) \qquad w_1 = \sup\{\rho: z_1^k > \rho \text{ i.o.}\},$$

of the sequence $\{z_1^k: k = 1, 2, ...\}$ of first components.

Now look at the intersection of C with the (closed) hyperplane

$$\{z: z_1 = w_1\}.$$

This is non-empty and compact. Its projection onto the first coordinate axis is w_1 and its projection onto the second coordinate axis is a non-empty compact set with largest element w_2, say. Next we look at the intersection of C with the (closed) subspace

$$\{z: z_1 = w_1, z_2 = w_2\};$$

this is non-empty and compact, its projection onto the $(1, 2)$-plane is (w_1, w_2) and its projection onto the third coordinate axis is a non-empty compact set with largest element w_3. And so on.

In this way we arrive at a point

$$w = \{w_n: n = 1, 2, ...\},$$

and I claim that w *belongs to* C. For look at the (non-empty) compact sets

$$\{z: z_1 = w_1, z_2 = w_2, ..., z_n = w_n, z \in C\} \quad (n = 1, 2, ...);$$

these form a descending nest and so from compactness have a common point, which can only be w, so that w belongs to each member of the nest, and therefore to C. We now make the

Definition. Let $\{z^k: k = 1, 2, ...\}$ be a sequence with track in $\bar{R}^\infty$. Then we shall write

$$\underset{k \to \infty}{\text{Lim Sup }} z^k$$

for the point w in the cluster set constructed in the manner just described.

It is obvious from the construction that property (*ii*) has been preserved, but the verification of property (*i*) is less straightforward. We shall achieve this by first making a second (equivalent) construction of w.

The new definition will be a recurrent one, starting with equation (16) as the definition of w_1, and then defining w_{n+1} in terms of $w_1, w_2, ..., w_n$ as follows:

(17) $$w_{n+1} = \inf_{r \geqslant 1} \sup \{\rho: \text{the sequence hits } E_{r,n}(\rho) \text{ i.o.}\}.$$

Here $E_{r,n}(\rho)$ denotes the following subset of $Z = \bar{R}^\infty$;

(18) $$E_{r,n}(\rho) = \{z: |z_j - w_j| < \frac{1}{r} \text{ for } j \leqslant n, \text{ and } z_{n+1} > \rho\}.$$

Let us show that the two definitions are equivalent; to this end we temporarily write $\bar{w}_{n+1}$ for the right-hand side of equation (17), and we suppose that we have already shown that $w_1 = \bar{w}_1, w_2 = \bar{w}_2, ..., w_n = \bar{w}_n$. As we have in fact defined $\bar{w}_1$ by equation (16), which forces it to coincide with w_1, this is an acceptable inductive hypothesis. Now we know that the (closed) subspace

$$\{z: z_1 = w_1, z_2 = w_2, ..., z_n = w_n\}$$

meets the cluster-set C in a non-empty compact set which projects into a non-empty compact set on the $(n+1)$st coordinate axis whose largest element is w_{n+1}. Accordingly there exists a cluster-point for the sequence whose first $n+1$ coordinates are $w_1, w_2, \ldots, w_{n+1}$, and the set (18) will be an open neighbourhood of this cluster point for the sequence, *if $\rho < w_{n+1}$.* When this happens, the sequence will hit $E_{r,n}(\rho)$ i.o., and so it is clear that the supremum which occurs in equation (17) cannot be less than w_{n+1}, whatever r may be. That is, we have shown that

$$(19) \qquad \qquad \bar{w}_{n+1} \geqslant w_{n+1}.$$

We cannot have the strict sign of inequality, however. We know that there is no cluster-point for our sequence in the compact set

$$(20) \qquad \qquad \{z : z_1 = w_1, \, z_2 = w_2, \ldots, z_n = w_n, \, z_{n+1} \geqslant \rho_0\},$$

for every $\rho_0 > w_{n+1}$. Thus the compact set (20) can be covered by finitely many basic open sets each of which is visited at most finitely often, and so the set $E_{r,n}(\rho_0)$ is visited at most finitely often, *for all sufficiently large r,* when $\rho_0 > w_{n+1}$. In other words,

$$\bar{w}_{n+1} \leqslant \rho_0 \quad \text{whenever } \rho_0 > w_{n+1},$$

and this precludes inequality at (19).

We can now use the second definition of $\lim \sup z^k$ to show that property (*i*) has been preserved. Let us make $\bar{R}$ and $Z = \bar{R}^\infty$ into measurable spaces by giving to each its σ-algebra of Baire (= Borel) sets, and let $(U, \mathscr{U})$ be an arbitrary measurable space and suppose that we have a sequence of measurable mappings

$$Y^k : U \to Z = \bar{R}^\infty \quad (k = 1, 2, \ldots).$$

Let us define

$$\tilde{Y} : U \to Z = \bar{R}^\infty$$

by

$$\tilde{Y}(u) = \operatorname*{Lim\,Sup}_{k \to \infty} Y^k(u);$$

it will be natural to call this new mapping $\operatorname{Lim\,Sup} Y^k$, and we have to show that it is measurable. Now if we introduce the component mappings associated with an arbitrary mapping $X : U \to Z = \bar{R}^\infty$ by

$$X_n(u) = n\text{th component of } X(u) \in Z = \bar{R}^\infty,$$

then we can say that X is a measurable mapping if and only if each X_n is a measurable mapping, in virtue of the way Baire sets are defined in this Z, for they can be taken to be just the measurable sets associated with the

Cartesian-product structure. But $\tilde{Y}_n(u)$ (for fixed u) can be identified with w_n in equations (16), (17) and (18), when the sequence $\{z^k \colon k = 1, 2, \ldots\}$ referred to in that definition is given by

$$z_n^k = Y_n^k(u) \quad (k = 1, 2, \ldots; n = 1, 2, \ldots).$$

From equations (17) and (18) (and the initiating definition (16)) it is now trivial, if tedious, to show recurrently that each of $\tilde{Y}_1, \tilde{Y}_2, \ldots$ is measurable, and so that $\tilde{Y} = \operatorname{Lim} \operatorname{Sup} Y^k$ is measurable, as required.

We now turn to the general situation where Z is *any* space satisfying Assumption (*a*). If $C(Z)$ denotes the Banach space of continuous real-valued functions over Z, we know that $C(Z)$ is separable and so we let Φ denote any countable dense set $(\varphi_1, \varphi_2, \ldots)$ of elements of $C(Z)$. This set Φ will be fixed in the following discussion. We now have a mapping

$$\Phi \colon Z \to \bar{R}^\infty,$$

which is clearly continuous and injective, and whose precise range is a compact subset K of $\bar{R}^\infty$. (Note that K depends on Φ.) By elementary topological arguments we see that K is homeomorphic with Z, and that Φ and Φ^{-1} establish a bijective mapping between Z and K which is continuous in each direction, and *so also measurable in each direction* if we continue to use Baire sets in Z, and agree to use Baire sets in K. The Baire ($=$ Borel) sets in K are of course just the traces of the Baire ($=$ Borel) sets in $\bar{R}^\infty$.

As in the preceding paragraph we suppose that with reference to some basic measurable space $(U, \mathscr{U})$ we have a sequence of measurable mappings

$$Y^k \colon U \to Z \quad (k = 1, 2, \ldots).$$

From what has just been said it will follow that the composed mappings

$$\Phi \circ Y^k \colon U \to \bar{R}^\infty \quad (k = 1, 2, \ldots)$$

are also measurable, and so we can proceed to construct the further measurable mapping

$$\operatorname{Lim} \operatorname{Sup}_{k \to \infty} \Phi \circ Y^k \colon \quad U \to \bar{R}^\infty.$$

We now observe that, for each separate u in U,

$$\operatorname{Lim} \operatorname{Sup}_{k \to \infty} \Phi \circ Y^k(u)$$

lies in the cluster-set of (and so in the closure of the track of) the Φ-image of the sequence

$$\{Y^k(u) \colon k = 1, 2, \ldots\}.$$

But $Y^k(u)$ is in Z, so $\Phi \circ Y^k(u)$ is in K, for each k, which is just to say that the track of the Φ-image of the sequence is in K (which is compact, and so closed). It follows that $\operatorname{Lim\,Sup} \Phi \circ Y^k(u)$ is also in K, and thus *we can define a unique element of Z by writing*

$$(21) \qquad \operatorname*{Lim\,Sup}_{k \to \infty} (Y^k(u); \Phi) = \Phi^{-1} \operatorname*{Lim\,Sup}_{k \to \infty} \Phi \circ Y^k(u).$$

I claim that the mapping

$$(22) \qquad \operatorname*{Lim\,Sup}_{k \to \infty} (Y^k(\cdot); \Phi) \colon U \to Z$$

is measurable. This is in fact an immediate consequence of the measurability of the mappings of which it is the composition. The significance of this result is that *we have constructed a measurable mapping which coincides with the partially defined mapping,*

$$\lim_{k \to \infty} Y^k \colon U \to Z,$$

wherever the latter exists.

Of course the mapping (22) is not a canonical object, for it depends on the particular choice of Φ, but for the applications (construction of standard modifications) this will never be important. It is in the nature of standard modifications not to be canonical; their function is to serve particular purposes, or to show by implication that some set of 'nice properties' is admissible for the name-class, and so is associated with a canonical extension of the canonical Daniell–Kolmogorov process. We shall here (in the matter of measurability) be concerned with standard modifications 'serving a particular purpose'.

We conclude this section by stating formally the following two results, each one of which is relative to a fixed dense denumerable family Φ of elements of $C(Z)$, where Z is a second-countable compact Hausdorff space.

Proposition 8. *Let Y^k for each $k = 1, 2, \dots$ be a Z-valued random variable defined over a given measurable space $(\Omega, \mathscr{F})$.*
Then

$$\operatorname*{Lim\,Sup}_{k \to \infty} (Y^k; \Phi)$$

is a random variable defined over $(\Omega, \mathscr{F})$.

Proposition 9. *Suppose that we have a sequence of name-classes of stochastic processes (labelled by $k = 1, 2, \dots$) with a common second-countable compact Hausdorff state-space Z and a common parameter-set A on which some σ-algebra $\mathscr{A}$ of measurable sets has been defined, and for each k let the* probability-space $(\Omega, \mathscr{F}, \mathrm{pr})$ *support a version* $\{Y_a^k: a \in A\}$ *of the process associated with the kth name-class. Then*

$$(\Omega, \mathscr{F}, \mathrm{pr}; Z, A; \underset{k \to \infty}{\mathrm{Lim\,Sup}}(Y_a^k; \Phi) \quad (a \in A))$$

is a stochastic process, and it will be measurable in the sense that the mapping

$$(\omega, a) \to \underset{k \to \infty}{\mathrm{Lim\,Sup}}(Y_a^k(\omega); \Phi)$$

is measurable relative to the σ-algebras

$$\mathscr{F} \otimes \mathscr{A} \quad \text{and} \quad \mathscr{B}_0(Z),$$

if each of the mappings

$$(\omega, a) \to Y_a^k(\omega)$$

is so measurable.

Proofs. These results follow directly from the preceding discussion. We have only to define the mapping Y^k by

$$Y^k(\omega, a) = Y_a^k(\omega) \quad (k = 1, 2, \ldots),$$

and the mapping Y by

$$Y(\omega, a) = (Y^1(\omega, a), Y^2(\omega, a), \ldots),$$

and then to check the measurability of the composition of the chain of mappings

$$\Omega \times A \xrightarrow{\quad Y \quad} Z^\infty \xrightarrow{\quad \Phi^\infty \quad} K^\infty \xrightarrow{\mathrm{Lim\,Sup}} K \xrightarrow{\quad \Phi^{-1} \quad} Z,$$

where

$$\Phi^\infty(z^1, z^2, \ldots) = (\Phi(z^1), \Phi(z^2), \ldots).$$

Of course the significance of Propositions 8 and 9 lies in the fact that, for each Φ, Lim Sup is *an* extension of lim.

9. We now replace Assumptions (*a*) and (*b*) by

Assumption (c). *Each of Z and A is a second-countable compact Hausdorff space.*

We do this because the proof of our concluding proposition will require some degree of compactness in the parameter set A. Assumption (*c*) is more than is really necessary, but most cases of interest can be reduced to this. For example if A were the real line we should compactify it to $R \cup \{-\infty, \infty\}$ and where necessary adjoin trivial random variables with parameters $\pm \infty$.

15

Our object will be to start with the canonical (*uncompleted!*) model (4) associated with a name-class which is continuous in probability, and to build on top of it a standard modification Y_a ($a \in A$) which is such that $(f, a) \rightarrow Y_a(f)$ is a measurable mapping, and which further has the property that $Y(f)$ is topologically D-separable as an element of the function-space Z^A, for every individual f. Here D is some dense denumerable set in A; it is quite arbitrary, but must be fixed at the outset, for our construction of the standard modification will depend on the choice of D. The construction will also depend on the choice of some dense denumerable set Φ of elements of $C(Z)$; this again is arbitrary, but must remain fixed throughout the construction.

We start by constructing the topological 'square' $A \times A$, and we observe that it (like A) will be second-countable, Hausdorff and compact. This being so, its 'diagonal',

$$\Delta = \{(a, a): a \in A\},$$

will be closed, and so compact, and so (since it is necessarily now a Baire set) it will be a countable intersection $\bigcap V_q$ of open sets V_q in $A \times A$. Indeed by the usual Urysohn argument we can choose a sequence of nested open sets V_q (each of which covers Δ) in such a way that

$$\Delta = \bigcap \bar{V}_q,$$

and we shall do this. We are of course already exploiting the additional topological restrictions placed upon A by Assumption (*c*).

As in the proof of Proposition 7, we introduce the q-nested open neighbourhoods $N^a(\varphi_j; q)$ of $a \in A$, and now we define

$$N^a(q) = N^a(\varphi_1; q) \cap N^a(\varphi_2; q) \cap \ldots \cap N^a(\varphi_q; q),$$

which will be an open neighbourhood of a.

For fixed q, the open sets $N^a(q)$ ($a \in A$) cover A, and so their squares cover Δ. Now each a in A satisfies both

$$(a, a) \in V_q$$

and

$$(a, a) \in N^a(q) \times N^a(q),$$

and as the products $U_r \times U_s$ of basic open sets for A form a countable basis for the open sets in $A \times A$, we see that with each $a \in A$ we can associate a member U^a of the countable basis for the A-topology in such a way that

$$a \in U^a,$$

$$U^a \times U^a \subset V_q$$

and

$$U^a \times U^a \subset N^a(q) \times N^a(q).$$

There may be more than one such U^a (for a given a); when there is a choice we agree to choose the 'first' such U^a (first with respect to some fixed positive-integer labelling of the basis $\{U_r: r = 1, 2, ...\}$). Obviously with this choice (U^a_q, let us call it) we shall have

$$a \in U^a_q \subset N^a(q) \subset N^a(\varphi_j; q) \quad (1 \leqslant j \leqslant q).$$

Now the squares of the basic open sets U^a_q form a covering (for fixed q) of the compact set Δ, and therefore the squares of finitely many of them do so; let such a finite covering of Δ be provided by the squares of

$$(23) \qquad\qquad U_{q,1}, U_{q,2}, ..., U_{q,r_q}.$$

Concerning the collection (23) we note the following facts.

(*i*) Each a in A belongs to at least one member of the collection (23).

(*ii*) Each member of the collection (23) is of the form U^b_q for some b in A, and so is covered by $N^b(q)$.

(*iii*) If a_1 and a_2 are distinct elements of A, then (a_1, a_2) lies outside $\bar{V}_q$ for all sufficiency large q, and so for all sufficiently large q the parameters a_1 and a_2 cannot lie together in the closure of a common member of the collection (23).

In (*iii*) we have made use of the fact that

$$U_{q,s} \times U_{q,s} \subset V_q \quad (s = 1, 2, ..., r_q),$$

which implies that

$$\overline{U_{q,s}} \times \overline{U_{q,s}} \subset \overline{V_q}.$$

If $a \in A$, then for each q we know that $a \in U_{q,s}$ for at least one value of $s \leqslant r_q$, and possibly more often; let $s(q, a)$ denote the smallest such s. We now call in the arbitrary (but fixed) dense denumerable subset D of A, and we suppose that it has already been labelled by the positive integers. We know that D must meet the open neighbourhood

$$W^a_q = U_{1,s(1,a)} \cap U_{2,s(2,a)} \cap ... \cap U_{q,s(q,a)}$$

of a, for each $q = 1, 2, ...$ Let d^a_q denote the 'first' value of D to lie in W^a_q.

For each q the elements a and d^a_q of A will both lie in the common member $U_{q,s(q,a)}$ of the collection (23). This member of the collection will be equal to $U^b_q \subset N^b(q)$ for some b in A, and by comparing $\varphi_j(f_d)$ via

$\varphi_j(f_b)$ with $\varphi_j(f_a)$ we see that we must have

$$\mathrm{pr}\left(\left\{f: |\varphi_j(f_d) - \varphi_j(f_a)| > \frac{2}{q}\right\}\right) < \frac{2}{q^3}$$

for $j = 1, 2, \ldots, q$, if $d = d_a^q$.

We write $B_0^{q,a}$ for the Baire set in Z^A on which

$$|\varphi_j(f_{d_a^q}) - \varphi_j(f_a)| \leqslant \frac{2}{q} \quad \text{for } j = 1, 2, \ldots, q.$$

As before, we see that the complement of this Baire set can be expressed as the union of q Baire sets each one of which has a pr-measure of $2/q^3$ at most, and so once again we see from the convergence of $\sum 2q/q^3$ that

$$\mathrm{pr}\left(\liminf_{q \to \infty} B_0^{q,a}\right) = 1.$$

In other words, outside a Baire set of probability zero we shall have

$$|\varphi_j(f_{d_a^q}) - \varphi_j(f_a)| \leqslant \frac{2}{q} \quad \text{for all } q \geqslant \max(j, Q(a,f)).$$

Thus on the complementary Baire set of probability one we must have

$$f_a = \lim_{q \to \infty} f_{d_a^q}.$$

Note that the Baire set of probability 1 on which this holds is dependent on a.

We therefore *define*

$$(24) \quad \begin{cases} (\mathbf{M}f)_a = Y_a(f) = \operatorname{Lim\,Sup}_{q \to \infty}(f_{d_a^q}; \Phi) & (a \in A \backslash D, f \in Z^A), \\ (\mathbf{M}f)_a = Y_a(f) = f_a & (a \in D, f \in Z^A). \end{cases}$$

By what we have just proved it is clear that $\{Y_a : a \in A\}$ is a standard modification of the coordinate-process built on top of the *uncompleted* canonical model, and moreover it has the desired measurability property. (Notice, in checking the latter statement, that each singleton $\{d\}$ in A is a Baire ($=$ Borel) set in A.)

To complete our task we must also show that for each fixed f, the new sample path

$$\{Y_a(f) : a \in A\}$$

is topologically D-separable. Notice first that in view of the second clause of the definition (24), we can write $Y(f)$ in place of f inside the Lim Sup. Next, remember that the Lim Sup always maps a sequence into an element of the cluster-set of that sequence.

Now

(25) $$\lim d_a^q = a \quad \text{for each } a \text{ in } A.$$

For let δ be any cluster-point of the sequence $\{d_a^q : q = 1, 2, ...\}$; then because a and $d_a^{\bar{q}}$ both belong to $W_{\bar{q}}^a$, it follows that they must both belong to

$$U_{q,s(q,a)}$$

for all values of $\bar{q} \geqslant q$, and so a and δ must both belong to

$$\overline{U_{q,s(q,a)}}$$

for every q. But this contradicts (*iii*), unless $\delta = a$, and equation (25) then follows.

Finally let U be any open neighbourhood of a, where $a \in A \backslash D$, and let G be any open set containing $Y_a(f)$. Then G must contain

$$Y_{d_a^q}(f) = f_{d_a^q}$$

for infinitely many values of q, and because of equation (25) we also know that d_a^q must lie in U for all but finitely many of those values of q. On referring back to (*ii*) of paragraph 2, it will be seen that we have established the D-separability of every modified sample path.

We have thus proved

Proposition 10. *If a name-class of stochastic processes with a state-space Z and a parameter-set A which are both second-countable compact Hausdorff spaces can be shown to be continuous in probability, then it will follow that a standard modification can be built on top of the canonical model in such a way as to obtain both process-measurability, and topologically D-separable sample paths. Here D is an arbitrary denumerable dense subset of A.*

In order to carry out the construction it is not necessary to complete the canonical model, and it is therefore possible to re-erect the standard modification, with both the properties stated above, on top of any version, complete or not, in the name-class.

In the $(\bar{R}, R)$ case there are extensions of the result which utilize the ordering of the real line which is the parameter-set, and these extensions are of great significance in, for example, the theory of Markov processes. We do not construct such extensions here, because for us A has no order structure, but it is likely that extensions for partly-ordered A could be constructed, if required.

We have now completed our survey, and note in conclusion that while the topics dealt with here may seem very remote from practical applications, they are in fact of importance there because of the way in which all mathematicians, pure or applied, habitually think. If we continue to find it convenient to make models of reality involving a time-axis which is a continuum, then we shall find ourselves talking about the supremum of a set of random variables Y_t, where t ranges through an interval, or possibly forming a Fourier transform,

$$\int_{-\infty}^{\infty} e^{ist} Y_t \, dt.$$

Such procedures are in general quite meaningless unless backed up by a formal study of separability (in the first case) or process-measurability (in the second). Thus, while as hewers of wood and drawers of water we can afford for most of the time to ignore these matters, we must not neglect them entirely if we care for the quality of the wood and the purity of the water.

REFERENCES

1. J. L. Doob, *Stochastic Processes*, Wiley, New York (1953).
2. R. M. Dudley, "On measurability over product spaces", *Bull. Am. math. Soc.* **77** (1971), 271–274.
3. P. R. Halmos, *Measure Theory*, Van Nostrand, Toronto (1950).
4. D. G. Kendall, "An introduction to stochastic analysis", 1.1 of this book.
5. ——, "Foundations of a theory of random sets", 6.2 of *Stochastic Geometry*, John Wiley, London (1973).
6. P. A. Meyer, *Probability and Potentials*, Blaisdell, Waltham (1966).
7. E. Nelson, "Regular probability measures on function space", *Ann. Math.* **69** (1959), 630–643.
8. J. Neveu, *Mathematical Foundations of the Calculus of Probability*, Holden Day, San Francisco (1965).
9. H. R. Pitt, "The definition of measures in function space", *Proc. Cam. Phil. Soc.* **46** (1950), 19–27.

Note added in proof. The results presented in this survey are not in all respects the strongest possible. Thus R. Borges in an early paper which I must apologize for having overlooked (*ZfW* **6** (1966), 125–128) shows that a separable standard modification exists (for some *unspecified D*) given that our Assumptions (*a*) and (*b*) hold, but without needing the assumption of completeness for the σ-algebra. (This throws some light on the fact that we escaped the need to assume completeness when proving Proposition 10.) It is also of interest to compare Proposition 10 with a paper by D. L. Cohn (*ZfW* **22** (1972), 161–165) which appeared while the present work was in the press; Cohn assumes continuity in probability and has Z compact metric (as here), but supposes that A is merely separable metric, and he proves the existence of a separable measurable modification (separability with respect to an *unspecified D*). Both writers work out 'measurable selection devices' of the type devised here in paragraph 8. It seems likely that, in the presence of continuity in probability (cf. our Proposition 7) both the Borges and Cohn theorems could be rephrased to permit the assertion of separability with respect to an *arbitrarily chosen* 'separating set' *D*, provided only that the latter is denumerable and dense in *A*, but I have not checked this. As remarked in the text, such freedom to choose *D* is of overriding importance in most practical applications.

6
CONCLUSION

6.1

Letter to F. Papangelou, Easter Day, 1970

R. DAVIDSON

Thank you very much for your farewell note. I'm sorry not to have written; of course things have been busy—I have had to write several lecture courses and am even now hammering out the syllabus for another—but it's no excuse. So how are you all and how is Ohio? My knowledge of the States is mostly derived from reading of the campaigns in the Civil War—and then the Columbus in Miss. was of greater strategic importance than that in Ohio.

Since you left I've been occupying my time as usual—semigroups and geometric stochastic processes. I tried to make progress on the problems left by my stay in Heidelberg—which are

(1) the main existence problem;

(2) a characterization of modulus-squared characteristic functions (*à la* Bochner's theorem).

(3) If you start off with a process of cars-with-velocities on a road, the joint distribution of positions and velocities being spatially homogeneous, and you then let the process run, can you say that the number of cars passing you in consecutive seconds does NOT converge to 0 in probability? This has bearings—vague—on (1), but is clearly of great interest—being so elementary—in its own right. I can do it if the velocities of the cars are independent.

Then I turned to semigroups with the idea of, if possible, extracting a Delphic kernel from a more or less arbitrary (cancellative commutative topological) semigroup. But the stumbling block was the production of enough homomorphisms (continuous, that is) into the unit disc. I only

447

realized rather late how beastly semigroups are (compared with groups)—there exists a compact commutative $\frac{1}{2}$ group with no non-trivial continuous homomorphisms, etc.

So then I moved off to Delphicize l.c.a. groups in the way I had used for laws on the line. This, as expected, works, i.e. we can, by quotienting out units and discarding elements with idempotent factors, reduce $M(G)$, where G is a 2nd countable l.c.a. group with a generating compact neighbourhood of e, to strongly Delphic form. I suppose I ought to write this up, but feel it is rather flogging a dead horse. The same ideas will work, of course, if we confine ourselves to measures concentrated on a fixed closed sub-$\frac{1}{2}$ group of G.

In geometric stochastic processes I have two problems, one posed by a physicist, the other by a doctor. In the physical problem, particles of gold are rained down on a quartz slab, where they migrate (presumably in independent Brownian motions) until they either evaporate, or meet other particles—when evaporation is arrested and the particles clump together; or they hit, and stick to, previous clumps. The physicist can only see clumps of a certain size, and wants to know the distribution of the number of these per unit area, after given time and with, I suppose, known rates of diffusion, evaporation and raining. There appear to be charming problems on the shape of clumps. I gave this one to my (unique) research student, but it is difficult to tear him away from the computer, and anyway he seems to like p-functions, etc., more.

The doctor's problem is a geometric continuous-state branching process (of arteries in the lung). He slices lungs and observes the distribution of small artery sizes, and wants to know as much as he can about the rules for forking of the arteries. So far I have thought of 2 models for him, and investigated the mean effects of them, but it is difficult to see how the models can be distinguished on the basis of his experiments.

I've just had an excellent week climbing in Snowdonia, and return refreshed to the construction of a new course of lectures on stochastic (Markov) processes for Part IB, partly on the lines of the Dynkin–Yushkevitch book which I read in Heidelberg—mostly at mealtimes. I'm very glad Klaus hurried through the translation, because the American edition costs \$15 against DM.14.80. . . . It is now extremely likely that I shall be visiting Russia (Ambartzumian, and Dynkin, anyhow, I think) this autumn, and I shall get D. and Y. in Russian, and also the excellent Armenian polylingual mathematical dictionary, if there are any left.

[*This letter was sent from Trinity College, Cambridge, to Papangelou (then in Ohio); the editors are very grateful to Professor Papangelou for permission to publish this survey of Davidson's research plans at that time.*]

6.2

Rollo Davidson: 1944-1970

E. F. HARDING and D. G. KENDALL

Rollo Davidson was born on 8 October 1944 in Bristol, and spent his childhood at Thornbury, Gloucestershire. Like his father, his father's father, both his mother's brothers and his own younger brother, he took a Scholarship to Winchester. Here his career was characterized by breadth (Ancient History as well as Mathematics at A-level) and speed (he won the Senior Mathematical Prize at the age of 16). One of his school papers which has been preserved is a Stewart McDowall Essay (1962) on 'The Appeal of Science to the Victorian Intelligentsia'.

He matriculated as a Scholar of Trinity College at the University of Cambridge in October 1962. He was awarded the Percy Pemberton prize as the Trinity undergraduate most distinguished in his studies in his first year, and at the end of his second year he was already a Wrangler. It would have been normal at that time to spend the third year working for Part III of the Mathematical Tripos, but with characteristic independence Davidson chose to take the course for the Diploma in Mathematical Statistics instead (in which he duly received distinction). It was this choice which determined the way he was to spend the next 5 years, in which brief period he made profound contributions to Probability Theory, a circumstance perhaps without parallel. It is idle to speculate on what he might have achieved, had he lived to attain full maturity in years and in his profession, but to his friends and colleagues there can be no doubt of the immense loss to learning, and to the society of scholars, occasioned by his tragically early death in 1970. By this time he had become a Smith's Prizeman (1967), a Research Fellow of Trinity College (1967), Ph.D. (1968), Assistant Lecturer in the Department of Pure Mathematics and Mathematical Statistics (1968), Lecturer in the Statistical Laboratory in the same Department (1969), and Fellow-elect of Churchill College (1970).

449

It is so usual for prodigies to be sophisticated, and perhaps even intolerable, that one must stress how far Davidson was from following this familiar pattern. Extremely diffident, overcoming a natural shyness by power of will alone, and far from self-confident in his mathematical powers, he did not at all realize until the last year or two that he had the capacity not merely to solve hard problems but to create new fields of enquiry, and to take the undisputed lead in them. In his work on Delphic semigroups he was the problem-solver, cracking one hard nut after another, but in his work on Stochastic Geometry, with which his name will always be linked, he quickly became, as Klaus Krickeberg has written, 'the reader one wrote for':

'I have systematized, made precise and generalized many things, but most of the basic intuitive ideas are his. I always assumed that we would go on like this for some time to come, pushing each other ahead and complementing each other, and while writing the article I thought of him as *the* reader of it, and imagined how he would be surprised and pleased with certain things.'

Davidson had an unpredictable variety of special interests outside mathematics, and any visitor to his rooms in Trinity would be struck by these. He was interested in trains, in physical geography, in old books (a 'regular' at David's stall on Saturday mornings, he collected especially old alpine books), and above all things he adored mountains. Mathematical excursions quickly became the passport to mountain wanderings. In June 1969 a 'Tagung' at Oberwolfach was devoted to 'Integral Geometry and Geometrical Probability'; Krickeberg and D.G.K. organized it, but it was the existence of Davidson's thesis which inspired the project. There were many happy rambles over the Black Forest hills during that week, which proved scientifically very fruitful, for it was followed by three weeks of collaboration in Heidelberg between Davidson and Krickeberg, which led to many advances.

We mention 'physical geography' because Davidson's papers contain a correspondence with the Hydrographer of the Navy on the existence (or not) of bores on certain Chinese rivers. This may have been linked with conversations with Trinity colleagues, though his own interest in bores undoubtedly came from growing up (by the Severn) near such a good one. But he may well have been equally interested in the observation of the Chinese bores by Commander W. Usborne Moore, R.N. (H.M.S. *Rambler*), in 1888–92, and he would certainly have leaped at the chance of an antiquarian contribution to physical science; he would have delighted in D. E. Cartwright's 'Tides and waves in the vicinity of Saint Helena'

(*Phil. Trans. Roy. Soc. A,* **270** (1971), 603–649) and in the utilization there of observations cited from 'Maskelyne, N. (1762*b*)'.

In August 1969 Davidson went to Canada to participate in the Twelfth Biennial Seminar of the Canadian Mathematical Society which was held in Vancouver. Here he went to the Cascade Pass with Daryl Daley, and on one of the weekends took a party to climb The Lions (getting within a few hundred feet of the summit). At about this time he was discussing the possible truth of the Riemann hypothesis with Littlewood, and lending a hand with Littlewood's psychophysical experiments and their statistical analysis.

From this point onwards Davidson's mathematical and mountaineering notes get rather mixed up. There is a 1969 entry of a visit to Snowdon (probably at Easter time) in characteristic style:

'0th day ... Bangor: Cod and Chips, Tea, Buns, 6s. 9d. Buns OK.
. . .

5th day ... Pen yr Olewen, C.D., C.L., F. Grach, down into
cwm below Craig Ysfa, up to Bwlch, Pen Helig ... very
fine: sun on tops of rocks.'

In June 1970 he went to Skye with E.F.H., climbing Sgurr Alasdair 'by something like Collie's route'. Later 'on the 15th we moved over to Sligachan and did the classic expedition: S. nan Gillean by the Pinnacle and West ridges. Very pleasant scrambling. . . .. The new Guide recommends descent on the right (W.) side of the 3rd Pinnacle, but this appears to be dictated by a desire to stop the holds on the usual route getting too smooth.' By avid reading he had absorbed the spirit of early British mountaineering, and in the Coolins he almost ritually enacted for himself the legendary traditions. The climbing apart, it was finding in the Glen Brittle Post Office log-book a Cambridge entry of the 1920s (with such names as Adrian, R.H. Fowler, Littlewood, and many others) that perhaps more than anything closed the link with the old days. Agile, swift and light as an elf on the mountains, and totally untroubled by 'exposure', he had advanced in skill and achievement in climbing as in everything else: his climbs at home were also stages in the symbolical ascent to the great peaks abroad, and he now had the Alps clearly in his sight.

By now, also, Davidson was preparing for a Royal Society sponsored visit to Erevan, in the Armenian Soviet Socialist Republic. Here he hoped especially to meet again and work with R.V. Ambartzumian, a distinguished young mathematician sharing both his mathematical and his mountaineering interests, but of course he had intended as well to visit the probabilists in Moscow, Leningrad and Kharkov. His notes on the proposed visit include the postscript: 'I'd like to cross a pass or two in the Caucasus'. But this expedition was not to take place.

In July he went to the Alps with the Cambridge University Mountaineering Club, accompanied by a Russian grammar and Paul-André Meyer's *Probabilités et Potentiel*, the latter decorated by mathematical comments in the margins and climbing notes on the fly leaf. After a very successful meet, the party broke up to go their several ways, Davidson remaining with Michael Latham (a gifted young mathematician from Gonville and Caius College) near Pontresina, to climb the Piz Bernina.

On 29 July 1970, while they were descending from the summit, an accident cost them both their lives.

David Williams expressed all our feelings at the time when he wrote in a letter to one of us:

> 'I feel so angry—if 'angry' is the word, (but anger *that* he died, not at *how*)—at such a loss for parents, friends, Cambridge, but, above all, for himself.'

From such feelings this book emerged. In the years that have gone to the making of it one has learned to see things in a slightly different perspective. Rollo's was a *magnificent life*; a flawless blend of personal relations, mathematics and mountain adventure. The hazards of the latter, never wholly to be avoided, are familiar to all, and to rail at its folly is to invite a reply which he himself might have made, in the words of one of the more sympathetic characters in contemporary fiction:

> 'If you always look over your shoulder, how can you still remain a human being?'

Author Index

Subject Index

SOCIAL SCIENCE LIBRARY

Manor Road Building
Manor Road
Oxford OX1 3UQ
Tel: (2)71093 (enquiries and renewals)
http://www.ssl.ox.ac.uk

This is a NORMAL LOAN item.

We will email you a reminder before this item is due.

Please see http://www.ssl.ox.ac.uk/lending.html
for details on:

- loan policies; these are also displayed on the notice boards and in our library guide.

- how to check when your books are due back.

- how to renew your books, including information on the maximum number of renewals.
Items may be renewed if not reserved by another reader. Items must be renewed before the library closes on the due date.

- level of fines; fines are charged on overdue books.

Please note that this item may be recalled during Term.